Lecture Notes in Computer Science

Edited by G. Goos and J. Hartmanis

403

S. Goldwasser (Ed.)

Advances in Cryptology – CRYPTO '88

Proceedings

Springer-Verlag

Berlin Heidelberg New York London Paris Tokyo Hong Kong

Editor

Shafi Goldwasser
Laboratory for Computer Science, Massachusetts Institute of Technology
545 Technology Square, Cambridge, MA 02139, USA

CR Subject Classification (1987): E.3

ISBN 0-387-97196-3 Springer-Verlag New York Berlin Heidelberg
ISBN 3-540-97196-3 Springer-Verlag Berlin Heidelberg New York

Printing and binding: Druckhaus Beltz, Hemsbach/Bergstr.
2145/3140-543210 – Printed on acid-free paper

CRYPTO '88

A Conference on the Theory and Application of Cryptography

held at the University of California, Santa Barbara,
August 21-25, 1988
through the cooperation of the Computer Science Department

Sponsored by:

International Association for Cryptologic Research

in cooperation with

The IEEE Computer Society Technical Committee
On Security and Privacy

General Chair
Harold Fredricksen, Naval Postgraduate School

Program Chair
Shafi Goldwasser, Massachusetts Institute of Technology

Program Committee

Eric Bach	University of Wisconsin
Paul Barret	Computer Security Ltd.
Tom Berson	Anagram Laboratories
Gilles Brassard	University of Montreal
Oded Goldreich	Technion Israel Institute of Technology
Andrew Odlyzko	Bell Laboratories
Charles Rackoff	University of Toronto
Ron Rivest	Massachusetts Institute of Technology

Foreword

The papers in this volume were presented at the CRYPTO '88 conference on theory and applications of cryptography, held August 21-25, 1988 in Santa Barbara, California. The conference was sponsored by the International Association for Cryptologic Research (IACR) and hosted by the computer science department at the University of California at Santa Barbara.

The 44 papers presented here comprise: 35 papers selected from 61 extended abstracts submitted in response to the call for papers, 4 invited presentations, and 6 papers selected from a large number of informal rump session presentations.

The papers were chosen by the program committee on the basis of the perceived originality, quality and relevance to the field of cryptography of the extended abstracts submitted. The submissions were not otherwise refereed, and often represent preliminary reports on continuing research.

It is a pleasure to thank many colleagues. Harold Fredricksen singlehandedly made CRYPTO '88 a successful reality. Eric Bach, Paul Barret, Tom Berson, Gilles Brassard, Oded Goldreich, Andrew Odlyzko, Charles Rackoff and Ron Rivest did excellent work on the program committee in putting the technical program together, assisted by kind outside reviewers.

Dawn Crowel at MIT did a super job in publicizing the conference and coordinating the activities of the committee, and Deborah Grupp has been most helpful in the production of this volume. Special thanks are due to Joe Kilian whose humor while assisting me to divide the papers into sessions was indispensable.

Finally, I wish to thank the authors who submitted papers for consideration and the attendants of CRYPTO '88 for their continuing support.

June 1989
Cambridge, MA

Shafi Goldwasser

Table of Contents

SHORT RUMP SESSION PRESENTATIONS
Chair: W. Diffie

Session 1

Cryptographic Primitives

Chair: S. Goldwasser, MIT

Weakening Security Assumptions and Oblivious Transfer

(Abstract)

Claude Crépeau*
Department of Computer Science
MIT

Joe Kilian†
Mathematics Department
MIT

1 Introduction

Our work is motivated by a recent trend in cryptographic research. Protocol problems that have previously been solved subject to intractability assumptions are now being solved without these assumptions. Examples of this trend include a new completeness theorem for multiparty protocols[BGW,CCD], and a protocol for byzantine agreement using private channels[FM]. These breakthroughs illustrate both the strengths and the weaknesses of using the cryptographic model. Devising first a protocol that uses cryptographic assumptions can give powerful intuition that later allows one to create a protocol that works without assumptions. However, there is a danger that the cryptographic assumptions one uses can become inextricably bound up in the protocol. It may take years before these assumptions can be ironed out of the final protocol.

One way to keep a firm grasp on ones cryptographic assumptions is to compartmentalize them into a small set of relatively simple primitives. One then attempts to build protocols on top of these primitives, without using any cryptographic assumptions in the high level design. The problem of eliminating cryptographic assumptions from the protocol is then reduced to that of implementing the primitives without cryptography.

In this abstract, we explore a particularly useful set of primitives, known as *oblivious transfers.* First introduced by Rabin, oblivious transfer protocols are games in which one player, Sam(the sender), can impart some information to another player, Rachel(the receiver), without knowing precisely what information he has imparted.

*Supported in part by an NSERC Postgraduate Scholarship. Some of this research was performed while visiting Bell Communication Research.

†Research supported in part by a Fannie and John Hertz foundation fellowship, and NSF grant 865727-DCR. Some of this research was performed while visiting Bell Communication Research.

Oblivious transfers come in a wide variety of flavors, and are not obviously reducible to each other. Following the work of Brassard, Crépeau, Robert[BCR], and Crépeau[C], we develop techniques for establishing equivalences between a wide variety of oblivious transfers.

We also investigate the properties of an ordinary noisy channel. By a noisy channel, we mean a communication line in which a transmitted bit is flipped with a certain fixed probability. This model has been extensively studied in coding theory, but relatively little was previously known about its cryptographic capabilities. We show that a noisy channel can be used to implement two-party cryptographic protocol without any intractability assumptions. In the forthcoming [CK] we also study a transfer mechanism we refer to as quantum transfer. This mechanism abstractly models a transfer mechanism based on quantum mechanics.

Weaker variants of two of the more standard forms of oblivious transfer are also studied. We investigate scenarios in which the security properties guarenteed by these mechanisms may be almost completely violated. We show that in many of these scenarios, it is still possible to achieve the full power of ordinary oblivious transfer.

The purpose of this abstract is to introduce the reader to the terminology and the statement of our results. To get the actual reductions and more detail on the application of the techniques described in this abstract, the reader should consult [CK].

Main Results

Our results may be summarized as follows. Before reading these theorems, we refer the reader to Section 2 of the paper, which provides the necessary terminology.

Theorem 1: α-1-2 slightly oblivious transfer is as powerful as 1-2 oblivious transfer.

Theorem 2: Noisy transfer is as powerful as 1-2 oblivious transfer.

Theorem 3: α-slightly oblivious transfer is as powerful as 1-2 oblivious transfer.

2 Definitions

In this section, we describe the various forms of information transfer mechanisms we will be considering. We define the two standard mechanisms, two weakened versions of the standard forms of oblivious transfer, and our nonstandard transfer mechanism.

2.1 Standard forms of oblivious transfer

There are two standard forms of oblivious transfer. We refer to these mechanisms as *oblivious transfer* and *1-2 oblivious transfer*.

Oblivious Transfer: In this protocol, Sam has a secret bit, b. At the end of the protocol, one of the following two events occurs, each with probability $\frac{1}{2}$.

1. Rachel learns the value of b.

2. Rachel gains no further information about the value of b (other than what Rachel knew before the protocol).

At the end of the protocol, Rachel knows which of these two events actually occurred, and Sam learns nothing.

Less formally, we can view this protocol as one in which Sam sends a letter to Rachel, which arrives exactly half the time.

1-2 Oblivious Transfer: In this protocol, Sam has two secret bits, b_0 and b_1. Rachel has a selection bit, s. At the end of the protocol, the following three conditions hold.

1. Rachel learns the value of b_s.

2. Rachel gains no further information about the value of b_{1-s}.

3. Sam learns nothing about the value of s.

Less formally, Sam has two secrets. Rachel can select exactly one of them, and Sam doesn't know which secret Rachel selected.

Dirtier Notions of Oblivious Transfer

In describing oblivious transfers, we make two distinct specifications. First, we specify what information is being transferred. Second, we impose a set of security conditions, specifying what information each party is guaranteed *not* to know at the end of the protocol, and specifying that certain events cannot be controlled by either party. The definitions of oblivious transfer and 1-2 oblivious transfer are particularly stringent in their security conditions. In oblivious transfer, Sam has no control over whether Rachel receives b. In 1-2 oblivious transfer, Sam gains no information about Rachel's selection s. We would like to be able to handle cases in which a malicious Sam can, thorough some form of cheating, violate these security conditions. This motivates the following definitions.

α-Slightly Oblivious Transfer: This protocol is the same as oblivious transfer, except that instead of Rachel learning bit b with probability $\frac{1}{2}$, she learns it with probability p. If Sam is nonmalicious, $p = \frac{1}{2}$. If Sam is malicious, he may choose any value of p he wishes, subject to $1 - \alpha \leq p \leq \alpha$.

α-1-2 Slightly Oblivious Transfer: This protocol is the same as $1 - 2$ oblivious transfer, except that at the conclusion of the protocol, a malicious Sam can guess Rachel's selection bit s with probability α.

In both these definitions, the interesting range for α is $\frac{1}{2} \leq \alpha < 1$.

2.2 Nonstandard transfer mechanism

We now consider our nonstandard transfer mechanism, motivated by coding theory.

Noisy Transfer: In this protocol. Sam has a secret bit, b. Rachel has no information about b. At the end of the protocol, Rachel receives a bit b'. With probability 3/4, $b' = b$, otherwise $b' = \bar{b}$. Sam learns nothing.

This protocol may be thought of as simulating a noisy communication channel, in which a bit is flipped with probability 1/4. We can parameterize the above definition by replacing the 3/4 with a probability ρ. We call this *ρ-noisy transfer*. In this paper, we only consider the "standard" noisy transfer, where $\rho = 3/4$.

Note that in these definitions, there is a careful distinction made between the powers of a malicious Sam verses the powers of a nonmalicious Sam. Since a malicious Sam is always more powerful than a nonmalicious Sam, it would at first seem natural to simply assume that Sam is malicious. However, we require that the protocols we build on top of these primitives meet the following two requirements: They must work when Sam is nonmalicious, and they must maintain their security conditions when Sam is malicious. So, for example, if one is building a protocol using a 3/4-slightly oblivious transfer subprotocol, one *cannot* require Sam to send 1000 bits, having at least 600 get through to Rachel. A malicious Sam could easily do this, but a nonmalicious Sam could not.

3 Making honest reductions more robust

In this section we sketch the ideas behind the technique for strengthening some of our reductions. Using this technique, we can write simple reductions which depend on the receiver being honest, and in a fairly routine fashion, convert them to protocols which are robust against cheating by the receiver. This technique will be crucial in our reductions from 1-2 oblivious transfer to α-oblivious transfer and noisy transfer.

3.1 The general scenario

We consider transfer mechanisms with the *verifiable obliteration* property. By this we mean that the transfer mechanism occasionally gives the receiver a value which is uncorrelated with the bit sent, and for which the receiver knows this fact. Two examples of such mechanisms are ordinary oblivious channel and α-oblivious transfer. Our intermediate goal is to implement some form or another of 1-2 oblivious transfer. Having accomplished this, we then try to apply the techniques leading to theorem 1 to implement standard 1-2 oblivious transfer.

For the complete description of this technique, consult [CK].

4 The power of noise

In this section we consider the cryptographic power of an ordinary noisy communication channel, i.e. one which inverts a transmitted bit with some fixed probability. We sketch the proof that this family of transfer mechanisms can be used to implement 1-2 oblivious transfer, and hence a wide variety of secure two-party protocols.

4.1 A philosophical remark

Noisy channels have been extensively studied in the field of coding theory, and it is interesting to see how our perspective differs from the more traditional one. Coding theory adopts the viewpoint that noise is a bad thing, to be eliminated as efficiently as possible. Given a noisy channel, a coding theorist tries to simulate a pristine, noiseless communication line.

From our point of view (following Wyner [W]), an ideal communication line is a sterile, cryptographically uninteresting entity. Noise, on the other hand, breeds disorder, uncertainty, and confusion. Thus, it is the cryptographer's natural ally. The question we consider is whether this primordial uncertainty can be sculpted into the more sophisticated uncertainty found in secure two-party protocols. The result outlined in this section answers this question in the affirmative.

4.2 An outline of our reduction

Our reduction consists of four main parts. We first show how to use a noisy transfer channel to simulate a very dirty transfer channel which has the total obliteration property. This allows us to start applying the techniques of Section 3. Using these techniques, we can show how to implement a version of 1-2 oblivious transfer similar to α-1-2 slightly oblivious transfer. We can then use the proof of Theorem 1 to get an almost pure 1-2 oblivious transfer channel. This channel may be used to simulate a pure 1-2 oblivious transfer channel.

Please consult [CK] for the details of the reduction.

5 Acknowledgments

We would like to acknowledge Gilles Brassard, Ernie Brickell, Ivan Damgård, Cynthia Dwork, Joan Feigenbaum, Shafi Goldwasser, and Silvio Micali for their valuable comments, ideas, and encouragement.

6 References

[BCR] Brassard, Gilles, Claude Crépeau, and Jean-Marc Robert. "Information Theoritic Reductions Among Disclosure Problems," *Proceedings of the* 27th *FOCS*, IEEE, 1986, 168–173.

[BGW] Ben-Or, Michael, Shafi Goldwasser, and Avi Wigderson, "Completeness Theorems for Noncryptographic Fault-tolerant Distributed Computation," *Proceedings of the* 20th *STOC*, ACM, 1988.

[C] Crépeau Claude, "Equivalence Between Two Flavours of Oblivious Transfer", *Proceedings of Crypto 87*, 1988,Springer-Verlag.

[CCD] Chaum David, Claude Crépeau and Ivan Damgård, "Multiparty unconditionally secure protocols," *Proceedings of the* 20th *STOC*, ACM, 1988.

[CK] Crépeau Claude, and Joe Kilian, "Achieving Oblivious Transfer Using Weakened Security Assumptions," to appear in *Proceedings of the* 29th *FOCS*, IEEE, 1988.

[EGL] Even S., Goldreich O., and A. Lempel, "A Randomized Protocol for Signing Contracts," *CACM*, vol. 28, no. 6, 1985, pp. 637-647.

[FM] Feldman, Paul, Silvio Micali. "Byzantine Agreement from Scratch," *Proceedings of the* 20th *STOC*, ACM, 1988.

[R] Rabin, M., "How to exchange secrets by oblivious transfer," Tech. Memo TR-81, Aiken Computation Laboratory, Harvard University, 1981.

[W] Wyner, A. D., "The Wire Tap Channel," Bell System Journal, 54, 1981, pp. 1355-1387.

Limits on the Provable Consequences of One-way Permutations

Russell Impagliazzo*
Computer Science Division
University of California at Berkeley
Berkeley, California 94720

Steven Rudich†
Computer Science Department
University of Toronto
Toronto, Canada M5S 1A4

Abstract

We present strong evidence that the implication, "if one-way permutations exist, then secure secret key agreement is possible", is not provable by standard techniques. Since both sides of this implication are widely believed true in real life, to show that the implication is false requires a new model. We consider a world where all parties have access to a black box for a randomly selected permutation. Being totally random, this permutation will be strongly one-way in a provable, information-theoretic way. We show that, if $P = NP$, no protocol for secret key agreement is secure in such a setting. Thus, to prove that a secret key agreement protocol which uses a one-way permutation as a black box is secure is as hard as proving $P \neq NP$. We also obtain, as a corollary, that there is an oracle relative to which the implication is false, i.e., there is a one-way permutation, yet secret-exchange is impossible. Thus, no technique which relativizes can prove that secret exchange can be based on any one-way permutation. Our results present a general framework for proving statements of the form, "Cryptographic application X is not likely possible based solely on complexity assumption Y."

1 Introduction

A typical result in cryptography will be of the form: With assumption X, we can prove that a secure protocol for task P is possible. Because the standard crypto-

*Research partially supported by NSF grant CCR 88-13632.
†Research partially supported by NSF grant CCR 88-13632 and an IBM doctoral fellowship.

graphic assumptions are, at present, unproved, many results focus on weakening the assumptions needed to imply that a given protocol is possible. As a consequence, we ask a new form of question: which assumptions are too weak to yield a proof that a secure protocol for P is possible?

The task we will study is secure secret-key agreement. Secret-key agreement is a protocol where Alice and Bob, having no secret information in common, agree on a secret-key over a public channel. Such a protocol is secure when no polynomial-time Eve listening to the conversation can determine part of the secret. Secure secret-key agreement is known to be possible under the assumption that trapdoor functions exist [DH76], [GM84]. However, researchers have been frustrated by unsuccessful attempts to base it on the weaker assumption that one-way permutations exist.

We provide strong evidence that it will be difficult to prove that secure secret-key agreement is possible assuming only that a one-way permutation exists. We model the existence of a one-way permutation by allowing all parties access to a randomly chosen permutation oracle. A random permutation oracle is provably one-way in the strongest possible sense. We show that any proof that secure secret-key agreement is possible in a world with a random permutation oracle would simultaneously prove $P \neq NP$. (Formally, $P = NP$ implies there is no secure secret-key agreement relative to a random permutation oracle.) We conclude that it is as hard to provably base a secure secret-key agreement protocol on an arbitrary one-way permutation as it is to prove $P \neq NP$. Furthermore, we can use the above result to construct an oracle relative to which one-way permutations exist, but for which secure secret-key agreement is impossible. (This oracle O is constructed by starting with an oracle for which P=NP, and adding on a random permutation oracle.) This means that any proof that the existence of a one-way function implies that of a secure secret-key agreement protocol cannot relativize (i.e., hold relative to any oracle). Non-relativizing proofs are few and far between not only in cryptography, but in complexity theory as a whole. Since the technique of examining complexity relative to an oracle was introduced in [BGS75], relativization results have been used to provide evidence for the difficulty of resolving questions in complexity theory [BG81]. (We will later briefly discuss the possibility that a non-relativizing proof basing secure secret-key agreement on a one-way permutation can be found.) Relativized complexity has not been frequently used in cryptography ([Bra, Bra83] is one exception to this rule); we hope the framework developed here will have wide applicability in separating the strengths of cryptographic assumptions.

Our result also has some implications for "black box" reductions between various other cryptographic assumptions and tasks. Instead of formalizing the notion of "black box" reducibility to an assumption or task, which would involve going into the specifics of these assumptions and tasks, we will use the phrase "A is black-box reducible to B" as an abbreviation for "If B holds relative to an oracle O, then A also holds relative to O". (In [I88], a general notion of "black box" proof is developed, and shown to be basically equivalent to that given in the preceding.) Since the definitions of the cryptographic tasks and assumptions mentioned here are lengthy, technical, and often not unique, to describe them formally would require a separate paper. (In fact,

A	B
One-way permutations exist	Secret-key agreement is possible
Signature schemes exist	Oblivious transfer is possible
Pseudo-random generators exist	Trapdoor functions exist
Private-key cryptosystems exist	Voting Schemes exist
Telephone coin flipping is possible	
Bit commitment with strong receiver is possible	
Bit commitment with strong sender is possible	
Collision free functions exist	

papers have been written describing various ways of formalizing some of the terms used here; for others, such papers do not presently exist but are greatly needed.) Therefore, rather than attempt to define these terms here, we will give references to papers introducing these concepts and/or papers clarifying them. We hope the reader not familiar with cryptography will still be able to follow the general idea of the following discussion.

Cryptographic tasks to be discussed here include: coin flipping by telephone ([Blu82]), electronic signatures ([DH76] , [GMR84]), private-key cryptography ([GM84, GGM84, LR86, Rac88]), bit-commitment (both the strong committer version ([GMW87]) and the strong receiver version ([BCC87])), identification ([DH76], [FFS86]), electronic voting ([Ben87]), oblivious transfer ([Blu81, Rab81]), and secret-key exchange itself ([DH76, Mer78]). General assumptions which have been used in cryptography include the existence of : one-way permutations ([P74]), pseudo-random generators ([BM84, Yao82]), trap-door permutations ([DH76]), and two-to-one collision-free functions. This last is a function f which is easily computable, is two-to-one on strings of length n, and where no polynomial-time algorithm, given n, can find strings x and y of length n with $f(x) = f(y)$.

Many reductions between the various assumptions and tasks listed above are known. In particular, it is known that the existence of a one-way permutation implies the following: pseudo-random generators exist ([Yao82]), private-key encryption is possible ([GM84, GGM84, LR86]), strong committer bit commitment is possible ([Yao82, GMW87]), telephone coin flipping is possible ([Blu82]), and electronic signatures are possible ([NY]). All of the preceding results relativize. We construct an oracle O relative to which one-way permutations exist, but for which no secret-key agreement protocol is possible. From relativized versions of these results, it follows that O will also have the property that, relative to O, pseudo-random generators exist, strong committer bit commitment is possible, etc. Thus, none of the preceding assumptions can imply that secure secret-key agreement is possible in a way which relativizes.

Furthermore, we can add to this list several statements which are not known to follow from the existence of a one-way permutation, but which O can be proved to satisfy because a truly random permutation is used in O's construction. For example, it is unknown whether a one-way permutation can be used to construct a two-to-one collision-free function, but it is easy to see that for a random permutation p, the function which outputs all but the last bit of p will be such a function. [NY] show that the existence of a two-to-one collision-free function suffices to construct a protocol for strong receiver bit commitment. Thus, neither the existence of two-to-one collision-free function nor that of a strong receiver bit commitment protocol suffices to construct a secret-key agreement scheme via a relativizing proof.

Similarly, if an assumption is sufficient to prove the possibility of secret-key agreement in a relativizing manner, it itself cannot be proven from the existence of a one-way permutation via a "black box" reduction. Examples of such assumptions include oblivious transfer ([Blu81, Rab81]), voting ([Ben87]), and trap-door functions ([DH76, GM84]).

To summarize:

There is an oracle O relative to which all A's hold, but all B's do not. (See above table.)

Some caution is needed in interpreting these results, since at least one non-relativizing construction in cryptography is known. In [I88] it is shown that the theorem proved in [GMW87], "the existence of a one-way permutation implies the existence of zero-knowledge protocols for all languages in NP", fails with respect to a random permutation. In contrast, the [FFS86] construction of an identification protocol based on any one-way function will *not* be possible with just a black box for a random permutation. Their construction is as follows. Every person chooses a random x, and announces publicly $f(x)$ as their I.D. To prove you are the person with I.D. I, you give a zero-knowledge proof of knowledge that you know an x with $f(x) = I$. However, to give a zero-knowledge

proof as in GMW and FFS that you know such an x, it is necessary to have an actual circuit that computes f, not just a black box which gives the value of f. In fact, if f is a random permutation, no such zero-knowledge proof will be possible. Thus, the [FFS86] scheme does not relativize. The [FFS86] protocol is exceptional even for those constructions involving zero-knowledge proofs. Most applications of zero-knowledge will in fact relativize, even though the literal statement of the [GMW87] theorem does not. In a world with a random permutation oracle, it *is* possible to give a zero-knowledge proof for any property *actually* in NP, as opposed to NP relativized to this oracle. It is only applications which attempt to "bootstrap", proving things concerning the values of the same function used to make the protocol zero-knowledge, which fail to relativize.

The above example is the only non-black box construction in cryptography known to the authors for a result in a general form (as opposed to results involving specific crypto-systems). Thus, it is fair to say that the result presented here shows that most of the standard techniques in cryptography cannot be used to construct a secret-key exchange protocol from a one-way permutation.

2 Notation and definitions

A *secret-key agreement protocol* is a pair of *PPTMs* called Alice and Bob. Each machine has a set of private tapes: a random-bit tape, an input tape, two work tapes, and a secret tape. In addition, they have a common communication tape that both can read and write. A run of the protocol is as follows: Alice and Bob both start with the same integer l written in binary on their input tapes; Alice and Bob run, communicating via the common tape; Alice and Bob both write an l-length string on their secret tape. If this string is the same, Alice and Bob are said to *agree*. The entire history of the writes to the communication tape is called ***the conversation***. $\alpha(l)$ will denote the probability that Alice and Bob agree on a secret of length l.

A *PPTM* Eve *breaks* a secret-ket agreement protocol if Eve, given only the conversation, can guess the secret with probability $\alpha(l)/poly(l)$. A protocol is *secure* if no Eve can break it. One could imagine far more stringent notions of security. For example, we might require that Eve can't even get one bit of the secret. However, in our scenario, we will be breaking secret-key agreement in the strong sense defined above, thus including the weaker notions of breaking that an applied cryptographer would use. (For example, a cryptographer would be happy to learn one bit of the secret.)

A *one-way permutation* is a 1-1, onto, polynomial-time computable function from n-bit strings to n-bit strings, where the inverse permutation is not computable in polynomial-time. In fact, for cryptography we require that no *PPTM* can expect to invert the function on more than a $1/poly(n)$ fraction of the inputs of length n.

We will abbreviate probabilistic polynomial-time Turing machine with the notation *PPTM*. The ***computation of a PPTM on a given input*** will be a trace of the entire run of the machine given the input. (The computations are indexed by the possible random tapes.) If the machine is an oracle machine, this would include all the queries and answers received during the computation. (In this case, each computation would be determined by a random tape, and by a finite set of query-answer pairs.) We use the notation *poly* to refer to some polynomial function. Thus, we can use the freewheeling arithmetic $poly * poly = poly$. A ***conversation*** between two *PPTMs* is the history of writes to the cells of a common communication tape.

We will use the following form of the *pigeonhole principle*:

Let M be a 0-1 matrix with a $1-\alpha$ proportion of 1s. For every $ab = \alpha$, a $1-a$ portion of the columns have at least a $1-b$ portion of 1s. (It suffices to note that the worst case is when the 0's are concentrated in an a by b rectangle.)

3 Uniform Generation

3.1 Polynomial-time relations

A relation, R, is polynomial-time if we can decide xRy in time polynomial in $\|x\| + \|y\|$. In this paper, we will only consider relations where the length of y is polynomially related to the length of x. *Is satisfied by* is an example of such a relation: x *is satisfied by* y iff x is a boolean formula and y is one of its satisfying assignments.

3.2 What is uniform generation?

Let R be the "is satisfied by" relation. We can ask two natural questions:

Existence Given x, does there exist a y such that xRy?
(Does a given formula has a satisfying assignment?)

Counting Given x, how many y exist such that xRy?
(How many satisfying assignments does a given formula have?)

The existence question, satisfiability, is NP-complete. The counting question, thought to be harder than satisfiability, is $\#P$-complete. Jerrum, Valiant, and Vazirani[JVV86] introduced a problem of intermediate complexity.

Uniform generation Given x, pick a y uniformly at random such that xRy.
(Given a formula, find a random satisfying assignment.)

More generally, let R be a polynomial-time relation. Let M be a $PPTM$ with a fixed (as opposed to expected) polynomial running time. We say M *uniformly generates* R if given x, M has at least a 50% chance of outputting a uniformly chosen y such that xRy; otherwise, M outputs "try again". If such a y does not exist, M will only output "try again". Notice that rerunning the algorithm when it fails to generate a random y will succeed in generating a random y in expected polynomial time.

3.3 $P = NP$ and uniform generation

Theorem 3.1 (JVV) *For any polynomial-time relation, there exists a $PPTM$ equipped with a $\sum_2^P$ oracle that uniformly generates it.*

Theorem 3.2 $P = NP \Longrightarrow$ *for any polynomial-time relation, there exists a $PPTM$ that uniformly generates it.*

Proof: $P = NP \Rightarrow$ the polynomial-time hierarchy collapses[CKS81] $\Rightarrow$ a polynomial-time machine can simulate a $\sum_2^P$ oracle $\Rightarrow$we can use previous theorem to uniformly generate. ■

Let M be a $PPTM$. There are possibly many different computations of M consistent with a given input and output. (Of course, there may be none.) The following corollary shows that if $P = NP$, we can efficiently pick a random element from the finite set of these computations.

Corollary 3.1 $P = NP \Longrightarrow$ *it is possible to generate a random computation for a given $PPTM$, M, with given input, I, and given output, O, in expected polynomial time.*

Proof: Checking that the trace of a computation is consistent with M, I, and O is a polynomial-time relation. ■

Corollary 3.2 $P = NP \Longrightarrow$ *given a conversation, C, between two $PPTMs$ M and N, we can uniformly generate a possible computation of M.*

Proof: Checking that C is consistent with a given computation of M is possible in polynomial-time. ■

3.4 An application to cryptography

Public-key cryptography relies on the assumption that **P≠NP**. The formal version of this fact, $P = NP$ implies secret-key agreement is not possible, is something one might see a rather technical proof of in a first-year course. We can use our results on uniform generation to give a particularly simple proof of the optimal result.

Theorem 3.3 $P = NP \Longrightarrow$ *Eve has an expected polynomial time algorithm to break any given secret-key agreement protocol in the strongest possible sense: Eve will find the secret with exactly the same probability that Alice and Bob agree on one.*

Proof: Fix a computation and resulting secret for Bob. We will show that the probability that Alice agrees with Bob is the same as the probability that Eve agrees with Bob. By corollary 3.2, Eve can generate a random computation of Alice consistent with the conversation. Alice's particular computation is, by definition, a random computation of Alice consistent with the conversation. Thus, Eve and Alice produce secrets with exactly the same probability distribution. They must, therefore, have exactly the same probability of agreeing with Bob. In other words, from Bob's point of view, Alice and Eve think alike; he will fool Eve with exactly the same probability that he will fool Alice. ∎

4 Random Oracles

4.1 Random function oracles

Let r be a random real between 0 and 1, chosen with the uniform distribution; express r in binary notation. A *random oracle* is the set induced from r as follows: $\{x :$ the xth binary digit of r is a 1 $\}$.

With each random oracle R, we can associate a function from n-bit strings to n-bit strings. $f(i)$ is defined by its length(i) binary digits; the jth digit is 1 iff $(2i+1)2^j \in R$. (Every natural is uniquely expressed as an odd times a power of 2.) Notice that as we vary over all possible R, we get all possible length-preserving functions, each one occurring with the same frequency. Furthermore, using R as an oracle, f is polynomial-time computable. Thus, a TM with a random oracle also has at its disposal an easy to compute length-preserving random function. The notions of a random oracle and a random function oracle will be used interchangeably. We state without proof a theorem a standard theorem concerning random functions:

Theorem 4.1 *For most oracles, the function associated with the oracle is one-way in the strongest possible sense: For every oracle PPTM, there exists a poly, such that the machine has expectation no more than $poly(n)/2^n$ of inverting the inputs of length n.*

4.2 Random oracles and uniform generation

Theorem 4.1 implies that uniform generation is impossible in a random world; it is impossible to uniformly generate an inverse to the function associated with the oracle. Our goal is, assuming $P = NP$, to break secret-key exchange in a random world. (In theorem 3.3, we saw how to break it in the real world.) Even though we can't hope for uniform generation in a random world (which would make life very easy), we can prove weak analogues of the uniform generation results, which will be helpful.

The idea is not to generate the computation of an oracle $PPTM$, M, with a particular random oracle, but rather, with a *random* random oracle; we want a random computation of the machine over all possible oracles. Let $M^{I,O}$ be the finite set of possible computations of M given input I, output O, using some oracle. (These computations are indexed by the random-bit tape, and the oracle query-answer pairs used during the computation.) A natural probability distribution to put on $M^{I,O}$ is to weight each computation by the probability that it occurs using a random oracle. We want to be able to pick a random element of the space $M^{I,O}$. Note: This time the distribution on the underlying set is not necessarily uniform. The probability of a computation with q queries being chosen is $2^{-q}/2^{-p}$ as likely as a computation with p queries being chosen.

Theorem 4.2 $P = NP \Longrightarrow$ *there exists a PPTM that picks a random element from the probability space* $M^{I,O}$ *in expected polynomial time.*

Proof: ¿From the oracle $PPTM\ M$, we construct a $PPTM\ M'$, such that a uniformly generated computation of M' given input I and output O, when suitably syntactically modified, yields a random element of the probability space $M^{I,O}$. Intuitively, M' is an oracle machine that makes up its own oracle on the fly.

Without loss of generality, assume the computation of M never makes the same oracle query twice; keep track of queries asked in a table, and use the oracle only when the table does not have the answer. Let $t(n)$ be a polynomial bound on the number of oracle queries M asks given an input of length n. M' starts its computation by writing down $t(n)$ random bits on a separate tape, called the *answer tape.* M' then proceeds as M would, except that when M asks the oracle for a query answer, M' answers the simulated query with the first unused bit from the answer tape. By corollary 3.2, we can generate a random computation m' of M', with input I and output O, in expected polynomial time. To make m' look like a random computation of M, strip away the answer tape, pretending that all answers came from an oracle; call the computation that remains m. The probability associated with an m asking q queries is proportional to 2^{-q}. Hence, m is a random element of $M^{I,O}$. ∎

We can strengthen our result slightly by fixing some finite portion of the oracles we wish to consider. Let E be a finite set of oracle addresses and their contents. An oracle is said to be *consistent* with E if the content-address pairs in E are also in the oracle. We define a space similar to $M^{I,O}$: $M_E^{I,O}$ is a finite set of computations of M given I and O, using oracles consistent with E. Each element in $M_E^{I,O}$ is weighed by the probability of it occurring using a random oracle consistent with E. Once again, we wish to pick a random element of the space.

Theorem 4.3 $P = NP \Longrightarrow$ *there exists a PPTM that picks a random element of the probability space* $M_E^{I,O}$ *in expected polynomial time.*

Proof: Same as the proof of the previous theorem with one important modification: Hardwire the answers to oracle queries in E into the finite state control of M'. When M' asks a query in E, *do not* use a bit from the answer tape. ∎

We can now prove the analogue of corollary 3.2 using **oracle** $PPTM$s Alice and Bob. In the case where oracle Alice and oracle Bob have conversation C, and E is a finite set of queries and answers, we define another similar space: A_E^C is the space of possible computations of oracle Alice consistent with the conversation C, where each computation is weighed by its probability of occurring with a random oracle consistent with E. The next theorem will be very important in the results on secret-key agreement.

Theorem 4.4 $P = NP \Longrightarrow$ *there exists a* $PPTM$ *that picks a random element of* A_E^C *in expected polynomial time.*

Proof: ¿From Alice's point of view a conversation is a set of inputs and outputs occurring at certain prescribed times during her computation. No further modification of the above proof technique is required. ■

5 Random Permutation Oracles

Random permutation oracles are similar to the random function oracles discussed in the previous section, except that the random functions must be 1-1 onto. A *random permutation oracle* Π is a random length-preserving function from the set of finite strings *onto* itself. Again, the function is chosen from the uniform distribution.

¿From the point of view of oracle $PPTM$s, there is no difference between the two types of oracles. We will formalize this in the spirit of pseudo-randomness.

A *tester* is an oracle $PPTM$ which, given n and a function oracle from n-bit strings to n-bit strings, outputs either 0 or 1. Let T be a tester. Let P_n be the probability that T will output a 0, when given n and a random function from n-bit strings to n-bit strings. Let P'_n be the probability that T will output a 0, when given n and a random permutation from n-bit strings to n-bit strings. Let $D_{T_n} = |P_n - P'_n|$. Thus, D_{T_n} measures how well the tester can distinguish between the two types of oracles.

Theorem 5.1 *For every tester* T, $D_{T_n} < poly(n)/2^n$

Proof: Assume T makes $q < poly(n)$ queries. In the case of a random function oracle, the answer to a previously unasked query is a random n-bit number, independent of the answers to previously asked queries. Thus, for each query made the probability that it gets the same answer as a previously made query is less than $q/2^n$. Summing, we conclude that the probability that two queries received the same answer is less than $q^2/2^n$. Next we observe that the distribution on possible query answers, given that all query answers are different, is the same for random function oracles and random permutation oracles; the probability that T will output a 0 given that all query answers are different, is the same for the two types of oracles. It follows that $D_{T_n} < q^2/2^n$. ■

The above theorem will allow us to first prove our results relative to a random oracle, and then extend them to a random permutation oracle.

It is a standard theorem that random permutations are very hard to invert.

Theorem 5.2 *Measure one of random permutation oracles are one-way in the strongest possible sense: For every oracle PPTM, there exists a poly, such that the machine has expectation no more than $poly(n)/2^n$ of inverting the inputs of length n.*

6 Cryptographic Lower Bounds

6.1 Introduction

We will show that the existence of a very strong one-way permutation is not an assumption likely to yield a proof that secure secret key agreement is possible. By theorem 5.2, we know that a random permutation oracle is one-way in the strongest possible sense. Therefore, we will use the availability of a random permutation oracle to model the existence of an ideal one-way permutation. We will show that it is as hard to prove secure secret key agreement is possible using a common random permutation oracle is it is to prove $P \neq NP$. The result will take the form of the contrapositive: $P = NP$ implies that any secret-key agreement protocol can broken even when a random permutation oracle is available to all parties.

Summarizing the results of this section: We first show that $P = NP$ implies there is no secret-key agreement protocol that is secure with measure $1/poly$ of random oracles (random function oracles). Theorem 5.1 will be used to extend the result to random permutation oracles. Further strengthening the result by swapping the quantifiers, we show $P = NP$ implies for measure one of oracles there is no secure secret-key agreement. A corollary of this result is the existence of an oracle relative to which one-way permutations exist, but secure secret-key agreement is impossible. We also distinguish between two strong senses of breaking a secret-key agreement protocol.

6.2 A normal form for secret-key agreement

To facilitate our analysis, we will assume that the secret-key agreement protocol has a normal form. Communication takes place in n rounds. Each round involves one person speaking and computing. Before each round, the party who is to speak asks the oracle a single query, and then does some computation. If Alice speaks first, the protocol would take the following form: Alice queries the oracle; Alice computes;

Alice speaks (i.e. writes on the communication tape); Bob queries the oracle; Bob computes; Bob speaks; Alice queries the oracle; Alice computes; Alice speaks; Bob queries the oracle; ...

Any protocol can be converted to normal form with only a polynomial blow-up in running time.

6.3 Notation and definitions

We wish to investigate a random world where Alice and Bob attempt to agree on an l-bit secret. In other words, we vary over runs of Alice, Bob, and Eve; and over oracles. Formally, a *world situation* is a five-tuple $< l, random_{Alice}, random_{Bob}, random_{Eve}, R >$. l, the input to Alice, Bob, and Eve, is the length of the secret being agreed upon. $random_{Alice}$, $random_{Bob}$, and $random_{Eve}$ are random bit tapes for Alice, Bob, and Eve to use during their computations (the random bit tapes are just long enough that they never get used up). R is a random oracle. Let WS_l be the set of all world situations where Alice and Bob attempt to agree on an l-length secret (l is the first entry of the five-tuple). We will also think of WS_l as a probability space with the uniform distribution. A world situation determines a random run of the protocol with a random oracle. With each world situation we can associate the following variables:

C_r, the conversation up to and including round r.

q_r, the query asked in round r.

A_r, the query-answer pairs Alice knows up to and including round r.

B_r, the query-answer pairs Bob knows up to and including round r.

If it is ambiguous which world situation C_r comes from, we write C_R^w to mean the conversation comes from world situation w.

World situation w *satisfies* C_r (written $w \models C_r$) means that the conversation between the machines in w is identical to C_r for the first r rounds. We will use the $\models$ notation with the other world situation variables as well.

Notice that none of the three polynomial time machines involved will be able to access the oracle past some very large address. Thus, without any loss, we can think of the oracle as finite. This means that the probability space WS_l is finite. Similarly any space we will discuss can be considered finite. This technical point will prevent the reader from suspecting any measure-theoretic fallacy.

6.4 Eve's sample space

We need to define the probability distributions Eve samples from during her algorithm. They have already been described in section 4.2, Theorem 4.4. We define them again here.

Call a random tape for Alice *consistent* with conversation C_r and oracle R if the run of Alice, determined by the random tape and input from Bob's portion of C_r, outputs Alice's portion of C_r. (What she does after round r does not matter.) Let E be a finite set of query-answer pairs.

Let $AS_E^{C_r}$ be the set of <oracle, random tape for Alice> pairs such that E is in the oracle and the random tape for Alice is consistent with C_r and the oracle. Eve will be sampling from the space $A_E^{C_r}$ of computations of Alice consistent with C_r and the query-answer pairs in E. The distribution on $A_E^{C_r}$ is induced from the uniform distribution on $AS_E^{C_r}$; sample a point in $AS_E^{C_r}$, that point corresponds to a computation of Alice: An <oracle, random tape for Alice> pair corresponds to a <finite portion of the oracle used during the computation, random tape for Alice pair>.

6.5 Eve's algorithm

We now give an algorithm for Eve to break a secret-key agreement protocol in a random world. This algorithm runs in polynomial time under the assumption that $P = NP$. S_l is a function of the form $1/poly$ (called a security parameter), which determines Eve's probability of failure. The smaller S_l, the longer Eve must run to break the protocol.

For each of n rounds of communication between Alice and Bob, Eve does $m = \lceil 3(n/S_l)\ln(2n/S_l)\rceil$ segments. Each segment has a simulate phase and an update phase. We will describe these phases in segment i for round r.

Without loss of generality, assume Alice speaks in round r. Let $E_{r,i-1}$ be the finite set of query-answer pairs that Eve knows about the oracle so far; $< q, a > \in E_{r,i-1}$ *iff* prior to round r, segment i, Eve has asked if q is in the oracle ($q \in R$?), and received answer a. Recall that C_r is the conversation that has occurred up to this round.

SIMULATION PHASE:

Using the method described in theorem 4.4, Eve picks a random run of Alice from the space $A_{E_{r,i-1}}^{C_r}$. (If Bob speaks in round r, Eve would instead simulate Bob.) Let $F_{r,i}$ be the set of queries that the simulated run of Alice asks her simulated oracle. (Note that so far in this segment, we have not asked any real oracle queries. Recall that when simulating a random Alice, we make up the answers to the oracle queries.)

UPDATING PHASE:

Eve asks all the queries in $F_{r,i}$ of the actual oracle R. Thus, $E_{r,i}$ equals $E_{r,i-1}$ union the new query-answer pairs Eve learned by asking $F_{r,i}$ of the oracle.

The following variables are also associated with any world situation:

$E_{r,i}$, the query-answer pairs Eve knows up to and including the ith segment of her simulation of round r.

$E_{r,0}$, the query-answer pairs Eve knows before she simulates round r. ($E_{r,0} = E_{r-1,m}$.)

$BPQ_{r,i}$, the query-answer pairs Bob knows and Eve does not, up to and including round r, segment i. BPQ stands for Bob's private queries. Note the relation: $BPQ_{r,i} = BPQ_{r,0} - E_{r,i}$.

6.6 Intersection queries and the secret

Intersection queries are the queries Alice and Bob ask in common during an execution of their protocol. A particular query becomes an intersection query, not when it is first asked by one party, but rather when it is later asked by the other party. For conceptual unity, we can assume without loss of generality that the secret is an intersection query; assume that as their final act Alice and Bob query the oracle at the location addressed by the secret.

The next Theorem will prove that with high probability Eve finds all the intersection queries. Thus, Eve will have a polynomial-length list containing the secret; Eve breaks the protocol.

6.7 The efficacy of Eve's algorithm

Theorem 6.1 *Suppose Alice and Bob attempt to agree on an l-length secret. The probability that Eve finds all the intersection queries is greater than $1 - S_l$. Formally,* $PROB_{x \in WS_l}[A_n \bigcap B_n \subseteq E_{n,m}] > 1 - S_l$.

Proof: (We show the stronger result that Eve probably anticipates (asks) a query before it becomes an intersection query.) Eve's algorithm has n rounds. If Eve fails to find all intersection queries, there must be a first round where she fails to anticipate an intersection query that occurs in the next round; there exists a first time $q \in A_r \bigcap B_r$ and $q \notin E_{r-1,m}$. To formalize the event that Eve fails for the first time to anticipate an intersection query in the next round, we write it as the conjunct of three events:

- Eve has, in previous rounds, anticipated all intersection queries about to happen. (Thus, Eve knows all intersection queries to date.)
- q_{r+1}, the query asked in the next round, is an intersection query.
  AND
- Eve fails to find q_{r+1}. ($q_{r+1} \notin E_{r,m}$.)

Lemma 6.1, the technical heart of the proof, will show this event has probability no more than S_l/n by showing that the complementary event has probability greater than $1 - S_l/n$. Thus, for each round the probability of failing for the first time to anticipate an intersection query in the next round is less than S_l/n. Summing the error probability for each round, we get a total error probability bounded by S_l. ∎

Lemma 6.1 *The probability that in round r, either*

- *In a previous round, Eve failed to anticipate the intersection query about to happen,*
- q_{r+1} *is not an intersection query,*
  OR
- *Eve finds* q_{r+1}

is greater than $1 - S_l/n$.

The proof of this lemma can be found in STOC'89

Theorem 6.2 *Theorem 6.1 is true relative to a random permutation oracle: Given any secret-key agreement protocol and a random permutation oracle, the probability that Eve finds all the intersection queries is greater than* $1 - S_l/2$.

Proof: Assume not. We will construct a tester to distinguish between a random function oracle and a random permutation oracle. We start with a protocol where Eve will find all the intersection queries with probability less than $1 - S_l/2$ if a random permutation oracle is used, and probability greater than $1 - S_l$ if a random function oracle is used. A tester can simulate runs of Alice, Bob, and Eve, counting the fraction of times Eve finds all the intersection queries. The essence of the situation is that the tester is flipping a coin with two possible biases: $1 - S_l/2$ and $1 - S_l$; the tester must guess which. If the tester flips the coin $1/S_l{}^2$ times, even a very weak form of the law of large numbers would tell us that Eve can guess the bias of the coin at least 99% of the time. This very strongly contradicts theorem 5.1. ∎

Notice the order of the quantifiers in the above result. We picked the protocol between Alice and Bob, then we picked the oracle (since the protocol is bound by definition to work with a random oracle). Then, we showed Eve can break the protocol. We prove a stronger result which reverses the quantifiers. First, we pick a random oracle; then a protocol for Alice and Bob (this time the protocol need not work properly on other oracles). Then, we show that Eve can break the protocol relative to the chosen oracle.

Theorem 6.3 $P = NP \Longrightarrow$ *relative to a random permutation oracle, any secret key agreement scheme can be broken.*

Proof: First, we argue that for every secret-key agreement protocol, there are only measure zero of oracles where it can't be broken. Fix a protocol. The $P = NP$ assumption allows us to use Eve's algorithm as before. Choose $S_l = 1/l^{2+\epsilon}$. Theorem

6.1 tells us that in $1 - S_l/2$ of world situations we succeed in breaking the protocol. By the pigeon-hole principle, for each length l, there are $1 - \sqrt{S_l/2}$ oracles relative to which there is a $1 - \sqrt{S_l/2}$ chance of Eve breaking the protocol. Call all such oracles good for length l. The probability that a random oracle fails to be good for length l is $\sqrt{S_l/2}$. $\sum_{l=0}^{\infty} \sqrt{S_l/2}$ converges; by the Borel-Cantelli lemma, measure one of oracles are good on all but finitely many lengths. For measure one of the oracles, past some length, Eve has a $1 - \sqrt{S_l/2}$ chance of breaking the protocol. (We can even non-uniformly boost Eve's ability to break protocols for finitely many lengths.) Thus, there are only measure zero oracles where the protocol can't be broken.

For each of the countably many protocols we throw out the measure zero of oracles where the protocol is secure. We have thrown out measure zero in all. Every protocol can be broken relative to the measure one of remaining oracles. ■

Corollary 6.1 *There exists an oracle relative to which a strongly one-way permutation exists, but secure secret-key agreement is impossible.*

Proof: Consider any oracle world where $P = NP$. Add a random permutation oracle to this world. Because all the techniques in our theorem relativize, we can conclude that secure secret-key agreement is not possible in the resulting world.

Construct an example of such an oracle as follows: The even numbers form an oracle for PSPACE (a PSPACE-complete problem), the odd numbers form a random permutation oracle. $P = NP$ relative to a PSPACE-complete oracle. We know the random permutation is one-way in the strongest possible sense. ■

The only other relativized result that we know in cryptography is Brassard[Bra83, Bra]. He explicitly constructs an oracle where secret-key agreement is possible.

So far, our sense of breaking a secret key agreement consists of finding a polynomial-sized list with the secret on it somewhere. The strongest sense of breaking secret key agreement is clearly to find the secret itself. We show how to extend Eve to actually find the secret. For the same reasons as before, the argument works equally well with both random oracles and random permutation oracles.

Eve's strategy can be extended as follows: Eve's final round will be her simulation of the $n-1$th round of the protocol. In each segment of her final round, Eve records her last query to the oracle. (Recall that the last query to the oracle should be thought of as the secret.) Of the final queries Eve has recorded, she outputs the one which occurs the majority of the time. (If there is no majority, output "failure".)

Theorem 6.4 *Suppose that Alice and Bob agree on a secret with probability at least* $1 - \alpha$ *over world situations in* WS_l*. Then, for every* $\delta > 0$*, there exists an Eve who can guess the secret with probability at least* $1 - \alpha(2 + \delta)$ *over world situations in* WS_l.

The proof of this theorem can be found in STOC'89

7 Related Work and Open Problems

In the work presented here, as in much of theoretical cryptography, we do not go into exactly how much time the adversary will take to break the protocol, as long as this time is polynomial. However, in real life, a protocol taking a large degree polynomial time to break may be almost as good as one secure against any polynomial time adversary. Merkle[Mer78] has suggested a protocol, based on any one-way function, the breaking of which would require an eavesdropper to take time quadratic in the time taken by the participants. (Here, time is measured as the number of calls to a black box for the one-way function.) We showed that for a protocol in normal form, an eavesdropper can always break the protocol in time $O(n^3 \log n)$; however, to put the protocol into normal form may square n, so our eavesdropper is actually taking time $O(n^6 \log n)$. This leaves open Merkle's question of whether his scheme is optimal.

Another general question brought up by this research is whether similar statements can be proved for other cryptographic applications. We have previously given a list of applications at least as strong as secret key agreement; that these are unlikely to be a consequences of the existence of a one-way permutation follows from the result here. However, it would be interesting to show that there is some natural application which cannot even be based on a much stronger assumption, such as the existence of a trapdoor permutation.

8 Acknowledgements

We are especially grateful to Manuel Blum and Amos Fiat, who asked us the question of whether a one-way permutation suffices for secret agreement, and presented a model in which it might be disproved. We would also like to thank Noam Nisan, Charlie Rackoff, and Umesh Vazirani. We give Manuel and Umesh a second helping of thanks for their support and encouragement during the many times we found a fatal flaw in what we thought was a proof.

References

[BGS75] T. Baker, J. Gill, and R. Solovay. Relativizations of the P=NP question. SIAM J. Comp., 4 (1975) pp. 431-442.

[BG81] C. H. Bennett and J. Gill. Relative to a random oracle A, $P^A neNP^A neCo - NP^A$ with probability 1. SIAM J. Comp. 10 (1981)

[BCC87] G. Brassard, D. Chaum, and C. Crépeau. Minimum disclosure proofs of knowledge. Technical Report PM-R8710, Centre for Mathematics and Computer Science, Amsterdam, The Netherlands, 1987.

[Ben87] J. Cohen Benaloh. *Verifiable Secret-Ballot Elections.* PhD thesis, Yale University, Sept 1987. YALEU/DCS/TR-561.

[Blu81] M. Blum. Three applications of the oblivious transfer: Part i: Coin flipping by telephone; part ii: How to exchange secrets; part iii: How to send certified electronic mail. Department of EECS, University of California, Berkeley, CA, 1981.

[Blu82] M. Blum. Coin flipping by telephone: A protocol for solving impossible problems. In *Proceedings of the 24th IEEE Computer Conference (CompCon)*, pages 133–137, 1982. reprinted in *SIGACT News*, vol. 15, no. 1, 1983, pp. 23–27.

[BM84] M. Blum and S. Micali. How to generate cryptographically strong sequences of pseudo-random bits. SIAM J. Comp. 13 (1984) pp. 850–864

[Bra] G. Brassard. An optimally secure relativized cryptosystem. *Advances in Cryptography, a Report on CRYPTO 81*, Technical Report no. 82-04, Department of ECE, University of California, Santa Barbara, CA, 1982, pp. 54–58; reprinted in *SIGACT News* vol. 15, no. 1, 1983, pp. 28–33.

[Bra83] G. Brassard. Relativized cryptography. *IEEE Transactions on Information Theory*, IT-19:877–894, 1983.

[CKS81] A.K. Chandra, D. Kozen, and L. Stockmeyer. Alternation. *JACM*, 28:114–133, 1981.

[DH76] W. Diffie and M. E. Hellman. New directions in cryptography. *IEEE Transactions on Information Theory*, IT-22:644–654, 1976.

[FFS86] U. Feige, A. Fiat and A. Shamir. Zero-knowledge proofs of identity. STOC, 1987.

[GGM84] O. Goldreich, S. Goldwasser, and S. Micali. How to construct random functions. In *Proceedings of the 25th Annual Foundations of Computer Science.* ACM, 1984.

[GMW87] O. Goldreich, S. Micali, and A. Wigderson. How to play any mental game or a completeness theorem for proto cols with honest majority. In *Proceedings of the 19th Annual Symposium on Theory of Computing.* ACM, 1987.

[GM84] S. Goldwasser and S. Micali. Probabalistic Encryption. *JCSS*, 28:270–299, 1984.

[GMR84] S. Goldwasser, S. Micali, and R. Rivest. A "paradoxical" solution to the signature problem. In *Proceedings of the 25th Annual Foundations of Computer Science.* ACM, 1984.

[I88] R. Impagliazzo Proofs that relativize, and proofs that do not. Unpublished manuscript, 1988.

[IY87] R. Impagliazzo and M. Yung. Direct minimum-knowledge computations. In *Proceedings of Advances in Cryptography.* CRYPTO, 1987.

[JVV86] Mark Jerrum, Leslie Valiant, and Vijay Vazirani. Random generation of combinatorial structures from a uniform distribution. *Theoretical Computer Science*, 43:169–188, 1986.

[LR86] M. Luby and C. Rackoff. How to construct pseudo-random permutations from pseudo-random functions. In *Proceedings of the Eighteenth Annual ACM Symposium on Theory of Computing*, 1986.

[Mer78] R. C. Merkle. Secure communications over insecure channels. *CACM*, 21(4):294–299, April 1978.

[NY] M. Naor and M. Yung. Universal One-Way Hash Functions and Their Applications. These precedings.

[P74] G. P. Purdy A high security log-in procedure. *CACM*, 17:442–445, 1974.

[Rab81] M. O. Rabin. How to exchange secrets by oblivious transfer. Technical Report TR-81, Harvard University, 1981.

[Rac88] C. Rackoff. A basic theory of public and private cryptosystems. Crypto 88.

[Yao82] A.C. Yao. Theory and applications of trapdoor functions. In *Proceedings of the 23rd Annual Symposium on Foundations of Computer Science*, pages 80–91. IEEE, 1982.

Generalized Secret Sharing and Monotone Functions

Josh Benaloh
University of Toronto

Jerry Leichter
Yale University

Abstract

Secret Sharing from the perspective of *threshold schemes* has been well-studied over the past decade. Threshold schemes, however, can only handle a small fraction of the secret sharing functions which we may wish to form. For example, if it is desirable to divide a secret among four participants A, B, C, and D in such a way that either A together with B can reconstruct the secret or C together with D can reconstruct the secret, then threshold schemes (even with weighting) are provably insufficient.

This paper will present general methods for constructing secret sharing schemes for *any* given secret sharing function. There is a natural correspondence between the set of "generalized" secret sharing functions and the set of monotone functions, and tools developed for simplifying the latter set can be applied equally well to the former set.

1 Introduction

The threshold schemes for secret sharing introduced by Blakley ([Blak79]) and Shamir ([Sham79]) have found many applications in recent years. There are, however, many secret sharing applications which do not fit into the model of threshold schemes.

In a recent paper ([ISN87]), Ito, Saito, and Nishizeki describe a general method of secret sharing whereby a secret can be divided among a set P of trustees such that any "qualified subset" of P can reconstruct the secret and such that unqualified subsets cannot. As they point out, it is most sensible to talk only about families of qualified subsets (or *access structures*) $\mathcal{A}$ which satisfy the property

$$A \in \mathcal{A}\,,\ A \subseteq A' \quad \Longrightarrow \quad A' \in \mathcal{A}.$$

It is hard to imagine a meaningful method of sharing a secret which does not satisfy this property.

The method of Ito, Saito, and Nishizeki can be roughly described as follows. For each of the (up to order $2^{|P|}$) sets of the access structure $\mathcal{A}$, divide the secret among each member of the set.[1] Thus, in the worst case, each of the n trustees may have to hold on the order of 2^n shares.

This paper gives a far simpler and more efficient method of developing a secret sharing scheme for any monotone access structure. The idea is to translate the access structure into a monotone formula.

Each variable in the formula is associated with a trustee in P, and the value of the formula is true if and only if the set of variables which are true corresponds to a subset of P which is in the access structure (i.e. the variables which are true correspond to a subset of trustees qualified to reconstruct the secret). This formula is then used as a template to describe how a secret is to be divided into shares.

Since every monotone function can implemented using just AND operators and OR operators, it is sufficient to show how to divide a secret "across" each of these two operators. It will be shown later how these formulae can be made more efficient by using general THRESHOLD operators and appealing to traditional threshold schemes.

Let p_1 and p_2 be trustees in P. To divide a secret s into shares such that p_1 and p_2 can reconstruct s, p_1 can be given a value s_1 and p_2 given a value s_2 such that $s = s_1 + s_2$. If s is selected from the range $0 \leq s < m$, then s_1 and s_2 can be chosen uniformly from this range subject to the constraint that $s = (s_1 + s_2) \bmod m$. In this case it can be shown in a very strong sense that neither p_1 nor p_2 can, without the other, obtain any information whatsoever about s.

To divide a secret s into shares such that p_1 or p_2 can reconstruct s, p_1 and p_2 can simply both be given the value s. With these two building blocks, it is easy to see how to construct a secret sharing scheme for any monotone access structure.

For instance, in the earlier example, a secret sharing scheme is sought for which either A together with B or C together with D can reconstruct the secret value s. The corresponding access structure can be written as $((A \wedge B) \vee (C \wedge D))$. Thus, to share a secret s according to this access structure, the secret is first moved across the OR yielding a situation in which the secret s now must be shared among AB and among CD. The value s is now moved across the two AND operators, yielding shares s_A, s_B, s_C, and s_D belonging respectively to A, B, C, and D such that $s_A + s_B = s$ and $s_C + s_D = s$. If the shares generated when a value is moved across an AND gate are random and independent of other selections, then it is not hard to show in a very strong sense that insufficient subsets of trustees obtain no information whatsoever about the original secret value.

There is, of course, no need to limit these gates to two inputs since both of the above operations generalize directly to gates with arbitrary fan-in. In general, a value can be moved across an arbitrary THRESHOLD operator by appealing to a traditional threshold scheme such as the Shamir scheme ([Sham79]). If some intermediate value s in a formula is to be moved across a threshold operator with n arguments and threshold k, the secret s is divided among the n arguments according to a (k, n)-threshold scheme, and these shares become the intermediate values for the next level of the formula.

[1] There is actually some minimization done as will be described later.

Since AND operators and OR operators are special cases of THRESHOLD operators, it would suffice to apply the Shamir threshold scheme to each operator of the formula. It is, however, often simpler to apply the direct methods above. Although the method of moving a secret across an OR operator described above does correspond exactly to Shamir's method of constructing a $(1, n)$-threshold scheme, the method given of moving a secret across an AND operator is computationally simpler than a Shamir (n, n)-threshold scheme. In addition, the threshold schemes given by Shamir and others have limitations which are not present in the scheme presented here. These limitations will be discussed later.

The method described by Ito, Saito, and Nishizeki in [ISN87] corresponds precisely to the case of minimal CNF-formulae in which conjunctions are formed by use of (n, n)-threshold schemes rather than by simple sums.

It is of course true that every monotone formula can be expressed as a CNF-formula and that there are a great many monotone formulae for which the CNF-formula is the smallest possible representation. However, there are also a great many cases in which the use of general monotone formulae (especially when arbitrary threshold operators are allowed) gives a much smaller formula than the CNF-formula. The number of shares which must be given to each trustee in these schemes as well as the complexity of reconstructing the secret from its shares are directly related to the size of the formula.

2 Preliminaries

To begin with, we must formally define the necessary access structures.

Definition Given a set P, a *monotone access structure* on P is a family of subsets $\mathcal{A} \subseteq 2^P$ such that

$$A \in \mathcal{A},\ A \subseteq A' \subseteq P \quad \Longrightarrow \quad A' \in \mathcal{A}.$$

Definition Let P be a set. The set V of *variables indexed by* P is the set $V = \{v_p : p \in P\}$.

Definition Given a monotone function F on variables indexed by a set P, the *access structure defined by* F is the set of subsets of A of P for which F is true precisely when the variables indexed by A are set to true.

It is clear that for every monotone function F, the access structure defined by F is a monotone access structure.

Definition For a given set P and an monotone access structure $\mathcal{A}$ on P, define $\mathcal{F}(\mathcal{A})$ to be the set of monotone formulae on $|P|$ variables such that for every formula $F \in \mathcal{F}(\mathcal{A})$, the output of F is true if and only if the true variables in F correspond exactly to a set $A \in \mathcal{A}$.

Note that $F, F' \in \mathcal{F}(\mathcal{A})$ implies that F and F' denote the same function. They may, however, represent entirely different formulae to express this function.

3 Generalized Secret Sharing

We can now begin to define secret sharing schemes. We start with a standard definition for threshold schemes.

Definition Given a set S of possible secret values, a (k,n)-threshold scheme on S is a (randomized) method of dividing each $s \in S$ into an array of shares $[s_1, s_2, \ldots, s_n]$ with each $s_i \in S$ such that

1. Given any set of k or more of the s_i, the secret value s is easily reconstructible.
2. Given any set of fewer than k of the s_i, the secret value s is completely undetermined in an information theoretic sense.

Shamir's polynomially based threshold scheme (see [Sham79]) satisfies the above definition whenever $|S|$ is a prime greater than n. It is not hard to remove the restriction that $|S|$ be prime by, for instance, factoring $|S|$ and using Chinese remaindering to encode secrets and shares. This kind of encoding, however, requires that all prime factors of $|S|$ be greater than n.

Other threshold schemes have been suggested by Blakley ([Blak79]), Asmuth and Bloom ([AsBl80]), and Kothari ([Koth84]), for example.

We want to show that no threshold scheme is sufficient to realize secret sharing on general monotone access structures. To do this, we show that there is no threshold scheme (even using weighting or multiple shares) such that the access structure $((A \wedge B) \vee (C \wedge D))$ can be achieved.

Theorem 1 *There exist monotone access structures for which there is no threshold scheme.*

Proof:

Consider the access structure $\mathcal{A}$ defined by the formula

$$((A \wedge B) \vee (C \wedge D)),$$

and assume that a threshold scheme is to be used to divide a secret value s among A, B, C, and D such that only those subsets of $\{A, B, C, D\}$ which are in $\mathcal{A}$ can reconstruct s.

Let a, b, c, and d respectively denote the weight (number of shares) held by each of A, B, C, and D. Since A together with B can compute the secret, it must be the case that $a + b \geq t$ where t is the value of the threshold. Similarly, since C and D can together compute the secret, it is also true that $c + d \geq t$.

Now assume without loss of generality that $a \geq b$ and $c \geq d$. (If this is not the case, the variables can be renamed.) Since $a + b \geq t$ and $a \geq b$, $a + a \geq a + b \geq t$. So $a \geq t/2$. Similarly, $c \geq t/2$. Therefore, $a + c \geq t$.

Thus, A together with C can reconstruct the secret value s. This violates the assumption of the access structure. ∎

Definition For a given threshold scheme, we use $\$_k(s; p_1, p_2, \ldots, p_n)$ to denote the random function which assigns shares $[s_1, s_2, \ldots, s_n]$ of a secret value s to trustees $p_1, p_2, \ldots, p_n$.

For certain access structures, every generalized threshold scheme *must* be able to assign multiple shares to each trustee (see theorem 3). In this case, we use $s_{i,j}$ to denote the jth share given to trustee p_i.

Definition Given a set P and a monotone access structure $\mathcal{A}$ on P, a generalized secret sharing scheme for $\mathcal{A}$ is a method of dividing a secret s into shares $s_{i,j}$ such that

1. When $A \in \mathcal{A}$, the secret s can be reconstructed from the shares $\bigcup_{i \in A} \bigcup_j s_{i,j}$.
2. When $A \notin \mathcal{A}$, the shares $\bigcup_{i \in A} \bigcup_j s_{i,j}$ give (in an information theoretic sense) no information whatsoever about the value of s.

We now define a generalized secret sharing scheme which satisfies the above definition.

Assume that the secret domain S is fixed to be the set $\{0, 1, \ldots, m-1\}$ for some positive integer m. We can now formally define the generalized secret sharing scheme described in section 1.

Let $\$(s, F)$ be the random function for $s \in S$ and a monotone formula F defined as follows.

- $\$(s, v_p)$ assigns the share s to trustee p.
- $\$(s, A \vee B) = \$(s, A) \cup \$(s, B)$.
- $\$(s, A \wedge B) = \$(s_1, A) \cup \$(s_2, B)$, where s_1 and s_2 are uniformly chosen from the secret domain S such that $s = (s_1 + s_2) \bmod m$.

If operators are allowed to have more than two arguments and if THRESHOLD operators are to be used, we add the following.

- $\$(s, \bigvee(F_1, F_2, \ldots, F_n)) = \bigcup_{1 \le i \le n} \$(s, F_i)$
- $\$(s, \bigwedge(F_1, F_2, \ldots, F_n)) = \bigcup_{1 \le i \le n} \$(s_i, F_i)$, where the s_i are chosen uniformly from S such that $s = (\sum_{i=1}^{n} s_i) \bmod m$.
- $\$(s, \text{THRESHOLD}_k(F_1, F_2, \ldots, F_n)) = \bigcup_{1 \le i \le n} \$(s_i, F_i)$, where $\$_k(s; p_1, p_2, \ldots, p_n) = [s_1, s_2, \ldots, s_n]$.

We now show that for every monotone access structure $\mathcal{A}$ and every monotone formula $F \in \mathcal{F}(\mathcal{A})$, the secret sharing scheme defined by $\$(s, F)$ satisfies the definition of a generalized secret sharing scheme.

Theorem 2 *Let P be a set and let $\mathcal{A}$ be a monotone access structure on P. Let F be a member of $\mathcal{F}(\mathcal{A})$, and let s be a secret value in $S = \{0, 1, \ldots, m-1\}$. The secret sharing scheme defined by $\$(s, F)$ is a generalized secret sharing scheme for $\mathcal{A}$.*

Proof:
It is easy to see that for any set $A \in \mathcal{A}$, the shares belonging to the members of A are sufficient to reconstruct the secret value s.

To see that if $A \notin \mathcal{A}$ then the shares belonging to the members of A give no information about the secret value s, we use induction on the number of operators of the formula F.

A formula with no operators consists of a single variable v_p. The access structure defined by v_p is the set of subsets of P which contain the trustee p. Thus, $\$(s, v_p)$ gives the secret value s to p alone and therefore allows only those sets of trustees which include p to determine s.

A monotone formula F with $d > 0$ operators can always be written in the form $\circ(F_1, F_2, \ldots, F_n)$ where $\circ$ is one of $\bigvee$, $\bigwedge$, and THRESHOLD_k, and where each of $F_1, F_2, \ldots, F_n$ is a monotone formula with less than d operators.

If the operator $\circ$ is $\bigvee$, then $\$(s, F_1, F_2, \ldots, F_n)$ is the union over i of $\$(s, F_i)$. By the inductive hypothesis, for each i, the members of a set A of trustees which is not in the access structure $\mathcal{A}$ can obtain no information whatsoever about the value of s from the values of the shares of $\$(s, F_i)$. Since for $i \neq j$, the shares of $\$(s, F_i)$ are chosen completely independently of the shares of $\$(s, F_j)$, no joint information is possible, and therefore, the shares of $\$(s, F)$ held by the members of an A not in $\mathcal{A}$ give no information at all about s.

If the operator $\circ$ is $\bigwedge$, then $\$(s, F_1, F_2, \ldots, F_n)$ is the union over i of $\$(s_i, F_i)$, where the s_i are chosen uniformly according the constraint that $s = (\sum_{i=1}^{n} s_i) \bmod m$. For each set A of trustees not in $\mathcal{A}$, there must be some i such that the shares of $\$(s_i, F_i)$ held my members of A give no information about s_i. (If this were not the case, then A would be in the access structure $\mathcal{A}$.) Since the shares given in each sub-formula are independent, this implies that the sum $s = (\sum_{i=1}^{n} s_i) \bmod m$ is completely undetermined by the shares held by the members of A.

Finally, if the operator $\circ$ is THRESHOLD_k, then $\$(s, F_1, F_2, \ldots, F_n)$ is the union over i of $\$(s_i, F_i)$, where the s_i are assigned according to the threshold scheme $\$_k$ by $\$_k(s; F_1, F_2, \ldots, F_n) = [s_1, s_2, \ldots, s_n]$. By assumption, a threshold scheme $\$_k$ allows sets of fewer than k shareholders to obtain no information at all about the value of s. If A is a set of trustees not in $\mathcal{A}$, then the members of A can obtain direct information about fewer than k of the s_i. Again by independence, the shares held by the members of A provide no information whatsoever about the value of s. ∎

Finally, we show that there are access structures which cannot be realized without giving multiple (or extra large) shares to some trustee.

Theorem 3 *There exists access structures for which any generalized secret sharing scheme must give some trustee shares which are from a domain larger than that of the secret.*

Proof:

Consider the access structure $\mathcal{A}$ defined by the formula

$$((A \wedge B) \vee (B \wedge C) \vee (C \wedge D)),$$

and fix a value a to be the share held by A.

Let $b_1, \ldots, b_m$ represent the set of possible shares available to B. Since A and B are together sufficient to compute the secret value s, each share b_i determines exactly one possible value s_i of the secret s. Also, since the share a alone is insufficient to give any information about the secret value s and since the number of possible values of s is equal to m (the number of possible values of the share held by B), every possible secret value s_i is determined by a together with exactly one value b_i.

Thus, for each a, we can construct a set of m pairs (s_i, b_i) which are consistent with a and such that each possible value of the secret and each possible value of B's share appear in *exactly* one such pair.

Now consider the possible value of the share held by C. Since B and C are together sufficient to compute the secret value s, and since each b_i can be matched with exactly one value c_i to form the secret s_i, there is exactly one value c_i consistent with each pair (s_i, b_i) in the set. (Note that the c_i are *not* necessarily distinct.)

If any two of these c_i are distinct, then considering the value held by A together with the value held by C would eliminate at least one of the possible consistent pairs and thereby eliminate at least one of the possible values of the secret s. But A and C are together *not* sufficient to determine *any* information about the value of the secret s. Thus, the value held by C must be completely determined by the value held by A.

Now since C and D are together sufficient to compute the secret value s, the value held by C together with the value held by D is sufficient to compute the secret value s. However, the value held by A completely determines the value held by C. Thus, the value held by A together with the value held by D is sufficient to compute s. This violates the premise that A and D are insufficient. ∎

4 Generalized Secret Sharing Homomorphisms

In [Bena86] and [Bena87], Benaloh describes a homomorphism property that is present in many threshold schemes which allows shares of multiple secrets to be combined to form "composite shares" which are shares of a composition of the secrets. Such secret sharing homomorphisms also apply to the generalized secret sharing scheme presented here.

For instance, if the shares of a secret value x (drawn from a fixed secret domain $S = \{0, 1, \ldots, m-1\}$) are added to the corresponding shares of a similarly chosen secret value y, then the sums represent shares of the value $(x + y) \bmod m$.

The applications of secret sharing homomorphisms includes fault-tolerant verifiable secret-ballot elections as well as verifiable secret sharing. The methods of verifiable secret sharing developed for threshold schemes in [Bena86] and [Bena87] and also by Feldman in [Feld87] can be used for generalized secret sharing too. The approach used by Feldman is actually somewhat better suited to these purposes. Here, the secret is distributed in such a way as to enable each trustee to, without further interaction, verify that its share is a well-formed and valid share of the secret.

The main requirement of these schemes is the presence of an appropriate homomorphism property, and the homomorphism property described above turns out to be sufficient.

5 Conclusions

This paper has shown how generalized secret sharing can be achieved in a method which is simpler and more efficient than in any previous scheme. There are, however, many cases in which this method is still unable to be applied efficiently.

For any given polynomial P, the number of n-variable monotone formulae of size no more than $P(n)$ is exponential in $P(n)$. However, the total number of monotone functions on n variables is doubly exponential in n. Therefore, most monotone access structures cannot be realized with a polynomially large number of polynomially sized shares.

Further methods of secret sharing which can efficiently realize additional access structures and an analysis of precisely what access structures can be efficiently realized are interesting areas for future research.

Acknowledgements

The authors would like to express their thanks to Ernie Brickell and to Steven Rudich for helpful discussions regarding this work.

References

[AsBl80] **Asmuth, C.** and **Bloom, J.** "A Modular Approach to Key Safeguarding." *Texas A&M University, Department of Mathematics,* College Station, TX (1980).

[Bena86] **Benaloh, J.** "Secret Sharing Homomorphism: Keeping Shares of a Secret Secret" *Proc. Crypto '86,* Santa Barbara, CA (Aug. 1986), 251–260. Published as *Advances in Cryptology,* ed. by A. Odlyzko in *Lecture Notes in Computer Science,* vol. 263, ed. by G. Goos and J. Hartmanis. Springer-Verlag, New York (1987).

[Bena87] **Benaloh, J.** "Verifiable Secret-Ballot Elections." Ph.D. Thesis presented at Yale University (Sep. 1987).

[Blak79] **Blakley, G.** "Safeguarding Cryptographic Keys." *Proc. AFIPS 1979 National Computer Conference,* New York, NY (June 1979), 313–317.

[Feld87] **Feldman, P.** "A Practical Scheme for Non-interactive Verifiable Secret Sharing." *Proc.* 28th *IEEE Symp. on Foundations of Computer Science,* Los Angeles, CA (Oct. 1987), 427–437.

[ISN87] **Ito, M.**, **Saito, A.**, and **Nishizeki, T.** "Secret Sharing Scheme Realizing General Access Structure." *Proc. Glob. Com,* (1987).

[Koth84] **Kothari, S.** "Generalized Linear Threshold Scheme." *Proc. Crypto '84,* Santa Barbara, CA (Aug. 1984), 231–241. Published as *Advances in Cryptology,* ed. by G. Blakley and D. Chaum in *Lecture Notes in Computer Science,* vol. 196, ed. by G. Goos and J. Hartmanis. Springer-Verlag, New York (1985).

[Sham79] **Shamir, A.** "How to Share a Secret." *Comm. ACM 22,* 11 (Nov. 1979), 612–613.

Session 2

Zero-Knowledge

Chair: C. Rackoff, University of Toronto

Everything Provable is Provable in Zero-Knowledge

Michael Ben-Or	Hebrew University
Oded Goldreich	Technion – Israel Institute of Technology
Shafi Goldwasser	M.I.T. Laboratory for Computer Science
Johan Håstad	Royal Institute of Technology, Sweden
Joe Kilian	M.I.T. Laboratory for Computer Science
Silvio Micali	M.I.T. Laboratory for Computer Science
Phillip Rogaway	M.I.T. Laboratory for Computer Science

Abstract

Assuming the existence of a secure probabilistic encryption scheme, we show that every language that admits an interactive proof admits a (computational) zero-knowledge interactive proof. This result extends the result of Goldreich, Micali and Wigderson, that, under the same assumption, all of *NP* admits zero-knowledge interactive proofs. Assuming envelopes for bit commitment, we show tht every language that admits an interactive proof admits a perfect zero-knowledge interactive proof.

1. Introduction

Suppose Bob is polynomially time-bounded, but Alice has unlimited computational resources. If ϕ is a satisfiable boolean formula, Alice can certainly convince Bob of this fact; she could send Bob a message y describing a satisfying truth assignment for ϕ, and Bob could check that y does indeed specify a satisfying truth assignment. In other words, the language L of satisfiable boolean formulas is in *NP*.

The interaction between Alice and Bob in this example is very simple: Alice sends a single message to Bob, and no other messages are sent between the two. If ϕ is satisfiable, there is some message y that Alice might send which will convince Bob to accept. But if ϕ is not satisfiable, then no message that Alice might send will convince Bob to accept.

In the paper of Goldwasser, Micali, and Rackoff [GMR], the authors extend the scenario above in two ways, to arrive at the notion of an *interactive proof for the language L*. First, the interaction between Alice and Bob is allowed to be more complicated, with Alice and Bob exchanging multiple messages. Secondly, Alice and Bob are taken to be *probabilistic*, and Bob may occasionally accept or reject erroneously. It is required that *if* an input is in L, then Alice can behave in such a way that Bob will almost always accept; but if an input is *not* in L, then, no mater what messages Alice sends, Bob will almost certainly reject.

A different notion of provability "beyond *NP*" was independently proposed by Babai [Bab]. This notion is called an *Arthur-Merlin protocol.* Babai's model is similar to that of [GMR], but is seemingly more limited, because the verifier is required to reveal to the prover all of his coin flips (right after making them). Though this loss of privacy seems an important restriction, Goldwasser and Sipser [GS] show that, in fact, the models are equivalent with respect to language recognition.

Let *IP* be the class of languages that admit interactive proofs. Clearly $NP \subseteq IP$, for an *NP*-interaction is a special type of *IP*-interaction, in which the prover (Alice) sends the one and only message, and the verifier (Bob) never errs. However, *IP* may be a much larger class of languages. For example, there is an interactive proof known for graph *non*isomorphism, even though there are not known to be succinct certificates for establishing that a pair of graphs are not isomorphic.

In this paper, we are concerned with *zero-knowledge* interactive proofs. A zero-knowledge interactive proof for a language L is an interactive proof for L for which, on any input in L, the prover divulges to the verifier no significant amount of information *except* that the input is in L. (The notion of zero-knowledge we refer to has sometimes been called *computational zero-knowledge*, to distinguish it from two other notions of zero-knowledge that appear in the literature, *perfect zero-knowledge* and *statistical zero-knowledge*. [Fo]) For example, a zero-knowledge interactive proof that ϕ is a satisfiable boolean predicate convinces the verifier that ϕ is satisfiable, but not, say, by exhibiting a satisfying truth assignment.

Though the models of [Ba] and [GMR] are equivalent with respect to language recognition, they are likely *not* the same with respect to zero-knowledge; zero-knowledge interactive proofs frequently make use of the verifier's ability to have secret coins.

It might seem that requiring an interactive proof be zero-knowledge is generally too much to hope for; one might expect that relatively few languages with interactive proofs admit zero-knowledge interactive proofs. This was shown to probably not be the case in a paper of Goldreich, Micali, and Wigderson [GMW1]. Here the authors show that, assuming the existence of a secure probabilistic encryption scheme, *every* language in *NP* admits a zero-knowledge interactive proof.

We generalize this result to establish that, under the same assumption, *every* language that admits an interactive proof admits a zero-knowledge interactive proof.

A brief note on the history of this theorem. The result was stated in [GMW1], attributed to Ben-Or; this proof was never published. A published sketch of a proof appears in the CRYPTO-87 paper of Chaum, Damgård, and van de Graaf [CDG]. However, the result seems to require a stronger assumption, such as a pair of claw-free trapdoor functions. We have been learned [I] that Russell Impagliazzo and Moti Yung independently and by different methods had a proof of this theorem, which will appear in journal-form shortly [Yu]; their work is sketched in [IY].

In this paper, we will point out some subtleties involved and formally prove the theorem. We then discuss the "physical model" in which a bit can be committed by putting it in an envelope, and we show how to obtain perfect zero-knowledge proofs for *IP* under this model. The technique used here is different from the method that employs encryption. (The proofs of [IY] and [CDG] can be adapted to the envelope model, as well, as pointed out by [Br] and [Yu].)

The paper is organized as follows. Section 2 gives the preliminaries needed to understand the main theorem, including both definitions and well-known or technical results. The reader familiar with this area might skim or skip this section. Section 3 is devoted to the proof of the main theorem. Section 4 shows how to obtain perfect zero-knowledge proofs for languages in *IP* in the model in which a bit can be committed by putting it in an envelope. The remainder of this section is an informal overview of the proof of the main theorem.

1.1. Overview of the construction

We wish to show that, if $(P \leftrightarrow V)$ is an interactive proof system for the language L, then P and V can be modified to P' and V', such that $(P' \leftrightarrow V')$ is an interactive proof system for L, *but* P' *is zero-knowledge over* L.

Suppose $(P \leftrightarrow V)$ is an interactive proof system for the language L. We would like to carry out the "same interaction" in a way that betrays essentially no information to V. To do this, we could have P encrypt each message that it sends to V. That is, P uses a secure encryption function, E. On the ith round, when P "would have" sent to V the string y_i, P instead sends to V the string $E(y_i, d_i)$, a random encryption of y_i. (We assume that $E(x, s) = E(y, t)$ implies $x = y$, and that from $E(x, s)$ and s one can efficiently compute x. Security is with respect to *nonuniform* poly-time computation.)

There are two immediate difficulties. First, how can V be expected to compute his responses to P, since he doesn't understand what P has sent? Second, how can V be convinced to accept the string x if, as far as he can tell, P has sent him complete gibberish?

The first problem—that V won't know what to do with the messages he's received—is answered as follows. By the result of Goldwasser and Sipser [GS], there is an *Arthur-Merlin* protocol, $(M \leftrightarrow A)$, for the same language, L. In $(M \leftrightarrow A)$, Arthur sends only his coin flips, so Arthur needn't understand the messages he's received in order to respond.

The second problem—that Arthur can't tell whether or not he ought accept—is answered as follows. If Arthur could *guess* the encryption keys, $d_1, d_2, \ldots$, he would have no problem knowing whether or not to accept, for he could decrypt each message sent by Merlin and accept or reject based on the same predicate he would have used had the conversation been carried out unencrypted. Of course, Arthur can't be expected to guess $d_1, d_2, \ldots$, but the statement "If you guessed $d_1, d_2, \ldots$, you'd accept x based on this interaction" is in *NP*. Since, by [GMW1], all of *NP* can be proven in zero-knowledge, there is a way for Merlin to convince Arthur of the validitity of this statement that is in zero-knowledge.

The construction just given has the following defect:

To show that $(M \leftrightarrow A)$ is a zero-knowledge proof system for L, we need to argue that *any* A^* learns essentially nothing by interacting with M. Suppose A^* cheats by flipping a biased coin in place of his random tape, with bias, say, $p = 3/4$. If A^* several times interacts with M on a common input $x \in L$, and if M usually convinces A^* to accept, then, intuitively, A^* has learned something: that most strings taken from this distribution lead to accepting x when used for A's random coins. This is "real knowledge," for it is entirely possible that even though M usually convinces A to accept if A uses a fair coin, M usually fails to convince A to accept if A uses a 3/4–biased coin.

(If we tried to prove that M is zero-knowledge, here's where we'd get stuck. The simulator M_{A^*} simulates the behavior of a "virtual prover," P^*, interacting with A^*. On common input $x \in L$, P^* sends random encryptions of the appropriate length string of 0's. When this phase of the interaction is finished, we append a simulated proof that Arthur would accept *if* he correctly guessed $d_1, d_2, \ldots$. But it is not always appropriate to append this simulated proof! For in the real interaction, $(M \leftrightarrow A^*)$, A^* sometimes rejects strings in L. Quite possibly, A^* has chosen a distribution of strings for which A would *usually* reject strings in L. So if P^* always appends the simulated proof that Arthur would accept, the resulting view may significantly differ from the real view of the interaction. But P^* has no way to know if it ought or ought not send the simulated proof.)

One possible fix is to use the result of Goldreich, Mansour, and Sipser [GMS], which says that we may, without loss of generality, take $(M \leftrightarrow A)$ to be *one-sided*. That is, we may assume that regardless of the coins that A employs on $x \in L$, A will be convinced to accept x. (Consequently, it will *always* be appropriate for P^* to convince A^* to accept in the simulated interaction.) This is the course that we shall follow.

Another possible fix is to use a "coin flip into the well" for A's coins ([Bl]). To make this proposal work, it is necessary that the coin flips that are agreed to are *statistically indistinguishable* ([Fo]) from truly random coin flips.

It is important that, in our formulation, independent encryption functions are, effectively, used on each round of the interaction. For example, we can not prove that it will be zero-knowledge for the prover to choose a public key encryption algorithm, E, with decryption algorithm D, send E to the verifier, and then send a random encryption of y_i under E, when y_i would otherwise be sent. (The verifier is convinced via the assertion, "If you guessed D, you'd accept the corresponding unencrypted conversation.")

(Here is the problem with such a scheme: Merlin, having received strings $x_1, \ldots, x_i$ from A^*, computes y_i by a probabilistic function of $(x, x_1, \ldots, x_i)$. Thus y_i is drawn from a probability space R_i which A^* has some influence over. The encryption function is secure, so $(x_1, \ldots, x_i)$ can be only slightly correlated to D. *But a weak correlation of $(x, x_1, \ldots, x_i)$ to D may result in R_i being strongly correlated to D, for M is not necessarily polynomial time.* In fact, there is no reason to assume that R_i is not precisely the space that A^* wants it to be. But we mustn't allow A^* to have such strong influence over R_i; what if A^* forces R_i to be the space, say, with unit probability mass on D? Then Merlin sends Arthur $E(D)$ (a random encryption of the decryption function). This possibly compromises the the encryption function. In general, we worry that A^* may be able to select a space R_i for Merlin such that taking y_i from R_i and encrypting y_i under E compromises E.)

In any case, requiring a public key cryptosystem is a *stronger* assumption than the commitment scheme $E(msg, rand)$ that we require.

2. Preliminaries

2.1. Interactive proof systems

The definition we give for an *interactive proof system* is essentially that of Goldwasser, Micali, and Rackoff [GMR]; see this paper for a more complete discussion of interacting Turing machines and interactive proofs. It does not significantly effect the model if one assumes that the prover never halts, the verifier sends the first message, and communication is done on a single communication tape. We build these assumptions into the definition:

An *interactive proof system*, $(P \leftrightarrow V)$, consists of a pair of probabilistic Turing machines, P and V, with common alphabet Σ. P and V each have distinguished *start* and *quiescent* states. V has distinguished *accept* and *reject* states, out of which there are no transitions. P and V operate on various one-way infinite tapes:

- P and V have a common read-only *input tape*.
- P and V each have a private *random tape*, and a private *work tape*.
- P and V have a common *communication tape*.
- V is polynomially time-bounded. This means that there is a polynomial p for which, on inputs of length n, V experiences at most $p(n)$ state transitions before it accepts or rejects. V does *not* transition when it is quiescent, and P is running.
- P is finite expected time. This means that there is a function f such that, on inputs of length n, P's expected computation time from start to quiescent states does not exceed $f(n)$, regardless of the messages P has received.
- There is a polynomial r such that P never writes more than $r(n)$ characters (including blanks) on the communication tape when the common input is of length n.

Execution begins with P in its quiescent state and V in its start state. V's entering its quiescent state arouses P, causing it to transition to its start state. Likewise, P's entering its quiescent state causes V to transition to its start state. Execution terminates when V enters its accept or reject state.

If P and V are given random tapes $\sigma, \tau \in \Sigma^\omega$, respectively, and are then run on input $x \in \Sigma^*$, with work tapes initially empty, then the final state that V enters is well-defined, and we say that $(P_\sigma \leftrightarrow V_\tau)(x)$ *accepts* or *rejects* accordingly. If we omit mention of σ and τ, then we may speak of the "probability that $(P \leftrightarrow V)(x)$ accepts," $Pr[(P \leftrightarrow V)(x) \text{ accepts}]$.

Definition. $(P \leftrightarrow V)$ is an *interactive proof system for the language L* if, for some $0 \leq \epsilon < 1/2$, we have both:

completeness: $\quad x \in L \implies Pr[(P \leftrightarrow V)(x) \text{ accepts}] \geq 1 - \epsilon,\quad$ and

soundness: $\quad (\forall P^*) \quad x \notin L \implies Pr[(P^* \leftrightarrow V)(x) \text{ accepts}] \leq \epsilon.$

$(P \leftrightarrow V)$ is a *one-sided interactive proof system* if, in place of completeness, we have:
perfect completeness: $x \in L \implies Pr[(P \leftrightarrow V)(x) \text{ accepts}] = 1.$

That is, a one-sided interactive proof *always* accepts x when $x \in L$, regardless of the contents of the random tapes.

The number ϵ in this definition is called the *error probability*. By the standard method of running the protocol multiple times, we may take the error probability to be any constant in $(0, 1]$—or even any error probability of the form $\epsilon(n) = 2^{-p(n)}$, where p is a polynomial.

In order to extend our discussion to speak of *knowledge*, we consider the possibility that V's work tape initially contains "some knowledge." Suppose P and V are given random tapes $\sigma, \tau \in \Sigma^\omega$, respectively, and are run on input $x \in \Sigma^*$, with $s \in \Sigma^*$ initially placed on V's work tape. Then not only is the final state of V well-defined, but so are:

- The *number of rounds*, $2m$, for which P and V interact. (The number of rounds is the number of messages sent between A and B.)
- The ith message sent from V to P, x_i. (A *message* is a prefix of the communication tape, from its left end to the first blank).
- The ith message sent from P to V, y_i.
- The (finite) prefix τ_0 of τ that V reads.

That is, $(P, V, \sigma, \tau, x, s)$ determine the number m and strings $\vec{x} = x_1; \ldots; x_m;$, $\vec{y} = y_1; \ldots; y_m;$, as above. We define from these the *public history* of the interaction and the *view* of the interaction:

$$(P_\sigma \overset{pub}{\longleftrightarrow} V_\tau)(x, s) = [x, \vec{x}, \vec{y}],$$
$$(P_\sigma \overset{view}{\longleftrightarrow} V_\tau)(x, s) = [x, s, \tau_0, \vec{y}].$$

Interpret the right hand side of each of these definitions as the binary encoding of the specified string, where '[', ';' and ',' are new (formal) symbols.

Informally, the public history is the interaction as it would be observed from the "outside;" the view is the interaction as seen from V's perspective.

If we omit mention of σ and τ, then $(P \overset{pub}{\longleftrightarrow} V)(x, s)$ and $(P \overset{view}{\longleftrightarrow} V)(x, s)$ are *probability spaces*. $(P \overset{pub}{\longleftrightarrow} V)$ and $(P \overset{view}{\longleftrightarrow} V)$ (no mention of x or s) are *families of probability spaces*.

2.2. Arthur-Merlin protocols

In the definition for an interactive proof, the verifier was not compelled to reveal his coins flips (the prefix of his random tape that he uses) to the prover. If the verifier does reveal his coin flips at each round, there is no reason for him to send anything else, since the prover himself could as well compute anything else the verifier would have sent. A seemingly weaker notion of an interactive proof, introduced by Babai [Ba] [BaM], is obtained by limiting the verifier's messages in this way.

Definition. An interactive proof for L, $(M \leftrightarrow A)$, is an *Arthur-Merlin protocol* if for some polynomials r and l, any interaction between M and A on an input of length n takes exactly $r(n)$ rounds, each message sent being of length $l(n)$. Moreover, the $r(n)$ messages sent by A, $x_1, \ldots, x_{r(n)}$, are precisely the prefix $\tau_0 = x_1 \ldots x_{r(n)}$ of A's random tape that A consumes.

The first condition alone is easily seen not to weaken the model from that of an interactive proof system. Surprisingly, the second condition does not weaken the model either:

Theorem 2.1. *If $(P \leftrightarrow V)$ is an interactive proof system for L, then there is a one-sided, Arthur-Merlin protocol $(M \leftrightarrow A)$ for L.*

The result without "one-sided" is due to Goldwasser and Sipser [GS]; it was extended to proofs of perfect completeness by Goldreich, Mansour, and Sipser [GMS]. Recently, J. Kilian [K] discovered a much simpler argument for Theorem 2.1.

2.3. Zero-knowledge

$C_n \in \Sigma^*$ $(n \in \mathbb{N})$, $\mathcal{C} = \{C_n\}$ is a *poly-size family of circuits* if there are polynomials p and q such that $|C_n| \leq p(n)$, and C_n encodes (via some fixed universal Turing machine) a (deterministic) algorithm which, on input $x \in \Sigma^*$, requires at most $q(|x|)$ steps before it outputs a bit, 0 or 1.

If C is an algorithm that outputs a bit, and R is a probability space, then we may speak of "the probability that C outputs a 1 on input drawn from R," $p_R^C = \sum_{\sigma \in \Sigma^*} Pr_R(\{\sigma\}) \cdot C(\sigma)$.

We define zero-knowledge in terms of families of probability spaces indexed by two variables, which are treated differently.

Definition. Let $R = \{R(x,s)\}$ and $S = \{S(x,s)\}$ be families of probability spaces, indexed by $\Sigma^* \times \Sigma^*$. Then R and S are *indistinguishable over* L if, for any poly-size family of circuits $\{C_n\}$, any polynomial q, and all sufficiently long x in L,

$$\left| p_{R(x,s)}^{C_{|x|}} - p_{S(x,s)}^{C_{|x|}} \right| < \frac{1}{q(|x|)}$$

for all $s \in \Sigma^*$. If R and S are indistinguishable over L, we write $R \equiv_L S$ to denote this.

Definition. P is *zero-knowledge over* L if, for any V, there exists an expected polynomial-time algorithm M_V such that $(P \overset{view}{\longleftrightarrow} V) \equiv_L M_V$.

Definition. $(P \leftrightarrow V)$ is a *zero-knowledge interactive proof system for* L if $(P \leftrightarrow V)$ is an interactive proof system for L, and P is zero-knowledge over L.

We have defined indistinguishability with respect to poly-size families of circuits. In the proof of the main theorem, it will be convenient to think of indistinguishability with respect to poly-size families of probabilistic polynomial time algorithms. As with circuits, this is a *non-uniform* concept; there may be no algorithm which, on input n, outputs the expected poly-time algorithm C_n. By an averaging argument, and exploiting nonuniformity, it is easy to see that the notion of indistinguishability is unchanged if we define indistinguishability with respect to poly-size families of circuits, or with respect to poly-size families of probabilistic polynomial time probabilistic algorithms.

2.4. Preliminary results

It is frequently convenient to assume that, when P and V interact, the interaction takes place for a fixed number of rounds, messages are of a fixed length, and V uses a fixed number of coin flips per round. The following proposition says that there is no loss of generality in making these assumptions. The proof is straightforward and has been omitted.

Proposition 2.2. *If $(P \leftrightarrow V)$ is a (zero-knowledge) (one-sided zero-knowledge) interactive proof system for L, then there exists an P', V', and polynomials r, l, t, such that $(P' \leftrightarrow V')$ is a (zero-knowledge) (one-sided zero-knowledge) interactive proof system for L, and on each input of length n, the interaction runs for exactly $r(n)$ rounds, each message exchanged of length $l(n)$, and V' flipping precisely $t(n)$ coins for each message that it sends.* ◇

With respect to language recognition, we may further assume that the prover is deterministic. This observation (actually, that *PSPACE* was enough for the prover), was first made by Feldman [Fe].

Proposition 2.3. *If $(P \leftrightarrow V)$ is an interactive proof system for L, then there is a deterministic P' for which $(P' \leftrightarrow V)$ is an interactive proof system for L. If $(P \leftrightarrow V)$ was an Arthur-Merlin protocol, then so will be $(P' \leftrightarrow V)$.* ◇

The next proposition depends on the fact that the "composition" of zero-knowledge interactive proofs remains zero-knowledge. A proof of Proposition 2.4 can be found in the paper of Tompa and Woll [TW].

Proposition 2.4. *If L admits a (one-sided) zero-knowledge interactive proof, then L admits a (one-sided) zero-knowledge interactive proof with error probability ϵ for any $0 < \epsilon \leq 1/2$.* ◇

To state the next lemma—which conceptually simplifies the argument of the main result—we define the ***composition*** of two interactive proof systems, $(P_1 \leftrightarrow V_1)$ and $(P_2 \leftrightarrow V_2)$. Let us assume that the former always uses $r(|x|)$ rounds on any input x, and that each message is of length $l(|x|)$. (By Proposition 2.2, this entails no loss of generality.) $((P_2 \circ P_1) \leftrightarrow (V_2 \circ V_1))$ is defined as follows: Initially, $P_2 \circ P_1$ and $V_2 \circ V_1$, acting on common input x, behave like P_1 and V_1, respectively, acting on common input x. This continues for the first $r(|x|)$ rounds. However, $P_2 \circ P_1$ and $V_2 \circ V_1$ each record the public history of the interaction during these $r(|x|)$ rounds. After that, $P_2 \circ P_1$ checks that the public history of the interaction is a public history that *could* arise in a $(P_1 \leftrightarrow V_1)$ interaction on x. If so, $P_2 \circ P_1$ behaves like P_2 acting on input of the public history of the preceeding interaction; if not, all future messages of $P_2 \circ P_1$ are the empty string. $V_2 \circ V_1$ continues by behaving like V_2, acting on input of the public history of the preceeding interaction.

The technical lemma we need is

Lemma 2.5. *Suppose $(P_2 \leftrightarrow V_2) \circ (P_1 \leftrightarrow V_1)$ is an interactive proof system for L. Suppose P_1 is zero-knowledge over L, and P_2 is zero-knowledge over L', where $L' = (P_1 \overset{pub}{\longleftrightarrow} V_1)(L)$ is the set of all public histories that might arise in a $(P_1 \leftrightarrow V_1)$ interaction about a string in L. Then $P_2 \circ P_1$ is zero-knowledge over L.*

Proof: Let r_1 be the polynomial such that $(P_1 \leftrightarrow V_1)$ uses $r_1(|x|)$ rounds on any input x, and let l be the polynomial such that each of these messages is of length $l(|x|)$.

Let W be any polynomial-time probabilistic algorithm that interacts with $P_2 \circ P_1$. We may assume that $(P_2 \circ P_1 \leftrightarrow W)$ uses $r(|x|)$ rounds on any input x, where r is a polynomial exceeding r_1 pointwise.

We must exhibit an expected polynomial time machine M (of two arguments) for which $(P_2 \circ P_1 \overset{view}{\longleftrightarrow} W) \equiv_L M$.

Begin by constructing from W machines W_1 and W_2 as follows. W_1, on input (x, s), behaves exactly like W would behave on (x, s), but only for $r_1(|x|)$ rounds. After that, W_1 immediately accepts (or rejects).

W_2 takes as input a pair (x, s), where we assume $s = [\ \hat{x}, \hat{s}, r, \vec{y}\ ]$, and $\vec{y}$ is of the form $y_1; \ldots; y_m;$. s is the view of part of a computation of W. W_2 runs W, to resurrect the state W would be in after the conversation indicated by s. After that, W_2 behaves like W, starting from this state.

Since P_1 is zero-knowledge over L, there is an expected poly-time M_1 such that $(P_1 \overset{view}{\longleftrightarrow} W_1) \equiv_L M_1$.

Since P_2 is zero-knowledge over L', there is an expected poly-time M_2 such that $(P_2 \overset{view}{\longleftrightarrow} W_2) \equiv_{L'} M_2$.

M is constructed by "composing" M_1 and M_2. On input (x, s), M first runs M_1, to compute a string which, without loss of generality, looks like $[x, s, \tau_1, \vec{y}_1]$. Next, M runs $M_2(\Lambda, [x, s, \tau_1, \vec{y}_1])$, producing a string $[\Lambda, [x, s, \tau_1, \vec{y}_1], \tau_2, \vec{y}_2]$. M outputs $[x, s, \tau_1\tau_2, \vec{y}_1\vec{y}_2]$.

$M(x, s)$ can be described as the probability space resulting from performing the following experiment:

$$\textit{Experiment } M: \quad \begin{array}{l} [x, s, \tau_1, \vec{y}_1] \leftarrow M_1(x, s). \\ [x, s, \tau_2, \vec{y}_2] \leftarrow M_2(\Lambda, [x, s, \tau_1, \vec{y}_1]). \\ \text{OUTPUT } [x, s, \tau_1\tau_2, \vec{y}_1\vec{y}_2]. \end{array}$$

$R(x, s) = (P_2 \circ P_1 \overset{view}{\longleftrightarrow} W)(x, s)$ can be viewed as the probability space associated with the following experiment:

$$\textit{Experiment } R: \quad \begin{array}{l} [x, s, \tau_1, \vec{y}_1] \leftarrow (P_1 \overset{view}{\longleftrightarrow} V_1)(x, s). \\ [x, s, \tau_2, \vec{y}_2] \leftarrow (P_2 \overset{view}{\longleftrightarrow} V_2)(\Lambda, [x, s, \tau_1, \vec{y}_1]). \\ \text{OUTPUT } [x, s, \tau_1\tau_2, \vec{y}_1\vec{y}_2]. \end{array}$$

We now have two families of probability spaces, R of "real" prover-verifier interactions, and M of simulated interactions. Let's introduce one more, H, of "hybrid" interactions, with the following experiment used to define the probability space $H(x, s)$:

$$\textit{Experiment } H: \quad \begin{array}{l} [x, s, \tau_1, \vec{y}_1] \leftarrow (P_1 \overset{view}{\longleftrightarrow} V_1)(x, s). \\ [x, s, \tau_2, \vec{y}_2] \leftarrow M_2(\Lambda, [x, s, \tau_1, \vec{y}_1]). \\ \text{OUTPUT } [x, s, \tau_1\tau_2, \vec{y}_1\vec{y}_2]. \end{array}$$

We now argue that $R \equiv_L M$. Suppose that this is not the case. Then there is some poly-size family $\mathcal{C} = \{C_n\}$ and some polynomial q such that

$$\left| p^{C_{|x|}}_{R(x,s_x)} - p^{C_{|x|}}_{M(x,s_x)} \right| \geq \frac{1}{q(|x|)}$$

for infinitely many $(x, s_x) \in L \times \Sigma^*$. Then either

$$\left| p^{C_{|x|}}_{R(x,s_x)} - p^{C_{|x|}}_{H(x,s_x)} \right| \geq \frac{1}{2q(|x|)} \tag{1}$$

or

$$\left| p^{C_{|x|}}_{H(x,s_x)} - p^{C_{|x|}}_{M(x,s_x)} \right| \geq \frac{1}{2q(|x|)} \tag{2}$$

for infinitely many $(x, s_x) \in L \times \Sigma^*$. We show that both of these cases are impossible.

Case 1. ((1) holds infinitely often.) Choose an (x, s_x) for which (1) holds. Single out the coin flips used by V_1, and the messages $y_1, \ldots, y_{r_1(|x|)}$ that P_1 might send to V_1 in the interactions defining $R(x, s_x)$ and $H(x, s_x)$. There is a ***particular*** sequence of coin flips σ for V_1 to use, and a ***particular*** vector of messages $\tau = y_1 \ldots y_{r_1(|x|)}$ for P_1 to use, such that

$$\left| p^{C_{|x|}}_{R_{\sigma,\tau}(x,s_x)} - p^{C_{|x|}}_{H_{\sigma,\tau}(x,s_x)} \right| \geq \frac{1}{2q(|x|)},$$

where $p^{C_n}_{R_{\sigma,\tau}(x,s_x)}$ is the probability that C_n outputs 1 when Experiment R is run with σ and τ used for V_1's coins and P_1's messages, respectively; likewise for $p^{C_n}_{S_{\sigma,\tau}(x,s_x)}$. Consequently, we may "hardwire" into C_n the values we obtain from $(P_1 \leftrightarrow V_1)$ interacting on (x, s_x) using σ and τ, to obtain a circuit which distinguishes $(P_2 \leftrightarrow V_2)(x, s_x)$ from $M_2(x, s_x)$ by at least $1/(2q(|x|)$. The existence of the family of circuits modified as specified here contradicts $(P_2 \leftrightarrow V_2)(\cdot,\cdot) \equiv_{L'} M_2(\cdot,\cdot)$.

Case 2, when (2) holds infinitely often, is handled analogously. ◇

For completeness, we state the following trivial proposition:

Proposition 2.6. *Let $L' \subseteq L$. If P is zero-knowledge over L, then P is zero-knowledge over L'.* ◇

2.5. Secure probabilistic encryption

The prover in our protocol will need the ability to securely commit a bit, and to convincingly decommit it. We formalize this by saying that a *secure probabilistic encryption scheme* is a function $E : \Sigma \times \Sigma^* \to \Sigma^*$ such that

(1) E is computable in polynomial time.

(2) *Unique decryption*: $E(\beta, x) = E(\beta', y)$ implies $\beta = \beta'$.

(3) Let $E_n(\beta)$ be the probability space obtained by setting $Pr(y) = 2^{-n} \cdot |\{x \in \Sigma^n : E(\beta, x) = y\}|$. We require that for any poly-size family of circuits $\mathcal{C} = \{C_n\}$, for any polynomial q, and for all sufficiently large n,

$$\left| p_{E_n(0)}^{C_n} - p_{E_n(1)}^{C_n} \right| < \frac{1}{q(n)}.$$

(Recall p_R^C is the probability that circuit C outputs 1 on input drawn from R. Note that to achieve the unique decryption condition with conventional encryption schemes, "certified primes" must be used [GK][AH].) We write $\{E_n(0)\} \equiv_{\mathbf{N}} \{E_n(1)\}$ to denote the security condition.

Without loss of generality, there is a polynomial q such that $|E(\beta, x)| = q(|x|)$ for all $x \in \Sigma^*$.

To encrypt a bit β with security parameter n, select a random n-bit string x and send $E(\beta, x)$. To decommit, reveal x. The unique decryption condition makes it impossible that the commited bit could be $1 - \beta$. Also, from x and $E(\beta, x)$ one can easily compute β.

To encrypt a string $m = \beta_1 \ldots \beta_\ell$ with security parameter n, send $E(\beta_1, x_1) \ldots E(\beta_\ell, x_\ell)$ for random n-bit strings $x_1, \ldots, x_\ell$. The encryption will be denoted $E_n(m, x)$, where $x = x_1 \ldots x_m$, and the corresponding probability space is denoted $E_n(m)$.

A secure encryption scheme exists if there are unapproximable predicates [GM], or if there are injective one-way functions [Ya][L][G]. (If f is injective one-way, then there are poly-time computable functions f' and a b such that $f'(x) = f'(y)$ implies $b(x) = b(y) \in \{0, 1\}$, and no poly-size circuit family can predict $b(x)$ given $f(x)$ by better than $1/2 + n^{-c}$, for any constant c. Given such f' and b, E as we have described it can readily be constructed.)

The crucial property we need of a secure probabilistic encryption scheme is the following:

Lemma 2.7. *Assume the existence of a secure probabilistic encryption scheme. Let $\{y_n\}$ be a collection of strings, where $|y_n| = l(n)$, for some nonconstant polynomial l. Then*

$$\{E_n(y_n)\} \equiv_{\mathbf{N}} \{E_n(0^{l(n)})\}.$$

Proof: Suppose to the contrary that there is a poly-size family $\mathcal{C} = \{C_n\}$ and a polynomial q such that

$$\left| p_{E_n(y_n)}^{C_n} - p_{E_n(0^{l(n)})}^{C_n} \right| \geq \frac{1}{q(n)}$$

for infinitely many $n \in \mathbf{N}$. Pick a particular n for which this holds. Define the strings y_n^i, for $0 \le i \le n$, by $y_n^i = y_n[1..i]0^{l(n)-i}$. Note $y_n^0 = 0^{l(n)}$, and $y_n^{l(n)} = y_n$. There exists, then, a j, $0 \le j < l(n)$, such that

$$\left| p^{C_n}_{E_n(y_n^j)} - p^{C_n}_{E_n(y_n^{j+1})} \right| \ge \frac{1}{l(n) \cdot q(n)}.$$

Note that y_n^j and y_n^{j+1} agree at all positions except the $(j+1)$–*st* where y_n^j is 1 and y_n^{j+1} is 0. Consequently, we may hardwire into C_n the values of y_n^j at each position except the $(j+1)$–*st*, to obtain a coin-flipping circuit which distinguishes encyptions of 0 from encryptions of 1 by at least $1/(l(n)q(n))$. Converting to a deterministic circuit, we contradict condition (3) about our encryption scheme. ◇

2.6. Zero-knowledge proofs for all of *NP*

The following lemma and theorem are due to Goldreich, Micali, and Wigderson [GMW1]. The proof of the first of these is omitted.

Lemma 2.8. *If secure probabilistic encryption is possible, then the language of (encodings of) 3-colorable graphs admits a (one-sided) zero-knowledge interactive proof.* ◇

Theorem 2.9. *If secure probabilistic encryption is possible, then any language in NP possesses a (one-sided) zero-knowledge interactive proof.*

Proof: Take $L \in NP$, and let M be a nondeterministic Turing machine for L. Fix a canonical transformation φ that takes any (M, x) (the encoding of a nondeterministic Turing machine and an input x) to a graphs G. φ is poly-time computable, and has the property that M accepts x iff $\varphi(M, x)$ is 3-colorable.

To prove $x \in L = L(M)$ in zero-knowledge, both the prover and the verifier compute the graph $G = \varphi(M, x)$, and engage in a zero-knowledge interactive proof (using Lemma 2.8) that G is 3-colorable. ◇

3. Proof of the main theorem

We now prove the main theorem of this paper:

Theorem 3.1. *Assuming a secure probabilistic encryption scheme exists, every language that admits an interactive proof admits a zero-knowledge interactive proof.*

Proof: Suppose L admits an interactive proof. Then, by Theorem 2.1, L admits a *one-sided, Arthur-Merlin* interactive proof $(M \leftrightarrow A)$. By Proposition 2.3, M may be assumed to be deterministic. By Proposition 2.4, we may take the error probability of $(M \leftrightarrow A)$ to be less than 1/5.

By Proposition 2.2, $(M \leftrightarrow A)$ may be assumed to always use $r(n)$ rounds, each message of length $l(n)$, when M and A interact with common input x of length n. (r and l are polynomials.)

We will construct from M and A a zero-knowledge, one-sided interactive proof system $(P \leftrightarrow V)$ for L.

Suppose Arthur's random tape contains a given infinite string. On input x of length n, Arthur only uses the $(l(n) \cdot r(n))$–bit prefix of this string, $x_1 \cdots x_{r(n)}$, where $|x_i| = l(n)$. Arthur sends Merlin $x_1, \cdots, x_{r(n)}$, receiving (interleaved with theses queries) the messages $y_1, \cdots, y_{r(n)}$. Then Arthur accepts or rejects according to the deterministic, poly(n)-time computable predicate

$$P_A(x, x_1, \cdots, x_{r(n)}, y_1, \cdots, y_{r(n)})$$

that he possesses.

To transform $(M \leftrightarrow A)$ into $(P \leftrightarrow V)$, a zero-knowledge interactive proof system for L, we will have the prover, P, behave like M, and the verifier, V, behave like A, with the following exceptions: The prover will encrypt each message y_i that he sends to the verifier, and then convince the verifier that he (V) would accept if he knew the corresponding encryption keys.

That is, the protocol runs in two phases. In the first phase, if P and V share input x of length n, then on round i, when Merlin "would have" sent to Arthur the string y_i, P instead randomly selects an $nl(n)$-bit string d_i and sends to V the string $\alpha_i = E_n(y_i, d_i)$. For the second phase, after all $r(n)$ rounds of the first phase are completed, the prover decides whether or not A would have accepted the corresponding unencrypted conversation, a fact which P can easily discern using P_A. If A would have accepted, then P convinces V that A would have accepted. That is, P convinces V of the validity of the NP-assertion

$$(\exists\, d_1, \ldots, d_{r(n)}, y_1, \ldots, y_{r(n)})\ [(E_n(y_i, d_i) = \alpha_i \text{ for all } i)\ \wedge$$
$$P_A(x, x_1, \ldots, x_{r(n)}, y_1, \ldots, y_{r_n})].$$

P convinces V of this assertion by computing a graph G which is 3-colorable if and only if the preceding assertion holds, and then convincing V that G is 3-colorable using the method of [GMW1]. Enough rounds are used in this protocol to convince V that G is 3-colorable with probability at least 4/5. Note G can be computed by a deterministic poly(n) time algorithm, φ, so both P and V "know" G after the first $r(n)$ interactions.

Let φ be the canonical map (appears in the proof of Theorem 2.9) that takes a tuple $(x, x_1, \ldots, x_{r(n)}, \alpha_1, \ldots, \alpha_{r(n)})$ to a graph G which is 3-colorable iff there is a guess $y_1, \ldots, y_{r(n)}$, $d_1, \ldots, d_{r(n)}$ for which $\alpha_i = E(y_i, d_i)$ and A would accept the corresponding unencrypted conversation according to P_A. We may assume that $|E(G)|$ is always a power of 2. Though we include details of Phase 2 for completeness, it can be viewed as a black box that accepts the public conversation with error probability $< 1/5$ whenever M would have accepted the corresponding unencrypted conversation.

Protocol for the prover, P (on input x of length n)

If $x \notin L$, all messages to the verifier are Λ. Otherwise ...

On rounds $1 \leq i \leq r(n)$: PHASE 1 ...
- Wait to receive a message x_i from the verifier.
- If $|x_i| \neq l(n)$, all future messages to the verifier are Λ. Otherwise ...
- Compute $y_i \leftarrow M(x, x_1, \ldots, x_i)$.
- Randomly select $d_i \in \Sigma^{nl(n)}$.
- Send $\alpha_i = E_n(y_i, d_i)$ to the verifier.

On round $i = r(n) + 1$: PHASE 2 ...
- Compute the graph $G = \varphi(x, x_1, \ldots, x_{r(n)}, \alpha_1, \ldots, \alpha_{r(n)})$, $V(G) = \{1, \ldots, \nu\}$.
- Compute a random (proper, vertex) 3-coloring of G, $\theta_i : V(G) \to \{01, 10, 11\}$.
- Randomly select $d_1^i, \ldots, d_\nu^i \in \Sigma^{2n}$.
- Send $E_n(\theta_i(1), d_1^i), \ldots, E_n(\theta_i(\nu), d_\nu^i)$ to the verifier.

For rounds $i \leftarrow r(n) + 2$ to ∞:
- Receive an edge $\{j, k\}$ from the verifier.
  All future messages are Λ if receive something not of this form.
- Send (d_j^{i-1}, d_k^{i-1}) to the verifier.
- Select a random 3-coloring of G, θ_i.
- Randomly select $d_1^i, \ldots, d_\nu^i \in \Sigma^{2n}$.
- Send $E_n(\theta_i(1), d_1^i), \ldots, E_n(\theta_i(\nu), d_\nu^i)$ to the verifier.

Protocol for the verifier, V (on input x of length n)

On round 1: PHASE 1 ...
- Read off first $l(n)$ bits of random tape into x_i.
- Send x_i to the prover.

On rounds $2 \leq i \leq r(n)$:
- Receive α_{i-1} from the prover.
- Read off next $l(n)$ bits of random tape into x_i.
- Send x_i to the prover.

On round $i = r(n) + 1$:
- Receive $\alpha_{r(n)}$ from the prover.
- Send Λ to the prover.

On round $i = r(n) + 2$: PHASE 2 ...
- Compute $G = \varphi(x, x_1, \dots, x_{r(n)}, \alpha_1, \dots, \alpha_{r(n)})$, $V(G) = \{1, \dots, \nu\}$.
- Receive $(\alpha_1^i, \dots, \alpha_\nu^i)$ from the prover.
- If not of this form, *reject*.
- Randomly select an edge $\{j, k\} \in E(G)$.
- Send $\{j, k\}$ to the prover.

For rounds $i \leftarrow r(n) + 3$ to $r(n) + 3 + 2m$:
- Receive (d_j^{i-1}, d_k^{i-1}) from the prover.
- If not of this form, or if it is not the case that for distinct $u, v \in \{01, 10, 11\}$ is $\alpha_j^{i-1} = E_n(u, d_j^{i-1})$, $\alpha_k^{i-1} = E_n(v, d_k^{i-1})$, *reject*.
- Receive $\alpha_1^i, \dots, \alpha_\nu^i$ from the prover.
- If not of this form, *reject*.
- Randomly select $\{j, k\} \in E(G)$, and send $\{j, k\}$ to the prover.

accept.

We have three things to check: that V accepts all strings in L; that V usually rejects strings not in L, even if P is replaced by some *other* probabilistic algorithm; and that P is zero-knowledge over L.

The first two of these claims are easy. Choose $x \in L$, where $|x| = n$. Then for *any* strings $x_1, \dots, x_{r(n)} \in \Sigma^{l(n)}$, we know that A interacting with M would accept when A sends messages $(x_1, \dots, x_{r(n)})$. Since the interactive proof for graph 3-colorability is one-sided, P will always always be able to convince V that A would accept $(x, x_1, \dots, x_{r(n)}, \alpha_1, \dots, \alpha_{r(n)})$ if A knew the corresponding endcyption keys. So, in fact, we retain perfect completeness.

Suppose V is interacting with a corrupt prover, P^*, and the common input is x, a string of length n, where $x \notin L$. The probability that V will accept a string which A would not have accepted when given the corresponding unencrypted messages is at most $1/5$. But for any $x \notin L$, A accepts with probability at most $1/5$. Thus V fallaciously accepts x with probability at most $2/5$, so the proof system is sound.

We now show that P is zero-knowledge over L. By Lemma 2.5, if we prove that P is zero-knowledge for the first phase of the interaction, we will be done: the whole interaction is the composition of $(P \leftrightarrow V)$ restricted to the first phase, with $(P \leftrightarrow V)$ restricted to the second phase, and the second phase is zero-knowledge over the output of the first phase.

Let P_1 be the protocol that carries out the first phase of the interaction. Inquiries beyond the $r(n)$-*th* are answered with the empty string.

Let W be a probabilistic poly-time algorithm that interacts with P_1. We may assume that W flips exactly $t(n)$ coins on each round, where t is a polynomial, and n is the length of the common input.

We may assume that $(P_1 \leftrightarrow W)$ always uses exactly $r(n)$ rounds, each message of length $l(n)$, when the common input is of length n.

M_W simulates a "virtual prover," $\hat{P}_1$, interacting with W. M_W uses its coins at odd positions for $\hat{P}_1$'s coins, and its coins at even positions for W's coins. M_W, after simulating the interaction, outputs the view of this interaction. Note that M_W is polynomial time.

Here is the protocol for $\hat{P}_1$:

On rounds $1 \leq i \leq r(n)$:
- Wait to receive a message x_i from the verifier.
- If $|x_i| \neq l(n)$, all future messages are Λ. Otherwise ...
- Randomly select $d_i \in \Sigma^{nl(n)}$.
- Send $E_n(0^{l(n)}, d_i)$ to the verifier.

On future rounds:
- Send Λ to the verifier.

We argue that $(P_1 \leftrightarrow W)(\cdot,\cdot) \equiv_L M_W(\cdot,\cdot)$. The auxiliary string plays no role in the proof (other than to be given to W), so we omit further mention of it. Denote the space $(P_1 \leftrightarrow W)(x)$ by $S_{r(|x|)}$, and $M_W(x)$ by $S_0(x)$. Assume for contradiction that these families of spaces are computationally ***distinguishable*** over L. That is, there exists a polynomial size family of circuits $\mathcal{C} = \{C_n\}$ and a polynomial h such that

$$\left| p^{C_{|x|}}_{S_0(x)} - p^{C_{|x|}}_{S_{r(|x|)}(x)} \right| \geq \frac{1}{h(|x|)}$$

for infinitely many x in L.

A "probability walk" is now used. Let $S_j(x)$ be the probability space obtained by using the real prover, P_1, for the first j rounds with V, and the virtual prover, $\hat{P}_1$, for the remaining rounds. That is, $S_j(x)$ is the probability space defined by the interaction between W and the following prover, $\tilde{P}_j$.

On rounds $1 \leq i \leq r(n)$:
- Wait to receive a message x_i from the verifier.
- If $|x_i| \neq l(n)$, all future messages are Λ. Otherwise ...
- Compute $y_i = \begin{cases} M(x, x_1, \ldots, x_i), & \text{if } i \leq j; \\ 0^{l(n)}, & \text{otherwise.} \end{cases}$
- Randomly select $d_i \in \Sigma^{nl(n)}$.
- Send $E_n(y_i, d_i)$ to the verifier.

On future rounds:
- Send Λ to the verifier.

Observe that this agrees with our previous definition of $S_0(x)$ and $S_{r(|x|)}(x)$, and that the defining algorithms for $S_j(x)$ and $S_{j+1}(x)$ differ in behavior only on the $(j+1)$-*st* round, at which point S_j uses $\hat{P}_1$ while S_{j+1} uses P_1.

By the triangle inequality, there are infinitely many x in L for which there is an associated i, $0 \leq i < r(|x|)$, such that

$$\left| p^{C_{|x|}}_{S_i(x)} - p^{C_{|x|}}_{S_{i+1}(x)} \right| \geq \frac{1}{r(|x|) \cdot h(|x|)}. \tag{1}$$

Using $\mathcal{C}$, we construct a poly-size family of expected polynomial time algorithms, $\mathcal{C}' = \{C'_n\}$, and an infinite collection of strings $\{z_n\}$, $|z_n| = l(n)$, such that C'_n effectively distinguishes the probability space $E_n(z_n)$ from the probability space $E_n(0^{l(n)})$.

Choose $x \in L$, $|x| = n$, and i, for which the bound in (1) holds. We show how to modify C_n to obtain C'_n. Let

$$\left| p^{C_n}_{S_i(x)} - p^{C_n}_{S_{i+1}(x)} \right| = \epsilon_n,$$

where $\epsilon_n \geq 1 \,/\, r(n)h(n)$.

Consider the first $i+1$ rounds between a prover and W. An $f(n) = (i+1)t(n)$-bit prefix, σ, of W's random tape, and strings $\alpha_1, \ldots, \alpha_i$ that the prover sends to W determine (1) the first $i+1$ messages, $x_1, \ldots, x_{i+1}$, that W sends; (2) W's state after this portion of the conversation; and (3) the string y_{i+1} that the prover will next encrypt and send to W (recall that the prover is deterministic on each round up to the point at which it encrypts).

Let $S_i^\sigma(x)$ be the probability space obtained by having $\tilde{P}_i$ interact with W on input x, where W's random tape has prefix σ. Then

$$p^{C_n}_{S_i(x)} = \sum_{\sigma \in \Sigma^{f(n)}} 2^{-f(n)} \cdot p^{C_n}_{S_i^\sigma(x)}.$$

Similarly, for S_{i+1}.

Now, by the triangle inequality,

$$\epsilon_n = \left| p^{C_n}_{S_i(x)} - p^{C_n}_{S_{i+1}(x)} \right| \leq \sum_{\sigma \in \Sigma^{f(n)}} 2^{-f(n)} \left| p^{C_n}_{S_i^\sigma(x)} - p^{C_n}_{S_{i+1}^\sigma(x)} \right|,$$

so there is a *particular* $\sigma \in \Sigma^{f(n)}$ which achieves

$$\left| p^{C_n}_{S_i^\sigma(x)} - p^{C_n}_{S_{i+1}^\sigma(x)} \right| \geq \epsilon_n.$$

Fix such a σ. For this σ, the prover induces a certain distribution on the first i messages it sends W, $\alpha = \alpha_1 \ldots \alpha_i$, $|\alpha| = g(n) = il(n)q(n)$, where $q(n)$ is the number of bits needed to encrypt a bit under E. Let λ_α be the probability that the prover's first i messages will be α, $\sum \lambda_\alpha = 1$. Define $S_i^{\sigma,\alpha}(x)$ as the probability space obtained by having $\tilde{P}_i$ interact with W on x, where W has random tape prefixed by σ, and $\tilde{P}_i$'s initial responses are α. Then

$$p^{C_n}_{S_i^\sigma(x)} = \sum_{\alpha \in \Sigma^{g(n)}} \lambda_\alpha \cdot p^{C_n}_{S_i^{\sigma,\alpha}(x)}.$$

Similarly, for S_{i+1}.

Now, as before

$$\epsilon_n \le \left| p^{C_n}_{S^\sigma_i(x)} - p^{C_n}_{S^\sigma_{i+1}(x)} \right| \le \sum_{\alpha \in \Sigma^{g(n)}} \lambda_\alpha \left| p^{C_n}_{S^{\sigma,\alpha}_i(x)} - p^{C_n}_{S^{\sigma,\alpha}_{i+1}(x)} \right|,$$

so there is a *particular* $\alpha \in \Sigma^{g(n)}$ which achieves

$$\left| p^{C_n}_{S^{\sigma,\alpha}_i(x)} - p^{C_n}_{S^{\sigma,\alpha}_{i+1}(x)} \right| \ge \epsilon_n.$$

Now σ and $\alpha = \alpha_1 \ldots \alpha_i$ determine $x_1, \ldots, x_{i+1}$ and y_{i+1} such that *if* the interaction determined by σ and α is executed, and *then* $d \in \Sigma^{nl(n)}$ is selected at random, and *then, either*
case 1: $E_n(y_{i+1}, d)$ is sent to W, *or*
case 2: $E_n(0^{l(n)}, d)$ is sent to W,

and *then*, $\hat{P}_1$ and W are allowed to continue their interaction (which will last another $r(n)-i-1$ rounds)—if all this is done, then the probability space associated with case 1 and the probability space associated with case 2 are distinguishable by C_n by at least ϵ_n.

C'_n is a probabilistic polynomial time algorithm that has σ and α "hardwired in." C'_n begins by bringing the state of W up to the state it would be in if its random tape began with σ, and it received messages $\alpha_1, \ldots, \alpha_i$ from the prover. The messages $x_1, \ldots, x_{i+1}$ that W would send are determined during this process, and they are recorded. C'_n expects a string α_{i+1} as input. This input is fed to W as its $(i+1)$-*st* message from the prover. From now on, C'_n uses its real random tape, and comes up with a query x_{i+2} for W to have made. However, C'_n answers its own queries using $\hat{P}_1$. This continues until C'_n has constructed a complete conversation, $((x_1, \alpha_1), \ldots, (x_{r(n)}, \alpha_{r(n)}))$, together with associated coin flips for W (which is σ with some random $(t(n)(r(n)-i-1)$-bit string appended). C'_n constructs the associated view of the conversation, and feeds this to C_n to obtain a bit, 0 or 1. C'_n outputs this bit.

The poly(n) length of α and σ guarantees that C'_n is expected polynomial time. And

$$\left| p^{C'_n}_{E_n(y_{i+1})} - p^{C'_n}_{E_n(0^{l(n)})} \right| \ge \epsilon_n$$

by our construction.

Set $z_n = y_{i+1}$. The family of probabilistic polynomial time algorithms $C' = \{C'_n\}$ (indexed by the same infinite set of naturals as in (1)) so constructed constitutes a poly-size family of probabilistic polynomial algorithms that distinguishes $\{(E_n(z_n)\}$ from $\{(E_n(0^{l(n)})\}$ by at least ϵ_n. By our remark that distinguishability by polynomial size families of probabilistic polynomial time algorithms implies distinguishability by poly-size families of circuits, we have contradicted Lemma 2.7. Our original assumption—that $(P_1 \leftrightarrow W)$ is distinguishable from M_W—is therefore in error.

That P itself is zero-knowledge follows from Lemma 2.5. The second phase of the interaction depends only on the public history of the first phase of the interaction. Recall that, by one-sidedness, whenever $x \in L$, A would accept when interacting with M, so the graph G generated following the interaction will always be 3-colorable. Since the second phase of the interaction is precisely the graph-isomorphism protocol applied to a deterministic poly-time computable function of the public history, Lemmas 2.8 and 2.6 tell us that the second phase of the interaction is zero-knowledge over the possible public histories. ◇

4. Notarized Envelopes: Description and Implementation

The interactive proof that a graph is 3-colorable ([GMW1]) can be implemented in perfect zero-knowledge using *envelopes* for committing strings. For each vertex, the prover puts into a vertex-labeled envelope a slip of paper giving the color of that vertex. These envelopes are placed before the verifier. The verifier chooses an edge and the prover allows the verifier to open the envelopes for the edge's endpoints. As a consequence of this protocol, all of *NP* can be implemented with envelopes in perfect zero-knowledge.

It is natural to ask if every language in *IP* can be proven in zero-knowledge using envelopes for commitment. The proof of the preceding section does not immediately give a solution to this problem. In this section, we answer this question in the affirmative.

4.1. Introduction to notarized envelopes

We now consider a stronger type of commitment scheme, known as *notarized envelopes*. Notarized envelopes allow one to commit and decommit a sequence of bits, $b_1, \ldots, b_n$, just as with ordinary envelopes. However, with notarized envelopes one can additionally prove any single *NP* assertion, $P(b_1, \ldots, b_n)$, during or after the commital stage. In our implementation using ordinary envelopes, this proof is in perfect zero-knowledge. If $P(b_1, \ldots, b_n)$ does not hold (or a poly-time bounded commitor does not have a witness of this fact), then the verifier will reject with probability at least $1/n^c$, where c is a constant which depending on P. This probability may be amplified arbitrarily by standard techniques.

A notarized envelope scheme may be thought of as a set of three protocols: A *commital* protocol, a *decommital* protocol, and a *zero-knowledge proof* protocol. Nearly all of the complexity of our implementation comes from the zero-knowledge protocol.

4.2. An implementation of notarized envelopes

Our reduction from notarized envelopes to ordinary envelopes is essentially a simplified version of Kilian's reduction from notarized envelopes (or, in his terminology, *commital with zero-knowledge proofs*) to oblivious transfer([K1]). However, our protocol has somewhat different properties from Kilian's, due to the fact that we are using envelopes instead of oblivious transfer. Using oblivious transfer, one can noninteractively commit bits with zero-knowledge proofs. Our scheme requires a constant number of rounds of interaction. It is not hard to show that any implementation based on ordinary envelopes must have some interaction, so our solution is optimal up to constant factors. Also, our implementation achieves perfect zero-knowledge, whereas Kilian's only achieves statistical zero-knowledge.

Commital and Decommital

We first present our protocols for committing and decommitting a set of bits, $b_1, \ldots, b_n$. In our protocols, we adopt the convention that Alice commits the bits, and Bob acts as the verifier.

Protocol Commit$(b_1, \ldots, b_n)$ /* Commit $b_1, \ldots, b_n$ */

1: Alice uniformly chooses bits $x_1, \ldots, x_{2n}$ subject to

$$b_i = x_i \oplus x_{i+n}.$$

2: Alice commits the x_i's to Bob, using ordinary envelopes. Bob is allowed to know which envelope is supposed to contain which x_i.

Protocol Decommit(i) /* Decommit b_i */

1: Alice opens the envelopes containing x_i and x_{i+n}. Bob computes $b_i = x_i \oplus x_{i+n}$.

Clearly, Bob gets no information about any of the b_i's from the commital protocol, and, on decommit, Bob only gains information that bit which is being decommitted.

Zero-knowledge Proofs

Our implementation of zero-knowledge proofs is somewhat more complicated. We first use the simple observation that it suffices to consider predicates, P, which are in NC^1 ([K1]). Furthermore, given an NC^1 predicate, $P(b_1,\ldots,b_n)$, the predicate $P'(x_1,\ldots,x_{2n})$ defined by

$$P'(x_1,\ldots,x_{2n}) = P(x_1 \oplus x_{n+1},\ldots,x_n \oplus x_{2n}),$$

will also be in NC^1. Now, if $P'(x_1,\ldots,x_{2n})$ is in NC^1, then by a theorem of Barrington [Ba], there is a polynomial sized width 5 permutation branching programs (W5PBP) for P'.

A branching program B may be thought of a a sequence of triples,

$$(i_1,\pi_1^0,\pi_1^1),\ldots,(i_m,\pi_m^0,\pi_m^1),$$

and a special element, a. For $j \in [1,m]$, we have $i_j \in [1,n]$, and $\pi_j^0,\pi_j^1 \in S_5$, where S_5 is the group of permutations on 5 elements. The special element a is also in S_5, and must not be equal to the identity. A branching program, B, realizes a predicate $P(b_1,\ldots,b_n)$ if the product

$$\prod_{j=1}^{m} \pi_j^{b_{i_j}} = \begin{cases} a & \text{when } P(b_1,\ldots,b_n) \text{ is true;} \\ I & \text{when } P(b_1,\ldots,b_n) \text{ is not true.} \end{cases}$$

Here, I represents the identity element for S_5. Given an NC^1 circuit for P', Barrington shows how to construct a canonical branching program which realizes P', which we denote by $B_{P'}$.

We can now describe our protocol for giving zero-knowledge proofs of some NC^1 predicate, P. We assume that Alice has committed bits $b_1,\ldots,b_n$, generating bits $x_1,\ldots,x_{2n}$.

Protocol Prove$(x_1,\ldots,x_{2n},P)$ /* prove $P(b_1,\ldots,b_n)$ */

1: Let $B_{P'}$ be a canonical W5PBP for P'. We write

$$B_{P'} = ((i_1,\pi_1^0,\pi_1^1),\ldots,(i_m,\pi_m^0,\pi_m^1),a)$$

Alice computes the sequence $A_1,\ldots,A_m$ by

$$A_j = \pi_j^{x_{i_j}}.$$

She then uniformly chooses $R_1,\ldots,R_{m-1}$, where $R_i \in S_5$. For convenience, we define $R_0 = R_m = I$. Finally, she computes a new sequence, $B_1,\ldots,B_m$, defined by

$$B_i = R_{i-1}A_iR_i^{-1}.$$

She then commits A_i, R_i, B_i, for $i \in [1,m]$, using ordinary envelopes.

2: Bob uniformly chooses one of the following three types of queries to make of Alice:

a: Bob asks Alice to reveal $B_1,\ldots,B_m$. He rejects if

$$\prod_{i=1}^{m} B_i \neq a.$$

b: Bob asks Alice to, for some $j \in [1,m]$, reveal A_i, R_{i-1}, R_i, B_i. He rejects if

$$B_i \neq R_{i-1}A_iR_i^{-1}.$$

The values of R_0, R_m are assumed to be I, and thus do not have to be revealed.

c: Bob asks Alice to, for some $j \in [1, m]$, reveal A_j and x_{i_j}. He rejects if

$$A_j \neq \pi_j^{x_{i_j}}.$$

Remark: The sequence of B_i's may be thought of as a randomized version of the sequence of A_i's. For any choice of $R_1, \ldots, R_{m-1}$ (assuming $R_0 = R_m = I$, we have

$$\prod_{i=1}^{m} B_i = \prod_{i=1}^{m} A_i.$$

Furthermore, it is not hard to show that if $R_1, \ldots, R_{m-1}$ are distributed uniformly, then the sequence $B_1, \ldots, B_m$ will be distributed uniformly over all sequences with the given product.

We claim that this protocol constitutes a perfect zero-knowledge proof system for P. First we show that this is indeed a proof system.

Lemma 4.1. *If $P'(x_1, \ldots, x_{2n})$ does not hold, then Bob will reject with probability at least $1/3m$.*

Proof: To simplify our argument, we assume that all of the x_i's are defined, that is, Alice never produced any empty or defective envelopes. Clearly, Alice gains nothing by such a tactic. Now, if

1. $\prod_{i=1}^{m} B_i = a$,
2. $(\forall i \in [1, m]) B_i = R_{i-1} A_i R_i^{-1}$, and,
3. $(\forall j \in [1, m]) A_j = \pi_j^{x_{i_j}}$,

then we have,

$$\prod_{j=1}^{m} \pi_j^{x_{i_j}} = \prod_{i=1}^{m} A_i$$

$$= \prod_{j=1}^{m} B_j \quad \text{(by the above remark)}$$

$$= a.$$

Therefore, if $P'(x_1, \ldots, x_n)$ does not hold, one of the above three equalities must not hold. If equality (1) does not hold, then test (a) will detect always detect this fact. If equality (2) does not hold, then test (b) will detect this fact with probability at least $1/m$. If equality (3) does not hold, then test (c) will detect this fact with probability at least $1/m$. Since each test will be invoked with probability $1/3$, the lemma follows. ◊

It is not hard to see that our protocol achieves perfect zero-knowledge. Bob is only allowed to make a single test, either (a), (b), or (c). If he makes test (a), all he sees is a random sequence of elements whose product is a. Tests (b) and (c) allow Alice to get information about x_{i_j}, for some value of j. However, this will give him no information about any of the bits b_i, since each is represented as an exclusive-or of two of the x_i's.

4.3. *IP* in perfect zero-knowledge with envelopes

The notarized envelope scheme just described gives us zero-knowledge proofs for all of *IP*:

Theorem 4.2. *Assuming envelopes for bit commitment, every language that admits an interactive proof admits a perfect zero-knowledge interactive proof.*

Proof: Let language L be in *IP*. Let $(M \leftrightarrow A)$ be a one-sided Arthur-Merlin protocol for L, rejecting strings not in L with probability at least $1/3$. We assume without loss of generality that there exist polynomials r, l such that on input $x \in L$, where $|x| = n$,

1. The protocol $(M \leftrightarrow A)$ takes $r(n)$ rounds.
2. Each of Arthur's messages, and Merlin's responses are $l(n)$ bits long.
3. Arthur's decision predicate is in NC^1. (This property is not really necessary, but simplifies our proof slightly.)

We adopt the following notation. The string A_i denotes Arthur's ith message, and M_i denotes Merlin's ith response. We denote by a_j^i the jth bit of Arthur's ith message, and by m_j^i the jth bit of Merlin's ith message. We denote by $\mathcal{A}(x, A_1, \ldots, A_{r(n)}, M_1, \ldots, M_{r(n)})$ the decision predicate computed by Arthur at the end of the protocol. Given some vector $\vec{A} = A_1, \ldots, A_{r(n)}$, and input x, of length n, we define $\mathcal{A}_{\vec{A},x}(M_1, \ldots, M_{r(n)})$ to be equal to $\mathcal{A}(x, A_1, \ldots, A_{r(n)}, M_1, \ldots, M_{r(n)})$. Any circuit for $\mathcal{A}$ can be trivially transformed into a circuit for $\mathcal{A}_{\vec{A},x}$ without increasing its size or depth.

We now exhibit a modified protocol $(M' \leftrightarrow A')$ which uses envelopes. We claim that this protocol will be in perfect zero-knowledge, and will also be a one-sided "weak" proof system for L. By "weak " we mean that for any $x \notin L$, A' will accept with probability at most $1 - 1/|x|^c$, for some fixed c.

Protocol $(\mathbf{M}' \leftrightarrow \mathbf{A}')(x)$

1: For $i \in [1, r(n)]$ Merlin and Arthur execute the following two steps. Arthur sends Merlin a random string A_i. Merlin computes his answer, M_i, and runs protocol **commit**$(m_1^i, \ldots, m_{l(n)}^i)$. The commital protocol will generate $Q = 2r(n)l(n)$ bits, which we denote by $x_1, \ldots, x_Q$.

2: Let $\vec{A} = A_1, \ldots, A_{r(n)}$. The prover executes protocol **prove**$(x_1, \ldots, x_Q, \mathcal{A}_{\vec{A},x})$. A' accepts iff he doesn't reject in protocol **prove**.

This protocol is clearly in perfect zero-knowledge, since the commital and proof protocols are in perfect zero-knowledge. Since the protocol is one-sided, the prover will always be able to execute protocol **prove**, so this doesn't give any information.

To see that this protocol remains a "weak" proof, we note that if $x \notin L$, then with probability at least $1/3$, $\mathcal{A}_{\vec{A},x}(M_1, \ldots, M_{r(n)})$ will not hold. This is due to the definition of $\mathcal{A}_{\vec{A},x}$, and the fact that $(M \leftrightarrow A)$ is a proof system. In this case, there is some c, depending on $\mathcal{A}$, such that A' will reject during the **prove** protocol with probability at least $1/n^c$. Hence, if $x \notin L$, then A' will reject with probability at least $1/3n^c$. This probability of rejection may be made exponentially close to 1, maintaining both the one-sideness and the perfect zero-knowledge properties, by the standard trick of running the above protocol many times in succession. ◇

It is interesting that the above proof never uses the ability to decommit notarized envelopes.

References

[AH] Adleman, L., and M. Huang, "Recognizing Primes in Random Polynomial Time," *Proceedings of the 19th STOC*, 1987, pp. 462–469.

[Bab] Babai, L., "Trading Group Theory for Randomness," *Proceedings of the 17th STOC*, 1985, pp. 421–429.

[Bar] Barrington, D., "Bounded Width Polynomial Size Branching Programs Recognize Exactly Those Languages in NC^1," *Proceedings of the 18th STOC*, 1986, pp. 1–5.

[BaM] Babai, L. and S. Moran, "Arthur-Merlin Games: A Randomized Proof System, and a Hierarchy of Complexity Classes," manuscript.

[Bl] Blum, M., "Coin Flipping by Telephone," *IEEE COMPCON*, 1982, pp. 133–137.

[BHZ] Boppana, R., J. Håstad, and S. Zachos, "Does co-NP Have Short Interactive Proofs?", Information Processing Letters, 1987, pp. 127–132.

[Br] Brassard, G., *personal communication,* Augest 1988.

[CDG] Chaum, D., I. Damgård, and J. van de Graaf, "Multiparty Computations Ensuring Privacy of Each Party's Input and Correctness of the Result," Proceedings of Crypto-87, pp. 87–119.

[Fe] Feldman, P., "The Optimum Prover Lives in *PSPACE*," manuscript.

[Fo] Fortnow, L., "The Complexity of Perfect Zero-Knowledge," *Proceedings of the 19th STOC*, 1987, pp. 204–209.

[G] Goldreich, O., "Randomness, Interactive Proofs and Zero-Knowledge (a survey)," Technion Technical Report, 1987.

[GK] Goldwasser, S., and J. Kilian, "Almost All Primes Can Be Quickly Certified," *Proceedings of the 18th STOC*, 1986, pp. 316–329.

[GM] Goldwasser., S., and S. Micali, "Probabilistic Encryption," *Journal of Computer and System Sciences*, Vol. 28, No. 2, 1984, pp. 270–299.

[GMR] Goldwasser, S., S. Micali, and C. Rackoff, "Knowledge Complexity of Interactive Proofs," *Proceedings of the 17th STOC*, 1985, pp. 291–305

[GMS] Goldreich, M., Y. Mansour, and M. Sipser, "Interactive Proof Systems: Provers that Never Fail and Random Selection," *Proceedings of the 28th FOCS*, 1987, pp. 449–461.

[GMW1] Goldreich, O., S. Micali, and A. Wigderson, "Proofs that Yield Nothing but their Validity and a Methodology of Cryptographic Protocol Design," *Proceedings of the 27th FOCS*, 1986, pp. 174–187.

[GMW2] Goldreich, O., S. Micali, and A. Wigderson, "How to Play Any Mental Game, or, A Completeness Theorem for Protocols with Honest Majority," *Proceedings of the 19th STOC*, 1987, pp. 218–229.

[GS] Goldwasser, S., and M. Sipser, "Arthur Merlin Games versus Interactive Proof Systems," *Proceedings of the 18th STOC*, 1986, pp. 59–68.

[I] Impagliazzo, R., *personal communications*, 1987.

[IY] Impagliazzo, R., and M. Yung, "Direct Minimum-Knowledge Computations," Proceedings of Crypto-87, pp. 40–51.

[K1] Kilian, J., "Founding Cryptography on Oblivious Transfer," *Proceedings of the 20th STOC*, 1988, pp. 20–31.

[K2] Kilian, J., "Primality Testing and the Cryptographic Complexity of Noisy Communications Channels," MIT Ph.D. Thesis (in preparation), 1988.

[L] Levin, L., "One-way Functions and Pseudorandom Generators," *Proceedings of the 17th STOC*, 1985, pp. 363–368.

[MRS] Micali, S., C. Rackoff and R. Sloan, "The Notion of Security for Probabilistic Cryptosystems," *SIAM Journal of Computing*, 17(2):412–426, April 1988.

[O] Oren, Y., "On the Cunning Power of Cheating Verifiers: Some Observations about Zero Knowledge Proofs," *Proceedings of the 28th FOCS*, 1987, pp. 462–471.

[TW] Tompa, M., and H. Woll, "Random Self-Reducibility and Zero Knowledge Interactive Proofs of Possession of Information," *Proceedings of the 28th FOCS*, 1987, pp. 472–482.

[Ya] Yao, A.C., "Theory and Applications of Trapdoor Functions," *Proceedings of the 23rd FOCS*, 1982, pp. 80–91.

[Yu] Yung, M., *personal communication*, Augest 1988.

A Perfect Zero-Knowledge Proof for a Problem Equivalent to Discrete Logarithm

Oded Goldreich *Eyal Kushilevitz*
Dept. of Computer Sc.
Technion
Haifa, Israel

ABSTRACT

An interactive proof is called *perfect zero-knowledge* if the probability distribution generated by any probabilistic polynomial-time verifier interacting with the prover on input a theorem ϕ, can be generated by another probabilistic polynomial time machine which only gets ϕ as input (and interacts with nobody!).

In this paper we present a *perfect* zero-knowledge proof system for a decision problem which is computationally equivalent to the Discrete Logarithm Problem. Doing so we provide additional evidence to the belief that *perfect* zero-knowledge proofs exist in a non-trivial manner (i.e. for languages not in BPP). Our results extend to the logarithm problem in any finite Abelian group.

1. INTRODUCTION

One of the most basic questions in complexity theory is how much knowledge should be yield in order to convince a polynomial-time verifier of the validity of some theorem. This question was raised by Goldwasser, Micali and Rackoff [GMR], with special emphasis on the extreme case where nothing but the validity of the theorem is given away in the process of proving the theorem. Such proofs are known as *zero-knowledge* proofs and have been the focus of much attention in recent years. Loosely speaking, whatever can be efficiently computed after participating in a zero-knowledge proof can be efficiently computed when just assuming the validity of the assertion.

The definition of zero-knowledge considers two types of probability distributions:

1) A distribution generated by a probabilistic polynomial-time verifier after participating in an interaction with the prover.

2) A distribution generated by a probabilistic polynomial-time machine on input the theorem.

Zero-knowledge means that for each distribution of type (1) there exists a distribution of type (2) such that these two distributions are ''essentially equal''. The exact definition of zero-knowledge depends on the exact interpretation of ''essentially equal'' distributions. Two extreme cases are of particular interest:

Research was partially supported by the Fund for Basic Research Administered by the Israeli Academy of Sciences and Humanities.

- *Perfect zero-knowledge.* This notion is derived when interpreting "essentially equal" in the most conservative way; namely, exactly equal.
- *Computational zero-knowledge.* This notion is derived when interpreting "essentially equal" in a very liberal way; namely, requiring that the distribution ensembles are polynomially indistinguishable. Loosely speaking, two distribution ensembles are polynomially indistinguishable if they can not be told apart by any probabilistic polynomial time test. For definition see [Y].

1.1. Known Results

Assuming the existence of one-way permutations, Goldreich, Micali and Wigderson showed that any language in *NP* has a computational zero-knowledge proof [GMW]. Using this result one can also show that whatever can be proven through an efficient interactive proof, can be proven through such a computational zero-knowledge proof [BGGHMR,IY]. Thus, assuming the existence of one-way permutations, the question of which languages have computational zero-knowledge proofs is closed.

Much less is known about perfect zero-knowledge. Clearly any language in *BPP* has a trivial perfect zero-knowledge proof (in which the prover is inactive...). Several languages believed not to be in *BPP* were shown to have perfect zero-knowledge proofs. These includes Quadratic Residuosity and Quadratic non-Residuosity [GMR], Graph Isomorphism and Graph non-Isomorphism [GMW], and membership and non-membership in a subgroup generated by a given group element [TW]. (It should be noticed that Tompa and Woll's proof of "possession of the Discrete Logarithm" is in fact a proof of membership in a subgroup generated by a primitive element. So are the proofs given by [CEGP,CG]).

The complexity of languages which have a perfect zero-knowledge interactive proofs was studied by Fortnow [F] and then by Aiello and Hastad [AH]. They prove that if a language L has a perfect zero-knowledge interactive proof system then both L and $\bar{L}$ have two-step interactive proofs. This implies that languages having perfect zero-knowledge proofs fall quite low in the polynomial time hierarchy (i.e. as low as $\Pi_2^P \cap \Sigma_2^P$). Using a result of Boppana, Hastad and Zachos [BHZ], such languages can also not be *NP*-complete, unless the polynomial time hierarchy collapses to its second level.

Perfect zero-knowledge proofs should not be confused with the perfect zero-knowledge *pseudo-proofs* presented by Brassard and Crepeau [BC]. By a *pseudo-proof* we mean that the verifier is convinced only if he believes that the prover is a polynomial-time machine with some auxiliary input (which is fixed before the protocol starts), and if some intractability assumption does hold. For example, *if factoring is intractable* then every NP language has a perfect zero-knowledge *pseudo-proof* [BC]. Brickell et. al. presented a perfect zero-knowledge *pseudo-proof* for a problem equivalent to the discrete logarithm problem, assuming the existence of any one-way permutation [BCDG]. It should be noted however that the class of languages having perfect zero-knowledge *pseudo-proof* does not seem to have the same complexity as the class of languages having perfect zero-knowledge proofs. Furthermore, assuming the intractability of factoring every language having an interactive proof has a perfect zero-knowledge pseudo-proof, and thus the class of languages having such proofs collides with the class of languages having computational zero-knowledge proofs.

1.2. Our Results

In this paper we present a *perfect* zero-knowledge proof system for a decision problem which is computationally equivalent to the Discrete Logarithm Problem. Doing so, we present a perfect zero-knowledge proof for a problem which is widely believed to be intractable. Thus, we provide additional evidence to the belief that *perfect* zero-knowledge proofs exist in a non-trivial manner (i.e. for languages not in BPP).

Let p be a prime, and g be a primitive element in the multiplicative group modulo p. The *Discrete Logarithm Problem* (DLP) is to find, given integers p, g and y, an integer x such that $g^x \equiv y \bmod p$. Solving DLP is considered intractable, in particular when $p-1$ has large prime factors. The best algorithms known for this problem run in subexponential time ($\exp\{O(\sqrt{\log p \, \log\log p}\,)\}$), see Odlyzko's survey [O]. It has been shown that determining* whether $x \le \frac{p-1}{2}$ is computationally equivalent to finding x, on inputs p, g and $g^x \bmod p$ [BM]. This is the case even if x is guaranteed to lie either in the interval $[1, \varepsilon p]$ or in the interval $[\frac{p-1}{2}+1, \frac{p-1}{2}+\varepsilon p]$, where $0<\varepsilon<1/6$ is a constant or a function bounded below by $(\log_2 p)^{-O(1)}$. This promise problem is hereby referred to as $DLP1$.

In this paper, we present a perfect zero-knowledge proof for $DLP1$. Using the computational equivalence with DLP, we have a perfect zero-knowledge proof for a problem considered computationally hard. Both our protocol and the computational equivalence of DLP and $DLP1$ extend to any finite Abelian group, in which the group operation can be implemented in polynomial-time and the order of the group is known (or can be efficiently found). (In the case of acyclic groups, one needs first to define the problems.)

It should be noted that DLP is always at least as hard as testing membership in a subgroup generated by an element of the group. In some cases, for example when $p-1=2q$ and q is prime, determining membership in a subgroup is easy (see Appendix), while solving DLP in the multiplicative group mod p is considered hard.

2. PRELIMINARIES

2.1. Promise Problems and Interactive Proofs

Loosely speaking a *promise problem* is a partial decision problem. That is, a decision problem in which only a subset of all possible inputs is being considered.
Formally a *promise problem* is a pair of predicates (Q, R). A Turing machine M *solves* the promise problem (Q, R) if for every z which satisfying $Q(z)$ machine M halts and it answer "yes" iff $R(z)$. When $\neg Q(z)$ we do not care what M does. This definition is originates from [ESY].

---

*) In fact, Blum and Micali proved a much stronger statement. Namely, that guessing this bit with success probability greater than $1/2+\varepsilon$ is as hard as retrieving x [BM].

We are going to extend the definition of interactive proofs given in [GMR] to promise problems. Intuitively, an interactive proof system for a promise problem (Q,R) is a two-party protocol for a "powerful" *prover* P and a probabilistic polynomial-time *verifier* V satisfying the following two conditions with respect to the common input, denoted z. If $Q(z) \wedge R(z)$ then with a very high probability the verifier is "convinced" of $R(z)$, when interacting with the prover. If $Q(z) \wedge \neg R(z)$ then no matter what the prover does, he cannot fool the verifier (into believing that "$R(z)$ is true"), except for with very low probability. When $\neg Q(z)$ nothing is required.

Definition 1: An *interactive proof for a promise problem* (Q,R) is a pair of interacting Turing-machines $<P,V>$, satisfying the following three conditions:

0) V is a probabilistic polynomial-time machine which share its input with P and they can communicate to each other using special communication tapes.

1) *Completeness condition*: For every constant $c>0$, and all sufficiently long z if $Q(z) \wedge R(z)$ then

$$Prob(V \text{ will accept } z \text{ after interacting with } P) \geq 1-|z|^{-c}.$$

2) *Soundness condition*: For every Turing machine P^*, every constant $c>0$, and all sufficiently long z if $Q(z) \wedge \neg R(z)$ then

$$Prob(V \text{ will reject } z \text{ after interacting with } P^*) \geq 1-|z|^{-c}.$$

2.2. Perfect Zero-Knowledge Proofs for Promise Problems

Here, again, we are going to extend the definition given by [GMR] to promise problems.

Definition 2: Let $<P,V>$ be an interactive proof system for a promise problem (Q,R), and V^* be an arbitrary verifier. Denote by $<P,V^*>(z)$ the probability distribution on all the read-only tapes of V^* when interacting with P (the prover) on common input z. We say that the proof system $<P,V>$ is a *perfect-zero-knowledge* for (Q,R) if for all polynomial-time verifier V^*, there exists a probabilistic machine M_{V^*} running in expected polynomial-time such that for every z satisfying $Q(z) \wedge R(z)$ the distributions $M_{V^*}(z)$ and $<P,V^*>(z)$ are equal.

2.3. The Discrete Logarithm Problem and a Related Promise Problem

Let p be a prime. The set of integers $[1,p-1]$ forms a cyclic group of $p-1$ elements under multiplication *mod p* which is denoted Z_p^*. The *Discrete Logarithm* problem (*DLP*) with input p,g and y is to find $x \in [1,p-1]$ such that $y \equiv g^x \bmod p$. (We use the notation $x = Dlog_g y$).

Let y be an element of Z_p^* and let g be a primitive element (a generator). We define the Half predicate H as follows:

$$H(p,g,y) \Leftrightarrow Dlog_g y \in [\frac{p-1}{2}+1, p-1]$$

Let $n = \log_2 p$ and let $\varepsilon(n) < \frac{1}{2}$ be a fraction bounded below by $\frac{1}{n^{O(1)}}$. We define the following predicate:

$$Q_\varepsilon(p,g,y) \Leftrightarrow g \text{ is a generator of } Z_p^* \text{ and } Dlog_g y \text{ in } [1, \varepsilon(n)(p-1)] \text{ or } [\frac{p-1}{2}+1, \frac{p-1}{2}+\varepsilon(n)(p-1)]$$

When it will be clear from the context we will shorten $H(p,g,y)$ and $Q_\varepsilon(p,g,y)$ by $H(y)$ and $Q(y)$ respectively.

The promise problem defined by the pair of predicates (Q_ε, H) will be called in this work *DLP* 1. Blum and Micali have shown that the *DLP* 1 is polynomially-equivalent to the original *DLP* in the group Z_p^* [BM].

2.4. Notations

1) Let s and t be two integers such that $1 \le s, t \le p-1$. $[s,t]$ is denoted the set of integers $\{s, s+1, \cdots, t-1, t\}$ in case $s \le t$ or $\{s, s+1, \cdots, p-1, 1, 2, ..., t\}$ in case $s > t$.
2) Let S be a set. The notation $r \in_R S$ means that r is chosen at random with uniform probability distribution among the elements of S.

3. THE PROTOCOL FOR DLP1 IN Z_p^*

In this section we will introduce a perfect zero-knowledge protocol for the promise problem *DLP* 1. In order to make the protocol more clear we will first introduce a protocol which is perfect zero-knowledge with respect to the honest verifier.

3.1. Protocol 1 - Perfect Zero Knowledge Proof with respect to the Honest Verifier

Here is a protocol for the promise problem $(Q_c(p,g,y), H(p,g,y))$ where c is a constant such that $0 < c \le 1/6$:

common input: The integers p, g and y as previously defined.

The following 3 steps are executed $n = log_2 p$ times (unless the verifier *rejects* previously), each time using independent random coin tosses.

V1) The verifier chooses at random a bit $b \in_R \{0,1\}$ and an integer $r \in_R [1, 2c(p-1)]$. The verifier computes $\alpha = y^b g^r$ and sends α to the prover.
P1) The prover computes $\beta = H(\alpha)$ and sends it to the verifier.
V2) If $\beta \ne b$, then the verifier *rejects*.

If all n rounds are completed without the verifier rejects then the verifier *accepts*.

Theorem 1: *Assuming $c < 1/6$ then protocol 1 constitutes an interactive proof system for DLP1.*

Proof: Recall that x denoted $Dlog_g y$ (i.e $y \equiv g^x \ mod\ p$).
Completeness: If $Q(y) \wedge H(y)$ then $x \in [\frac{p-1}{2}+1, \frac{p-1}{2}+c(p-1)]$ and then, according to the ranges in which b and r are chosen from, in each round $\beta = H(\alpha) = H(y^b \cdot g^r) = H(g^{bx+r}) = b$.
Soundness: If $Q(y) \wedge \neg H(y)$ then we have $x \in [1, c(p-1)]$. Therefore if the verifier chooses $b=0$ then $Dlog_g \alpha \in [1, 2c(p-1)]$ and if he chooses $b=1$ then $Dlog_g \alpha \in [x+1, x+2c(p-1)]$. In this case, for any prover P' we are looking for the probability that V does not reject in a single round:
Prob(V does not reject)= $Prob(P'(\alpha)=b)$

$$= Prob(b=0) \cdot Prob(P'(\alpha)=b \mid b=0) + Prob(b=1) \cdot Prob(P'(\alpha)=b \mid b=1)$$

$$=\frac{1}{2}\cdot Prob(P'(g^r)=0) + \frac{1}{2}\cdot Prob(P'(yg^r)=1)$$

$$=\frac{1}{2}\cdot\frac{1}{2c(p-1)}\ \Big(\sum_{i=1}^{x} Prob(P'(g^i)=0) + \sum_{i=x+1}^{2c(p-1)} \big(Prob(P'(g^i)=0)+Prob(P'(g^i)=1)\big)$$

$$+\sum_{i=2c(p-1)+1}^{2c(p-1)+x} Prob(P'(g^i)=1)\ \Big)$$

$$\le\frac{1}{2}\cdot\frac{1}{2c(p-1)}\cdot\left[x+(2c(p-1)-x)+x\right]$$

$$=\frac{1}{2}\cdot\left[\frac{x}{2c(p-1)}+1\right]$$

$$\le\frac{1}{2}\cdot\left[\frac{c(p-1)}{2c(p-1)}+1\right]=\frac{3}{4}$$

Therefore in n iterations the probability that the verifier will not reject this input is exponentially low. (i.e. $(3/4)^n$) □

Remark: It is clear that this protocol is perfect zero-knowledge with respect to the honest verifier V. The simulator M_V chooses the random tape for V, and therefore knows the b which V will choose and can compute $\beta=H(\alpha)=b$.

The interactive proof for DLP1 presented above is probably not zero-knowledge with respect to arbitrary verifiers: a cheating verifier interacting with the prover may send α's which he wants to know $H(\alpha)$, he could also choose $r \notin [1,2c(p-1)]$ and get in this way some additional information about x. The way to prevent this, is to let the verifier first "prove" to the prover that he "knows" $H(\alpha)$. This is done in the following protocol.

3.2. Protocol 2 - Perfect Zero Knowledge Proof with respect to Any Verifier

The previous protocol will be modified. The modification follows an idea of [GMR] used also in [GMW] and simplified by [Bh]. However in our case the modification is more complex.

In the following protocol we provide an interactive proof to the promise problem $(Q_{\frac{c}{n^2}}(p,g,y),H(p,g,y))$ where c is a constant such that $0<c\le\frac{1}{12}$.

common input: The integers p,g and y as previously defined.

The following 5 steps are repeated $n=\log_2 p$ times (unless the verifier *rejects* previously), each time using independent random coin tosses.

V1) The verifier chooses at random a bit $b\in_R\{0,1\}$ and an integer $r\in_R[\left\lfloor\frac{p-1}{4}\right\rfloor+1,\left\lfloor\frac{p-1}{4}\right\rfloor+c(p-1)]$. The verifier computes $\alpha=y^b g^r$ and sends α to the prover. In addition to α he computes n pairs of integers. The $i-th$ pair is denoted α_i and is constructed in the following way: The verifier chooses at random $\gamma_i\in_R\{0,1\}$ and $r_{i,0},r_{i,1}\in_R[1,c(p-1)]$. He computes $\alpha_{i,0}=y^{\gamma_i}\cdot g^{r_{i,\gamma_i}}$ and $\alpha_{i,1}=y^{\gamma_i+1 \bmod 2}\cdot g^{r_{i,\gamma_i+1 \bmod 2}}$ and at last sets $\alpha_i=(\alpha_{i,0},\alpha_{i,1})$. The verifier sends the list of pairs to the prover.

P1) The prover chooses at random, a subset $I \subseteq \{1,2,...,n\}$ with uniform probability distribution among all 2^n subsets. The prover sends I to the verifier.

V2) If I is not a subset of $\{1,2,...,n\}$ then the verifier halts and *rejects*. Otherwise, the verifier replies with $\{(\gamma_i, r_{i,0}, r_{i,1}) : i \in I\}$ and $\{(\gamma'_i = \gamma_i + b + 1 \ mod\ 2,\ r'_i = r + r_{i,b+1\ mod\ 2}) : i \in \bar{I}\}$. (where $\bar{I} = \{1,2,...,n\} - I$).

P2) For every $i \in I$ the prover checks that α_i is constructed according to the protocol. (i.e. $r_{i,0}, r_{i,1} \in [1, c(p-1)]$ and $\alpha_i = (y^{\gamma_i} \cdot g^{r_{i,\gamma}},\ y^{\gamma_i+1\ mod\ 2} \cdot g^{r_{i,\gamma+1\ mod\ 2}})$). He also checks for every $i \in \bar{I}$ that $r'_i \in [\left\lfloor \frac{p-1}{4} \right\rfloor + 2, \left\lfloor \frac{p-1}{4} \right\rfloor + 2c(p-1)]$ and $y \cdot g^{r'_i} = \alpha \cdot \alpha_{i,\gamma'_i}$. If either conditions is violated the prover stops. Otherwise, the prover computes $\beta = H(\alpha)$ and sends it to the verifier.

V3) If $\beta \neq b$, then the verifier *rejects*. Otherwise he continues.

If all n rounds are completed without the verifier rejects then the verifier *accepts*.

Theorem 2: *Protocol 2 constitutes a perfect zero-knowledge interactive proof system for DLP1.*

Proof: We will first prove that Protocol 2 is an interactive proof for DLP1 and then we will show that it is perfect zero-knowledge. Recall again that $x = Dlog_g y$.

Completeness: Similar to the completeness in theorem 1.

Soundness: We are going to prove that although α and the list of pairs $S = \{\alpha_1 \cdots \alpha_n\}$ can give information to the prover, there is a big enough probability that α and S will not give him anything that will help him to convince V that $x \in [\frac{p-1}{2}+1, \frac{p-1}{2}+\frac{c(p-1)}{n^2}]$ when in fact $x \in [1, \frac{c(p-1)}{n^2}]$.

We call α *good* if it is constructed using $r \in [\left\lfloor \frac{p-1}{4} \right\rfloor + \frac{c(p-1)}{n^2} + 1, \left\lfloor \frac{p-1}{4} \right\rfloor + c(p-1) - \frac{c(p-1)}{n^2}]$. Otherwise α is *bad*. Intuitively, when α is good the prover can not learn anything about b from α, for any $x \in [1, \frac{c(p-1)}{n^2}]$ (since in this case $Prob(b=0 \mid y^b \cdot g^r = \alpha) = \frac{1}{2}$). The probability that α is bad is $\frac{2}{n^2}$.

Similarly we will call a pair α_i *good* if both $r_{i,0}$ and $r_{i,1}$ are in $[\frac{c(p-1)}{n^2}+1, c(p-1)-\frac{c(p-1)}{n^2}]$. Otherwise α_i is *bad*. The list of pairs S is *good* if every α_i is good, and is *bad* otherwise. The probability that a pair α_i is bad is less than $\frac{4}{n^2}$ and the probability that S is bad is therefore less than $\frac{4n}{n^2}$.

We remark here that since P' has infinite power we can assume without loss of generality that P' is deterministic. Therefore for any α and S the prover P' always chooses the same subset I, denoted $f(\alpha, S)$.

Our first claim is the following:

$\forall$ *good* α $\forall$ *good* S $\forall I\ \forall r'_i\ \forall \gamma'_i$

$$Prob(b{=}0 \mid y^b g^r{=}\alpha \ \wedge f(\alpha,S){=}I \ \wedge \ \forall i \in \bar{I}\ (y \cdot g^{r'_i}{=}\alpha \cdot \alpha_{i,\gamma_i} \wedge r'_i{=}r{+}r_{i,\gamma_i})) \ = \ \frac{1}{2}$$

The reason is that when α and S are good then assigning any value to b yields a unique values to all the other variables r , $r_{i,0}$ and $r_{i,1}$. Thus, assuming $I{=}f(\alpha,S)$, there are only two elements in the conditional probability space, one corresponds to $b{=}0$ and the other to $b{=}1$. Using this claim we will show now that the probability that P' will convince V in single round is low:

$$Prob(P'(\alpha, S, \{\gamma_i, r_{i,0}, r_{i,1} : i \in f(\alpha,S)\}, \{\gamma'_i, r'_i : i \notin f(\alpha,S)\}) = b)$$

$$\le Prob(P'(\alpha, S, \{\gamma_i, r_{i,0}, r_{i,1} : i \in f(\alpha,S)\}, \{\gamma'_i, r'_i : i \notin f(\alpha,S)\}) = b \mid \alpha \text{ and } S \text{ are good})$$

$$+Prob(\alpha \text{ is bad or } S \text{ is bad})$$

$$\le \frac{1}{2} + \frac{4n+2}{n^2}$$

Therefore the probability that P' will mislead V (i.e. provide correct β's) in all n rounds is exponentially low.

Zero-knowledge: For every interactive machine V^*, we will present a machine M_{V^*} so that for every input satisfying $Q(p,g,y) \wedge H(p,g,y)$ then $M_{V^*}(p,g,y) = \langle P, V^* \rangle (p,g,y)$. The machine M_{V^*} uses V^* as a subroutine.

The idea of the simulator M_{V^*} is to cause V^* to yield all the information needed for calculating $H(\alpha)$. This is done by executing V^* several times with the same random tape, so that V^* will send the same α and S. Machine M_{V^*} will try to get for one of the pairs α_i the information $\{\gamma_i, r_{i,0}, r_{i,1}\}$ in one round and $\{\gamma'_i, r'_i\}$ in another. If this information is constructed according to the protocol (M_{V^*} will check it) then this is enough for calculating $H(\alpha)$.

Following is a detailed description of M_{V^*}. Machine M_{V^*} starts by choosing a random tape $r \in_R \{0,1\}^q$ for V^*, where $q{=}poly(|p,g,y|)$ is a bound on the running time of V^* on the current input (Clearly, V^* reads at most q bits from its random tape). M_{V^*} places r on its record tape and proceeds in n rounds as follows.

Round j:

S1) M_{V^*} initiates V^* on the input (p, g and y) and random tape r, and reads from the communication tape of V^* the pairs α and $\alpha_1 \cdots \alpha_n$. M_{V^*} chooses a random subset I and places it on the communication tape of V^*. M_{V^*} also appends I to its record tape.

S2) M_{V^*} reads from the communication tape of V^* $\{(\gamma_i, r_{i,0}, r_{i,1}) : i \in I\}$ and $\{(\gamma'_i, r'_i) : i \in \bar{I}\}$. For every $i \in I$ machine M_{V^*} checks whether $\gamma_i \in \{0,1\}$, $r_{i,0}, r_{i,1} \in [1, c(p-1)]$ and whether $\alpha_i {=} (y^{\gamma_i} \cdot g^{r_{i,\gamma_i}},\ y^{\gamma_i+1 \bmod 2} \cdot g^{r_{i,\gamma_i+1 \bmod 2}})$. It also checks for every $i \in \bar{I}$ whether $r'_i \in [\left\lfloor \frac{p-1}{4} \right\rfloor + 2, \left\lfloor \frac{p-1}{4} \right\rfloor + 2c(p-1)]$ and $y \cdot g^{r'_i} {=} \alpha \cdot \alpha_{i,\gamma_i}$. If either conditions is violated M_{V^*} outputs its record tape and stops. Otherwise, M_{V^*} continues to step (S3).

S3) The purpose of this step is to find $H(\alpha)$. This is done by repeating the following procedure (until $H(\alpha)$ is found):

(S3.1) Machine M_{V^*} chooses at random a subset $K \subseteq \{1,2,...,n\}$ not equal to I. Machine

M_{V^*} initiates V^* on the same input and the same (!) random tape r and places K as the first message on the read-only communication tape of V^*. Consequently, machine M_{V^*} reads from the communication tape of V^* $\{(\delta_i, s_{i,0}, s_{i,1}): i \in K\}$ and $\{(\delta'_i, s'_i): i \in \bar{K}\}$.

(S3.2) M_{V^*} checks whether the information he received is ok. (The same tests as he does for the answers to I). If it is not ok he returns back to step (S3.1). Otherwise M_{V^*} finds i such that $i \in I \cap \bar{K}$ or $i \in \bar{I} \cap K$. Such an i exists since $I \neq K$, without loss of generality we assume that $i \in I \cap \bar{K}$. Machine M_{V^*} sets $\beta = (\gamma_i + \delta'_i + 1) mod 2$.

(S3.3) In parallel to (S3.1) and (S3.2), try to find $H(\alpha)$ by exhaustive search. (Make one try per each invocation of V^*.)

S4) Once β is found, machine M_{V^*} appends β to its record tape, thus completing round j.

If all rounds are completed then M_{V^*} outputs its record tape and halts.

We now have to prove the validity of the construction. First, we will prove that the simulator M_{V^*} indeed terminates in expected polynomial-time. Next, we will prove that the output distribution produced by M_{V^*} does equal the distribution over V^*'s tapes (when interacting with P). Once these two claims are proven, the Theorem follows.

Claim 1: Machine M_{V^*} terminates in expected polynomial time.

Proof: We consider the expected running time on a single round with respect to a particular random tape r. We call a subset $I \subseteq \{1,2,...,n\}$ *good* if V^* answers properly on message I with random tape r. Denote by g_r the number of good subsets with respect to random tape r. Clearly, $0 \le g_r \le 2^n$. We will compute the expected number of times V^* is invoked in round j as a function of g_r. We need to consider three cases:

Case 1 ($g_r \ge 2$): In case the subset I chosen in step (S1) is good, we have to consider the probability that another subset K is also good. In case the set I chosen in step (S1) is bad, the round is completed immediately. Thus, the expected number of invocations is

$$\frac{g_r}{2^n} \cdot \left[\left(\frac{g_r - 1}{2^n - 1} \right)^{-1} + 1 \right] + \frac{2^n - g_r}{2^n} \cdot 1 < \frac{g_r}{g_r - 1} + 1 \le 3$$

Case 2 ($g_r = 1$): With exponentially small probability (i.e. 2^{-n}) the subset I chosen in step (S1) is good. In this case we find β by exhaustive search (in stage (S3.3)). Otherwise, the round is completed immediately. Thus, the expected complexity of M_{V^*} in case 2 is bounded by one invocation of V^* and an additional $(p-1) \cdot 2^{-n} \le 1$ step.

Case 3 ($g_r = 0$): The subset I chosen in step (S1) is always bad, and thus M_{V^*} invokes V^* exactly once and then halts.

The claim follows by additivity of expectation and the fact that V^* is polynomial-time. □

Claim 2: The probability distribution $M_{V^*}(p, g, y)$ is identical to the distribution $\langle P, V^* \rangle (p, g, y)$.

Proof: Both distributions consists of a random r, and sequence of elements, each being either (I, β) (with good I) or a bad I, with random I. In $\langle P, V^* \rangle (p, g, y)$ we have $\beta = H(\alpha)$ we need to show that this is the case also in $M_{V^*}(p, g, y)$. i.e. we will prove that when I is good then M_{V^*} succeeds in finding $H(\alpha)$. But this is true because either he finds $H(\alpha)$ by exhaustive search or find an i in which $\gamma_i, r_{i,0}, r_{i,1}, \delta'_i$ and s'_i are all correct. (i.e. $r_{i,0}, r_{i,1} \in [1, c(p-1)]$, $s'_i \in [\left\lfloor \frac{p-1}{4} \right\rfloor +2, \left\lfloor \frac{p-1}{4} \right\rfloor +2c(p-1)]$, $\alpha_{i,j} = y^{\gamma_i + j \bmod 2} \cdot g^{r_{i,\gamma_i+j \bmod 2}}$ and $y \cdot g^{s'_i} = \alpha \cdot \alpha_{i,\delta'_i}$). In this case we have:

$$
\begin{aligned}
H(\alpha) &= H(yg^{s'_i} \cdot (\alpha_{i,\delta'_i})^{-1}) \\
&= H(yg^{s'_i} \cdot (y^{\gamma_i+\delta'_i \bmod 2} \cdot g^{r_{i,\gamma_i+\delta'_i \bmod 2}})^{-1}) \\
&= H(y^{\gamma_i+\delta'_i+1 \bmod 2} \cdot g^{s'_i - r_{i,\gamma_i+\delta'_i \bmod 2}}) \\
&= H(y^{\gamma_i+\delta'_i+1 \bmod 2}) = \gamma_i + \delta'_i + 1 \bmod 2 \ \square
\end{aligned}
$$

The Theorem follows. □

Remark 1: It is not hard to see that instead of executing the protocol sequentially, we can execute all the rounds in parallel.

Remark 2: Let s and t be two integers such that $1 \le s, t \le p-1$. We define $dist(s,t)$ to be the minimal distance between s and t over the circle of numbers $[1, p-1]$. Consider the following promise problem hereby referred to as *DLP* 2: Promised that $x \in [s,t]$ or $x \in [(s+\frac{p-1}{2}) \bmod (p-1), (t+\frac{p-1}{2}) \bmod (p-1)]$ and $dist(s,t) < \frac{p-1}{12n^2}$ does $x \in [(s+\frac{p-1}{2}) \bmod (p-1), (t+\frac{p-1}{2}) \bmod (p-1)]$? An easy modification to protocol 2 yields a perfect zero-knowledge interactive proof system for *DLP* 2:

Protocol 3

common input: p, g and y as before and also s.

1) P and V both perform $y' := y \cdot g^{-s+1}$
2) P and V perform protocol 2 on input p, g and y'.

Theorem 2': *Protocol 3 constitutes a perfect zero-knowledge interactive proof system for DLP2.*

Proof: Since it is promised that $Dlog_g y \in [s,t]$ or $Dlog_g y \in [(s+\frac{p-1}{2}) \bmod (p-1), (t+\frac{p-1}{2}) \bmod (p-1)]$ and that $dist(s,t) < \frac{p-1}{12n^2}$ then after executing step 1 we have $Dlog_g y' \in [1, \frac{p-1}{12n^2}]$ or $Dlog_g y' \in [\frac{p-1}{2}+1, \frac{p-1}{2}+\frac{p-1}{12n^2}]$ and now our theorem follows from theorem 2.

4. EXTENSIONS

4.1. Generalization of the Protocol to other Cyclic Groups

Let G be an arbitrary cyclic group such that the following conditions holds:

1) The group operation of G can be implemented in polynomial-time.
2) The order of G (to be denoted N) is either given or can be computed in polynomial-time.

We can extend the definitions of the *DLP* and the *DLP* 1 in the obvious way. The needed modifications are to replace any multiplication *mod p* by the group-operation of G and to replace $p-1$ by the group order (N).

With the same modifications our protocol will be a perfect zero-knowledge proof for the promise problem *DLP* 1 in G (since the protocol does not make any use of the special structure of Z_p^*, but merely its being cyclic). What we still have to show is that the *DLP* 1 is polynomially equivalent to the *DLP* itself in any cyclic group. The Blum-Micali proof (used in Z_p^*) extends easily only to groups in which N is even and both testing quadratic-residuosity and taking square-root can be performed in polynomial time. Unfortunately, this does not seem to be the case in all groups and a different argument is needed. We present a proof for the equivalence of *DLP* and *DLP* 1 based on ideas of Kaliski [Ka].

We define the oracle LOG_G as follows:

$LOG_G(g,y,s,d)=0$ if $Dlog_g y \in [s,(s+d) \bmod N]$

$LOG_G(g,y,s,d)=1$ if $Dlog_g y \in [(s+\left\lfloor \frac{N}{2} \right\rfloor) \bmod N, (s+\left\lfloor \frac{N}{2} \right\rfloor+d) \bmod N]$

In any other case the answer of the oracle LOG_G is unexpected.

It should be noticed that when $d<\frac{N}{12n^2}$ this oracle solves the promise problem *DLP* 2 for which protocol 3 is a perfect zero-knowledge proof.

Theorem 3: The following 2 problems are polynomially equivalent for any cyclic group G of N elements and a generator g:

1) . Given $g,y \in G$ such that g is a generator of G find x such that $x=Dlog_g y$. (*DLP*)

2) Given $y \in G$, a generator $g \in G$, $0<s<N$ and d such that $0<d<\frac{N}{12n^2}$ compute $LOG_G(g,y,s,d)$. (*DLP* 2)

Proof: It is obvious that if we know to solve the first problem we can solve the second one. We will prove the other direction by presenting an algorithm that solves the *DLP* using the oracle $LOG_G(g,y,s,d)$. The algorithm is based on the following elementary lemma:

Lemma 1: For any cyclic group G of order N and for every $y \in G$:

If $Dlog_g y^2 \in [s,t]$ then $Dlog_g y$ is in $[\left\lceil \frac{s}{2} \right\rceil, \left\lfloor \frac{t}{2} \right\rfloor]$ or in $[\left\lceil \frac{s+N}{2} \right\rceil, \left\lfloor \frac{t+N}{2} \right\rfloor]$

Proof (of the Lemma): Let $x \in [s,t] \subseteq \{0,1,\cdots,N-1\}$ and try to find a number w such that $x=2 \cdot w$. We deal with two cases:

Case 1: N is odd. Since N is odd there exists a unique number $2^{-1} \bmod N$. In this case one can easily verify that if x is even then $w=\frac{x}{2} \in [\left\lceil \frac{s}{2} \right\rceil, \left\lfloor \frac{t}{2} \right\rfloor]$ and if x is odd then $w=\frac{x}{2} \in [\left\lceil \frac{s+N}{2} \right\rceil, \left\lfloor \frac{t+N}{2} \right\rfloor]$.

Case 2: N is even. Since N is even $2^{-1} \bmod N$ not exists. In this case only for even x's we have such w. Actually we have two such numbers: $w_1=\frac{x}{2}$ and $w_2=\frac{x+N}{2}$. It is easy to verify that

$w_1=\frac{x}{2} \in [\left\lceil \frac{s}{2} \right\rceil, \left\lfloor \frac{t}{2} \right\rfloor]$ and $w_2=\frac{x+N}{2} \in [\left\lceil \frac{s+N}{2} \right\rceil, \left\lfloor \frac{t+N}{2} \right\rfloor]$.

Now taking $x=Dlog_g y^2$ the lemma follows for every N. □

Note that if the interval in which $Dlog_g y^2$ is found is of size d then the intervals in which $Dlog_g y$ can be found are of size $\lceil d/2 \rceil$. This is used in the following algorithm.

Algorithm 1: (The input is $y \in G$ and a generator g)

(1) Let $n=\log_2 N$

(2) Compute $y_1=y$, $y_2=y_1^2$, $y_3=y_2^2$ ... $y_n=y_{n-1}^2$. /* $y_i=y^{2^{i-1}}$ */

(3) Let $s=0$.

(4) Let $d=\frac{N}{12n^2}$

(5) For $k=n$ to 1 do
  If ($LOG_G(g, y_k, s, d)=0$) then $s'=s$
  else $s'=(s+\left\lfloor \frac{N}{2} \right\rfloor) \bmod N$
  $s=\lceil s'/2 \rceil$
  $d=\lceil d/2 \rceil$
 end

(6) If ($g^s=y$) output s
 else if ($g^{s+1}=y$) output $s+1$
 else $s=s+\frac{N}{12n^2}$; goto (4)

The idea is that we are trying to find an s such that $Dlog_g y_n=Dlog_g y^{2^n}$ is in the range of size $\frac{N}{12n^2}$ starting from s. Assume that we are in the right interval then according to the lemma in each round in step (5) we reduce by a factor of 2 the size of interval in which we are looking for $Dlog_g y_k$. Therefore at the end after $n=log_2 N$ rounds we are looking for $Dlog_g y_1=Dlog_g y$ in an interval of size 2. Now, we check which of the two numbers in the interval is $Dlog_g y$. If both are not fitted then the current s is wrong and we increase it by $\frac{N}{12n^2}$ and try again. After at most $\frac{N}{d}=12n^2$ iterations we should find the right s. Therefore the number of times we will have to execute steps (4-6) is $O(n^2)$. Now, assuming that LOG_G is polynomial-time and recall the assumptions about G (i.e. N is known or can be computed in polynomial-time and the group operation can also be implemented in polynomial time) then this algorithm is also polynomial-time. □

4.2. Generalization of the Results to Acyclic Groups

In an acyclic group which is finite and Abelian we do not have a generator but a *generating-tuple* $\bar{g}=(g_1,g_2,\cdots,g_k)$. Any element $y \in G$ can be uniqely expressed as $y=g_1^{x_1}\cdots g_k^{x_k}$. The order of each g_i is denoted N_i and the number of elements in the group is

$N = N_1 \cdot N_2 \cdots N_k$. The *DLP* and the *DLP* 1 are defined with respect to g_1. (For example the *DLP* in such a group is: Given y - find x_1 such that $\exists x_2 \cdots x_k \mid y = g_1^{x_1} \cdots g_k^{x_k}$).

Our protocol with some modifications will work here too. We have to assume that we know (or can compute in polynomial-time) not only the group size N but also N_1. We should replace every occurance of N in the previous protocol by N_1 and also everything done with respect to g has to be done with respect to g_1. In addition we should randomize everything by elements chosen at random from the subgroup generated by $(g_2, g_3, \cdots, g_k)$. For example in step (V1) of the protocol the verifier should compute $\alpha = y^b \cdot g_1^{r_1} \cdot g_2^{r_2} \cdots g_k^{r_k}$, where $r_1 \in_R [\left\lfloor \frac{N_1}{4} \right\rfloor + 1, \left\lfloor \frac{N_1}{4} \right\rfloor + cN_1]$ and $r_2 \cdots r_k \in_R [1, N]$.

Using the same modifications described above we can also modify theorem 3 to show that the *DLP* and the *DLP* 1 are still equivalent in an acyclic group.

APPENDIX: Determining Membership in a Subgroup - Special Case

In this Appendix we consider the problem of determining membership in a subgroup generated by an element g in Z_p^*, when $p-1=2q$ and q is prime. We will show that in this special case, testing membership in a subgroup is easy. This should be contrasted with the believed intractability of *DLP* also for this case.

One can readily verify that if $p-1=2q$ with q prime then Z_p^* has $q-1$ primitive elements (i.e. elements of order $p-1$), $q-1$ elements of order q, one element of order 2, and one element of order 1 (i.e. the identity). Furthermore, all the elements of order q and the identity element form a subgroup which is generated by any of the elements of order q. Thus, the question of whether a is in the subgroup generated by g reduces (in this case!) to testing the order of both a and b (a is in the subgroup generated by b iff the order of a divides the order of b). Finally note that testing the order of an element is easy (in this case!).

REFERENCES

[AH] Aiello, W., and J. Hastad, "Perfect Zero-Knowledge Languages can be Recognized in Two Rounds", *Proc. 28th FOCS*, 1987, pp. 439-448.

[BGGHMR] Ben-or, M., O. Goldreich, S. Goldwasser, J. Hastad, S. Micali, and P. Rogaway, This proceedings.

[Bh] Benaloh, (Cohen), J.D., "Cryptographic Capsules: A Disjunctive Primitive for Interactive Protocols", *Crypto86*, Abstract #21, Santa Barbara, California, August 1986.

[BM] Blum, M., and Micali, S., "How to Generate Cryptographically Strong Sequences of Pseudo-Random Bits", *SIAM Jour. on Computing*, Vol. 13, 1984, pp. 850-864.

[BHZ] Boppana, R., J. Hastad, and S. Zachos, "Does Co-NP Have Short Interactive Proofs?", *IPL*, 25, May 1987, pp. 127-132.

[BC] Brassard, G., and C. Crepeau, "Non-Transitive Transfer of Confidence: A Perfect Zero-Knowledge Interactive Protocol for SAT and Beyond", *Proc. 27th FOCS*, pp. 188-195, 1986.

[BCDG] Brickell E.F., D. Chaum, I. Damgard, and J. van de Graaf, "Gradual and Verifiable Release of a Secret", preprint, 1987.

[CEGP] Chaum, D., J.H. Evertse, J. van de Graaf, and R. Peralta, "Demonstrating Possession of a Discrete Logarithm without Revealing It", *Crypto86*, Abstract #20, Santa Barbara, California, August 1986.

[CG] Chaum, D. and J. van de Graaf, "An Improved protocol for Demonstrating Possession of a Discrete Logarithm without Revealing It", *Eurocrypt87*, Amsterdam, The Netherlands, April 1987, IV-15 to IV-21.

[ESY] Even, S., A.L. Selman, and Y. Yacobi, "The Complexity of Promise Problems with Applications to Public-Key Cryptography" *Information and Control*, Vol. 61, 1984, pp. 159-173.

[F] Fortnow, L., "The Complexity of Perfect Zero-Knowledge", *Proc. of 19th STOC*, pp. 204-209, 1987.

[GMW] Goldreich, O., S. Micali, and A. Wigderson, "Proofs that Yield Nothing But their Validity and a Methodology of Cryptographic Protocol Design", *Proc. 27th FOCS*, pp. 174-187, 1986. Submitted to *JACM*.

[GMR] Goldwasser, S., S. Micali, and C. Rackoff, "Knowledge Complexity of Interactive Proofs", *Proc. 17th STOC*, 1985, pp. 291-304. To appear in *SIAM Jour. on Computing*.

[IY] Impagliazo, R., and M. Yung, "How to show that IP has Zero-Knowledge Proofs", private communication.

[Ka] Kaliski, B., "A Pseudo-Random Bit Generator Bases on Elliptic Logarithms", Ph.D. thesis, MIT, in preparation.

[O] Odlyzko, A., "Discrete Logarithm in finite fields and their cryptographic significance", Preprint.

[TW] Tompa, M., and H. Woll, "Random Self-Reducibility and Zero-Knowledge Interactive Proofs of Possession of Information", *Proc. 28th FOCS*, 1987, pp. 472-482.

[Y] Yao, A.C., "Theory and Applications of Trapdoor Functions", *Proc. of the 23rd IEEE Symp. on Foundation of Computer Science*, 1982, pp. 80-91.

Zero-Knowledge With Finite State Verifiers
(Extended Abstract)

Cynthia Dwork Larry Stockmeyer

IBM Almaden Research Center
San Jose, CA 95120

Abstract. We initiate an investigation of interactive proof systems (IPS's) and zero knowledge interactive proof systems where the verifier is a 2-way probabilistic finite state automaton (2pfa). Among other results, we show:

1. There is a class of 2pfa verifiers and a language L such that L has a zero knowledge IPS with respect to this class of verifiers, and L cannot be recognized by any verifier in the class on its own;

2. There is a language L such that L has an IPS with 2pfa verifiers but L has no zero knowledge IPS with 2pfa verifiers.

1. Introduction

Issues in complexity theory and cryptography motivated Babai [1] and Goldwasser, Micali, and Rackoff [7] to introduce the concept of an interactive proof system. Speaking informally, an *Interactive Proof System* (IPS) for membership in a language L is a two-party protocol whereby a "prover" convinces a "verifier" that elements $x \in L$ are actually in L. The concept is interesting only if the verifier is not itself sufficiently powerful to recognize L.

To date, almost all research in interactive proof systems has dealt with the case that the verifier is a probabilistic Turing machine (ptm) which runs in polynomial time. Due to the present lack of understanding of the power of polynomial time computation, many previous results depend on unproven assumptions, typically that a certain problem is not in P or that a certain cryptosystem cannot be broken in polynomial time. If the given assumptions are false, then either the proof becomes invalid or the result becomes trivial. For example, the important and powerful result that any language in NP has a zero knowledge IPS [6] would become unproven if secure probabilistic encryption schemes do not exist, and would become vacuous if P = NP.

The ability to prove lower bounds is crucial to understanding the structure of the class of languages with interactive proof systems. We therefore restrict the class of verifiers, namely, to 2-way probabilistic finite state automata (2pfa). We have obtained a number of results on 2pfa's and IP(2pfa), the class of languages with interactive proof systems in which the verifier is a 2pfa, examining public coins, private coins, and zero knowledge proof systems. ([4] contains a preliminary report of these results, including all proofs.) For the remainder of this abstract we restrict our attention to zero knowledge interactive proofs, noting only that the class IP(2pfa) is quite rich, despite the restricted power of the verifier, containing, for example, any language recognizable by a deterministic exponential time Turing machine.

2. Definitions

Our definition of an interactive proof parallels the one used in previous papers on interactive proofs where the verifier is a polynomial-time bounded ptm, for example, [7,6], and the one given by [3] in a more general setting. The main difference in our case is that the verifier is a 2-way probabilistic finite state automaton (2pfa). A 2pfa consists of a probabilistic finite state control and a 2-way head which reads the input string. Transition probabilities are assumed to be rational. In addition, the verifier can communicate with a prover which sees the same input. In our case, the communication is done via a single communication cell which can hold a single symbol from some finite communication alphabet. The prover writes a symbol in the cell only in response to a symbol written by the verifier. At some point in the interactive computation, the verifier can halt and either accept or reject. The prover-verifier pair (P, V) is an *interactive proof system for the language L with error probability ε* if

1. for all $x \in L$, $(P, V)(x)$ accepts with probability at least $1 - \varepsilon$, and
2. for all $x \notin L$ and all provers P^*, $(P^*, V)(x)$ rejects with probability at least $1 - \varepsilon$.

Let IP(2pfa) be the class of languages L such that L has an interactive proof system with error probability $\varepsilon < 1/2$.

Let 2PFA denote the class of languages recognized by 2pfa's with error probability $\varepsilon < 1/2$. Equivalently, 2PFA is IP(2pfa) restricted to IPS's (P, V) where P and V do not communicate (so the prover can be empty).

In some results we will want to talk about the expected or worst-case *time complexity* of an IPS (P, V), defined to be the expected (averaged over all random choices made by V and P) or worst-case number of steps taken by the verifier before halting and measured as a function $T(n)$ of the length n of the input.

A *sweeping 2pfa* is a 2pfa restricted so that the input head can switch direction only when reading an endmarker. In any computation, the input head alternately sweeps across the input from left to right, then from right to left, and so on.

We shall also use a more general form of recognition called separation. Let M be an IPS or a 2pfa, and let A and B be sets of words with $A \cap B = \emptyset$. Then *M separates A and B* if there is some constant $\varepsilon < 1/2$ (the error probability) such that, for all $x \in A$, $M(x)$ accepts with probability at least $1 - \varepsilon$, and for all $x \in B$, $M(x)$ rejects with probability at least $1 - \varepsilon$ (we do not care about the behavior of M on inputs not in A or B).

We temporarily defer the definition of "zero knowledge" interactive proof system.

2.1. An Example

If x is a string, let x^R be x written backwards. Define

$$Palindromes = \{\, x \in \{0,1\}^* \mid x = x^R \,\}.$$

We describe an IPS (P, V) for *Palindromes* with error probability ε for any constant $\varepsilon > 0$. If x is a palindrome, the interaction involves k iterations, where $k = \lceil \log_2(1/\varepsilon) \rceil$. On each iteration, the prover P sends x to the verifier one symbol at a time. At the start of each iteration, the verifer V (privately) tosses a fair coin. Letting w denote the string received from the prover during this iteration, if the outcome of the coin toss is "heads" then V checks that $w = x$ and rejects if not. If the outcome is "tails" then V checks that $w = x^R$ and rejects if not. If the check succeeds for all k iterations, then V accepts. It is easy to see that (P, V) is an IPS for *Palindromes* with error probability ε. This shows:

Theorem 2.1. *Palindromes* $\in$ *IP(2pfa). Moreover, for any error probability* $\varepsilon > 0$, *there is an IPS for Palindromes where the verifier is a sweeping 2pfa which runs in worst-case time* $O(n)$.

This theorem contrasts with the following impossibility result.

Theorem 2.2. *Palindromes* $\notin$ *2PFA.*

In fact, we prove a somewhat stronger result, from which the theorem follows. Theorem 2.2 is particularly interesting in light of Freivalds' result [5] that 2PFA contains certain nonregular sets, such as $\{ 0^n 1^n \mid n \geq 1 \}$.

3. Zero Knowledge Interactive Proof Systems

3.1. Old and New Definitions

Informally, an interactive proof system (P, V) for a language L is *zero knowledge* if for any input $x \in L$ and any verifier V^*, the only information which V^* can get from P during their interaction is the single bit of information that x belongs to L. Previous papers, e.g. [7], considered zero-knowledge only for ptime-ptm verifiers; we generalize the definition to an arbitrary class of verifiers as follows. Fix some class $\mathcal{V}$ of verifier machines, for example, 2pfa's or polynomial-time ptm's. Let $(\emptyset, \mathcal{V})$ be the subclass of machines in $\mathcal{V}$ that do not communicate with the prover (the symbol $\emptyset$ in this notation should be a reminder that the prover is empty). The interactive computation of $(P, V^*)(x)$ defines a distribution of conversations between P and V^*. The IPS (P, V) is *zero knowledge* if for any verifier $V^* \in \mathcal{V}$ there is an $M_{V^*} \in (\emptyset, \mathcal{V})$ such that, for all $x \in L$, $M_{V^*}(x)$ produces a distribution of conversations which is "close" to the distribution produced by $(P, V^*)(x)$.

At first glance, it would appear that the IPS (P, V) for palindromes described above is perfect zero knowledge according to this definition. On input x, the conversation consists of the prover sending x to the verifier several times, and obviously a 2pfa can produce this conversation alone. On an intuitive level, however, this IPS is clearly not zero knowledge for the following reason. Let A be the set of "double palindromes", i.e., the set of palindromes of the form ww^R where w is itself a palindrome, and let B be the set of palindromes not in A. It is not hard to see that there is a 2pfa V^* such that (P, V^*) separates A and B. On input x, V^* first checks that $|x|$ is even and rejects if not. Then starting from the left endmarker, V^* moves its head two to the right for every symbol sent to it by the prover until the right endmarker is reached. At this point, P has finished sending w and is ready to send w^R to V^*, where $x = ww^R$. So V^* is now in a position to compare w with w^R. Since we can show that no 2pfa separates A and B, it is clear that P is giving V^* some extra information which it cannot get by itself.

This suggests the following definition of zero knowledge which we call "recognition zero knowledge" to distiguish it from previous definitions.

Let $\mathcal{V}$ be a class of verifier machines. Let (P, V) be an IPS for the language L where $V \in \mathcal{V}$. Then (P, V) is a *recognition zero knowledge IPS for L with* $\mathcal{V}$ *verifiers* if, for any $V^* \in \mathcal{V}$ and any $A, B \subseteq L$ with $A \cap B = \emptyset$ such that (P, V^*) separates A and B, there is an $M_{V^*} \in (\emptyset, \mathcal{V})$ such that M_{V^*} separates A and B.

This is a fairly weak definition, in the sense that if a language has no recognition zero knowledge IPS then it has no zero knowledge IPS in a strong intuitive sense.

3.2. Languages Having No Zero Knowledge IPS

We first consider the palindrome language *Palindromes* defined in §2.1. We are able to show that the ability of a V^* to get extra information from the prover is not a property just of the particular IPS (P, V) described in §2.1. It is an inherent property of *Palindromes*.

Theorem 3.1. *There is no recognition zero knowledge IPS for Palindromes with 2pfa verifiers. This remains true with 2pfa verifiers which run in either polynomial worst-case time or polynomial expected time.*

By a similar proof, we can show that the graph isomorphism problem has no recognition zero knowledge IPS with 2pfa verifiers. This result contrasts with the situation for polynomial-time ptm verifiers, where graph isomorphism does have a (recognition) zero knowledge IPS [6]. We remark that the graph isomorphism problem does have an IPS with a 2pfa verifier.

3.3. A Language With a Recognition Zero Knowledge IPS

That the graph isomorphism problem has no (recognition) zero knowledge IPS with 2pfa verifiers suggests that techniques which have been used to obtain zero knowledge IPS's with ptime-ptm verifiers will not extend to 2pfa verifiers. In fact, we have no example of a language $L \notin$ 2PFA which has a recognition zero knowledge IPS with 2pfa verifiers. With 2pfa verifiers restricted to a certain class $\mathcal{R}$, however, we do have such an example. Let $\mathcal{R}$ denote the class of sweeping 2pfa's that halt in polynomial expected time.

Theorem 3.1, showing that there is no recognition zero knowledge IPS for palindromes, also holds with $\mathcal{R}$ verifiers. It is interesting to contrast this latter result with the result obtained next, that the unary version of palindromes has a recognition zero knowledge IPS with $\mathcal{R}$ verifiers. The unary version of palindromes is the language

$$Upal = \{\, 0^n 1^n \mid n \geq 1 \,\}.$$

Greenberg and Weiss [8] show that *Upal* cannot be recognized by any 2pfa which runs in polynomial expected time; in particular, *Upal* is recognized by no machine in $\mathcal{R}$, so the next result is not vacuous.

Theorem 3.2. *There is a recognition zero knowledge IPS for Upal with $\mathcal{R}$ verifiers.*

Actually, we prove a stronger result from which Theorem 3.2 follows immediately. We describe an IPS (P, V) for *Upal* with the following property. For any V^* and any $\varepsilon < 1/2$, let A (B) be the set of integers n such that $(P, V^*)(0^n 1^n)$ accepts (rejects) with probability at least $1 - \varepsilon$. Then there is a set C of integers such that C separates A and B (i.e., $A \subseteq C$ and $B \cap C = \emptyset$) and $\{\, 1^n \mid n \in C \,\}$ is regular. Our proof of this fact differs from previous proofs of zero knowledge in a significant way. Whereas previous proofs involved a simulation which used V^* as a "black box", our proof uses the internal structure of V^* in an essential way. This proof draws upon several facts from the theory of Markov chains.

4. Related Work

Other results on interactive proof systems with restricted verifiers appear in [2] and [3]. In these papers Condon and Ladner considered the case in which the verifier is restricted to run in space logarithmic in the length of the input, but they did not address the question of zero knowledge.

More recently, Kilian [9], adopting a defintion of zero knowledge based on the one presented here, has shown that, for verifiers which use logarithmic space and polynomial time, every language which has an IPS also has a zero knowledge IPS; no unproved assumptions are needed to obtain this result.

Note

The authors thank the program committee of CRYPTO '88 for inviting this paper to the conference.

References

[1] L. Babai, Trading group theory for randomness, *Proc. 17th ACM Symp. on Theory of Computing* (1985), pp. 421–429.

[2] A. Condon, Computational models of games, Ph.D. Thesis, Tech. Report 87-04-04, Computer Science Dept., University of Washington, Seattle, WA, 1987.

[3] A. Condon and R. Ladner, Probabilistic game automata, *Proc. Conference on Structure in Complexity Theory, Lecture Notes in Computer Science*, Vol. 223, Springer-Verlag, New York, 1986, pp. 144–162.

[4] C. Dwork and L. Stockmeyer, Interactive proof systems with finite state verifiers (preliminary report), IBM Research Report RJ 6262 (1988).

[5] R. Freivalds, Probabilistic two-way machines, *Proc. International Symposium on Mathematical Foundations of Computer Science, Lecture Notes in Computer Science*, Vol. 118, Springer-Verlag, New York, 1981, pp. 33–45.

[6] O. Goldreich, S. Micali, and A. Wigderson, Proofs that yield nothing but their validity and a methodology of cryptographic protocol design, *Proc. 27th IEEE Symp. on Foundations of Computer Science* (1986), pp. 174–187.

[7] S. Goldwasser, S. Micali, and C. Rackoff, The knowledge complexity of interactive proof systems, *Proc. 17th ACM Symp. on Theory of Computing* (1985), pp. 291–304.

[8] A. G. Greenberg and A. Weiss, A lower bound for probabilistic algorithms for finite state machines, *J. Comput. Syst. Sci.* 33 (1986), pp. 88–105.

[9] J. Kilian, Zero-knowledge with log-space verifiers, *Proc. 29th IEEE Symp. on Foundations of Computer Science* (1988), to appear.

Session 3

Number Theory

Chair: A. Odlyzko, Bell Laboratories

Intractable Problems in Number Theory

Eric Bach
Computer Sciences Department
University of Wisconsin
Madison, WI 53706

Abstract. This paper surveys computational problems related to integer factorization and the calculation of discrete logarithms in various groups. Its aim is to provide theory sufficient for the derivation of heuristic running time estimates, and at the same time introduce algorithms of practical value.

0. Introduction

Several problems in number theory are believed to be computationally intractable, a property that is potentially of great use in cryptography. Included in this category are problems related to integer factorization and the evaluation of discrete logarithms in various groups. The purpose of this paper is to summarize current knowledge about them, from a theoretical viewpoint.

In line with the long-term goals of complexity theory we should like to settle the question of whether these problems are really difficult, in the sense of having no probabilistic polynomial time algorithms. However, two features of this program seem inappropriate to the present context. First, a concentration on the asymptotic behavior of algorithms may be too restrictive, as a designer of public-key cryptosystems has to make compromises between efficiency and security and so must consider problems of a fixed size. Second, a restriction to algorithms that can be rigorously analyzed is too stringent if one wishes to design a system that will resist all known attacks. Since currently we cannot even prove asymptotic lower bounds on the complexity of these problems, design decisions must be based on what we believe to be the best algorithms. Such has been the state of affairs ever since the invention of public-key cryptology; it seems unlikely to change soon.

Preparation of this paper was supported by the National Science Foundation, via grants DCR-8504485 and DCR-8552596.

Of course, there *have* been improvements in our ability to solve these problems, most strikingly for factorization. A paper written in the early 1980's [Pomerance 1982] noted that the available algorithms could factor numbers up to 50 digits; the record now stands at 100 digits [Lenstra and Manasse 1988]. Thus the size of numbers whose factorization is feasible has doubled in ten years, and more advances are sure to follow. Certainly, some of this progress has come from the use of more powerful computers; what may not be so evident is the impact of new techniques, most notably the elliptic curve [Lenstra 1987] and quadratic sieve [Pomerance 1984] algorithms. Both of these algorithms are easy to parallelize on currently available machines.

Given an algorithm, one should always try to find the most general structure to which it applies. Thus, to highlight similarities and hide details, I have used algebraic language wherever possible. Sometimes the level of abstraction is greater than that needed merely to describe an algorithm. I would argue, however, that from this vantage point one can see clearly how the algorithms arise from the basic ideas. Necessarily, some details are lost; for more complete descriptions I refer to the surveys in the reference list (marked with a "*") as well as to the original papers.

In considering running times the reader should equate "step" with "bit operation."

1. Problems related to factoring

The problem of factorization makes sense in any unique factorization domain, of which the most basic example is the ordinary integers $\mathbb{Z}$. Thus we wish to compute the prime divisors of a number n presented as input.

If n is prime, then the problem is easy, as there are efficient randomized algorithms to test primality. With no more work than that of evaluating a power modulo n — an $O(\log n)^3$ process — one can tell if a number is prime, with an error probability of at most 1/4 [Rabin 1980]. If certainty is needed, then a more complicated deterministic algorithm [Adleman et. al. 1983] will *prove* that n is prime in at most $(\log n^{\log\log\log n})^{c+o(1)}$ steps. This algorithm also has a randomized version that is likely to find such a proof within the same time bound; for this it is conjectured that $c = 1/\log 2 \cong 1.442....$ Finally, a new test due to Atkin and based on complex multiplication has been recently implemented [Morain 1988]; this has proved useful for testing numbers up to 571 digits but it has not yet been analyzed.

In a statistical sense, we understand quite well how numbers factor. One can imagine that a random number n has prime factors whose *lengths* are selected by a "random bisection" process: choose a prime p whose length is uniformly distributed in the interval $(0, \log n)$, replace n by n/p and repeat, and so on. From this one gets intuition about how typical numbers factor as well as an efficient method for generating random numbers together with their factorizations [Bach 1988].

However, we do not know a polynomial time algorithm for factoring, even if we use randomness or make a reasonable assumption such as the extended Riemann hypothesis. We do not even know how to efficiently produce any useful information about the factors of a number. For instance, one might ask (from a formal analogy with polynomials) if extracting the squarefree part of a number, or just deciding if it is squarefree, takes less time than computing the full factorization; no such result is known. Neither can we count the prime factors of a number in any way better than finding them all.

One often finds factorization problems represented as equation-solving problems. For instance, an algorithm to solve the congruence

$$x^2 \equiv a \pmod{n} \tag{1.1}$$

can be used to efficiently factor n [Rabin 1979]. One could make a formal analogy with (1.1) and speculate that for e relatively prime to the Euler function $\phi(n)$, the congruence

$$x^e \equiv a \pmod{n} \tag{1.2}$$

cannot be efficiently solved without finding information from which one could easily factor n. The security of the RSA cryptosystem [Rivest et. al. 1978] relies on this conjecture as well as on the belief that factoring is difficult.

There is also an *existence* problem related to square roots modulo n: decide whether

$$\exists x [x^2 \equiv a \pmod{n}]. \tag{1.3}$$

This was used in the design of a probabilistic encryption method [Goldwasser and Micali 1982]. A necessary but not sufficient condition for (1.3) to hold is that the Jacobi symbol $(a \mid n)$ equals 1; this is computable in $O(\log n)^2$ time [Collins and Loos 1982]. Problem (1.3) clearly has *some* relation to factoring, for if a is a quadratic residue modulo n, then for each p dividing n, a is a square modulo p. By quadratic reciprocity, the factors of n are restricted to certain arithmetic progressions. However, recovering the factors from this information seems not to be easy. There is also a relationship between deciding (1.3) and computing $\omega(n)$, the number of distinct prime factors of n, since for odd n, the fraction of quadratic residues in $(\mathbb{Z}/n\mathbb{Z})^*$ is $2^{-\omega(n)}$; however, this does not immediately imply a polynomial-time equivalence between these problems.

More generally, one might wish to decide if, for a number e not prime to $\phi(n)$,

$$\exists x [x^e \equiv a \pmod{n}]. \tag{1.4}$$

This problem has been applied to the design of election protocols [Cohen and Fischer 1985]. It has been argued on heuristic grounds that an efficient algorithm to solve (1.4) for general e and n would lead to an algorithm for factoring that, although not polynomial time, would outperform any currently known on certain numbers [Adleman and McDonnell 1983].

Problems (1.1)-(1.4) are all solvable in random polynomial time for prime moduli and hence (by the Chinese remainder theorem and Hensel's lemma) for moduli whose factorization is known. The first two might be called ''zero-dimensional'' problems, for the analogous equations over the complex numbers have only finitely many solutions. Despite our intuition that increasing the dimension increases the complexity, similar one-dimensional problems *are* efficiently solvable. In particular, there is an efficient algorithm [Pollard and Schnorr 1987] to solve

$$x^2 - dy^2 \equiv a \pmod{n} \qquad (1.5)$$

as well as efficient algorithms for related problems in algebraic number rings [Adleman et. al. 1987].

All of the problems (1.1) - (1.4) make sense if $\mathbb{Z}$ is replaced by a ring and n is replaced by an ideal of finite index. Such generalizations appear not to have been studied much, although cryptographic schemes similar to the RSA have been proposed using algebraic numbers [Williams 1986].

2. Problems related to discrete logarithms

Just as the factorization problem is concerned with rings, the discrete logarithm problem is concerned with groups. Thus let G denote a finite cyclic group, in which the equality predicate, group multiplication, and inverses can be efficiently computed. If g is a generator of G and a another element of G, we wish to solve

$$g^x = a; \qquad (2.1)$$

this is the *discrete logarithm problem.* (The restriction to cyclic groups is no constraint because the group generated by an element is always cyclic.) If G has order m, then

$$G \cong \mathbb{Z}/m\mathbb{Z}. \qquad (2.2)$$

One can efficiently compute the reverse direction ($g^x \leftarrow x$) of this isomorphism by repeated squaring, with $O(\log x)$ group multiplications. The discrete logarithm problem is that of computing the forward direction. Of course $\mathbb{Z}/m\mathbb{Z}$ has a natural ring structure, and one might ask if the multiplication operation can be transplanted to G; that is, if one can efficiently

$$\text{compute } g^{xy} \text{ given } g^x, g^y. \qquad (2.3)$$

This is the *Diffie–Hellman* problem; clearly an algorithm to compute the forward direction of (2.2) (that is, solve (2.1)) can be used to solve it. For most groups of interest, it is unknown if the converse holds, although this has been shown in certain cases for $(\mathbb{Z}/p\mathbb{Z})^*$ [den Boer 1988].

Various groups have been suggested in cryptographic applications of problems (2.1) and (2.3). The original key-exchange proposal [Diffie and Hellman 1978] suggested $(\mathbb{Z}/p\mathbb{Z})^*$ where p is prime; one might also use $\mathbb{F}_q^*$, the multiplicative group of a finite field. There are also possible applications where the ambient group is non-cyclic, employing the unit group $(\mathbb{Z}/n\mathbb{Z})^*$ [Shmuely 1985, McCurley 1987], class groups of imaginary quadratic fields [Buchmann and Williams 1988], and various algebraic groups such as elliptic curves [Miller 1985, Koblitz 1987], abelian varieties [Koblitz 1988], and matrix groups [Varadharajan 1986].

With such an abundance of examples, one might well ask how far the generalization can be pushed. It seems that nothing about (2.1) or (2.3) requires that the group be finite, or even that inverses be computable; perhaps one could use semigroups instead of groups.

3. Algorithms

Remarkably, many of the best algorithms for the problems discussed above have apparent running times that are moderate powers of the following function:

$$L(n) = e^{\sqrt{\log n \log\log n}} \tag{3.1}$$

(here n is the number to be factored or the size of the group and $\log n$ is its natural logarithm). Before presenting algorithms, it will be worthwhile to discuss this function and how it arises.

$L(n)$ is often called a subexponential function because it grows more slowly than n^{ε} for any $\varepsilon > 0$; the appellation "subexponential" is apt because n^{ε} is an exponential function of the length $\log n$. However, most of our intuition deals with polynomial time algorithms, so it is convenient to pretend that $L(n)$ is a polynomial in $\log n$ with a slowly growing exponent, and define $E(n)$ by $L(n) = (\log n)^{E(n)}$. The following values hold:

n	10^{50}	10^{100}	10^{200}	10^{500}	10^{1000}
$\log n$	115	230	460	1151	2303
$E(n)$	4.9	6.5	8.7	12.8	17.2

From the above chart, if an algorithm requires $L(n)^c$ steps, a small reduction in c will have a large effect on its running time.

$L(n)$ arises from considerations of smoothness (a number is *smooth* with respect to a bound M if all its prime factors are less than or equal to M). Briefly, there is a tradeoff between making smooth numbers plentiful (M should be large) and making smooth numbers easy to recognize (M should be small).

To quantify this, we can use the random bisection heuristic cited above to get a plausible estimate for the "probability" $P(\alpha)$ that a random number near q is composed of prime factors less than q^{α}. Conditioning on the first factor's relative length x (which is presumed to be uniformly distributed),

$$P(\alpha) = \int_0^{\alpha} P\left(\frac{\alpha}{1-x}\right) dx;$$

after the change of variable $\lambda = 1/\alpha$ this becomes

$$\rho(\lambda) = \frac{1}{\lambda} \int_{\lambda-1}^{\lambda} \rho(t)\, dt.$$

This equation, together with the initial condition $\rho(\lambda) = 1$ for $0 < \lambda < 1$, defines the *Dickman rho-function*. As a rule of thumb, $\rho(\lambda) \cong \lambda^{-\lambda}$; consequently,

$$\Pr[x \le q \text{ is } q^{\alpha}\text{-smooth}] \cong \alpha^{1/\alpha}. \tag{3.2}$$

This can be used in a simple argument that underlies many running time calculations. Consider a two-phase procedure that first assembles a set of M-smooth numbers (with some desired properties) and then processes this set further to complete the algorithm. The first phase simply chooses random candidates (of size roughly q) and adds them to the set if they are smooth. To find the work for this phase, multiply the requisite number of smooth numbers by the work necessary to check a number for smoothness, and divide by the probability that a random number near q is smooth. If first two factors combined produce a term around M^k and the second phase of the algorithm takes M^l steps, then by the approximation (3.2), the total time is roughly

$$T = T_1 + T_2 \cong M^k \lambda^\lambda + M^l \tag{3.3}$$

where $\lambda = (\log q)/(\log M)$. If $T_1 \gg T_2$, we can minimize $\log T_1$ by setting its derivative to zero and find that asymptotically

$$\lambda \cong \sqrt{(2k \log q)/\log\log q}\,,$$

so

$$T = T_1 + T_2 \cong L(q)^{\sqrt{2k}} + L(q)^{\sqrt{l^2/2k}} \tag{3.4}$$

(the first term dominates if $2k \geq l$). Evidently we would like q, k, and l to be small; in fact, much of the progress in factorization and discrete logarithms has come from reducing these parameters.

Naturally, one would like to justify calculations such as the above, but this can be rigorously done only for certain algorithms. The problem is not with the approximation (3.2) — which can be sharpened — but with the tacit assumption that the numbers constructed by the algorithm are smooth with the same probability as random numbers of comparable size. Because in many important cases we are unable to prove this, there has arisen a notion of "heuristic" running time bounds for such algorithms. Thus we distinguish between proofs that an algorithm uses or expects to use only a certain number of steps (so-called "rigorous" bounds) and plausibility arguments for such assertions that always rely on unproved *ad hoc* assumptions. Of course, we can always try out a factoring or discrete logarithm algorithm and see if it works, since any answer produced can be quickly checked. For this reason, heuristic arguments are very useful, even if they are mathematically suspect.

In the descriptions below all running times will be heuristic, unless otherwise noted (the asymptotic notations 'O' and 'o' are reserved for proved results). Furthermore, the calculations are what might be called "first-order": they are only accurate enough to derive the correct value of c in an estimate of the form $L(n)^c$. In particular, they ignore relatively small factors such as powers of $\log n$.

Algorithms for factoring

Most factorization algorithms rely on what might be called a "functorial" approach. The idea is to associate with each ring $\mathbb{Z}/n\mathbb{Z}$ an object X_n in a generic fashion, so that the factorization given by the Chinese remainder theorem transfers to a factorization of X_n, thus:

$$\begin{array}{ccc} \mathbb{Z}/n\mathbb{Z} & \cong & \mathbb{Z}/p\mathbb{Z} \times \mathbb{Z}/q\mathbb{Z} \\ & \downarrow & \\ X_n & \cong & X_p \times X_q \end{array} \tag{3.5}$$

(in this section assume that n has two distinct prime factors p and q). We then use the factorization of X_n to recover the factors of n, usually by constructing special elements of X_n. The easiest way to guarantee that (3.5) occurs is to define X_n with polynomial equations modulo n, though this may not be the only way to proceed.

The best algorithms for factoring numbers composed of two equally large primes are the *quadratic residue* family of algorithms. These algorithms work with the group $X_n = \{x : x^2 \equiv 1 \bmod n\}$, for any element of X_n that is not congruent to $\pm 1 \bmod n$ (at least half of the elements of X_n have this property) will allow us to factor n as $\gcd(x-1, n)$. Equivalently, we can homogenize and seek numbers x and y for which $x^2 \equiv y^2$ but $x \not\equiv \pm y$. The algorithms in this family all do this by performing three basic steps:

1) Generate many quadratic residues mod n.
2) Try to factor them using primes $p \le M$, to construct congruences of the form $\prod_{p \le M} p^{e_p} \equiv r^2$.
3) Using linear algebra on the exponents modulo 2, combine the congruences multiplicatively to find x and y with $x^2 \equiv y^2$.

The *continued-fraction* factoring algorithm [Morrison and Brillhart 1970] generates residues around $\sqrt{n}$ in size by evaluating the continued fraction of $\sqrt{n}$, factors them by trial division, and uses Gaussian elimination for the linear algebra. Since roughly M linear equations are needed, we can take $q = n^{1/2}$, $k = 2$, and $l = 3$ in (3.4) to find that the running time is approximately $L(n)^{\sqrt{2}}$.

The *quadratic sieve* algorithm [Pomerance 1984] dispenses with the need for trial division, by using values of a polynomial to form residues around $\sqrt{n}$ in size. Instead of factoring each residue separately, the algorithm processes polynomial arguments one prime at a time, only examining those for which the corresponding value will be divisible by that prime. Neglecting log factors, the amortized cost of factorization per residue may be taken as constant. Using the notation of (3.3), the number of polynomial arguments processed must be the number of smooth residues needed (M) times the inverse smoothness probability (λ^λ). If Gaussian elimination is used for the linear algebra, then the running time is the result of taking $k = 1$ and $l = 3$ in (3.3). A good choice for M is obtained by balancing T_1 (the cost of sieving) and T_2 (the cost of equation solving), which leads to a running time of approximately $L(n)^{\sqrt{9/8}}$.

Since a number m has no more than $\log_2 m$ prime factors, the running time of this and similar algorithms can be improved by exploiting the sparsity of the linear equations.

A randomized algorithm based on shift-register synthesis [Wiedemann 1984] will solve an $M \times M$ linear system of equations over a finite field with $O(Mw)$ field operations, if there are w nonzero coefficients. Therefore, for theoretical purposes we may take $l = 2$ in analyzing the Gaussian elimination phase of the quadratic sieve algorithm; this leads to the improved estimate $L(n)$ for the running time.

If one wishes to factor a number with a known or suspected small prime factor p, the algorithm of choice is the *elliptic curve* method [Lenstra 1987]. This takes X_n to be the set of solutions to $y^2 \equiv x^3 + ax + b \pmod n$. By the Chinese remainder theorem, $X_n \cong X_p \times X_q$, but X_p has some additional structure. Augmented by an additional "point at infinity" (0:1:0), it forms an abelian group $\bar{X}_p$ with (0:1:0) as the identity (this group is written additively). The group operations are given by rational functions, which can be evaluated mod n. By the Riemann hypothesis for finite fields, $p + 1 - 2\sqrt{p} \le |\bar{X}_p| \le p + 1 + 2\sqrt{p}$, and the group order can be randomized within this interval by varying a and b. If we are lucky and find an M-smooth group (that is, one of M-smooth order), then any element must become the identity when multiplied by $E = \prod_{p \le M} p^{\lfloor \log_p M \rfloor}$. Of course, no rational operations can produce the point at infinity, so a factor is detected when one attempts this multiplication and divides by a non-unit in $\mathbb{Z}/n\mathbb{Z}$. For success, we expect to need only *one* M-smooth group, but by the prime number theorem, multiplication by E requires roughly M operations. The running time is therefore estimated by taking $q = p$, $k = 1$ and $l = 0$ in (3.4); one expects to extract p in approximately $L(p)^{\sqrt{2}}$ steps.

A related algorithm — it does not fit the paradigm (3.5)! — is based on *class groups* [Schnorr and Lenstra 1984]. Here one chooses a random small multiplier μ, and forms a group from the invertible ideals modulo similarity of a subring A of $\mathbb{Q}(\sqrt{-\mu n})$. In the simplest case, $-\mu n$ is the field discriminant, whose divisors are exactly the ramified primes. Solutions to $x^2 = 1$ in the class group lead in a straightforward way to these primes. (Factors can also extracted from square roots of 1 in the general case, but the theory is more complicated). If the group order h depends "randomly" on μ, as suggested by heuristic considerations [Cohen and Lenstra 1984], we may try many values of μ and hope that one of the resulting groups is M-smooth. If so we can annihilate the odd part of the group by brute force, then square repeatedly to find solutions to $x^2 = 1$. Since $h \cong \sqrt{n}$, we can evaluate the running time by taking $q = \sqrt{n}$, $k = 1$ and $l = 0$ in (3.4) and find it to be roughly $L(n)$.

The above discussion cites *three* factorization methods with a conjectured running time near $L(n)$, and one might suspect that this is the true complexity of factoring. However, the algorithms are all based on similar ideas, so it is equally plausible that the $L(n)$ running times are simply a consequence of this similarity. Of these algorithms, the quadratic sieve is the best algorithm in practice (unless we think the number to be factored might have a small prime divisor). It is superior because a typical step in its execution is a single-precision subtraction; a step of the elliptic curve algorithm must evaluate a pair of rational functions (at a cost of $O(\log n)^2$), and a step of the class group algorithm must perform a gcd calculation followed by a 2-dimensional lattice reduction (again, an $O(\log n)^2$ operation).

The *cyclotomic* family of factoring algorithms takes $X_n = (A/nA)^*$, where A is a ring of algebraic integers. In these cases, X_p is a direct sum of finite fields, each of order p^k-1 for some k, and we can easily factor n when any algebraic factor of p^k-1 is smooth [Bach and Shallit 1985]. The practically important cases are $k=1,2$; that is, the method is useful when $p \pm 1$ is smooth. For example, if $A = \mathbb{Z}$, then the unit group modulo p has order $p-1$, and by raising to a large enough power E we can annihilate this group, factoring n with $\gcd(x^E - 1, n)$ [Guy 1976]. The $p+1$ method [Williams 1982] works in a similar fashion with the group of elements in the finite field $\mathbb{F}_{p^2}$ that have norm 1. Both methods have a refinement in which the running time is proportional to the square root of the smoothness bound [Montgomery 1987]; they are useful as preliminary steps in factorization, before a complicated method like the quadratic sieve is used.

Some attention has also been paid to the effects of "second-order" smoothness, that is, smoothness of the automorphism group of $(A/pA)^*$. For example, if the map $x \rightarrow x^e$, an automorphism of $(\mathbb{Z}/p\mathbb{Z})^*$, has a small order t, then we can split n with $\gcd(x^{e^t} - x, n)$. This leads to a requirement that $\phi(p-1)$, the order of the automorphism group, have at least one large factor if p is going to be difficult to remove from n. Similarly, by considering automorphisms of the group of norm-1 elements in $\mathbb{F}_{p^2}$, we see that $\phi(p+1)$ should be chosen to have a large factor.

By properly building primes, the methods of the previous two paragraphs are easy to defend against. What appears to be more difficult is constructing a number that resists the elliptic curve or class group factorization methods. No one knows how to make the smoothness of the groups that occur in these algorithms less likely than the smoothness of random numbers of a comparable size.

A few words should be said here about rigorous analyses of factorization algorithms. Surprisingly, the best known running time for a deterministic factoring algorithm is $n^{1/4+o(1)}$ [Pollard 1974]; this can be lowered to $n^{1/5+o(1)}$ if the Extended Riemann Hypothesis is assumed [Schoof 1982]. The best randomized algorithm for factoring takes expected time $L(n)^{\sqrt{4/3}+o(1)}$ [Vallée 1988], although assuming the ERH, a randomized algorithm related to the class group method has an expected running time of $L(n)^{1+o(1)}$ steps [Lenstra 1987].

Contrasted with the variety of factoring algorithms, very little seems to be known about direct attacks on the RSA encryption scheme (1.2) or the residue problems (1.3) and (1.4). It has been shown that an algorithm to find or guess individual bits of a solution to (1.2) could be used to efficiently find complete solutions [Chor 1986], and that the cost of obtaining individual solutions to (1.2) can be reduced by accumulating other solutions [Desmedt and Odlyzko 1986], but no method to attack these problems has surfaced that is substantially better than factorization. Unfortunately, we cannot rule out the possibility that one exists.

Algorithms for discrete logarithms

The complexity of the discrete logarithm problem depends very much on the group considered. The most general algorithms are "canonical" in the sense that they use only the group operations; however their running times are exponential. In several important cases, though, we know methods with subexponential running times, equal to or better than those of the best factorization algorithms. However, these methods require the group to be specified as part of a larger structure.

The *baby-step/giant-step* algorithm [Shanks 1971] works in any group, as follows. Assume that $|G| \le t^2$, then a solution to $g^x = a$ can be written $x_0 + x_1 t$ with $0 \le x_i < t$. By computing the $2t$ elements g^{x_0} and $a \cdot g^{-x_1 t}$ and looking for a match (one can either sort or use hashing), x can be found in roughly $|G|^{1/2}$ steps (the space requirement is comparable; if $|G|$ is known, this can be reduced with a variant of the "rho" algorithm [Pollard 1978]).

This idea can be extended [Pohlig and Hellman 1978] if G is smooth in the sense of having a long chain factorization, where $|G_i/G_{i-1}| = p$:

$$1 = G_0 \subset G_1 \subset G_2 \subset \cdots \subset G_k = G.$$

Then the index x is expressible as $\sum x_i p^i$, $0 \le x_i < p$, and via the homomorphism $G_i \to G_1$ (raise to the power p^{i-1}), computation of the x_i's reduces to the solution of k discrete logarithm problems in G_1. Using the above algorithm, the complexity is roughly $k\sqrt{p}$.

Finally, assume that the factorization of $m = |G|$ is known: $m = \prod m_i$, where the m_i's are relatively prime. This induces a factorization of G into groups of relatively prime order, and if the m_i's are small we can solve the discrete log problem by going counterclockwise around the following diagram:

$$\begin{array}{ccc} G & \to & \mathbb{Z}/m\mathbb{Z} \\ \downarrow & & \uparrow \\ \prod G_i & \to & \prod \mathbb{Z}/m_i\mathbb{Z} \end{array}$$

(to project G into G_i, raise to the power m/m_i, to go across, solve the problem in each group G_i, and to go up, use the Chinese remainder theorem).

By combining the last two algorithms one sees that, except for a factor that is polynomial in $\log|G|$, the discrete log problem for a p-smooth group is solvable in time roughly $\sqrt{p}$.

In certain groups one can use the *index-calculus* family of algorithms, which work essentially by doing factorization on the left of (2.2) and linear algebra on the right. To use these algorithms G must be specifiable in the following way: start with a ring A that has unique factorization (or more generally, unique ideal factorization), take the free

Abelian group generated by the primes (certain "exceptional" primes may be omitted), and form the quotient group modulo a set of multiplicative identities. If G is represented as such a group, then factorizations in A lead to identities in G, which can be exploited to compute discrete logarithms. An important feature of this family of algorithms is that once one logarithm is computed, others can be found relatively quickly (typically in time equal to the square root of that needed to compute the first logarithm).

For example, take $G = (\mathbb{Z}/p\mathbb{Z})^*$ and $A = \mathbb{Z}$ [Adleman 1980]. For a smoothness bound M, roughly M smooth numbers of the form g^x will serve to tell us the discrete logarithms of all primes up to M. To find them, we try to factor random powers of g using primes less than or equal to M; each successful factorization gives a linear equation in $\mathbb{Z}/(p-1)\mathbb{Z}$ for the logarithms. The time required to construct this "database" can be estimated by taking $k=1$ and $l=2$ in (3.3), assuming that a subexponential factorization algorithm and sparse matrix techniques (generalized to finite rings) are used. This gives a time of roughly $L(p)^{\sqrt{2}}$ for the first phase of the algorithm. Once this is completed, computing the logarithm of a requires *one* smooth number of the form $a \cdot g^r$; if r is chosen at random, this will succeed after approximately λ^λ trials, in approximately $L(p)^{\sqrt{2}/2}$ steps.

This method can be modified so that it uses smooth numbers near $\sqrt{p}$ rather than near p [Coppersmith et. al. 1985]. In the analysis one has to replace p by $p^{1/2}$ in the above formulas; if this is done one finds that roughly $L(p)$ steps are needed to find the logarithms of small primes, and the work per additional logarithm is close to $L(p)^{1/2}$.

Similar methods are available for $\mathbb{F}_q^*$ when $q = 2^n$ (or, more generally, a power of a small prime); they have been exhaustively surveyed [Odlyzko 1985]. To study them, one needs an analog of (3.3) for polynomials (since elements of $\mathbb{F}_{2^n}$ are represented in this fashion). Calling a polynomial (over $\mathbb{F}_2$) d*-smooth* if all of its irreducible factors have degree at most d, the analogous approximation to (3.2) is

$$\Pr[\, f \text{ of degree } d \text{ is } \alpha d-\text{smooth}\,] \cong \alpha^{1/\alpha}. \tag{3.6}$$

Assume that the algorithm requires a collection of m-smooth polynomials, each of degree roughly d. Again, the work in assembling them is the size of the collection times the work required to test a candidate (estimated as 2^{mk}) times the inverse smoothness probability λ^λ. Taking $\lambda = d/m$, and assuming a second phase of complexity 2^{ml}, the total time is

$$T = T_1 + T_2 \cong 2^{mk}\lambda^\lambda + 2^{ml} \tag{3.7}$$

which is minimized asymptotically for $m = \sqrt{(d \log d)/(2k \log 2)}$, and leads to

$$T_1 + T_2 \cong M(d)^{\sqrt{2k\log 2}} + M(d)^{\sqrt{l^2 \log 2 / 2k}} \tag{3.8}$$

where $M(d) = e^{\sqrt{d \log d}}$.

The basic index-calculus algorithm in $\mathbb{F}_{2^n}^*$ first tries to find m-smooth polynomials g^x which have degree n. Ignoring log factors, roughly 2^m polynomials are needed. Taking $d=n$, $k=1$ and $l=2$ in (3.8) (the time to factor can be neglected [Berlekamp 1967], and as usual the linear equations are sparse), we find the time for the first phase to be roughly $M(n)^{\sqrt{2\log 2}}$, and the time to extract additional logarithms to be about $M(n)^{\sqrt{\log 2/2}}$. As with $(\mathbb{Z}/p\mathbb{Z})^*$, this can be improved by working with smooth polynomials whose degree is a constant fraction of n [Odlyzko 1985].

The asymptotically fastest algorithm for discrete logarithms in $\mathbb{F}_{2^n}^*$ is an extension of the index-calculus idea that works with smooth polynomials of degree around $n^{2/3}$ [Coppersmith 1984]. It requires time roughly $K(n)^c$, where $K(n) = \exp(n^{1/3}\log^{2/3} n)$ and $c \cong 1.41$ (*not* the square root of 2!).

The above algorithms will compute discrete logarithms in $\mathbb{F}_{p^m}^*$ when p is small or m is 1. Perhaps due to a lack of applications, there are no algorithms known to be efficient when both m and p vary. The basic algorithm can be generalized by replacing $\mathbb{Z}$ by a ring of algebraic integers [ElGamal 1986]; this handles $\mathbb{F}_{p^m}^*$ when m is fixed, but it is unclear how it can be generalized to take account of all cases.

One can use the index-calculus method to find logarithms in $(\mathbb{Z}/n\mathbb{Z})^*$ (this was used in Desmedt and Odlyzko's attack on the RSA scheme), but there is a simpler approach: just factor the group (by factoring n), and solve the problem in each group separately. In some sense, this is the best possible method, because an algorithm to solve arbitrary discrete logarithms modulo n can be used to efficiently factor n. It can also be shown that discrete logarithms in $(\mathbb{Z}/p^e\mathbb{Z})^*$ reduce in polynomial time to discrete logarithms in $(\mathbb{Z}/p\mathbb{Z})^*$, via p-adic logarithms [Bach 1984]. The group $(\mathbb{Z}/n\mathbb{Z})^*$ does have one advantage: we *know* that the Diffie-Hellman problem (2.3) for this group is difficult, if factoring is hard [Shmuely 1985]; this holds in some cases even if the generator g is fixed [McCurley 1987].

There is also an index-calculus algorithm for the class group of an imaginary quadratic field of discriminant $-\Delta$, if the class number h is known [McCurley 1988]. In this case, an ideal $\mathbf{A}$ is called M-smooth if each prime ideal $\mathbf{p}$ dividing it satisfies $N\mathbf{p} \le M$ (the number of prime ideals of norm at most M is roughly the number of ordinary primes at most M, by the prime ideal theorem). Each ideal class contains an ideal of norm at most $\sqrt{\Delta}$, and we can attempt to find the indices of all small prime ideals in the group generated by $\mathbf{g}$ by factoring enough M-smooth ideals of the form $\mathbf{g}^x$ (factorization of an ideal reduces to factorization of its norm in $\mathbb{Z}$), and using linear algebra in $\mathbb{Z}/h\mathbb{Z}$. Analogously to (3.2),

$$\Pr[\mathbf{A} \text{ with } N\mathbf{A} \le q \text{ is } q^{\alpha}-\text{smooth}] \cong \alpha^{1/\alpha}, \tag{3.9}$$

[Hazlewood 1977], so that the asymptotic complexity of the first stage can be found by taking $q = \sqrt{\Delta}$, $k = 1$, and $l = 2$ in (3.3); it is roughly $L(\Delta)$. To solve $\mathbf{g}^x = \mathbf{A}$ given logarithms of all small prime ideals requires one smooth ideal (of the form $\mathbf{g}^r\mathbf{A}$), therefore time roughly $L(\Delta)^{1/2}$.

Discrete logarithms in elliptic curves and abelian varieties have also been considered [Miller 1985, Koblitz 1987, Koblitz 1988]. These groups have the advantage that the index-calculus algorithm appears not to generalize to them, and if the order of the group is properly chosen, the exponential-time algorithms outlined earlier in this section can be made very expensive.

Since all the discrete logarithm algorithms (except for the baby-step/giant-step procedure) require knowledge of the group order, it is worthwhile to summarize how difficult this is to compute. For $\mathbb{F}_q^*$, the group order is just $q-1$. The orders of the last three groups are more refractory. It is known that any algorithm to compute $\phi(n)$, the order of $(\mathbb{Z}/n\mathbb{Z})^*$, allows one to easily factor n [Miller 1976]; a similar result holds for the class number [Shanks 1971], although $h(-\Delta)$ can be computed in roughly $L(\Delta)^{\sqrt{9/8}}$ steps [McCurley 1988]. Finally, although the number of solutions to $y^2 = x^3 + ax + b \pmod p$ can be found in $O(\log p)^8$ steps [Schoof 1985], the degree of this bound is too high for the algorithm to be practical.

Perhaps because the problem lacks the notoriety of factoring, the rigorous analysis of discrete logarithm procedures has not received as much attention. The exponential-time algorithms are easy to analyze; the index-calculus methods, relying on smoothness, are not. However, there are randomized algorithms for discrete logarithms in $(\mathbb{Z}/p\mathbb{Z})^*$ and $\mathbb{F}_{2^n}^*$ whose expected running times can be proved to be $L(p)^{\sqrt{2}+o(1)}$ and $M(n)^{\sqrt{2\log 2}+o(1)}$, respectively [Pomerance 1987].

In contrast to factorization, there is also not much known about the special cases in which discrete logarithms are easy to compute. If the group is smooth, then one can use the factorization of the group to advantage as explained above. In particular, taking $G = (\mathbb{Z}/p\mathbb{Z})^*$, discrete logarithms can be easily found if $p-1$ is smooth. No one knows if the smoothness of $p+1$ (or higher cyclotomic polynomials) helps in this case.

An intriguing unanswered question asks if the complexity of the discrete logarithm problem in $(\mathbb{Z}/p\mathbb{Z})^*$ equals that of the factorization problem. More generally, one would like to classify these and similar problems into degrees of difficulty; although partial results along these lines are known [Shallit and Shamir 1985, Woll 1987, Landau 1988], a complete theory has not yet been developed.

4. Practical considerations

From the above discussion, if one wishes to concoct difficult instances of a factorization or discrete logarithm problem, one must avoid smoothness. In particular, not only must the original structure not be smooth, but neither must any related structures have this property. Unfortunately, without any good lower bounds on computational complexity, we are uncertain exactly what structures count as related. In addition, all of the algorithms discussed in this paper are in some sense algebraic, but this does not eliminate the possibility that methods of a more combinatorial nature could be useful.

In using the heuristic running times developed above, it is important to recognize that first-order formulas like (3.3) tend to overestimate running times, often by several orders of magnitude. For example, evaluating $L(n)^{\sqrt{9/8}}$ (the running time of the quadratic sieve algorithm with Gaussian elimination) at $n = 10^{92}$ gives 3×10^{15} operations, or almost a year if an operation takes 8 nanoseconds. However, an actual 92-digit factorization [te Riele 1988] took 3 days on an NEC SX-2, a machine whose cycle time *is* 8 nanoseconds.

For factorization and discrete logarithms, it would be useful to have a simple "second-order" theory accurate enough to account for such discrepancies. This has yet to be worked out in any detail, but some techniques for improving the estimates can be suggested.

First, although the rough estimate $\rho(\lambda) \equiv \lambda^{-\lambda}$ is surprisingly useful for values of practical interest (if $5 \le \lambda \le 10$, it overestimates ρ, by a factor of 4 at most), it is not hard to get better estimates. For example, if ξ denotes the positive root of $e^{\xi} - 1 = \lambda\xi$, and $\mathrm{Ei}(\xi)$ denotes the exponential integral function (that is, the Cauchy principal value of $\int_{-\infty}^{\xi} t^{-1} e^{t}\, dt$), then as $\lambda \to \infty$,

$$\rho(\lambda) \sim \frac{1}{\sqrt{2\pi\lambda}} \cdot \frac{1}{\xi} \cdot e^{-\lambda\xi + \mathrm{Ei}(\xi)}$$

[de Bruijn 1951]. This is already quite accurate; when $\lambda = 5$ the error is only 10%. To get more precision, one can replace the integral in the definition of ρ by an approximation such as Simpson's rule and solve for $\rho(\lambda)$ in terms of "previous" values; this gives an iterative scheme from which it is easy to compute ρ numerically [van de Lune and Wattel 1969]. There is also an asymptotic expression for ρ in terms of elementary functions ($\lambda^{-\lambda}$ is its dominant factor), but it is not very precise unless λ is large.

From the published data [Schnorr and Lenstra 1984] it appears that the probability of smoothness is estimated very well by the Dickman rho-function; this conclusion is also supported by the asymptotic theory [Canfield et. al. 1983]. Once one has a good method to estimate this function, it is not hard to restore the "missing" log factors in formulas such as (3.3) and find a good value for λ numerically. This has been done at least for the continued fraction algorithm [Wunderlich 1985].

For polynomials over $\mathbb{F}_2$, analogs to the Dickman rho-function have been tabulated and the running times of various discrete logarithm algorithms worked out [Odlyzko 1985]. However, much less is known about the accuracy of estimates such as (3.9), which give the smoothness probability of ideals in algebraic number fields and therefore affect running time estimates for computing discrete logarithms in class groups and extension fields of $\mathbb{Z}/p\mathbb{Z}$.

Of course, one can simply try out algorithms and see how they perform on a variety of machines. The most comprehensive such experiments have been performed with factoring algorithms, most notably using benchmark numbers from the Cunningham project [Brillhart et. al. 1983]. For algorithms such as the quadratic sieve that collect many rows of a matrix, one expects by the law of large numbers that the running time can be extrapolated from the time needed to find a few rows. The elliptic curve and similar algorithms are more chancy; since only one smooth group is required, there is no reliable way to predict when the algorithm will finish.

Finally, some mention should be made of the parallel versions of these algorithms. Algorithms such as the elliptic curve and class group factorization method have a straightforward parallelization: give each processor its own group to try. For the quadratic sieve, much benefit can be gained by using the multiple-polynomial version

[Silverman 1987]. This has the theoretical advantage that the residues sieved are smaller than those of the unadorned algorithm, as well as the practical advantage that each processor can be given its own polynomial from which to generate values for sieving. This was the algorithm that factored the 100-digit Cunningham number $(11^{104}+1)/(11^{8}+1)$.

5. Acknowledgements

I would like to thank Josh Benaloh, Susan Landau, Kevin McCurley, François Morain, Andrew Odlyzko, Carl Pomerance, and Jeffrey Shallit for helpful comments. The support of the National Science Foundation is also gratefully acknowledged.

6. References

L.M. Adleman, A subexponential algorithm for the discrete logarithm problem with applications to cryptography, in Proceedings of the 1980 IEEE Symposium on Foundations of Computer Science, New York: IEEE (1980).

L.M. Adleman and R. McDonnell, An application of higher reciprocity to computational number theory, in Proceedings of the 23rd IEEE Symposium on Foundations of Computer Science, New York: IEEE (1983).

L.M. Adleman, C. Pomerance, and R.S. Rumely, On distinguishing prime numbers from composite numbers, Annals of Mathematics 117 (1983), pp. 173-206.

* L.M. Adleman and K.S. McCurley, Open problems in number theoretic complexity, in Discrete Algorithms and Complexity (Proceedings of the Japan-US Joint Seminar), London: Academic Press (1987).

L.M. Adleman, D.R. Estes, and K.S. McCurley, Solving bivariate quadratic congruences in random polynomial time, Mathematics of Computation 48 (1987), pp. 17-28.

E. Bach, Discrete logarithms and factoring, University of California at Berkeley Computer Science Division Report 84/186 (1984).

E. Bach and J. Shallit, Factoring with cyclotomic polynomials, Mathematics of Computation 52 (1989).

E. Bach, How to generate factored random numbers, SIAM Journal on Computing 17 (1988), pp. 179-193.

E.R. Berlekamp, Factoring polynomials over finite fields, Bell System Technical Journal 46 (1967), pp. 1853-1859.

B. den Boer, Diffie-Hellman is as strong as discrete log for certain primes, preprint, Centre for Mathematics and Computer Science, Amsterdam (1988).

J. Brillhart, D.H. Lehmer, J.L. Selfridge, B. Tuckerman, and S.S. Wagstaff, Jr., Factorizations of $b^n \pm 1$, $b = 2,3,5,6,7,10,11,12$ up to High Powers, Providence: American Mathematical Society (1983).

N.G. de Bruijn, The asymptotic behavior of a function occurring in the theory of primes, Journal of the Indian Mathematical Society 15 (1951), pp. 25-32.

J. Buchmann and H.C. Williams, A key-exchange system based on imaginary quadratic fields, Journal of Cryptology 1 (1988).

E.R. Canfield, P. Erdös, and C. Pomerance, On a problem of Oppenheim concerning "Factorisatio Numerorum," Journal of Number Theory 17 (1983), pp. 1-28.

B.-Z. Chor, Two Issues in Public Key Cryptography, Cambridge: MIT Press (1986).

H. Cohen and H.W. Lenstra, Jr., Heuristics on class groups of number fields, in Number Theory (Lecture Notes in Mathematics 1068), Berlin: Springer (1984).

J.D. Cohen and M.J. Fischer, A robust and verifiable cryptographically secure election scheme, in Proceedings of the 26th Annual ACM Symposium on Foundations of Computer Science, New York: IEEE (1985).

G. Collins and R. Loos, The Jacobi symbol algorithm, SIGSAM Bulletin 16 (1982), pp. 12-16.

D. Coppersmith, Fast evaluation of logarithms in fields of characteristic two, IEEE Transactions on Information Theory 30 (1984), pp. 587-594.

D. Coppersmith, A.M. Odlyzko, and R. Schroeppel, Discrete logarithms in GF(p), Algorithmica 1 (1986), pp. 1-15.

Y. Desmedt and A.M. Odlyzko, A chosen text attack on the RSA cryptosystem and some discrete logarithm schemes, in Proceedings of CRYPTO '85 (Lecture Notes in Computer Science 218), Berlin: Springer (1986).

W. Diffie and M. Hellman, New directions in cryptography, IEEE Transactions on Information Theory 22 (1978), pp. 644-654.

T. ElGamal, On computing logarithms over finite fields, in Proceedings of CRYPTO '85 (Lecture Notes in Computer Science 218), Berlin: Springer (1986).

S. Goldwasser and S. Micali, Probabilistic encryption, Journal of Computer and System Sciences 28 (1984), pp. 270-299.

* R.K. Guy, How to factor a number, in Proceedings of the Fifth Manitoba Conference on Numerical Mathematics (1976).

D.G. Hazlewood, On ideals having only small prime factors, Rocky Mountain Journal of Mathematics 7 (1977), pp. 753-768.

N. Koblitz, Elliptic curve cryptosystems, Mathematics of Computation 48 (1987), pp. 203-209.

N. Koblitz, A family of Jacobians suitable for discrete log cryptosystems, Proceedings of CRYPTO '88, Berlin: Springer (1989).

S. Landau, Some remarks on computing the square parts of integers, Information and Computation 78 (1988), pp. 246-253.

A.K. Lenstra, Fast and rigorous factorization under the generalized Riemann hypothesis, University of Chicago Computer Science Department Report 87-007 (1987) [to appear, Indagationes Mathematicae].

* A.K. Lenstra and H.W. Lenstra, Jr., Algorithms in number theory, to appear in Handbook of Theoretical Computer Science, Amsterdam: North-Holland.

A.K. Lenstra and M. Manasse, 100 digit factorization, announcement (1988).

H.W. Lenstra, Jr., Factoring integers with elliptic curves, Annals of Mathematics 126 (1987), pp. 649-673

J. van de Lune and E. Wattel, On the numerical solution of a differential-difference equation arising in analytic number theory, Mathematics of Computation 23 (1969), pp. 417-421.

K.S. McCurley, A key distribution system equivalent to factoring, preprint, IBM Almaden Research Center (1987).

K.S. McCurley, Cryptographic key distribution and computation in class groups, to appear in Proceedings of the NATO Advanced Study Institute on Number Theory and Applications (Banff, May 1988), Dordrecht: Reidel. [Available as IBM Almaden Research Center Technical Report #6433.]

G.L. Miller, Riemann's hypothesis and tests for primality, Journal of Computer and System Sciences 13 (1976), pp. 300-317.

V. Miller, Use of elliptic curves in cryptography, in Proceedings of CRYPTO '85 (Lecture Notes in Computer Science 218), Berlin: Springer (1986).

P.L. Montgomery, Speeding the Pollard and elliptic curve methods of factoring, Mathematics of Computation 48 (1987), pp. 243-264.

F. Morain, Implementation of the Goldwasser-Kilian-Atkin primality testing algorithm, University of Limoges / INRIA Report (1988).

M.A. Morrison and J. Brillhart, A method of factoring and the factorization of F_7, Mathematics of Computation 29 (1975), pp. 183-205.

J.M. Pollard, Theorems on factorization and primality testing, Proceedings of the Cambridge Philosophical Society 76 (1974), pp. 521-528.

J.M. Pollard, Monte Carlo methods for index computation (mod p), Mathematics of Computation 32 (1978), pp. 918-924.

J.M. Pollard and C.-P. Schnorr, An efficient solution of the congruence $x^2+ky^2 \equiv m \pmod{n}$, IEEE Transactions on Information Theory IT-33 (1987), pp. 702-709.

S. Pohlig and M. Hellman, An improved algorithm for computing logarithms over GF(p) and its cryptographic significance, IEEE Transactions on Information Theory 24 (1978), pp. 106-110.

* C. Pomerance, Analysis and comparison of some integer factoring algorithms, in Computational Methods in Number Theory (v. 1), edited by H.W. Lenstra, Jr., and R. Tijdeman, Amsterdam Mathematical Centre Tract #154 (1982).

C. Pomerance, The quadratic sieve factoring algorithm, in Proceedings of EUROCRYPT 84 (Lecture Notes in Computer Science 209) Berlin: Springer (1985).

C. Pomerance, Fast rigorous factorization and discrete logarithm algorithms, in Discrete Algorithms and Complexity, Proceedings of the Japan-US Joint Seminar, London: Academic Press (1987).

* A.M. Odlyzko, Discrete logarithms and their cryptographic significance, Proceedings of EUROCRYPT '84 (Lecture Notes in Computer Science 209), Berlin: Springer (1985).

M.O. Rabin, Digitalized signatures and public-key functions as intractable as factorization, MIT Laboratory for Computer Science Report TR-212 (1979).

M.O. Rabin, Probabilistic algorithm for testing primality, Journal of Number Theory 12 (1980), pp. 128-138.

H. te Riele, W. Lioen and Dik Winter, New factorization records, announcement (1988).

R. Rivest, A. Shamir, and L. Adleman, A method for obtaining digital signatures and public-key cryptosystems, Communications of the ACM 21 (1978), pp. 120-126.

D. Shanks, Class number, a theory of factorization, and genera, in Proceedings of Symposia in Pure Mathematics 20, Providence: American Mathematical Society (1971).

C. Schnorr and H.W. Lenstra, Jr., A Monte Carlo factoring algorithm with linear storage, Mathematics of Computation 43 (1984), pp. 289-311.

R. Schoof, Quadratic fields and factorization, in Computational Methods in Number Theory (v. 2), edited by H.W. Lenstra, Jr., and R. Tijdeman, Amsterdam Mathematical Centre Tract #155 (1982).

R. Schoof, Elliptic curves over finite fields and the computation of square roots mod p, Mathematics of Computation 44 (1985), pp. 483-494.

J. Shallit and A. Shamir, Number-theoretic functions which are equivalent to number of divisors, Information Processing Letters 20 (1985), pp. 151-153.

Z. Shmuely, Composite Diffie-Hellman public-key systems are hard to break, Technion Computer Science Department Report 356 (1985).

R.D. Silverman, The multiple polynomial quadratic sieve, Mathematics of Computation 48 (1987), pp. 329-339.

B. Vallée, Quasi-uniform algorithms for finding small quadratic residues and application to integer factorisation, preprint, Université de Caen (1988) [Presented at 1988 AMS Computational Number Theory Conference].

V. Varadharajan, Trapdoor rings and their use in cryptosystems, in Proceedings of CRYPTO '85 (Lecture Notes in Computer Science 218), Berlin: Springer (1986).

D.H. Wiedemann, Solving sparse linear equations over finite fields, IEEE Transactions on Information Thery 32 (1986), pp. 54-62.

H.C. Williams, A $p+1$ method of factoring, Mathematics of Computation 39 (1982), pp. 225-234.

H.C. Williams, An M^3 public-key encryption scheme, in Proceedings of CRYPTO '85 (Lecture Notes in Computer Science 218), Berlin: Springer (1986).

* H. Woll, Reductions among number-theoretic problems, Information and Computation 72 (1987), pp. 167-169.

M. Wunderlich, Implementing the continued fraction factoring algorithm on parallel machines, Mathematics of Computation 44 (1985), pp. 251-260.

A Family of Jacobians Suitable for Discrete Log Cryptosystems

Neal Koblitz, Dept. of Mathematics GN-50,
Univerity of Washington, Seattle WA 98195

Abstract. *We investigate the jacobians of the hyperelliptic curves* $v^2 + v = u^{2g+1}$ *over finite fields, and discuss which are likely to have "almost prime" order.*

1. The discrete logarithm problem in a finite abelian group A consists in finding for given $a, b \in A$ an integer m such that $a = mb$, if such m exists. In cases when the discrete log problem appears to be intractable in A, one can construct certain public key cryptosystems in which taking large multiples of a group element is the trapdoor function. The first examples of A that were considered were the multiplicative groups of finite fields. However, because some special techniques for attacking the discrete log problem are available in that case, it is useful to study other sources of finite abelian groups. In [6] we investigated the use of the jacobians of hyperelliptic curves defined over finite fields.

In the present article we consider an especially simple family of such curves. We first give an algorithm for the group law for this family. Next, we recall how to compute the number of points in terms of jacobi sums. In order for the discrete log problem to be intractible, we would like the number of points on the jacobian to be "almost prime" in the sense of [6]. Some necessary conditions for this are given, and some examples are tabulated.

2. For each positive integer g (the *genus*) we consider the hyperelliptic curve $v^2 + v = u^{2g+1}$ defined over the field $\mathbf{F}_p$ of p elements, where p is a prime not dividing $2g + 1$. Let $K = \mathbf{F}_{p^n}$. A *K-divisor* is a finite formal sum $D = \sum m_i P_i$ of $\overline{K}$-points on the curve which is fixed by any $\sigma \in \mathrm{Gal}(\overline{K}/K)$. Its *degree* is $\sum m_i$. The finite abelian group of K-points of the jacobian, denoted $\mathbf{J}(K)$, is the quotient of the group of K-divisors of degree zero by the subgroup of divisors of rational functions (defined over K) on the curve. Every element $D \in \mathbf{J}(K)$ is uniquely associated to a pair of functions $a, b \in K[u]$ for which $\deg a \leq g$, $\deg b < \deg a$, and $b(u)^2 + b(u) - u^{2g+1}$ is divisible by $a(u)$; namely, D is the equivalence class of the g.c.d. of the divisors of the functions

$a(u)$ and $b(u)-v$. The element D of $\mathbf{J}(K)$ is then denoted $\mathrm{div}(a,b)$. For more details, see [6] and [2].

To add two elements $\mathrm{div}(a_1,b_1)$, $\mathrm{div}(a_2,b_2)\in\mathbf{J}(K)$, one proceeds in two stages. First, let $d=d(u)$ be the g.c.d. of the three polynomials $a_1(u)$, $a_2(u)$ and $b_1(u)+b_2(u)+1$; and choose $s_1(u)$, $s_2(u)$ and $s_3(u)$ to be polynomials in u such that $d=s_1a_1+s_2a_2+s_3(b_1+b_2+1)$. Next, set $a=a_1a_2/d^2$ and

$$b(u)=(s_1(u)a_1(u)b_2(u)+s_2(u)a_2(u)b_1(u)+s_3(u)(b_1(u)b_2(u)+u^{2g+1}))/d(u) \pmod{a(u)}.$$

In stage 2, if $\deg a>g$, we replace the pair (a,b) by the equivalent pair (a',b') defined by setting $a'(u)=(u^{2g+1}-b(u)^2-b(u))/a(u)$ and $b'(u)=-b(u)-1 \ (mod\ (a'(u))$. Since $\deg a'<\deg a$, successive application of this procedure leads to a pair $\mathrm{div}(a'',b'')$ with $\deg a''\le g$ such that $\mathrm{div}(a'',b'')=\mathrm{div}(a_1,b_1)+\mathrm{div}(a_2,b_2)$. This concludes the description of the group law in $\mathbf{J}(K)$.

3. Let g be a fixed positive integer, let $\mathbf{J}(K)$ denote the K-points of the jacobian of the curve $v^2+v=u^{2g+1}$ defined over $\mathbf{F}_p$, where the degree $d=2g+1$ is prime to p, and let N_n denote $\#(\mathbf{J}(\mathbf{F}_{p^n}))$. As explained in [6], the zeta-polynomial $Z(T)=Z_g(T)$

$$Z(T)=\prod_{j=1}^{g}(T-\alpha_j)(T-\overline{\alpha}_j)$$

of the curve $v^2+v=u^{2g+1}$ is related to N_n as follows:

$$N_n=\prod_{j=1}^{g}|1-\alpha_j^n|^2.$$

The polynomial $Z(T)$ is computed from the number of $\mathbf{F}_{p^n}$-solutions of $v^2+v=u^{2g+1}$ for $n=1,2,\ldots,g$, and the result is as follows (see, e.g., [13]).

For simplicity, we shall henceforth suppose that $d=2g+1$ is prime. In practice, this is the only case we shall be interested in, because of Theorem 4(1a) below. Let f denote the multiplicative order of p modulo d, so that $d|p^f-1$, and let h denote $2g/f$. Let χ be a fixed character of $\mathbf{F}_{p^f}^*$ of order d, i.e., $\chi(\rho)=e^{2\pi i/d}$ for some generator ρ of $\mathbf{F}_{p^f}^*$. Let m_j, $1\le j\le h$, run through a set of representatives of $(\mathbf{Z}/d\mathbf{Z})^*$ modulo the subgroup $\{p, p^2, \ldots, p^f\}$, and let χ_j denote the character χ^{m_j}. For $j = 1, 2, \ldots, h$ let J_j denote the jacobi sum

$$J_j = \sum_{x \in \mathbf{F}_{p^f}} \chi_j(x)\chi_j(1-x).$$

Then J_j is a complex number of absolute value $p^{f/2}$, and

$$Z(T) = \prod_{j=1}^{h} (T^f + J_j).$$

In what follows we shall suppose that n is prime to f, in which case the preceding formula for $Z(T)$ implies that

$$N_n = \prod_{j=1}^{h} (1 + (-1)^{n+1} J_j^n).$$

For cryptographic purposes, we wish to choose g and n so that N_n is "almost prime" in the sense of [6]. For n prime this means that $N_n/N_1 = \prod_{j=1}^{g} |(1-\alpha_j^n)/(1-\alpha_j)|^2$ is prime. Clearly this is possible only if n is prime to f. A second necessary condition is that $Z_g(T)$ not factor over the rational numbers. The theorem that follows describes classes of g which must be avoided, and also a class of g for which $Z_g(T)$ is irreducible.

4. Theorem. *Let $g > 1$ be an integer. Then:*

(1) *the polynomial $Z_g(T)$ factors over the rationals* (a) *if $d = 2g+1$ is composite;* or (b) *if $d = 2g+1$ is prime and either* (i) *p is a quadratic nonresidue modulo d, or else* (ii) *p has order g modulo d and g is even.*

(2) *the polynomial $Z_g(T)$ is irreducible over the rationals if $d = 2g+1$ is a prime, g is odd, and p has order g modulo d.*

The proof of this theorem is straightforward, and will be omitted.

Corollary. *For $p = 2$ and $g < 100$, the polynomial $Z_g(T)$ is irreducible over* $\mathbf{Q}$ *for $g = 1, 3, 11, 15, 23, 35, 39, 51, 83, 95, 99$, and is reducible over* $\mathbf{Q}$ *for all other values except possibly for $g = 36, 44, 56, 63, 75$.*

5. Thus, in order to find examples of almost prime $\#\mathbf{J}(\mathbf{F}_{p^n})$, we must choose g so as not to fall in cases (1a) or (1b) of Theorem 4, and choose n prime to f.

For $p=2$, here are the first few values of g with irreducible $Z_g(T)$:

$$Z_1(T) = T^2+2$$

$$Z_3(T) = T^6-2T^3+8$$

$$Z_{11}(T) = T^{22}-48T^{11}+2048$$

$$Z_{15}(T) = T^{30}-6T^{25}-16T^{20}+352T^{15}-512T^{10}-6144T^5+32768$$

In the case $p=2$ and $g=3$, we tested $\#\mathbf{J}(\mathbf{F}_{2^n})$ for all primes $n<50$, and found the following list of all the almost prime cases, i.e., where this number is 7 times a prime. (We wish to thank Andrew Odlyzko for verifying primality of the three large unfactored integers below, using the Cohen-Lenstra algorithm.)

n	$\#\mathbf{J}(\mathbf{F}_{2^n})$
2	$7\cdot 11$
13	$7\cdot 78536756663$
29	$7\cdot 221060721300991678702 83191$
47	$7\cdot 39822759283090398466982419047946 0780961207$

6. Remarks. 1. If $\mathbf{J}$ is the jacobian of $v^2+v=u^d$ with $d=2g+1$ prime, it is not hard to show that $d|\#\mathbf{J}(\mathbf{F}_p)$. This prevents $\#\mathbf{J}(\mathbf{F}_p)$ from being prime for all but very small values of p and d (since $\#\mathbf{J}(\mathbf{F}_p)\sim p^g$). However, $\#\mathbf{J}(\mathbf{F}_p)/d\sim p^g/d$ may be prime. For example, in the first table above, for $g=15$, $d=31$, $p=2$ we have $\#\mathbf{J}(\mathbf{F}_2)=Z_{15}(1)=31\cdot 853$.

2. For fixed prime p, part (2) of Theorem 4 gives us a source of jacobians over $\mathbf{F}_p$ with irreducible $Z_g(T)$: the curves $v^2+v=u^d$ with d a prime $\equiv 3\ (mod\ 4)$ for which p is the square of a primitive root modulo d. For fixed p, the frequency with which such d occur is given by a (generalization of a) conjecture of E. Artin, according to which there is a positive constant probability that a prime $d\equiv 3\ (mod\ 4)$ has p as the square of a primitive root. For example, when $p=2$ (in which case $d\equiv 7\ (mod\ 8)$, since 2 must be a quadratic residue modulo d), the number of $d<x$ with the desired property is conjecturally asymptotic to

$$c\frac{x}{4\log x},\qquad c=\prod_{\text{primes } l\geq 3}\left(1-\frac{1}{l(l-1)}\right)\approx 0.746\cdots.$$

More information about Artin's conjecture can be found in [11, p. 80-83 and 222-225].

3. In searching for suitable jacobians of curves over finite fields $\mathbf{F}_{p^n}$, one can take several points of view. (a) One can fix the genus g and the field (i.e., p and n), and let the coefficients of the curve's equation vary. One expects, roughly speaking, that as these coefficients vary the number of points on the jacobian will be nearly uniformly distributed in an interval of the form $(p^{gn} - cp^{(g-1/2)n}, p^{gn} + cp^{(g-1/2)n})$. This has been studied in detail in the cases $g = 1$, $n = 1$, p large (see [8]) and $g = 2$, $n = 1$, p large (see [1]).

(b) One can fix a curve with rational coefficients, and consider the jacobian of its reduction modulo p (i.e., over $\mathbf{F}_p$) as p varies. In the case $g = 1$, conjectural formulas for the probability that the corresponding elliptic curve has a prime number of points are given in [5].

(c) One can fix $\mathbf{F}_p$ (or a finite extension of $\mathbf{F}_p$) and also fix a curve with coefficients in that field. One then considers $\mathbf{J}(\mathbf{F}_{p^n})$, i.e., the group of points of $\mathbf{J}$ with coordinates in a finite extension of the field of definition, which is chosen so that $\#\mathbf{J}(\mathbf{F}_{p^n})$ is "almost prime" in the sense of [6]. For this, the curve must have been chosen so that its zeta-polynomial $Z(T) = \prod_{j=1}^{2g}(T-\alpha_j)$ is irreducible over $\mathbf{Q}$, i.e., all of the α_j are conjugates of $\alpha = \alpha_1$. Suppose, for example, that the curve is defined over $\mathbf{F}_p$, it has irreducible zeta-polynomial, and one considers extensions $\mathbf{F}_{p^n}$ of prime degree n. In that case one is interested in primality of the norm of the algebraic integer $(\alpha^n - 1)/(\alpha - 1)$ as n varies. This is a generalization of the Mersenne prime problem, and most likely the frequency of occurrence of prime values is predicted by a heuristic estimate of the same form as in the classical Mersenne case (see [12]).

The point of view (c) is illustrated in the second table above.

(d) One can fix the field of definition $\mathbf{F}_{p^n}$ and examine a family of curves of varying genus. This was the point of view in the first table above. Even if p^n is small, the size of the group of points will grow rapidly with the genus, since it is of order p^{gn}. If one wants $\#\mathbf{J}$ to be a prime number or the product of a large prime and a small factor, then a necessary condition is that the zeta-polynomial be irreducible.

One advantage of point of view (d), in addition to the possible desirability of having one more parameter to vary (the genus g), is that one can limit oneself to curves with special symmetry properties (e.g., the family considered in this report), and this seems to make it possible to compute the number of points much more rapidly (and also carry out the algorithm for finding multiples of points somewhat faster) than in the case of a general curve.

In conclusion, we recall that, because index calculus type algorithms for finding discrete logs in $\mathbf{F}_{p^n}^*$ apparently do not carry over to elliptic curves (see [9]) or hyperelliptic curves, the only known algorithm for finding discrete logs in $\mathbf{J}(\mathbf{F}_{p^n})$ takes time roughly proportional to the square root of the largest prime factor in $\#\mathbf{J}(\mathbf{F}_{p^n})$. Thus, as far as we know, discrete log cryptosystems using $\mathbf{J}(\mathbf{F}_{p^n})$ seem to be secure for relatively small p^n (even when $p = 2$). From the standpoint of implementation, this feature may outweigh the added time required to compute the more complicated group operation.

References

[1] L. M. Adleman and Ming-Deh A. Huang, "Recognizing primes in random polynomial time," preprint.

[2] D. Cantor, "Computing in the jacobian of a hyperelliptic curve," *Math. of Computation,* **48** (1987), 95-101.

[3] N. Koblitz, *A Course in Number Theory and Cryptography*, Springer-Verlag, New York, 1987.

[4] N. Koblitz, "Elliptic curve cryptosystems," *Math. of Computation*, **48** (1987), 203-209.

[5] N. Koblitz, "Primality of the number of points on an elliptic curve over a finite field," *Pacific J. Math.*, **131** (1988), 157-165.

[6] N. Koblitz, "Hyperelliptic cryptosystems," to appear.

[7] N. Koblitz and D. Rohrlich, "Simple factors in the jacobian of a Fermat curve," *Can. J. Math.*, **30** (1978), 1183-1205.

[8] H. W. Lenstra, Jr., "Factoring integers with elliptic curves," Report 86-18, Mathematisch Instituut, Universiteit van Amsterdam, 1986.

[9] V. Miller, "Use of elliptic curves in cryptography," *Advances in Cryptology – Crypto '85*, Springer-Verlag, New York, 1986, 417-426.

[10] A. M. Odlyzko, "Discrete logarithms and their cryptographic significance," *Advances in Cryptography: Proceedings of Eurocrypt 84*, Springer-Verlag, New York, 1985, 224-314.

[11] D. Shanks, *Solved and Unsolved Problems in Number Theory*, 3rd ed., Chelsea, New York, 1985.

[12] S. S. Wagstaff, Jr., "Divisors of Mersenne Numbers," *Math. of Computation*, **40** (1983), 385-397.

[13] A. Weil, "Numbers of solutions of equations in finite fields," *Bull. Amer. Math. Soc.*, **55** (1949), 497-508.

Computation of Approximate L-th Roots Modulo n and Application to Cryptography

Marc Girault
Service d'Etudes communes des Postes et Télécommunications
BP 6243, 14066 Caen Cedex, France.

Philippe Toffin, Brigitte Vallée
Département de mathématiques
Université, 14032 Caen Cedex, France.

ABSTRACT

The goal of this paper is to give a unified view of various known results (apparently unrelated) about numbers arising in crypto schemes as RSA, by considering them as variants of the computation of approximate L-th roots modulo n. Here one may be interested in a number whose L-th power is "close" to a given number, or in finding a number that is "close" to its exact L-th root. The paper collects numerous algorithms which solve problems of this type.

This work was supported by a convention between SEPT and University of Caen.

I. INTRODUCTION

That a lot of public-key cryptosystems or digital signature schemes are based on the computation of L-th roots modulo n is today a very well known fact. Roughly speaking, and assuming that n is a large integer (say, at least, a 320-bit long one), this computation is easy when n is prime or when all its prime factors are known, hard when n is composite and its factors unknown. The cryptographic validity of the famous system RSA [RSA] (as well as many other systems) is based on this dissymmetry.

But very often in public-key cryptology, the problem is raised of extracting *approximate* L-th roots modulo n, in a sense that will be stated more precisely in next section. As this problem is weaker than the problem of extracting *exact* L-th roots modulo n, we may hope that it will be solved even when the factors of n remain hidden. As shown below, that hope is often fulfilled, provided that we do not demand a "too good" approximation.

For example, in the Morrison-Brillhart factorization algorithm [MB], the most consuming part is the quest of integers x such that x^2 (mod n) is as small as possible (hoping that it is "smooth"), where n is the number to be factored. The continued fraction algorithm allows us to find such values of x, but most of the time, x^2 (mod n) has still too large factors to be useful. Fortunately, from time to time, one of them is smooth enough to be factored in the so-called factor base, and will contribute to discovering a factor of n. But this factorization algorithm would become much more efficient if another method was discovered, which finds square roots modulo n of still smaller integers.

Another example is provided by Okamoto and Shiraishi's digital signature scheme [OS]. In this scheme, the signature of the message m is an integer s such that s^2 (mod n) is close to h(m) -where h is a one-way hash-function- instead of being exactly equal, as in the Rabin scheme (the "square root variant" of RSA). The claimed advantage of this scheme was a very fast signature computation compared to the computation time necessary to extract an exact square root modulo n. But Brickell and Delaurentis broke the scheme by showing that s can be efficiently computed, even when the factors of n are hidden [BD]. Now, it remains an open question: can their attack be generalized to the version of Okamoto and Shiraishi's scheme in which the signature s is such that s^L # h(m) (mod n) with $L \geq 4$?

This paper aims at collecting the results already established concerning these questions and improving them whenever possible. First (section II), we state the problems we are going to deal with. Then (section III), we recall how such problems naturally arose in public-key cryptology and briefly indicate how they were solved (or not...). Finally (section IV), we describe most of the algorithms sketched in section III, generally with enough details to effectively implement them.

II. THE PROBLEMS

What do we mean by approximate L-th roots modulo n? In fact, this includes a lot of various questions amongst which we will consider the following ones (n, L and y_0 are three given positive integers, with $L \geq 2$ and $y_0 < n$):

Firstly, we wish to find an integer whose L-th power modulo n is close to the given integer y_0. We subdivide this problem into three ones:

(1): Find x such that $x^L \# y_0$ (mod n) (no matter where x stands).
(1a): Given x_0 such that $y_0 = x_0{}^L$ (mod n), find $x \# x_0$ ($x = x_0$) such that $x^L \# y_0$ (mod n).
(1b): Given x_0, find $x \# x_0$ such that $x^L \# y_0$ (mod n).

Secondly, we wish to get some information about an (existing but unknown) exact L-th root x_0 of the given integer y_0. We subdivide this problem into both which are complementary :

(2a): Find x such that $x \# x_0$.
(2b): Given x such that $x \# x_0$, find x_0.

(Note that problems (2a) and (2b) cannot be both efficiently solved with the same order of approximation, or there is an efficient algorithm which finds exact L-th roots modulo n.)

Of course, the symbol # may have various significations, upon which depends whether the problem is efficiently solvable or not. In order to be more specific, we state again the above problems by replacing "$x \# x_0$" with "$x = x_0 + O(n^a)$ (mod n)" and "$x^L \# y_0$ (mod n)" with "$x^L = y_0 + O(n^b)$ (mod n)", where a and b are real numbers picked in the interval]0,1[. Note that, if some of these problems are easily solved when prime factors of n are known, this knowledge apparently does not help to solve other ones, for example (1a) and (1b).

III. HOW PROBLEMS AROSE

III-1. Problem (1) and its variants

As already noted in the introduction, problem (1) was considered by Morrison and Brillhart (using an idea from Lehmers & Powers) with L=2 and y_0=0 [MB]. By computing continued fractions of $n^{1/2}$, one obtains values of x such that x^2(mod n) = $O(n^{1/2})$. But only a few of them are useful to factorize n, because the quadratic residues modulo n which are required are generally much smaller than $n^{1/2}$. Unfortunately, no efficient algorithm is known, which solves (1) with b<1/2. On the other hand, we show in section IV that continued fraction algorithm can still be used to solve (1) for small exponents greater than 2, but with b growing rapidly with L.

The case {L=2; any y_0} was solved by Brickell and Delaurentis [BD], when they cryptanalysed Okamoto and Shiraishi's signature scheme [OS]. In this scheme, the signature of the message m (or rather of its hashed version h(m)) is an integer s, not too small in absolute value, and such that s^2 (mod n) = h(m) + $O(n^{2/3})$. The public modulus n of the signer is in the form $n=p^2q$ (p and q distinct primes), because this permits a very fast computation of s when p and q are known, which was the claimed advantage of this scheme. Unfortunately for it, Brickell and Delaurentis showed that s can be computed with the same efficiency without knowing the factors of n and no matter what form n takes! We will see that their method can be easily extended to any exponent b greater than 1/2 (hence solving the problem (1) for L=2, y_0=0 and b>1/2). Moreover, it solves the variant (1b) with L=2 and b>1/2, but only if a+b/2>5/4. The Brickell and Delaurentis method can also be used for L=3 [BO] but does not seem to work for L≥4.

One year later, the cryptanalysis of some hash-function using modulo-n operations due to Davies and Price is reduced in [G] to solving (1a) for L=2 and a=b=7/8. First, the problem is linearized by putting x = x_0 + u. It can then be stated as follows: find a "very small" u such that $2ux_0$ # 0 (mod n). Now, the equation $2ux_0$ = v (mod n) with small unknowns u and v is solvable by developing $2x_0/n$ in continued fractions, i.e. by applying the extended Euclid's algorithm to the integers $2x_0$ and n. As the solutions provided are shown to be such that |uv|<n, the problem (1a) appears to be solved for L=2, a+b≥1 and b≥2/3.

Does this last method allow us to solve (1b) (and consequently the problem (1) itself) for the same values of L, a and b? In this case, since we have no more y_0 = x_0^2 (mod n), we are led to an "affine" problem rather than a linear one; explicitly: find a very small u such that $2ux_0$ # z_0 (mod n), where z_0 = x_0^2 - y_0 (mod n). In [OS], is presented a variant of Euclid's algorithm which allows to do that. Ironically enough, Shamir used it to give another cryptanalysis of the OS signature scheme discussed above and presented in the same paper, as well as the cryptosystem proposed one year after by Okamoto in [O1]!

Recently, the authors have described in [VGT1] (or [VGT2]) a different technique powerful enough to solve also variant (1b) for L=2, a+b>1 and b>2/3. First, the problem is linearized as above. Then, the linear equation (E): $2ux_0 = v \pmod n$ is interpreted as the equation of the integer lattice R spanned by $(1,2x_0)$ and $(0,n)$. This point of view allows us to transform the problems into lattice ones ("find a short vector in R" or "find a point of the lattice close to a given point"), for which algorithmic solutions are known in all dimensions, based on the LLL basis reduction algorithm [LLL]. Moreover, this method can be refined in order to find x in a "quasi-uniform way", leading to a factorisation algorithm [V] whose proven complexity is smaller than Dixon's one [D].

III-2. **Problems (2a) and (2b)**

Problem (2a) with L=2 was first solved by Blum, Blum and Shub [BBS], when they analysed the left-unpredictability of the so-called $x^2 \pmod n$ generator. In their paper, n is a Blum integer so that the mapping "squaring modulo n" is a permutation over the set of quadratic residues modulo n. So, by working in this set, the square root of a quadratic residue is defined in an inambiguous manner. It is shown in [VV] that the location of x_0 (equal to 0 if $x_0<n/2$, 1 if not) cannot be guessed, even with a very small advantage, unless factoring is easy. It follows that (2a) with L=2 is not efficiently solvable for any a<1 since even the location of x_0 cannot be found.

The same problem is solved for all the L which are coprime to phi(n) -the RSA context- by (first) Goldwasser, Micali and Tong [GMT], then many others until Alexi, Chor, Goldreich and Schnorr [ACGS], when they studied the security of the RSA bits. It can be proved that the location of x_0 cannot be guessed, even with a small advantage, unless inverting RSA (i.e. finding x_0 in full) is easy.

The problem (2b) has been partly solved by Shamir when he cryptanalysed the first version of Okamoto's cryptosystem [O1]. In this cryptosystem, the public modulus n is in the form $n=p^2q$, as for Okamoto and Shiraishi's signature scheme. Moreover, the public key contains another integer x, itself of a very particular form. With the notations of (2b), x_0-x plays the role of the plaintext and y_0 is the ciphertext. Shamir found two cryptanalysis for this system. The first one, based on the OS-variant of Euclid's algorithm, does not make use of the form of n but is valid only for L=2. The second one works for any L but does make use of the particular form of n and x. At this stage, the problem (2b) appears to be solved for {L=2, a≤1/3} and only in very particular cases if L>2.

In [VGT1], the authors, using the lattice technique already mentioned in section III.1, specify an efficient algorithm solving (2b) for {L=2, a≤1/3} and more generally for any reasonably small L, sufficiently large n and sufficiently small a (in the order of about $1/L^2$). So, an L-th root modulo n can always be calculated, if we are given a sufficiently good approximation of this root. Moreover, the technique is general enough to apply to other types of approximations, such as used in the second version of Okamoto's cryptosystem [O2]. This version hence appears to be broken too.

These results have incidences on the predictibility of congruentiel pseudo-random generators. In particular, the truncated x^2 (mod n) generator, obtained by removing the 1/3 least significant bits of the sequence, is right-predictible (in a sense section V will make clear).

IV. THE ALGORITHMS

We now describe the algorithms with more details. Frow now on, n is a positive integer and Z(n) denotes the ring of the integers modulo n that we identify with the interval of length n centered at 0. For any u in Z(n), |u| denotes the absolute value of u, i.e. the maximum of u and -u (for example, |21 (mod 25)|=|-4|=4).

The symbols x, x_0, y, y_0, u denote elements of Z(n) whilst a and b denote real numbers in]0,1[. The notation O(f(n)) stands for any function g(n) such that |g(n)| ≤ kf(n) for some integer k and any sufficiently large n.

For L a positive integer, we ask if the equation:

$$x^L = y \bmod n \qquad (E)$$

admit solutions (x,y) which satisfy some closeness requirements, and if we can discover them. For example, can we find (x,y) such that x is "close" to a given integer x_0 and y "close" to another one y_0? Or without conditions on x but such that y is exactly equal to y_0?

Of course, we can (and will often have to...) reduce our ambitions and claim our satisfaction if we succeed in inferring some partial information about such solutions, or, conversely, if some additional information about these solutions permit us to recover them entirely. The most famous case is the computation of an exact square root modulo n (n a large integer), where two extreme situations are possible. Either the factors of n are known and such a computation is child's play; or they are hidden and we can infer almost nothing about the solution.

It must be noticed that most of the algorithms which are presented below do not work for all the values of their inputs. Only in some cases (e.g. algorithms VGT1 and VGT2), the set of values for which they fail has been carefully analysed in the original papers. But the fact they provide solutions *most of the time* is satisfactory enough for cryptologic applications. This is particularly true for cryptanalytic ones, which only require that the algorithm work for a non-negligible fraction of the input values.

IV-1. Finding roots with small residual

IV-1.1. Without conditions for x

We first consider the problem of finding solutions of (E) where y is close to y_0:

<u>Pb (1)</u>: Given n, L, b and y_0, find x such that $|x^L - y_0 \pmod n| \leq O(n^b)$.

Of course, if b is big enough, there is a straight-forward way of finding some of them, which consists in detecting the elements of $[y_0\text{-}0(n^b), y_0\text{+}0(n^b)]$ which are *true* L-th powers (as opposed to L-th powers *modulo n*) of an integer x. For instance, if L=2 and $b \geq 1/2$, $x=[y_0^{1/2}]$ (where [z] denotes the closest integer to the real number z) is a trivial solution of our problem since:

$$x^2 = (y_0^{1/2} + \beta)^2 \qquad \text{with } |\beta| \leq 1/2$$
$$= y_0 + 2\beta y_0^{1/2} + \beta^2 \implies |x^2 - y_0| \leq y_0^{1/2} + 1 < n^{1/2} \leq n^b.$$

In case b is a little bit too small and $[y_0\text{-}0(n^b), y_0\text{+}0(n^b)]$ does not contain true L-th powers, other intervals $[y_0\text{+}kn\text{-}0(n^b), y_0\text{+}kn\text{+}0(n^b)]$ can be tried, for k=1,2,.... Solutions which are found in this way may be considered as trivial ones. As they are necessarily small, with the same order of magnitude of $n^{1/L}$, we may define trivial solutions of (1) as solutions $x = O(n^{1/L})$. It may occur that the algorithms which are presented below provide trivial solutions, but they (generally) provide also non-trivial ones.

The quadratic version ($L=2$) of Pb (1) has been solved in the general case by Morrison and Brillhart for $y_0=0$ [MB]. They find approximations in the order of magnitude of $n^{1/2}$. More precisely:

Algorithm MB

**Input*: n

**Output*: some x such that $|x^2 \pmod n| \leq 2n^{1/2}$

**Method*: develop $n^{1/2}$ in continued fractions; call x_i/y_i the convergents of $n^{1/2}$; output the x_i.

**Proof*: from a well-known inequality of continued fractions theory, we have:

$|n^{1/2} - x_i/y_i| \leq 1/(y_i y_{i+1})$

then:

$|n^{1/2} + x_i/y_i| \leq 2n^{1/2} + 1/(y_i y_{i+1})$

$|n - x_i^2/y_i^2| \leq 2n^{1/2}/(y_i y_{i+1}) + 1/(y_i y_{i+1})^2$

$|y_i^2 n - x_i^2| \leq 2n^{1/2} y_i/y_{i+1} + 1/y_{i+1}^2 \leq 2n^{1/2} \Longrightarrow |x_i^2 \pmod n| \leq 2n^{1/2}$.

**Remark*: when the periodicity of the development of $n^{1/2}$ is small (for example if $n=m^2+1$), this algorithm only provides trivial solutions; if not, there may be a lot of (non-trivial) solutions.

We now show that the MB idea can be extended to small exponents other than 2, but with less efficiency because only first convergents have a good chance to lead to success :

Algorithm MB'

**Input*: n, L (small integer = O(log n) and ≥ 3), ß (real $\leq 1/L$)

**Output*: nothing or some x such that $|x^L \pmod n| \leq Ln^{(L-1)/L+\beta}$

**Method*: develop $n^{1/L}$ in continued fractions; call x_i/y_i the convergents of $n^{1/L}$; output the x_i until (say) $y_i > n^{\beta/(l-2)}$.

**Proof*: from $|n^{1/L} - x_i/y_i| \leq 1/(y_i y_{i+1})$ we deduce:

$n^{(L-1)/L} + n^{(L-2)/L}(x_i/y_i) + \dots + n^{1/L}(x_i/y_i)^{L-2} + (x_i/y_i)^{L-1} \leq Ln^{(L-1)/L} + L^2 n^{(L-2)/L}/y_i y_{i+1}$

then:

$|n - x_i^L/y_i^L| \leq Ln^{(L-1)/L}/(y_i y_{i+1}) + L^2 n^{(L-2)/L}/(y_i y_{i+1})^2$

hence:

$|y_i^L n - x_i^L| \leq Ln^{(L-1)/L} y_i^{(L-1)}/y_{i+1} + L^2 n^{(L-2)/L} y_i^{(L-2)}/y_{i+1}^2 \leq Ln^{(L-1)/L} y_i^{L-2}$

(because the second term of the sum is easily shown to be smaller than $Ln^{(L-1)/L}(y_i^{L-2}-y_i^{L-1}/y_{i+1})$ except perhaps in some very particular cases)
finally:
$|x_i^L \pmod n| \leq Ln^{(L-1)/L+ß}$.

But continued fractions do not seem to help in solving the problem when y_0 is non zero. In [BD], Brickell and Delaurentis describe an algorithm which solves the case {L=2; any y_0} with b=2/3 (i.e. approximations of $n^{2/3}$). But, as their original algorithm can be easily extended to b=1/2+e, for 0<e<1/2, it appears to be almost as efficient than algorithm MB, and much more general.

The idea is the following one: not only $[y_0^{1/2}]$ is a solution of Pb (1) but also $k[z^{1/2}]$ where $z = y_0k^{-2} \pmod n$, for any positive integer $k \leq n^{e/2}$, as shown in the proof below. These new solutions are not trivial ones but are in $O(n^{1/2+e/2})$. In order to find solutions of any magnitude, Brickell and Delaurentis proceed as follows:

<u>Algorithm BD</u>

**Input*: n, y_0, e (0<e<1/2)
**Output*: some x such that $|x^2 - y_0 \pmod n| = O(n^{1/2+e})$

**Step 1*: find k (coprime with n) and x' such that $k=O(n^{e/2})$ and $2kx' \pmod n = O(n^e)$;
**Step 2*: calculate $y = y_0 - x'^2 \pmod n$
$z = yk^{-2} \pmod n$
$t = [z^{1/2}] = z^{1/2} + ß$ with $|ß| \leq 1/2$;
**Step 3*: output x = x' + kt.

**Remark*: for step 1, it suffices to choose x' in one of the intervals I_i centered in [ni/2k] of radius $[n^e/2k]$.

**Proof*: in a straightforward manner (all the equalities standing modulo n):

$$\begin{aligned} x^2 &= x'^2 + k^2t^2 + 2kx't \\ &= x'^2 + k^2(z^{1/2}+ß)^2 + 2kx't \\ &= x'^2 + k^2z + 2k^2z^{1/2}ß + k^2ß^2 + 2kx't \\ &= x'^2 + y_0 - x'^2 + 2k^2z^{1/2}ß + k^2ß^2 + 2kx't \\ &= y_0 + O(n^{1/2+e}) + O(n^e) + O(n^{1/2+e}) \\ &= y_0 + O(n^{1/2+e}) \end{aligned}$$

Does algorithm BD extend to $L \geq 2$? One can remark that the above proof does not work for L=3, except if one makes some very particular choices, namely : $x' = [n/3]$ and $k = 1$. Then, choosing $y = y_0 - x'^3 \pmod n$ and t = nearest integer to $y^{1/3}$ divisible by 3, leads to success for b = 2/3; details may be found in [BO]. For $L \geq 4$, the method does not seem to work at all.

IV-1.2. With conditions for x

We now come to algorithms which not only solve Pb (1) but provide solutions which are themselves close to a given integer x_0. This new problem can be subdivided into two subproblems. In first one, Pb (1a), y_0 is nothing but the L-th power of x_0; in other words, we have to find a solution (x,y) of (E), close to another already known solution. In second one, Pb (1b), y_0 and x_0 are any two elements of Z(n). It is clear that this last problem is harder than both Pb (1) and Pb (1a). Note also that knowledge of the factors of n completely solves Pb (1) but does not seem to help to solve Pb (1a) and Pb (1b).

Pb (1a): Given n, L, a, b, x_0 and y_0 such that $y_0 = x_0^L \pmod n$, find $x = x_0$ such that $|x - x_0| \leq O(n^a)$ and $|x^L - y_0 \pmod n| \leq O(n^b)$.

Pb (1b): Given n, L, a, b, x_0 and y_0, find x such that $|x - x_0| \leq O(n^a)$ and $|x^L - y_0 \pmod n| \leq O(n^b)$.

First, a closer look at BD-algorithm shows that it solves (but not very well) Problem (1b):

Algorithm BD'

Input: n, x_0, y_0, a, b (s.t. $a+b/2>5/4$ and $b>1/2$)

Output: some x such that $|x - x_0| \leq O(n^a)$ and $|x^2 - y_0 \pmod n| \leq O(n^b)$

Method: as in algorithm BD with $e=b-1/2$, by choosing k close to $n^{e/2}$ and x' in the interval I_i which is the closest one to x_0.

Proof: the distance between two consecutive intervals I_i and I_{i+1}, defined in the remark of algorithm BD, is smaller than $n/2k = O(n^{1-e/2})$.

Better solutions to Pb (1b) are obtained by linearizing it, as will be explained in subsection IV.1.1.2. Beforehand, we have to make a digression into Euclid's algorithm and some of its extensions.

IV-1.2.1. A Euclidean digression

We consider here the equation:

$$dx = y \pmod n \qquad (E')$$

where d is a positive integer smaller than n, and ask if there are solutions (x,y) close to a given pair (x_0,y_0): $x = x_0 + O(n^a)$ and $y = y_0 + O(n^b)$.

Let us start with the case $x_0 = y_0 = 0$. It is proven in [DC] (or [G]) that such solutions certainly exist if a+b = 1. More precisely, for any pair (X,Y) whose product is greater than n, there is at least one solution (x,y) such that $|x|<X$ and $|y|<Y$. In order to discover it (or them), it is useful to remark that finding small (x,y) satisfying (E') comes to finding a good approximation of the fraction d/n. So, here again, we (almost) always find such a solution by developing it in continued fractions i.e. applying Extended Euclid's algorithm to d and n:

Algorithm EE

**Input*: n, d, a, b (s.t. $a+b\geq 1$)

**Ouput*: nothing or some x such that $|x| \leq n^a$ and $|dx \pmod n| \leq n^b$

**Method*: apply Extended Euclid's algorithm to n and d; one obtains coefficients l_i and m_i such that $l_i n + m_i d = r_i$ where the r_i are the successive remainders (the last non-zero remainder being equal to the greatest common divisor of n and d); output the smallest (in absolute value) m_i such that $n^{1-b} \leq |m_{i+1}|$ (the case "such a m_i does not exist" is very rare).

**Proof*: the fractions $|l_i/m_i| = -l_i/m_i$ are in fact the convergents of the development of d/n in continued fractions; hence:
$|d/n + l_i/m_i| \leq 1/|m_i m_{i+1}|$ ==> $|dm_i + nl_i| \leq n/|m_{i+1}|$ ==> $|dm_i \pmod n)| \leq n/|m_{i+1}| \leq n^b$.
Moreover, $|m_i| \leq n^{1-b} \leq n^a$ since $a+b \geq 1$.

Now, what happens if x_0 and y_0 are non-zero? Of course, it is enough to solve the problem for $x_0 = 0$ and any y_0 (if x_0 is not equal to zero, it suffices to replace y_0 with y_0-$dx_0 \pmod n$). Okamoto and Shiraishi provide in [OS] an extension of Extended Euclid's algorithm which very often solves this problem. We hope that we do not deform it too much by presenting it as follows:

Algorithm OS

*$Input$: n, d, a, b (s.t. $a+b\geq 1$), y_0

*$Ouput$: nothing or some x such that $|x| \leq n^a$ and $|dx - y_0 \pmod n| \leq n^b$

*$Step\ 1$: apply Extended Euclid's algorithm to d and n (as in algorithm DC);

*$Step\ 2$: introduce a sequence y whose first term is y_0 and following ones are defined by: $y_i = y_{i-1} - q'_i r_i$ where q'_i is the quotient in the division of y_{i-1} by r_i;

*$Step\ 3$: introduce also the sequences h_i and k_i whose first terms h_0 and k_0 are zero and following ones are defined by: $h_i = h_{i-1} + q'_i l_i$ and $k_i = k_{i-1} + q'_i m_i$;

*$Step\ 4$: output k_i such that $n^{1-b} \leq |k_i| \leq n^a$ (mind: its existence is questionable, especially if a+b is close to 1).

*$Proof$: From: $l_i n + m_i d = r_i$, we easily deduce:

$h_i n + k_i d = h_{i-1} n + k_{i-1} d + (y_{i-1} - y_i)$; then :

$h_i n + k_i d = 0 + (y_0 - y_1) + (y_1 - y_2) + ... + (y_{i-1} - y_i) = y_0 - y_i \implies k_i d \pmod n = y_0 - y_i$.

Moreover, when it can be shown that $|k_i| y_i < n$ (it is very often the case), we have : $|k_i| \geq n^{1-b} \implies y_i \leq n^b$.

IV-1.2.2. Come back to our problems

In [G], one of the authors shows that the quadratic version of (1a) can be solved with $a+b\geq 1$ and $b\geq 2a$ (which is equivalent to: $a+b\geq 1$ and $a\leq 1/3$ or: $a+b\geq 1$ and $b\geq 2/3$). The idea consists in reducing the problem to a linear one by taking advantage of the fact we already know a solution of (E):

Algorithm G

*$Input$: n, a, b (s.t. $a+b\geq 1$ and $b\geq 2a$), x_0, y_0 (s.t. $y_0 = x_0^2 \pmod n$)).

*$Output$: nothing or some x such that $0 < |x - x_0| \leq n^a$ and $|x^2 - y_0 \pmod n| \leq 2n^b$)

*$Method$: perform algorithm EE with inputs n, $2x_0$, a, b; output $x = x_0 + m_i$.

*$Proof$: $x^2 = (x_0 + m_i)^2 = x_0^2 + m_i^2 + 2m_i x_0^2 \pmod n$

$= y_0 + m_i^2 + 2m_i x_0 \pmod n$

We know from previous section that: $|2m_i x_0 \pmod n| \leq n^b$.

Moreover, $|m_i| \leq n^a \implies m_i^2 \leq n^{2a} \leq n^b$ since $b\geq 2a$.

It follows that $|x^2 - y_0 \pmod n| \leq 2n^b$.

Let us now consider Pb (1b). If, in algorithm G, we substitute algorithm OS to Euclid's one, we obtain a new algorithm that Shamir used to cryptanalyse OS signature scheme:

Algorithm S1

Input: n, a, b (s.t. $a+b\geq 1$ and $b\geq 2a$), x_0, y_0

Output: nothing or some x such that $|x - x_0| \leq n^a$ and $|x^2 - y_0 \pmod n| \leq 2n^b$

Method: perform algorithm OS with inputs n, $2x_0$, a, b, $z_0 = y_0 - x_0^2 \pmod n$; output $x = x_0 + k_i$.

Proof: $x^2 = (x_0 + k_i)^2 = x_0^2 + k_i^2 + 2k_ix_0 \pmod n$

We know from previous section that, very often: $|2k_ix_0 - z_0 \pmod n| \leq n^b$.

Moreover, $|k_i| \leq n^a \Longrightarrow k_i^2 \leq n^{2a} \leq n^b$, since $b\geq 2a$.

It follows that $|x^2 - y_0 \pmod n| \leq 2n^b$.

Another point of view has been recently considered by the authors in [VGT1 or 2]. Using the theory of lattice basis reduction, they present an algorithm which solves Pb (1b) in the same conditions as algorithm S1 but is more adapted to generalizations (see IV.2). The starting point is identical: we want $x = x_0 + u$, and $x^2 \pmod n = y_0 + v$ with u and v small. These two equalities imply: $2ux_0 = z_0 + w \pmod n$ with $z_0 = y_0 - x_0^2 \pmod n$ and $w = v - u^2 \pmod n$. But the set of vectors (u,u') such that $2ux_0 = u' \pmod n$ may be seen as the lattice $R(x_0)$ spanned by vectors $(1,2x_0)$ and (0,n). If we find a point (u,u') of $R(x_0)$ close to $(0,z_0)$ in that $u = O(n^a)$ and $u' = z_0 + O(n^b)$, then $v = w+u^2 = O(n^b) + O(n^{2a}) = O(n^b)$, since $b\geq 2a$, and the problem is solved. We now see how to find such a point (u,u'):

Algorithm VGT1

Input: n, a, b (s.t. $a+b>1$ and $b\geq 2a$), x_0, y_0

Output: nothing or some x such that $|x - x_0| \leq O(n^a)$ and $|x^2 - y_0 \pmod n| \leq O(n^b)$

Step 1: consider the lattice $M(x_0)$ spanned by vectors $(k,2k'x_0)$ and $(0,k'n)$ where $k = [n^{(b-a)/2}]$ and $k' = 1/k$.

Step 2: use LLL algorithm [LLL] to find a point (t,t') of $M(x_0)$ very close to the point $(0,k'z_0)$ with $z_0 = y_0 - x_0^2 \pmod n$.

Step 3: output $x = x_0 + t/k$ (mind: its existence is questionable, especially if a+b is close to 1).

**Proof*: let $e = (a+b-1)/2$. Except in some exceptional cases (see details in [VGT1]), the shortest vector of the lattice $M(x_0)$ is not too small, of length $\geq n^{1/2-e}$. The lattice theory tells us that any ball $B(P,r)$ with $r > (2H_2)^{1/2}n^{1/2+e}$, where H_2 is Hermite's constant in dimension 2 (for definition see e.g. [LLL] or [VGT2]), contains at least one point of the lattice. Moreover, LLL algorithm (i.e. Gauss' algorithm in dimension 2) allows us to find such a point, say $T(t,t')$. Let P be the point $(0,k'z_0)$. Since the distance between T and P is smaller than r, we have $|t|$ and $|t'-k'z_0| \leq (2H_2)^{1/2}n^{1/2+e}$. It remains to put $u=t/k$ and $u'=t'/k'$ (remark that u and u' are necessarily integers). Then: $|u| \leq (2H_2)^{1/2}n^{1/2+e}n^{(a-b)/2} \leq O(n^a)$ and $|u'-z_0| \leq (2H_2)^{1/2}n^{1/2+e}n^{(b-a)/2} \leq O(n^b)$, the inequalities we wanted.

IV-2. Finding something about exact roots

We now consider the problem of finding x_0, an exact L-th root modulo n of a given integer y_0. Here again, there are situations in which the problem can be considered as trivial: when y_0 is a *true* L-th power or when factorization of n is known. At the opposite, the problem is specially hard in almost all other cases, since one (presently) does not know how extract L-th roots modulo n without factors of n. Between these two extreme situations, we may consider intermediary ones. First, can we at least infer some partial information about where stands x_0? Or, on contrary, if we are given some information about location of x_0, can we recover it entirely?

IV-2.1. Inferring some partial information about location of x_0

We first consider the following problem:

Pb (2a): Given n, L, a and y_0 (known to be the L-th power modulo n of an integer x_0), find x such that $|x - x_0| \leq O(n^a)$.

This question has been widely discussed between 1982 and 1984, since it is related to security of RSA (or Rabin) bits [BBM],[GMT],[VV],[ACGS]. The conclusion was that Pb (2a) has definitely no solution at all, even when a is very close to 1 (we refer the reader to the introduction).

IV-2.2. Finding x_0 with some help

Let us come now to our last problem:

Pb (2b): Given n, L, a, y_0 (known to be the L-th power modulo n of an integer x_0), and x

Pb (2b): Given n, L, a, y_0 (known to be the L-th power modulo n of an integer x_0), and x such that $|x - x_0| \le O(n^a)$, find x_0.

In [O2], Okamoto presents an algorithm due to Shamir which solves the quadratic version of this problem (it is the little sister of algorithm S1):

Algorithm S2

Input: n, y_0, x (s.t. $|x - x_0| \le O(n^{1/3})$)
Output: nothing or x_0

Method: apply algorithm OS to n, $2x_0$, 1/3, 2/3, $z_0 = x^2 - y_0 \pmod n$; output $x = x_0 + k_i$ for $k_i \# n^{1/3}$; check that $y_0 = x_0^2 \pmod n$.

Proof: let $x = x_0 + u$; then $2x_0 u = z_0 - u^2 \pmod n$ and we are reduced to finding $u = O(n^{1/3})$ such that $|2x_0 u - z_0 \pmod n| \le O(n^{2/3})$. Such an u is likely to be one of the k_i close to $n^{1/3}$, provided by algorithm OS.

In [VGT1], the authors, using the lattice technique inroduced in IV-1. solve Pb (2b) for any (reasonably small) L and a in the order of about $2/L^2$. We only state here a weak version of this result (more generally, x_0 can be found even if y_0 is not exactly known, provided the approximation on y_0 is in the order of about 2/L) and we suppose n square-free for simplicity:

Algorithm VGT2

Input: n (square-free), a=2/[L(L+1)], y_0, x (s.t. $|x - x_0| \le O(n^a)$)
Output: nothing or x_0

Sketch of the method: similar to algorithm [VGT1] (but mind: now x is known and x_0 is the unknown!).

Sketch of the proof : similar to VGT1; the property used here is the following one: if the shortest vector of M(x) is not too small, of length $\ge n^{1/L-\epsilon}$, then the ball B(P,r) with $r < n^{(1/L-\epsilon)}/2$ contains at most one point of the lattice. This point is found, if it exists and lies in a slightly smaller ball, by LLL algorithm.

To be more explicit, let us consider the case L=3. In that case, the lattice R(x) is the one spanned by the vectors $(1,0,3x^2)$, $(0,1,3x)$ and $(0,0,n)$ and the lattice M(x) is obtained by multiplying the first (resp. second, resp. third) coordinate by k (resp. k', resp. k") such that $kn^a = k'n^{2a} = k''n^b$ and kk'k"=1.

The general result of [VGT1] can interestingly be applied to pseudo-random number generators (PRNG). Consider for example the case L=2 (hence a=1/3) and the sequence : $s_{i+1} = s_i^2 \pmod n$ where s_0 is a secret seed. Let t_i be the number obtained by removing the $[(\log_2 n)/3]$ least significant bits of s_i, for $i \geq 1$. It results from [VGT1] that this PRNG is not secure, since we can recover s_1 and s_2 (hence all the s_i, hence all the t_i!) from t_1 and t_2.

V. CONCLUSION

We have shown how various known algorithms may be considered as variants of the computation of approximate L-th roots modulo n, and have given a unified description of all these (often revisited) algorithms. Except for the most complicated ones, for which given references should be consulted, enough details are provided to implement them.

On the way, we have improved or extended some of these algorithms (see algorithm MB', extension of algorithm BD to any small exponent e, algorithm BD', and "new look" of algorithm OS).

Some questions remain open amongst which we point out:

1) can we solve $x^L \# y_0 \pmod n$ for $L \geq 4$ with an approximation on y of order $n^{2/3}$ (this problem is related to generalized Okamoto-Shiraishi signature scheme) ?

2) can we solve $x^2 \# y_0 \pmod n$ with an approximation on y of order less than $n^{1/2}$ (this problem is related to Morrison-Brillhart factorization algorithm) ?

3) can we solve $x \# x_0$ and $x^2 \# y_0 \pmod n$ with an approximation on x and y of order $n^{1/2}$ (this problem is related to quadratic congruentiel pseudo-random number generator) ?

VI. ACKNOWLEDGEMENTS

We thank one member of the program committee for having inspired the redaction of our abstract.

VII. REFERENCES

[ACBG] W. Alexi, B. Chor, O. Goldreich and C.P. Schnorr, "RSA and Rabin functions: certain parts are as hard as the whole", SIAM J. Comp., Vol. 17, pp. 194-209, 1988.

[BBS] L. Blum, M. Blum and M. Shub, "Comparison of two pseudo-random number generators", Advances in Cryptology, Proc. of Crypto 82, Plenum press, New York, 1983, pp.61-78.

[BO] E. Brickell and A. Odlyzko, "Cryptanalysis: a survey of recent results", Proc. of the IEEE, Vol. 76, n°5, May 1988, pp. 578-593.

[BD] E. Brickell and J. Delaurentis, "An attack on a signature scheme proposed by Okamoto and Shiraishi", Proc. of Crypto '85, LNCS, Vol. 218, Springer-Verlag, 1986, pp.10-14.

[D] J.D. Dixon, "Asymptotical fast factorization of integers", Math. Comp., Vol. 36, 1981, pp. 255-260.

[DC] W. De Jonge and D. Chaum, "Attacks on some RSA signatures", Advances in Cryptology, Proc. of Crypto '85, LNCS, Vol. 218, Springer-Verlag, 1986, pp.18-27.

[G] M. Girault, "Hash-functions using modulo-n operations", Proc. of Eurocrypt '87, LNCS, Vol. 304, Springer-Verlag, 1988, pp. 217-226.

[GMT] S. Goldwasser, S. Micali and P. Tong, "Why and how to establish a private code on a public network", Proc. of the 23rd IEEE FOCS, 1982, pp. 134-144.

[LLL] A.K. Lenstra, H.W. Lenstra and L. Lovasz, "Factoring polynomials with integer coefficients", Mathematische Annalen, 1982, Vol. 261, pp. 513-534.

[MB] M.A. Morrison and J. Brillhart, "A method of factorization and the factorization of F_7", Math. Comput., Vol. 29, 1975, pp. 183-205.

[O1] T. Okamoto, "Fast public-key cryptosystem using congruent polynomial equations", Electronics Letters, 1986, Vol.22, pp. 581-582.

[O2] T. Okamoto, "Modification of a public-key cryptosystem", Electronics Letters, 1987, Vol. 23, pp.814-815.

[OS] T. Okamoto and A. Shiraishi, "A fast signature scheme based on quadratic inequalities", Proc. of the 1985 Symposium on Security and Privacy, Apr.1985, Oakland, CA.

[RSA] R.L. Rivest, A. Shamir and L. Adleman, "A method for obtaining digital signatures and public-key cryptosystems", CACM, Vol. 21, n°2, Feb. 1978, pp. 120-126.

[V] B. Vallée "Integer factorisation with quasi-uniform generation of small quadratic residues", presented at "Computational Number Theory" Conference, Bodwin College, Jul. 88; submitted to Compte Rendus de l'Académie des Sciences de Paris (preprint available from the author).

[VGT1] B. Vallée, M. Girault and P. Toffin, "How to break Okamoto's cryptosystem by reducing lattice bases", Proc. of Eurocrypt '88, to appear.

[VGT2] B. Vallée, M. Girault and P. Toffin, "How to guess L-th roots modulo n by reducing lattice bases", Proc. of Conference of ISSAC-88 and AAECC-6, Jul. 88, to appear.

[VV] U.V. Vazirani and V.V. Vazirani, "Efficient and secure pseudo-random number generation", Advances in Cryptology, Proc. of Crypto '84, LNCS, Vol. 196, Springer-Verlag, 1985, pp.193-202.

Session 4

Cryptanalysis

Chair: A. Odlyzko, Bell Laboratories

On the McEliece Public-Key Cryptosystem

Johan van Tilburg

Dr. Neher Laboratories, Department of Applied Mathematics

P.O. Box 421, 2260 AK Leidschendam, the Netherlands

Abstract

Based on an idea by Hin, the method of obtaining the original message after selecting k of n coordinates at random in the McEliece public-key cryptosystem is improved. The attack, which is more efficient than the attacks previously proposed, is characterized by a systematic method of checking and by a random bit swapping procedure. An optimization procedure similar to the one proposed by Lee and Brickell is used to improve the attack. The attack is highly suitable for parallel and pipelined implementation. The work factor and the values, which yield 'maximum' security for the system are given.

It is shown that the public-key can be reduced to $k \times (n-k)$ bits.

1 Introduction

At Crypto'87 Adams and Meijer [1] presented a paper in which the 'optimum' values for the parameters of the McEliece public-key cryptosystem [9] are given. As shown in [1] these values improve the cryptanalytic complexity of the system and increase the information rate. As noted in [4,9] there are several ways of attacking McEliece's cryptosystem. Of the known attacks, the one which requires the least effort is based on decoding a more or less arbitrary linear code containing correctable errors. It has been proved in [2] that the general decoding problem for linear codes is NP-complete, so one certainly expects that for sufficiently large code parameters, the minimal effort for this attack will become computationally infeasable. The best known attack is based on selecting and solving k of n equations obtained from the (publicly known) encryption matrix and the cryptogram. Thereafter it is necessary to verify whether the obtained solution is unique and gives the correct plaintext. If the solution is not correct, then a new set of k equations has to be selected etc. For the attack it was shown in [1] that for a suitable choice of the parameters this minimal effort can be maximized.

This paper gives an improved method to obtain the original message after selecting k of n cryptogram bits. A bit swapping procedure is used to randomly renew the set of k-bits one bit at a time. A fast validation wether the selected k-bits are error-free and the corresponding columns of the publicly known encryption matrix are linearly independent is part of the algorithm.

At the same time when this paper was accepted for presentation at Crypto'88, Lee and Brickell [7] presented an elegant attack on the McEliece public-key cryptosystem at Eurocrypt'88. Their attack is based on a generalization of two well known attacks and includes a systematic method for checking whether the obtained message agrees with the original message and is closely related to our attack.

Sections 2 and 3 describe the public-key cryptosystem and some well known attacks on this system. Section 4 discusses the basics of the proposed attack including the way of validation. The algorithm, based on a bit swapping procedure, is subsequently given in the next section. Section 6 considers in more details the bit swapping procedure. In section 7 the work factor is discussed and in Section 8 an optimization similar to the one proposed by Lee and Brickell is used to improve the attack. Finally we note in Section 9, that the public-key can be reduced to $k \times (n-k)$ bits without affecting the security of the system.

2 McEliece's Cryptosystem

The McEliece public-key cryptosystem can be easily understood from the following description. Let C be a linear $[n,k,d]$ code over $GF(2)$ with code length n, dimension k and minimum distance d. Let the $k \times n$ matrix G be a generator matrix of C and let the $(n-k) \times n$ matrix H be a parity check matrix of C. The publicly known encryption matrix E is defined by

$$E = SGP, \tag{1}$$

where S is a $k \times k$ non-singular binary matrix over $GF(2)$ and P is an $n \times n$ permutation matrix. The scheme also uses a subset Z of $GF(2)^n$ with the property that the Hamming weight $w_H(\underline{z})$ of the vectors $\underline{z} \in Z$ is less or equal than $t = (d-1)/2$. Generally $w_H(\underline{z}) = t$.

A k-message $\underline{m}$ is encrypted into the n-bit ciphertext $\underline{e}$ as follows

$$\underline{e} = \underline{m}E + \underline{z} = \underline{c} + \underline{z}, \tag{2}$$

where $\underline{c}$ is a n-bit permuted codeword from C.

Decryption is straightforward. An enciphered message $\underline{m}$ is *formally* decrypted by the following steps.

1. Compute $\underline{e}' = \underline{e}P^T$ and obtain the error pattern $\underline{z}' = \underline{z}P^T$
   Let $\underline{c}' = \underline{e}' - \underline{z}'$.

2. Calculate $\underline{m} = \underline{m}' \times S^{-1}$, where $\underline{m}'$ represents the first k-bits of $\underline{c}'$. The result is the plaintext $\underline{m}$.

This encryption scheme must satisfy the properties introduced by Diffie and Hellman [3] to become a public-key cryptosystem. Therefore the decryption process must be fast if the private-keys S, P and G are known and the decryption process must be infeasible if only the public-key E is known. Furthermore the encryption process must be fast if one has only knowledge of the public-key E. McEliece based his cryptosystem on the existence of Goppa codes, which meet the conditions for a public-key cryptosystem and can easily be generated.

We note that Goppa codes are in general not maximum distance separable codes (MDS). The only binary MDS codes are the trivial ones which are of no use in the (binary) McEliece scheme. More details about Goppa codes can be found in e.g. [8].

3 Cryptanalysis of the McEliece Cryptosystem

In this section we will discuss some general and well known attacks on the McEliece scheme. We shall not pay attention to special cases for which fast cryptanalysis exist.

3.1 Factoring the encryption matix

Let G_s denote the generator matrix G in systematic form and let the encipher matrix E be SG_sP. The number of non-singulier matrices S is given by $0.29 \times 2^{k^2}$. There exist approximately $2^{mt}/t$ generator matrices G_s for a binary irreducible t-error correcting Goppa code. And there are $n!$ possible permutation matrices. Moreover as shown by Adams and Meijer [1] the only transformation which transforms the encryption matrix E into a generator matrix G which algebraic structure allows us to use a fast decoding algorithm, is the original transformation i.e. $G = S^{-1}EP^{-1}$. Therefore we may conclude that for sufficiently large parameters it will be infeasible to obtain the private-keys S, G and P by an exhaustive search.

3.2 Recover message from cryptogram and encryption matrix

McEliece states in [9] that probably the most promising attack on his scheme consists of actually solving the basic problem, i.e. decoding a more or less arbitrary $[n, k, d]$ linear code containing t correctable errors. As it has been proved that the general decoding problem is NP-complete [2], one certainly expects that for large code parameters this attack will be infeasible.

A straightforward approach is based on a brute force *distance* search; comparing the cryptogram $\underline{e}$ to each permuted codeword $\underline{c} = \underline{m}E$. If the Hamming distance result is: $d_H(\underline{e}, \underline{c}) \leq t$, then $\underline{m}$ is the original message. However this method has a work factor of about $O(2^k)$. For $k = 654$ this becomes $2^{654} \approx 10^{197}$, which is astronomically large.

Another approach is based on a brute force search for a correct *syndrome*. Let D be the matrix HP. Clearly $ED^T = 0$. Find an error vector $\underline{z}$ with minimum weight for which $\underline{e}D^T = \underline{z}D^T$. However, it seems to be necessary to search through all solutions of this equation to find the desired $\underline{z}$ of minimum weight and has a work factor of about $O(n^t)$.

McEliece proposes in [9] to select randomly k of n ciphertext bits from $\underline{e}$ in the hope that none of the k selected bits are in error, and based on this assumption, to obtain the correct plaintext $\underline{m}$. The probability p_k of no error in the chosen k-bits of $\underline{e}$, however, is equal to

$$p_k = \frac{\binom{n-t}{k}}{\binom{n}{k}} = \prod_{i=0}^{k-1}(1 - \frac{t}{n-i}). \tag{3}$$

Selecting k-bits, which are not in error, does not guarantee that the corresponding $k \times k$ sub-matrix of E is non-singular. This only holds for maximum distance separable codes (MDS, [8]). In case of an MDS code every k columns of the encryption matrix are linearly independent. Since the Goppa codes used in the McEliece scheme are not MDS, we will have k linearly independent columns with a probability $q_k > 0$. This also holds for the encryption matrix E, since S works on the message space and P permutes the code words. Clearly, q_k can not be estimated by assuming that E is a random matrix.

The amount of work involved in solving k simultaneous equations in k unkown is k^a (e.g. $a = 2.8$ [6]). Let V_k be the average work factor if k columns are linearly dependent. Hence, before finding the message $\underline{m}$ with this attack one expects a work factor of

$$W = \gamma \times [(1 - q_k)V_k + q_k k^a] \times q_k^{-1} \times p_k^{-1}. \tag{4}$$

We can use the Hamming distance to check whether the obtained message $\underline{m}$ is correct plaintext. If the result of the Hamming distance is: $d_H(\underline{e}, \hat{\underline{m}}E) \leq t$, then $\hat{\underline{m}}$ is the original message $\underline{m}$. The additional cost to validate each message $\hat{\underline{m}}$ is therefore $O(nk)$.

Adams and Meijer [1] established by exhaustive search that for values of 'a' between 2 and 3, the maximal work-factor (without validation, $\gamma = 1$ and $q_k = 1$) is reached at $t = 37$. In this case for $a = 3$ the work-factor is approximately $2^{84.1}$, while for $t = 50$ this becomes $2^{80.7}$. As a consequence of this improvement, the value of k is increased from 524 to 654; i.e. the information rate $R = k/n$ is increased from 0.51 to 0.64.

4 Main Idea

A straightforward approach is based on a brute force distance search as mentioned in the previous section. Despite the high work factor this approach has the advantage that there are no additional validation costs, because the validation is part of the attack itself. As suggested by Hin [5], this attack can be improved by taking the constraints imposed by the cryptogram into account. For this reason we have to restate the above attack in terms of the cryptogram $\underline{e}$ instead of the message $\underline{m}$.

For the attack to be described in the next section we need a decomposition of the encryption matrix E in the following form

$$E = S_a[I_k|A_a]P_a^T, \tag{5}$$

where I_k is the $k \times k$ identity matrix and A_a is a $k \times (n-k)$ binary matrix. Since every linear code is equivalent to a systematic code, this decomposition is always possible.

If we apply a permutation matrix P_a to $\underline{e}=\underline{m}E+\underline{z}$, then we obtain the relation

$$\underline{e}P_a = \underline{c}P_a + \underline{z}P_a, \tag{6}$$

which will be denoted as $\underline{e}_a = \underline{c}_a + \underline{z}_a$.

The function $FKB(\underline{x})$ is defined as

$$FKB(\underline{x}) = FKB(x_1, x_2, \ldots, x_k, x_{k+1}, \ldots, x_n) = x_1, x_2, \ldots, x_k.$$

Hence, $FKB(\underline{x})$ selects the first k-bits from a n-bit vector $\underline{x}$.

We are now able to prove the next theorem.

Theorem 1 *If P_b is a permutation matrix for which $\underline{e}=\underline{m}E+\underline{z}$ can be written as*

$$\underline{e}_a P_b = \underline{m} S_a S_b [I_k|A_b] + \underline{z}_a P_b,$$

then

$$w_H[FKB(\underline{z}_a P_b)] = 0 \iff d_H[FKB(\underline{e}_a P_b)[I_k|A_b], \underline{e}_a P_b] \leq t. \tag{7}$$

Proof. We have

$$d_H[\underline{m} S_a S_b [I_k|A_b], \underline{e}_a P_b] = w_H[(\underline{z}_a P_b)] \leq t.$$

From which it follows that

$$d_H[FKB(\underline{e}_a P_b - \underline{z}_a P_b)[I_k|A_b], \underline{e}_a P_b] \leq t.$$

If $w_H[FKB(\underline{z}_a P_b)] = 0$, then it follows that $d_H[FKB(\underline{e}_a P_b)[I_k|A_b], \underline{e}_a P_b] \leq t$. On the other hand, if $d_H[FKB(\underline{e}_a P_b)[I_k|A_b], \underline{e}_a P_b] \leq t$, then $FKB(\underline{e}_a P_b)[I_k|A_b]$ must be the codeword with whom $\underline{e}_a P_b$ corresponds to. Therefore the $FKB(\underline{e}_a P_b)$ must be error-free, i.e. $w_H[FKB(\underline{z}_a P_b)] = 0$.

□

Observe that this theorem describes McEliece attack (with validation) in a more general form. During the initial phase of the attack, k cryptogram bits are randomly selected (without replacement) from $\underline{c}$. The k selected bits form the set $\mathcal{A}$ and the remaining $(n-k)$-bits are assigned to set $\mathcal{B}$. Selecting k new bits in the McEliece attack is replaced by a permutation P_b which swaps *at most* k new bits from set $\mathcal{B}$ for k-bits from set $\mathcal{A}$. The permutation is only succesful if the corresponding columns of the encryption matrix E are linearly independent as has been mentioned in section 3. The theorem states that the obtained solution is unique and gives the correct message $\underline{m}$ if the distance verification is positive. In the next section we will describe an attack based on this kind of bit swapping.

5 One Bit Swapping Attack

The McEliece attack can be considered as a k-bit swapping attack. To obtain a low complexity and to determine in a fast way if a given permutation fulfils, we will present an algorithm for a one bit swapping procedure only.

The algorithm for a one bit swapping attack consists of the following 5 steps.

Step 1 - *initialisation.*
Decompose the encipher matrix E, i.e. calculate a permutation matrix P_a and a matrix A_a such that $E = S_a[I_k|A_a]P_a^T$. Set-up a pointer table: FOR $i := 1$ TO n DO $Ptable[i] := i$.
Calculate $\underline{c}_a = \underline{c}P_a$ and up-date the pointer table.

Step 2 - *checking.*
Check if it holds that $d_H(FKB(\underline{c}_a)[I_k|A_a], \underline{c}_a) \leq t$. This can be done by checking whether there are no more than t errors with respect to $FKB(\underline{c}_a)A_a$. If there are t errors or less in the *last* $(n-k)$-bits of $\underline{c}_a$, then proceed to step 5.

Step 3 - *swapping.*
The algorithm PRP produces a pseudo-random permutation P_b. The permutation P_b swaps one column, say i, from the I_k part of the matrix $[I_k|A_a]$ for one column, say j, from the A_a part. The swapping procedure is as follows.

```
REPEAT
        Select permutation P_b from PRP.
        IF column j has not an '1' as i-th entry
           THEN P_b does not fulfil
           ELSE P_b fulfils
                  swap(Ptable[i], Ptable[j]);
                  c_a := c_a P_b
        FI;
UNTIL P_b fulfils.
```

Step 4 - *up-dating.*
Compute $[I_k|A_a]P_b$ into the form $S_b[I_k|A_b]$. The new 'stripped' generator matrix will be defined as $[I_k|A_a] := [I_k|A_b]$. Compute $\underline{e}_a := \underline{e}_a P_b$.
Proceed to step 2.

Step 5 - *calculate plaintext.*
At this stage there are no errors in $FKB(\underline{e}_a)$ and consequently $FKB(\underline{z}_a)$ is 0. The first k positions of the pointer table ($Ptable$) show locations in $\underline{e}$ without error. Select the corresponding columns of the encryption matrix E which are guaranteed linearly independent and calculate the plaintext $\underline{m}$.

6 Number of Swaps

A ciphertext $\underline{e}$ is obtained by adding an error vector $\underline{z}$ with Hamming weight t to a permuted codeword $\underline{c} = \underline{m}E$. Therefore there are t 'disturbed' bits in the cryptogram $\underline{e}$ which differ from the permuted codeword bits in $\underline{c}$. In the attack bits are repeatedly swapped in order to obtain k non-disturbed ciphertext bits. During the initial phase of the attack, k cryptogram bits are randomly selected (without replacement) from $\underline{e}$. The k selected bits form the set $\mathcal{A} = \{e_\nu\}$ and the remaining $(n-k)$-bits are assigned to set $\mathcal{B} = \{e_\mu\}$. The procedure $swap(e_\nu, e_\mu)$, which swaps a bit from set $\mathcal{A}$ for a bit from set $\mathcal{B}$, has one of the following values

- $s = \ \ 0$ if a (non-)disturbed bit e_ν is swapped for a (non-)disturbed bit e_μ,
- $s = -1$ if a disturbed bit e_ν is swapped for a non-disturbed bit e_μ,
- $s = +1$ if non-disturbed bit e_ν is swapped for a disturbed bit e_μ.

For the conditional probability $Pr\{i+s|i\}$, i.e. the probability that an event with i disturbed bits e_ν in $\mathcal{A}$ is followed after a swap by an event with $i+s$ disturbed bits e_ν in $\mathcal{A}$, we find

$$\sum_s Pr\{i+s|i\} = 1,\ Pr\{i-1|i\} = \frac{i(n-k-t+i)}{k(n-k)} \text{ and } Pr\{i+1|i\} = \frac{(k-i)(t-i)}{k(n-k)} \quad (8)$$

If $N_i = N_i(n,k,t)$ denotes the average work factor for a state with i errors, then N_{i-1} follows from

$$N_{i-1} = \sum_{r=0}^{\infty}\sum_{i=0}^{r} \binom{r}{i} \Big[Pr\{i-1|i-1\}^{i+1} Pr\{i|i-1\}^{r-i} + \\ + Pr\{i-1|i-1\}^{i} Pr\{i|i-1\}^{r-i+1}(N_i+2)\Big]$$

$$= \frac{Pr\{i-1|i-1\} + Pr\{i|i-1\}\cdot(N_i+2)}{Pr\{i-2|i-1\}} \quad (9)$$

Table 1 The average number of swaps N_i for state i.

n=1024			k=624			t=39			
N_1	N_2	N_3	N_4	N_5	N_6	N_7	N_8	N_9	N_{10}
$2^{59.4}$	$2^{53.3}$	$2^{48.3}$	$2^{43.9}$	$2^{39.9}$	$2^{36.3}$	$2^{33.0}$	$2^{30.0}$	$2^{27.3}$	$2^{24.7}$

The average number of random swaps (with replacement) $N(n,k,t)$ depending on all the possible $\binom{n}{t}$ initial states is given by

$$N(n,k,t) = \sum_{j=1}^{t} \frac{\binom{k}{j}\binom{n-k}{t-j}}{\binom{n}{t}} \sum_{i=1}^{j}(N_i + j). \tag{10}$$

7 Work factor

Let W_i denotes the average work factor of step i. With a probability of approximately one half ($q_k \approx \frac{1}{2}$) a permutation P_b is found in step 3 which can be used. The permutation P_b can be generated and validated in a fast way and independent from the main algorithm. Steps 2 and 4 are only executed when a correct permutation P_b is determined. Therefore we can neglect W_3 ($V_k \approx 0$ in equation 4). Since steps 1 and 5 are executed only once, we can neglect W_1 and W_5 in view of the on average $N(n,k,t)$ repeated steps 2 and 4. Therefore the main algorithm has an average work factor

$$W = (W_2 + W_4) \times N(n,k,t). \tag{11}$$

$M(j,i)$ is a notation for j simultaneous i-bit multiplications and similarly $A(j,i)$ denotes j simultaneous i-bit additions. If simultaneous i-bit operations are left out of consideration, then e.g. $M(j,i)$ becomes, with a little ambiguity, $jM(1,i) = jM(i)$. Moreover if only 1 bit operations are considered, then this notation reduces to $M(j,i) = ji$.

For W_2 we find that

$$W_2 = M(n-k,k), \tag{12}$$

$$(t+1) \cdot M(k) \le W_2 \le (n-k) \cdot M(k). \tag{13}$$

$$\text{On average } W_2 \text{ will be } 2(t+1) \cdot M(k). \tag{14}$$

For W_4 we obtain

$$W_4 = A(k-1, n-k-1), \tag{15}$$

$$A(n-k-1) \leq W_4 \leq (k-1) \cdot A(n-k-1). \tag{16}$$

On average W_4 will be $\frac{k-1}{2} \cdot A(n-k-1)$. (17)

In general the work factor (11) becomes

$$W = [M(n-k,k) + A(k-1,n-k-1)] \times N(n,k,t). \tag{18}$$

If we use for example the average values (14) and (17) in (11), then we obtain the following work factor

$$W = [2(t+1) \cdot M(k) + \frac{k-1}{2} \cdot A(n-k-1)] \times N(n,k,t). \tag{19}$$

This way we find for the overall average work factor (without parallelism etc.)

$$W = \frac{1}{2} \cdot [4k(t+1) + k(n-k) - (n-1)] \times N(n,k,t). \tag{20}$$

The maximum value of W is approximately $2^{76.8}$ for $t = 39$. The average number of swaps is in this case $2^{59.4}$.

8 Further Improvements

At Eurocrypt'88, Lee and Brickell [7] presented a generalized attack on the McEliece scheme. Briefly, the attack is as follows: a set of k-bits is selected at random from the cryptogram. The set is tested by an exhaustive search for an error pattern with no more than j errors. In case an error pattern is found with j or less errors, the algorithm stops, otherwise a new set of k-bits is selected. For $j = 0$ the *traditional* attack is obtained and a *brute force* distance search for $j = t$. Lee and Brickell have found (with some assumptions) that the optimum j which minimizes the maximum work factor is 2 for all values of useful code parameters.

8.1 Search for one correctable error

Lee and Brickell propose in [7] a random update of only one bit instead of all the k-bits at the same time. This bit swapping is actually one of the basics of our method. From section 6 it follows that the last steps, i.e. removing the last j errors, dominate the work factor. An optimization procedure similar to the Lee-Brickell method is used to speed-up our attack. While in our case there is a *trade-off* between the swap-complexity and the complexity of the exhaustive search with checking, the optimum j which minimizes the maximum work factor is found to be 1. This low optimum is due to the low complexity of the swap-procedure, which is $O(k \times (n-k))$. For a single error pattern search a new step has to be added to the attack described in section 5. If $\underline{u}_i$ is the i-th unit vector, then the new step becomes

Step 2a - *search for one correctable error*
For $1 \leq i \leq k$, check if it holds that $d_H(FKB(\underline{e}_a - \underline{u}_i)[I_k|A_a], \underline{e}_a) \leq t$. This can be done by checking whether there are no more than t errors with respect to $FKB(\underline{e}_a - \underline{u}_i)A_a$. If for a certain i there are t or less errors in the *last* $(n-k)$-bits of $\underline{e}_a$, then correct bit $Ptable[i]$ in $\underline{e}$ and proceed to step 5.

For the average work factor W_{2a} for step $2a$ we find that

$$W_{2a} = 2(t+1)k. \tag{21}$$

The maximum overall work factor W is approximately $2^{71.1}$ for $t = 39$. The average number of swaps is in this case $2^{53.4}$.

8.2 Partial search for two correctable errors

Since the value of W_{2a} is small compared to $(W_2 + W_4)$, a partial search for patterns with two errors can be considered additionally. The number of partial search patterns used in step $2b$ below is denoted by n_s.

Step 2b - *partial search for two correctable errors*
For $1 \leq i < k$ and $i < j \leq k$, check if it holds that $d_H(FKB(\underline{e}_a - \underline{u}_i - \underline{u}_j)[I_k|A_a], \underline{e}_a) \leq t$. This can be done by checking whether there are no more than t errors with respect to $FKB(\underline{e}_a - \underline{u}_i - \underline{u}_j)A_a$. If for certain i and j there are t or less errors in the *last* $(n-k)$-bits of $\underline{e}_a$, then correct bit $Ptable[i]$ and $Ptable[j]$ in $\underline{e}$ and proceed to step 5. If $\#\{(i,j)\} = n_s$ then proceed to step 3.

For the average work factor W_{2b} for step $2b$ we find that

$$W_{2b} = 2(t+1) \cdot n_s. \tag{22}$$

If we assume a uniform distribution of the error patterns, then the probability of succes follows from

$$Pr\{\text{Succes Partial Search}|i = 2\} = n_s / \binom{k}{2} \tag{23}$$

The average work factor N_i for states 3 to t follows from equation 9. The average work factor N_2 for state $i = 2$ becomes

$$N_2 = \frac{Pr\{i = 2|i = 2\} + Pr\{i = 3|i = 2\} \cdot (N_3 + 2)}{Pr\{i = 1|i = 2\} + Pr\{\text{Succes Partial Search}|i = 2\}} \tag{24}$$

The maximum overall work factor W is approximately $2^{69.7}$ for $t = 39$ and $n_s = 5769$. The average number of swaps is in this case $2^{50.3}$.

8.3 General Attack

Let P_b be a permutation matrix which swaps *at most* i-columns from the I_k part for i-columns in the A_a part of the $[I_k|A_a]$ matrix. Let S be a subset of $GF(2)^k$ with the property that the Hamming weight $w_H(\underline{s})$ of the vectors $\underline{s} \in S$ is less or equal than j. If *all* vectors $\underline{s}$ with Hamming weight equal to j are used during one search, then the attack is called *complete* otherwise *partial.*

The general **[i,j] – swap attack** follows from

1. *initialisation*
   - decompose encipher matrix: $E = S_a[I_k|A_a]P_a^T$
   - calculate $\underline{e}_a = \underline{e}P_a$
   - set-up pointer table
2. *checking*
   - Check if there exists an $\underline{s} \in S$ such that $d_H[(FKB(\underline{e}_a) - \underline{s})[I_k|A_a], \underline{e}_a] \leq t$. If there exists such an $\underline{s} \in S$, then correct $\underline{e}$ with $\underline{s}$ using the pointer table and proceed to step 4.
3. *swapping*
   - select a permutation P_b which fulfils
   - P_b swaps at most i-columns from the I_k part for i-columns in the A_a part of the $[I_k|A_a]$ matrix
   - Transform $[I_k|A_a]P_b$ into $S_b[I_k|A_b]$
   - let $[I_k|A_a] := [I_k|A_b]$ and $\underline{a}_a := \underline{a}_a P_b$
   - up-date the pointer table and proceed to step 2
4. *calculate plaintext*
   - the first k positions of the pointer table show locations in $\underline{e}$ without error
   - select the corresponding columns of the encryption matrix E which are guaranteed linearly independent
   - calculate the plaintext $\underline{m}$

For a *complete* $[i,j]$-swap attack with $i < j$ all search patterns $\underline{s} \in S$ have to be used during the initial round. However for the subsequent rounds only the search patterns with $(j-i) \leq w_H(\underline{s}) \leq j$ have to be considered, since there are at least $(j-i)$ errors after each i-swap. For a *partial* attack this becomes $(j-i-1) \leq w_H(\underline{s}) \leq j$.

9 Reduced Public-Key

From the attack described in section 5 and the fact that the existence of more than one trapdoor in the system is unlikely [1], it follows that although a factorisation of E is found, no information about the original S, G and P matrices is revealed. For this reason a $k \times (n-k)$ matrix A of a decomposition of the encryption matrix $E = SGP = S'[I_k|A]P'$ can be published instead of E. Encryption can be done in the following way

- Use a publicly known seed $\underline{s}$. The seed $\underline{s}$ generates a new non-singular binary matrix S^* . The encryption scheme becomes

$$\underline{e} = \underline{m}E + \underline{z} = \underline{m}S^*[I_k|A] + \underline{z} = \underline{w}[I_k|A] + \underline{z}. \tag{25}$$

- Use an publicly known invertible function f which transforms a message $\underline{m} \in GF(2)^k$ into a word $\underline{w} \in GF(2)^k$. The function f may also depend on the error vector $\underline{z}$. In this case the following encryption scheme is obtained

$$\underline{e} = \underline{m}E + \underline{z} = f(\underline{m}, \underline{z}).[I_k|A] + \underline{z} = \underline{w}[I_k|A] + \underline{z}. \tag{26}$$

To keep the seed $\underline{s}$, used to generate a non-singulier matrix S^*, secret does not increase the security of the system. Since a chosen-plaintext attack by majority voting of each position of a row of the encipher matrix E will be successful and reveal $[S^*|S^*A]$ and consequently S^*.

In both cases it follows that

- **The sender** generates an error vector $\underline{z}$, computes $\underline{w}$ and calculates a cryptogram $\underline{e} = \underline{w}[I_k|A] + \underline{z}$.
- **The receiver** determines the error pattern $\underline{z}$, removes it from $\underline{e}$, computes $\underline{w} = FKB(\underline{e} - \underline{c})$ and calculates the message $\underline{m} = \underline{w}S^{*-1}$ or $\underline{m} = f^{-1}(\underline{w}, \underline{z})$.

It follows that the public-key can be reduced to $n \times (n-k)$-bits. For $n = 1024$ and $t = 39$ the reduced key becomes 399 kbits.

Acknowledgement

The author wishes to thank Paul Hin for the numerous discussions we had about cryptosystems based on coding theory during our final years at the Delft University of Technology and for reading this manuscript. I wish to thank IACR for the support given which made it possible to present this paper at Crypto'88. Finally, I wish to thank Jean-Paul Boly for his comments.

References

[1] C. Adams and H. Meijer, "Security-Related Comments Regarding McEliece's Public-Key Cryptosystem", in: *Advances in Cryptology - CRYPTO '87*, Carl Pomerance ed., Lecture Notes in Computer Science # 293, Springer-Verlag, pp. 22 4-228, 1988.

[2] E.R. Berlekamp, R.J. McEliece and H.C.A. van Tilborg, "On the inherent intractability of certain coding problems", *IEEE Trans. Inform. Theory*, vol. IT-24, pp. 384-386, 1978.

[3] W. Diffie and M.E. Hellman, "New Directions in Cryptography", *IEEE Trans. Inform. Theory*, vol. IT-22, pp. 644-654, 1976.

[4] P.J.M. Hin, "Channel-Error-Correcting Privacy Cryptosystems", M.Sc. Thesis, Delft University of Technology, 1986 (in Dutch).

[5] P.J.M. Hin, Private Communication, December 1986.

[6] D.E. Knuth, *The art of computer Programming, Vol.2 / Seminumerical Algorithms*, Addison-Wesley Publishing Company, 1981.

[7] P.J. Lee and E.F. Brickell, "An observation on the Security of McEliece's Public-Key Cryptosystem", Presented at Eurocrypt'88, Davos, May 1988.

[8] F.J. MacWilliams and N.J.A. Sloane, *The Theory of Error-Correcting Codes*, Amsterdam: North-Holland, 1978.

[9] R.J. McEliece, "A Public-Key Cryptosystem Based on Algebraic Coding Theory", DSN Progress Report 42-44, JPL Pasadena, pp. 114-116, 1978.

A Constraint Satisfaction Algorithm for the Automated Decryption of Simple Substitution Ciphers

Michael Lucks
Department of Computer Science and Engineering
Southern Methodist University
Dallas, Texas 75275

Abstract

This paper describes a systematic procedure for decrypting simple substitution ciphers with word divisions. The algorithm employs an exhaustive search in a large on-line dictionary for words that satisfy constraints on word length, letter position and letter multiplicity. The method does not rely on statistical or semantical properties of English, nor does it use any language-specific heuristics. The system is, in fact, language independent in the sense that it would work equally well over any language for which a sufficiently large dictionary exists on-line. To reduce the potentially high cost of locating all words that contain specified patterns, the dictionary is compiled into a database from which groups of words that satisfy simple constraints may be accessed simultaneously. The algorithm (using a relatively small dictionary of 19,000 entries) has been implemented in Franz Lisp on a Vax 11/780 computer running 4.3 BSD Unix. The system is frequently successful in a completely automated mode -- preliminary testing indicates about a 60% success rate, usually in less than three minutes of CPU time. If it fails, there exist interactive facilities, permitting the user to guide the search manually, that perform very well with minor human intervention.

1. Introduction

Despite its relative insecurity compared to modern encryption techniques, the simple substitution cipher remains a classical problem that has defied reliable automated decryption. Human cryptanalysis of substitution ciphers is usually begun by obtaining a trial entry to the code, i.e. guessing the decodings one or more letters. The initial guesses may be based on a variety of simple techniques, such as n-gram frequencies, doubled letters or short word patterns. The partial decryption yielded by the entry may then be used deduce full words through visual recognition and by observing syntactic and semantic patterns. The guessed words, in turn, yield further letter decodings and the process is repeated until the entire is message is deciphered. Some of the automated systems have attempted to imitate this method. Carroll and Martin [CM86], for instance, have developed a microcomputer-based program which utilizes expert system methodology to capture the knowledge and heuristics that an experienced cryptanalyst might employ in both the entry and deduction phases. Schatz [S77] uses singular value decomposition of a cipher's digram matrix to obtain a prediction of a cryptogram's vowels. Using the vowels and some special clues (e.g. one-letter words and apostrophes) as an entry, Schatz's program performs a heuristic search for words guided by a small vocabulary and a database of rules which reflect statistical properties of the English language. A very different method, proposed by Peleg and Rosenfeld [Peleg-Rosenfeld], employs a relaxation algorithm to determine all of the plaintext letters in parallel by iteratively updating the joint probabilities for the decoding of each ciphertext letter, with respect to its two nearest neighbors. The above systems assume that the plaintext conforms to various statistical properties of English. For long cryptograms this is a reasonable assumption, however messages that are short in length or contain uncommon combinations of letters (e.g. acronyms), are particularly difficult, if not impossible for such systems to solve.

An exhaustive search that generates all 26! keys is a reliable, but clearly impractical decryption method. A more reasonable (but still exhaustive) approach is to conduct the search at the word level, rather than at the letter level, using a large on-line dictionary. For each word in the ciphertext, the dictionary is searched for all of words that satisfy some known constraints. Since the ciphertext contains word divisions, word length is always a known constraint. Multiple occurrences of the same letter in the same word a second important pattern constraint. If the dictionary is complete, then each plaintext word must appear somewhere in the corresponding list of constrained words. If we examine all possible combinations from the constrained lists, the correct translation of the entire message must eventually appear. The search for the correct combination is conducted as a depth-first tree walk, in which each branch in the search tree corresponds to a guess for the decoding of a particular word in the ciphertext. Although the search space is initially very large, it is greatly reduced during the course of the search because each time a word is chosen as a possible decryption it imposes additional constraints upon other word that shares one or more of its letters. Hence, as a choices are made for each word, the set of possible choices for the other words becomes progressively smaller. Backtracking is performed whenever there are remaining words for which the set of potential decryptions is empty. Hence, if the dictionary is complete, the search will eventually find a set of choices for the ciphertext words which mutually satisfy all known constraints. With high probability, this set of words is very close to the correct plaintext. Even if some plaintext words are not in the dictionary, the constraints imposed by those that are may be sufficient to provide an unambiguous decryption that is apparent by visual inspection. Wall [W80] describes such a procedure, but claims it is feasible only if special purpose hardware (a content addressable memory) is used to support parallel lookup of words from the dictionary. Wall actually implemented this method, simulating the parallel hardware via APL vector operations and excluding lookup time from the performance analysis. Our approach is much the same as Wall's, but with the following improvements:

1) no special hardware is required; instead the dictionary is compiled into a database designed to facilitate efficient lookup;
2) a control strategy is employed to guide the search toward promising paths;
3) the use of letter multiplicity (i.e. multiple occurrences of the same letter in the same word) as a constraint results in a much smaller search space;
4) certain guesses for words are recognized as yielding inconsistencies, and hence immediately rejected instead of being propagated further in the search.

2. The Database

Our system is based on an exhaustive search for pattern words in a dictionary of over 19,000 entries. The word search entails determining the set of words in the dictionary that satisfy specified constraints on word length, letter position and letter multiplicity. An example of such a pattern is the set all words with six letters having *e* in position 2 and *w* in position 5. A more complicated example is the set of all eight letter words ending in *t* in which the same letter occurs in positions 1, 5 and 7. Extracting such information by repeatedly scanning the dictionary for pattern matches would be impractically slow. Instead, the dictionary is compiled into a database that is partitioned according letter, word length and letter position. Associated with each letter in the alphabet is a list of numbered properties 1, 2,...m, where m is the maximum length of any word in the dictionary. The value of each property j is a vector V_j, indexed from 1 to j. If we want to find all words of length n containing the letter l in position i, we look on the property list of l and access the ith element of the vector found on property n. For instance, all 10 letter words containing r in position 6 are found by looking in the sixth entry of the vector found in property 10 on the property list of r. For simplicity, the database may also be viewed as a three dimensional array D, indexed by word length, letter and letter position, in which the entries are lists of words. For parameters i, j and k, an entry $D(i,j,k)$ would contain the list of all words of length i in which letter j occurs in position k. Words satisfying more complicated patterns are found by computing the union and intersection of one-letter patterns. The intersection of $D(9,b,4)$ and $D(9,w,7)$, for example, would be the set of all 9 letter words containing b in position 4 and w in position 7. To get all 6 letter words containing the same letter in positions 5 and 6, we take the union of all words having letter l in positions 5 and 6, where l ranges from a thru z.

The dictionary is compiled in a Franz Lisp session, separate from the execution of the decryption program. The resulting Lisp image may then be stored on disk to permit fast loading of the system. The Lisp image, including the database of approximately 19,700 words, occupies about 2.9 megabytes of disk space.

3. The Search Technique

Viewing a cipher as a list of words $[w_0, w_1, \ldots, w_n]$, our decryption process amounts to a state-space search in which each state T_i is a pair $[P_i, S_i]$. P_i is a list $[p_{i_0}, p_{i_1}, \ldots, p_{i_n}]$ where each p_{i_j} is itself a list containing all possible decryptions for word w_j in the ciphertext. S_i is a the current substitution list, i.e. a list of pairs of letters $([C_1,d_1], [C_2,d_2], \ldots, [C_m,d_m])$ indicating that letter d_k is currently assumed to be the decoding of the ciphertext letter C_k. Each node in the search tree represents a modified state which reflects the constraints imposed by a new guess for some ciphertext word. At the root node T_0, S_0 is empty and P_0 is obtained by searching the dictionary for the possible decryptions of each word, subject to the constraints of word length and multiple occurrences in a word of the same letter. For example, consider the following cryptogram taken from *The Dallas Morning News*:

MZDDTK CJQLAPZZ DKDM CJQLNZPQ TZJKDA MPQBPQB TNT MNQBM

Possible decryptions of the first ciphertext word are words that satisfy the pattern MZDDTK, i.e. all six-letter words having the same letter in positions 3 and 4. In this case, the word search routine returns a list of 320 words, [babble,bobbin,...,sizzle]. The same procedure is then repeated for each word in the ciphertext. Table 1 summarizes these initial possibilities.

Word	Ciphertext	Possible Decryptions	# Possibilities
w_0	MZDDTK	[babble,...,sizzle]	320
w_1	CJQLAPZZ	[absentee,...,megawatt]	90
w_2	DKDM	[afar,...,vivo]	39
w_3	CJQLNZPQ	[academia,...,bayberry]	130
w_4	TZJKDA	?	?
w_5	MPQBPQB	[alfalfa]	1
w_6	TNT	[ala,...,wow]	28
w_7	MNQBM	[aloha,...,widow]	93

Table 1. Initial possible decryptions of ciphertext words

The entries for the word TZJKDA are left blank because it has no multiple occurrences of any letter. Its possible decipherments (all six-letter words) are so numerous (2,850 words) that it is best to postpone evaluation of this word, as will be discussed later.

Word	Ciphertext	Possible Decryptions	# Possibilities
w_0	MZDDTK	[cobble,...,sizzle]	246
w_1	CJQLAPZZ	[divorcee,...,kilowatt]	32
w_2	DKDM	[afar,...,vivo]	32
w_3	CJQLNZPQ	[charisma,...,petulant]	42
w_4	TZJKDA	?	?
w_5	MPQBPQB	-	0
w_6	TNT	[ala,...,wow]	28
w_7	MNQBM	[aloha,...,widow]	79

Table 2. Reduced possible decryptions of ciphertext words

The size of these initial lists of possibilities may be reduced considerably by removing ***inconsistent*** words, i.e. words that imply an ambiguous decryption key. For instance, ***babble*** is inconsistent with MZDDTK because it implies that both M and D translate to b. (In Wall's algorithm, such inconsistencies are not recognized.) Table 2 displays the reduced possibility lists in which inconsistent words have been extracted. Note that MPQBPQB has no possible decryption, i.e. the plaintext for the word doesn't appear in the dictionary.

The initial state of the search at the root node of the search tree is $T_0 = [P_0, S_0]$, where S_0 is an empty list and P_0 corresponds to the lists of possible decryptions in Table 2. For instance, p_{0_0} is the list of candidate decryptions for the first word, i.e. p_{0_0} = [cobble,...,sizzle]. Similarly, p_{0_1} = [divorcee,...,kilowatt], p_{0_2} = [afar,...,vivo], ... , p_{0_7} = [aloha,...,widow]. Each descendant node in the tree may be viewed as a guess for some word in the ciphertext. To expand the root node, a particular word is chosen from some p_{0_i} as a trial decryption of w_0. The successor state is $T_1 = [P_1, S_1]$, where S_1 is list of letter substitutions implied by the choice and P_1 is equal to $[P_{1_1},...,P_{1_{(i-1)}},P_{1_{(i+1)}},...,P_{1_7}]$, where each P_{1_j} is the subset of P_{0_j} whose words do not violate the new constraints imposed by S_1. (Note that P_1 does not contain the possible decryptions for w_i, since w_i is the word for which a guess is being made.) If *divorcee* is selected as trial decryption of w_1, for example, then S_1 = $[(C,d),(J,i),(Q,v),(L,o),(A,r),(P,c),(Z,e)]$. Each P_{1_i} is now filtered to remove words which conflict with S_1. The filter succeeds in two ways. In one case, words are discarded because the same ciphertext letter has two decryptions, e.g. *charisma* is dismissed as a possible decryption for w_3 (CJQLNZPQ) because the implied decoding (C,c) conflicts with the assumption from S_1 that C decodes to d. The second way that filtering works is to eject words that require two different ciphertext letters to decode to the same plaintext. For example, *cobble* is no longer a possible decryption of w_1 because the required substitution of c for M conflicts with the constraint (P,c) in S_1. The new constraints also provide additional information about w_4 (TZJKDA), the word which had not been previously evaluated. Under the constraints (Z,e) and (D,c), the possible decryptions for for TZJKDA is now the set of all six-letter words containing e in position 2 and c in position 5. The result is a list of 21 words [deduce,...,select] (all of which get filtered out). If we had evaluated w_4 earlier, we would have to filter the entire list of 2,850 six-letter words in the dictionary.

The search space is greatly reduced by the seven constraints of S_1, as indicated in Table 3 which corresponds to P_1.

Word	Ciphertext	Possible Decryptions	# Possibilities
w_0	MZDDTK	[bellum]	1
w_2	DKDM	[alan,...,sash]	4
w_3	CJQLNZPQ	-	0
w_4	TZJKDA	-	0
w_5	MPQBPQB	-	0
w_6	TNT	[ala,...,tat]	8
w_7	MNQBM	-	0

Table 3. Possible decryptions at state $T_1 = [P_1,S_1]$
with w_1 decoded as CJQLAPZZ = divorcee

The node corresponding to state T_1 may now be expanded by choosing among the 13 possible decodings for the w_0, w_2 and w_6. If *bellum* is chosen for w_0, the resulting additional constraints in state T_2 filter out all of the remaining possibilities for w_2 and w_6, so we have reached a dead end in the search.

When a dead end is encountered, the trial plaintext under the current set of constraints is evaluated to decide whether or not the constraints yield a likely decryption of the ciphertext. The main criterion considered by the evaluation routine is the number of words in the ciphertext which are completely determined. The evaluation function awards points to any completed word, whether or not its decipherment is in the dictionary -- the mere fact that all of the letters can be unambiguously decoded is a positive sign. Greater weight, of course, is given to words found in the dictionary and longer words are assigned more points than shorter ones. Extra credit is given to completed words that were not among those selected for expansion, i.e. words that were filled in as a result of other selections. If the score returned by the evaluation function is sufficiently high and is equal to or greater than the previous highest score, the current state is considered to be a possible solution and the trial plaintext is displayed to the user. In any case, the search continues by backtracking to the previous node and re-expanding with a new word choice.

In the present example, the constraints derived from the first selected word are sufficient to shrink the search space to a manageable level after the expansion of only one node -- fortunately the selected word provides sufficient constraints. This is not always the case, however. For instance, if we had selected *ala* as a decryption for w_6 (TNT) at state T_0 (instead of *divorcee* for w_1), P_1 would contain far more possibilities, as shown in Table 4. Here the number of combinations of remaining possible decryptions is 3,456 (6× 6× 16× 6) rather than 32 (4× 8) as in Table 3.

Word	Ciphertext	Possible Decryptions	# Possibilities
w_0	MZDDTK	[giddap,...,hurray]	6
w_1	CJQLAPZZ	[divorcee,...,princess]	6
w_2	DKDM	[vivo,...,cock]	16
w_3	CJQLNZPQ	[preclude]	1
w_4	TZJKDA	-	0
w_5	MPQBPQB	-	0
w_7	MNQBM	[elide,...,plump]	6

Table 4. Possible decryptions at state $T_1 = [P_1, S_1]$ with w_6 decoded as TNT = *ala*.

The striking contrast is due, of course, to the difference in the number of constraints imposed by the two choices. Word w_1 has 7 distinct letters, yielding 7 constraints, as opposed to only 2 constraints produced by the 2 distinct letters in w_6. Short words (i.e. words with less than 5 letters) hence pose a problem for our algorithm. Not only do they fail to provide the desired constraints on other words, they also are less likely to be filtered out themselves because there are fewer possibilities for letter conflicts. If there are several unresolved short words, the combinatorics involved in checking all possible combinations rapidly gets out of hand. Since most cryptograms contain a high percentage of such words, the full tree may be extremely bushy and a complete traversal usually cannot be executed in a reasonable amount of time. It is therefore advisable that the traversal be directed toward cheap, promising paths and steered away from expensive, dubious ones, so that a satisfactory solution may be displayed to the user at a relatively early stage of the search. Fortunately, it is usually possible to achieve this goal if we are careful in the selection of nodes to be expanded.

To deal with the short word problem, the ciphertext is separated into two groups. Group A contains the longer words in the message, i.e. words of six or more letters, while group B contains the rest. (If there are not three or more long words in the message, the definition of "long" is dynamically redefined so that there are at least three.) No words from group B are considered for expansion until all of the words in group A have been either been expanded or have no remaining possible decryptions. By examining longer words first we hope that the search space will already be somewhat constrained before the short words are processed. When only words from group B remain, the current state is evaluated and a decision is made whether to continue on the current path or to backtrack. The node is expanded only if there is some evidence that the current path looks promising or if the cost of expansion is relatively small. The primary measure for evaluating the promise of a path is the number of completely deciphered words which are also found in the dictionary, particularly words that were not chosen as guesses. A secondary measure is the number of letters remaining to be deciphered -- the fewer the better. If a path is not found to be promising by the above criteria, the next node may still be expanded if it can be done cheaply, i.e. if the number of successors is small and the tree is already of sufficient depth.

Another useful heuristic for optimizing the search is Wall's suggestion that the most constrained word, i.e. the word with the fewest number of possible decryptions, should be expanded first. If a word found in the dictionary happens to be highly constrained at the root node, expanding it right away will almost always yield a speedy correct decryption because the search converges very fast once the right path is found. (This rule should be subordinate to the short word heuristics, however -- a short word should not be expanded prior to a long one even if it is more constrained.)

The workings of the search may be illustrated by completing the decryption of our example. (An abbreviated trace of the search is found in the appendix.) The words in group A are w_0, w_1, w_3, w_4 and w_5. In the initial state (Table 2) the most constrained long word is w_1, with 32 possibilities, hence CJQLAPZZ is chosen to be expanded first. From its list of possible decryptions, *kilowatt* is selected as the first trial word. Since there are no possible decryptions for any of the other long words under the new constraints, this choice is rejected and the search immediately backtracks and the word *waitress* is tried. This choice is rejected for the same reason, as are the next 10 choices for w_1. The first trial guess for w_1 that is considered promising is *buckaroo*. This path is considered worthy to pursue because it allows another long word (w_0) to be deciphered into a word appearing in the dictionary, namely *sodden*. After *sodden* is selected to expand w_0, the state is considered promising enough to warrant expansion of a short word, so *eye* is chosen for TNT. At this point a dead end is reached. Since there has been no previous solution offered, the current decryption is the best available so far, hence it is displayed to the user as:

```
sodden buckaroo dnds buckyorc eounda src-rc- eye syc-s
MZDDTK CJQLAPZZ DKDM CJQLNZPQ TZJKDA MPQBPQB TNT MNQBM
```

As shown in the appendix, the search now backtracks to *sodden* and selects *ewe* for TNT. This yields a solution which appears equally good as the first, so it too is displayed. After 13 possible solutions involving *buckaroo* are discovered, the search backtracks to top level and other choices are tried for w_1. Several other paths are explored, but the depth of the search never exceeds 2. Eventually *mandrill* is selected for w_1, which happens to be the correct decryption. This leads immediately to *mandolin* for w_3. The only word that now satisfies the constraints for w_4 is *slater* (*player* is not in the dictionary). This path terminates in the solution

```
-leest mandrill ete- mandolin slater -in-in- sos -on--
MZDDTK CJQLAPZZ DKDM CJQLNZPQ TZJKDA MPQBPQB TNT MNQBM
```

which is still very obscure. Backtracking to the *mandolin* level, *sleety* is now tried for w_0, yielding the somewhat intelligible

```
sleety mandrill eyes mandolin tlayer sin-in- tot son-s
MZDDTK CJQLAPZZ DKDM CJQLNZPQ TZJKDA MPQBPQB TNT MNQBM
```

The next choice for w_0 is *sleepy* which yields the correct answer

```
sleepy mandrill eyes mandolin player sin-in- pop son-s
MZDDTK CJQLAPZZ DKDM CJQLNZPQ TZJKDA MPQBPQB TNT MNQBM
```

The full plaintext is obvious by inspection, however there is no way for the system to determine that B decodes to *g* because neither *singing* nor *songs* is in the dictionary and only these words contain *g*. (Since *player* is not in the dictionary, the score of 4100 is no better than the score of the previous decryption.)

4. Interactive Mode

When the system fails in the fully automated mode, a backup interactive mode is provided through which the user may analyze the cipher and supply his/her own guesses for letters. Commands exist which permit the user to display first order statistics, to add and delete guesses for letters, and to simultaneously display the message and its partial decryption. With some guesses for letters, the automated search may then be repeated, this time guided by the user-supplied constraints. In many cases where the automated system fails, a successful decryption is achieved via correct guesses for only one or two letters.

5. Extensions

The current system might be improved in a variety of ways that have yet to be attempted. An ability to recognize plural, prefixed and suffixed forms as words, for instance, would take care of the majority of examples that the present system can't handle automatically. Whether these forms should be added to the dictionary (at the cost of a significantly larger search space) or detected by a separate routine is under investigation. A second desirable extension would be to integrate the various heuristics and statistical approaches found in [S77], [PR79], [CM86], [A84] and [A86]. The information obtained from the statistical analyses might be valuable both in guiding the automated search as well as aiding the interactive user. Finally, an moderate improvement in performance would almost certainly result from a careful editing of the dictionary, which currently contains a many extremely rare words and omits many common ones. It would also be desirable to order the words in the database, so that more frequently used words are considered first. These tedious tasks have not yet been undertaken.

6. Performance

The system has been implemented in Franz Lisp on a Vax 11/780 computer. In tests on more than 100 examples chosen at random from newspapers and magazines, the system was successful in a completely automated mode about 60% of the time. Usually the solution was obtained in less than three minutes of CPU time. In approximately 30% of the trials, the program required rather trivial human intervention, such as the guessing of a common short word such as *the* or *and*. Failure most commonly occurred on examples in which none of the longer words in the plaintext were present in the dictionary. This situation occurs, for instance, when all of the long words are plurals or suffixed, since these forms are not likely to be found in our limited dictionary. When this happens, the system is forced to use small words as trial entries, thereby establishing few constraints and hence greatly expanding the search space. The second most common cause of failure was that none of the words in the plaintext contained any repeated letters. In this case, the program is unable to proceed (unless there are some one-letter words) because there are no entry candidates. This situation is most likely to arise in very short messages or in examples composed mostly of short words.

7. Conclusions

We have described an automated method for decrypting simple substitution ciphers based on exhaustive search and controlled thru constraints imposed by word patterns. No statistical analyses or language-specific heuristics are employed. Although quite successful in its own right, we believe that the technique could be used as a driver to an even more powerful system in which heuristics and statistical information would assist in directing the search. This hybrid approach would exploit the somewhat unstructured methods of the human cryptanalyst while retaining the systematic character of the exhaustive search that enables successful automation.

Acknowledgement

I would like to thank Dr. James G. Dunham, SMU Dept. of Electrical Engineering, for his advice and assistance in the development of this project.

References

[A84] Roland Anderson, "Finding Vowels in Simple Substitution Ciphers by Computer", Cryptologia, vol. 8, no. 4, Oct. 1984, pp. 348-358.

[A86] Roland Anderson, "Improving the Machine Recognition of Vowels in Simple Substitution Ciphers", Cryptologia, vol. 10, no. 1, Jan. 1986, pp.10-33.

[CM86] John H. Carroll and Steve Martin, "The Automated Cryptanalysis of Substitution Ciphers", Cryptologia, vol. 10, no. 4, Oct. 1986, pp. 193-209.

[PR79] Shmuel Peleg and Azriel Rosenfeld, "Breaking Substitution Ciphers Using a Relaxation Algorithm", CACM, vol. 22, no. 11, Nov. 1979, pp. 598-605.

[S77] Bruce R. Schatz, "Automated Analysis of Cryptograms", Cryptologia, vol. 1, no. 2, April 1977, pp. 116-142.

[W80] Rajendra Wall, "Decryption of Substitution Cyphers with Word Divisions Using a Content Addressable Memory", Cryptologia, vol. 4, no. 2, April 1980, pp. 109-115.

Appendix

Program Execution with Trace of Word Search

The current depth of the search tree is indicated by the number on the left. The ciphertext of the word currently being examined is denoted in upper case, while the trial decryption for the word is in lower case. (A portion of the trace has been omitted to save space.)

```
MZDDTK CJQLAPZZ DKDM CJQLNZPQ TZJKDA MPQBPQB TNT MNQBM

0 CJQLAPZZ kilowatt
0 CJQLAPZZ waitress
0 CJQLAPZZ ruthless
```

```
0 CJQLAPZZ princess
0 CJQLAPZZ marquess
0 CJQLAPZZ giantess
0 CJQLAPZZ dutchess
0 CJQLAPZZ congress
0 CJQLAPZZ compress
0 CJQLAPZZ baroness
0 CJQLAPZZ buckaroo
|1 MZDDTK sodden
||2 TNT eye

*** -- Solution #1
       Score = 3100
sodden buckaroo dnds buckyorc eounda src-rc- eye syc-s
MZDDTK CJQLAPZZ DKDM CJQLNZPQ TZJKDA MPQBPQB TNT MNQBM

||2 TNT ewe

*** -- Solution #2
       Score = 3100
sodden buckaroo dnds buckworc eounda src-rc- ewe swc-s
MZDDTK CJQLAPZZ DKDM CJQLNZPQ TZJKDA MPQBPQB TNT MNQBM

||2 TNT eve

*** -- Solution #3
       Score = 3100
sodden buckaroo dnds buckvorc eounda src-rc- eve svc-s
MZDDTK CJQLAPZZ DKDM CJQLNZPQ TZJKDA MPQBPQB TNT MNQBM

|1 MZDDTK toddle
|1 MZDDTK soffit
|1 MZDDTK joggle
|1 MZDDTK toggle
|1 MZDDTK pollen
||2 TNT eye

*** -- Solution #4
       Score = 3100

pollen buckaroo lnlp buckyorc eounla prc-rc- eye pyc-p
MZDDTK CJQLAPZZ DKDM CJQLNZPQ TZJKDA MPQBPQB TNT MNQBM
                         .
                         .
                         .

{ To save space, the next 15 trial solutions are omitted }
                         .
                         .
                         .

0 CJQLAPZZ nutshell
|1 MZDDTK bloody
||2 TNT did
```

```
*** -- Solution #20
       Score = 3100
bloody nutshell oyob nutsilet dluyoh bet-et- did bit-b
MZDDTK CJQLAPZZ DKDM CJQLNZPQ TZJKDA MPQBPQB TNT MNQBM

||2 TNT dad

*** -- Solution #21
       Score = 3100
bloody nutshell oyob nutsalet dluyoh bet-et- dad bat-b
MZDDTK CJQLAPZZ DKDM CJQLNZPQ TZJKDA MPQBPQB TNT MNQBM

|1 MZDDTK gloomy
0 CJQLAPZZ mandrill
|1 CJQLNZPQ mandolin
||2 TZJKDA slater

*** -- Solution #22
       Score = 3100
-leest mandrill ete- mandolin slater -in-in- sos -on--
MZDDTK CJQLAPZZ DKDM CJQLNZPQ TZJKDA MPQBPQB TNT MNQBM

||2 MZDDTK sleety

*** -- Solution #23
       Score = 4100
sleety mandrill eyes mandolin tlayer sin-in- tot son-s
MZDDTK CJQLAPZZ DKDM CJQLNZPQ TZJKDA MPQBPQB TNT MNQBM


||2 MZDDTK sleepy

*** -- Solution #24
       Score = 4100
sleepy mandrill eyes mandolin player sin-in- pop son-s
MZDDTK CJQLAPZZ DKDM CJQLNZPQ TZJKDA MPQBPQB TNT MNQBM
```

Session 5

Pseudorandomness

Chair: E. Bach, University of Wisconsin

On the Existence of Pseudorandom Generators

Oded Goldreich
Dept. of Computer Sc.
Technion
Haifa, Israel

Hugo Krawczyk
Dept. of Computer Sc.
Technion
Haifa, Israel

Michael Luby
Dept. of Computer Sc.
University of Toronto
Ontario, Canada

ABSTRACT – Pseudorandom generators (suggested and developed by Blum and Micali and Yao) are efficient deterministic programs that expand a randomly selected k-bit seed into a much longer pseudorandom bit sequence which is indistinguishable in polynomial time from an (equally long) sequence of unbiased coin tosses. Pseudorandom generators are known to exist assuming the existence of functions that cannot be efficiently inverted on the distributions induced by applying the function iteratively polynomially many times. This sufficient condition is also a necessary one, but it seems difficult to check whether particular functions, assumed to be one-way, are also one-way on their iterates. This raises the fundamental question whether the mere existence of one-way functions suffices for the construction of pseudorandom generators.

In this paper we present progress towards resolving this question. We consider ***regular*** functions, in which every image of a k-bit string has the same number of preimages of length k. We show that if a regular function is one-way then pseudorandom generators do exist. In particular, assuming the intractability of general factoring, we can now prove that pseudorandom generators do exist. Other applications are the construction of pseudorandom generators based on the conjectured intractability of decoding random linear codes, and on the assumed average case difficulty of combinatorial problems as subset-sum.

1. INTRODUCTION

In recent years randomness has become a central notion in the theory of computation. It is heavily used in the design of sequential, parallel and distributed algorithms, and is of course crucial to cryptography. Once so frequently used, randomness itself has become a resource, and economizing on the amount of randomness required for an application has become a natural concern. It is in this light that the notion of pseudorandom generators was first suggested and the following fundamental result was derived: the number of coin tosses used in any practical application (modeled by a polynomial time computation) can be decreased to an arbitrarily small power of the input length.

The key to the above informal statement is the notion of a pseudorandom generator suggested and developed by Blum and Micali [BM] and Yao [Y]. A *pseudorandom generator* is a deterministic polynomial time algorithm which expands short seeds into longer bit sequences, such that the output ensemble is polynomially-indistinguishable from the uniform probability distribution. More specifically, the generator (denoted G) expands a k-bit seed into a longer, say $2k$-bit, sequence so that for every polynomial time

Research done while the third author was visiting the Computer Science Department of the Technion. First author was supported by grant No. 86-00301 from the United States - Israel Binational Science Foundation (BSF), Jerusalem, Israel. Third author was partially supported by a Natural Sciences and Engineering Research Council of Canada operating grant No. A8092 and by a University of Toronto grant.

algorithm (distinguishing test) T, any constant $c>0$, and sufficiently large k

$$\left| Prob\left[T(G(X_k))=1\right] - Prob\left[T(X_{2k})=1\right] \right| \leq k^{-c},$$

where X_m is a random variable assuming as values strings of length m, with uniform probability distribution. It follows that the strings output by a pseudorandom generator G can substitute the unbiased coin tosses used by any polynomial time algorithm A, without changing the behavior of algorithm A in any noticeable fashion. This yields an equivalent polynomial time algorithm, A', which randomly selects a seed, uses G to expand it to the desired amount, and then runs A using the output of the generator as the random source required by A. The theory of pseudorandomness was further developed to deal with function generators and permutation generators and additional important applications to cryptography have emerged [GGM, LR]. The existence of such seemingly stronger generators was reduced to the existence of pseudorandom (string) generators.

In light of their practical and theoretical value, constructing pseudorandom generators and investigating the possibility of such constructions is of major importance. A necessary condition for the existence of pseudorandom generators is the existence of one-way functions (since the generator itself constitutes a one-way function). However, it is not known whether this necessary condition is sufficient. Instead, stronger versions of the one-wayness condition were shown to be sufficient. Before reviewing these results, let us recall the definition of a one-way function.

Definition 1: A function $f:\{0,1\}^* \rightarrow \{0,1\}^*$ is called *one-way* if it is polynomial time computable, but not "polynomial time invertible". Namely, there exists a constant $c>0$ such that for **any** probabilistic polynomial time algorithm A, and sufficiently large k

$$Prob\left[A(f(x),1^k) \notin f^{-1}(f(x))\right] > k^{-c}, \quad (*)$$

where the probability is taken over all x's of length k and the internal coin tosses of A, with uniform probability distribution.
(Remark: The role of 1^k in the above definition is to allow algorithm A to run for time polynomial in the length of the preimage it is supposed to find. Otherwise, any function which shrinks the input by more than a polynomial amount would be considered one-way.)

1.1. Previous Results

The first pseudorandom generator was constructed and proved valid, by Blum and Micali, under the assumption that the discrete logarithm problem is intractable on a non-negligible fraction of the instances [BM]. In other words, it was assumed that exponentiation modulo a prime (i.e. the 1-1 mapping of the triple (p,g,x) to the triple $(p,g,g^x \bmod p)$, where p is prime and g is a primitive element in Z_p^*) is one-way. Assuming the intractability of factoring integers of the form $N=p\cdot q$, where p and q are primes and $p\equiv q\equiv 3 \bmod 4$, a simple pseudorandom generator exists [BBS, ACGS] (1). Under this assumption the permutation, defined over the quadratic residues by modular squaring, is one-way.

Yao has presented a much more general condition which suffices for the existence of pseudorandom generators; namely, the existence of one-way permutations [Y] (2).

Levin has weakened Yao's condition, presenting a necessary and sufficient condition for the existence of pseudorandom generators [L]. Levin's condition, hereafter referred to as *one-way on iterates*, can be derived from Definition 1 by substituting the following line instead of line (*)

$$(\forall i, 1 \le i < k^{c+2}) \; Prob\left[A(f^{(i)}(x), 1^k) \notin f^{-1}(f^{(i)}(x)) \right] > k^{-c},$$

where $f^{(i)}(x)$ denotes f iteratively applied i times on x. (As before the probability is taken uniformly over all x's of length k.) Clearly, any one-way permutation is one-way on its iterates. It is also easy to use any pseudorandom generator in order to construct a function which satisfies Levin's condition.

Levin's condition for the construction of pseudorandom generators is somewhat cumbersome. In particular, it seems hard to test the plausibility of the assumption that a particular function is one-way on its iterates. Furthermore, *it is an open question whether Levin's condition is equivalent to the mere existence of one-way functions*.

1.2. Our Results

In this paper we present progress towards resolving the above open problem. We consider "regular" functions, in which every element in the range has the same number of preimages. More formally, we use the following definition.

Definition 2: A function f is called *regular* if there is a function $m(\cdot)$ such that for every n and for every $x \in \{0,1\}^n$ the cardinality of $f^{-1}(f(x)) \cap \{0,1\}^n$ is $m(n)$.

Clearly, *every* 1-1 function is regular (with $m(n)=1$, $\forall n$). Our main result is

Main Theorem: *If there exists a regular one-way function then there exists a pseudorandom generator.*

A special case of interest is of 1-1 one-way functions. The sufficiency of these functions for constructing pseudorandom generators does not follow from previous works. In particular, Yao's result concerning one-way permutations does not extend to 1-1 one-way functions.

1) A slightly more general result, concerning integers with all prime divisors congruent to 3 mod 4, also holds [CGG].

2) In fact, Yao's condition is slightly more general. He requires that f is 1-1 and that there exists a probability ensemble Π which is invariant under the application of f and that inverting f is "hard on the average" when the input is chosen according to Π.

Regularity appears to be a simpler condition than the intractability of inverting on the function's iterates. Furthermore, many natural functions (e.g. squaring modulo an integer) are regular and thus, using our result, a pseudorandom generator can be constructed assuming that any of these functions is one-way. In particular, if factoring is weakly intractable (i.e. every polynomial time factoring algorithm fails on a non-negligible fraction of the integers) then pseudorandom generators do exist. This result was not known before. (It was only known that the intractability of factoring a special subset of the integers implies the existence of a pseudorandom generator.) Using our results, we can construct pseudorandom generators based on the (widely believed) conjecture that decoding random linear codes is intractable, and on the assumed average case difficulty of combinatorial problems as subset-sum.

The main theorem is proved essentially by transforming any given regular one-way function into a function that is one-way on its iterates (and then applying Levin's result [L]).

It is interesting to note that not every (regular) one-way function is "one-way on its iterates". To emphasis this point, we show (in Appendix A) that from a (regular) one-way function we can construct a (regular) one-way function which is easy to invert on the distribution obtained by applying the function *twice*. The novelity of this work is in presenting *a direct way to construct a function which is one-way on its iterates from any regular one-way function (which is not necessarily one-way on its iterates).*

1.3. Subsequent Results

Recent results of Impagliazzo, Levin and Luby extend our results in two directions [ILL]. First, they generalize the regularity condition deriving a necessary and sufficient condition for the existence of pseudorandom generators. The new condition requires that the function f is one-way on a distribution induced by a function h, while the distribution induced by $f \circ h$ has almost the same entropy as the distribution induced by h. Second, they show that using non-uniform definitions of one-way functions and pseudorandom generator, yields their equivalence.

2. MAIN RESULT

2.0. Preliminaries

In the sequel we make use of the following definition of *strongly* one-way function. (When referring to Definition 1, we shall call the function *weak* one-way or simply one-way).

Definition 3: A polynomial time computable function $f : \{0,1\}^* \rightarrow \{0,1\}^*$ is called *strongly one-way* if for any probabilistic polynomial time algorithm A, any positive constant c, and sufficiently large k,

$$Prob\left[A(f(x),1^k) \in f^{-1}(f(x))\right] < k^{-c},$$

where the probability is taken over all x's of length k and the internal coin tosses of A, with uniform probability distribution.

Theorem (Yao [Y]): There exists a strong one-way function if and only if there exists a (weak) one-way function. Furthermore, given a one-way function, a strong one can be constructed.

It is important to note that Yao's construction preserves the regularity of the function. Thus, we may assume without loss of generality, that we are given a function f which is strongly one-way and regular.

For the sake of simplicity, we assume f is *length preserving* (i.e. $\forall x, |f(x)|=|x|$). Our results hold also without this assumption (see subsection 2.6).

Notation: For a finite set S, the notation $s \in_R S$ means that the element s is randomly selected from the set S with uniform probability distribution.

2.1. Levin's Criterion: A Modified Version

The proof of the Main Theorem relies on the transformation of a function which is one-way and regular into a function which satisfies a variant of Levin's condition (i.e., being one-way on iterates). The modified condition, relating to functions which leave the first part of their argument unchanged, requires that the function is one-way on a number of iterates which exceeds the length of the second part of its argument. (Levin has required that the function is one-way on a number of iterations exceeding the length of the entire argument.) A precise statement can be found in Lemma 1 bellow. Before proving the sufficiency of the modified condition for constructing pseudorandom generators, we recall the basic ideas behind Levin's condition.

Levin's condition is motivated by Blum-Micali scheme for the construction of pseudorandom generators [BM]. This scheme uses two basic elements. The first, a (strongly) one-way function f, and the second, a boolean predicate $b(\cdot)$ called a "hard-core" of the function f. (Roughly speaking, a Boolean function $b(\cdot)$ is a *hard-core predicate of* f, if it is polynomial time computable, but no polynomial time probabilistic algorithm given $f(x)$, for randomly selected x, can compute the value of $b(x)$ with a probability significantly better than 1/2). A pseudorandom generator G is constructed in the following way. On input x (the seed), the generator G applies iteratively the one-way function $f(\cdot)$ on x for t $(= poly(|x|))$ times (i.e. $f(x), f^{(2)}(x), \ldots, f^{(t)}(x)$). In each application of f, the predicate $b(f^{(i)}(x))$ is computed and the resultant bit is output by the generator. That is, G outputs a string of length t. Blum and Micali show that the above sequence of bits is unpredictable when presented in reverse order (i.e. $b(f^{(t)}(x))$ first and $b(f^{(1)}(x))$ last), provided that the boolean function $b(\cdot)$ is a hard-core predicate on the distribution induced by the iterates $f^{(i)}, 0 \le i \le t$. The unpredictability of the sequence is proved by showing that an algorithm which succeeds to predict the next bit of the sequence with probability better than one half can be transformed into an algorithm for "breaking" the hard-core of the function f. Finally applying Yao's Theorem [Y] that unpredictable sequences are pseudorandom we get that the above G is indeed a pseudorandom generator.

The hard part of the proof of Levin's Theorem (namely, that the existence of a function f being one-way on iterates implies the existence of pseudorandom generators) is in showing that the existence of a one-way function implies the existence of a hard-core predicate on the iterates of another function.[3] In order to construct this bit, the original function f is modified into a new one-way function f', and the hard-core predicate $b(\cdot)$ is constructed with respect to the new f'. The function $f'(x)$ consists of the parallel application of the original f on many copies, i.e. $f'(x_1,...,x_s)=(f(x_1),...,f(x_s))$. The x_i's are of equal size, say n, and Levin's construction uses a number of copies $s(n)$ which is any function that grows faster than $c \cdot \log n$, for any constant c. For constructing a pseudorandom generator, following Blum-Micali scheme, f' should be iterated on a seed of length k for at least $k+1$ iterations [4]. Recall that the seed has the form $(x_1, \ldots, x_{s(n)})$. Thus in order to have f' which is one-way for $n \cdot s(n)+1$ iterations, we need that the original function f is one-way for this number of iterations when applied to the substrings x_i of length n. Let $\tau(n)$ be a function which is an upper bound on the function $n \cdot s(n)+1$. For simplicity we may assume $\tau(n)=n^2$. Thus, we get that in order to construct a pseudorandom generator it suffices to have a function f which is strongly one-way for $\tau(n)$ iterations when applied to strings of length n. This is Levin's sufficient (and necessary) condition for the existence of pseudorandom generators. (Observe that Levin's condition as presented in section 1.1 refers to weak one-way functions, and then a greater number of iterations is required).

In our work we use the concept of "one-wayness on iterates" in a slightly modified way. We consider a function $F(\cdot,\cdot)$ defined as

$$F(h,x)=(h, F_0(h,x)) \qquad (*)$$

That is, F applies a function F_0 on its arguments and concatenates the first argument h to this result. The advantage of considering this kind of functions is that in order to construct a pseudorandom generator based on this function, it suffices to require that the function F is strongly one-way for $\tau(|x|)$ iterations, instead of the $\tau(|h|+|x|)$ iterations required by the straightforward application of Levin's result. The way we prove the sufficiency of this condition is as follows. First, we use Levin's modification of the function F into a new function $F'(h',x')$ for which a hard-core predicate does exist. (This is the same as the transformation from f to f' in the above description of Levin's construction). An important and simple observation is that F' preserves the form (*). Then, the

3) This part of the proof can be avoided using a recent result of Goldreich and Levin [GL]. This result states that any function $f'(x,r)=(f(x),r)$, where $|x|=|r|$, has a hard-core predicate for the uniform distribution on r and any distribution on x for which f is one-way.

4) A notable property of pseudorandom generators is that in order to have a generator which expands strings to any polynomial length, it suffices to construct a generator which expands strings of length k into strings of length $k+1$. This generator can be iteratively applied for polynomially many times without harming the pseudorandomness of its output [GrM].

function F' is applied by the generator G for at least $|x'|+1$ iterations. Note that F' remains one-way for all these iterations, as the original $F(h,x)$ is one-way for $\tau(|x|)$ iterates, and then $|x'|+1$ pseudorandom bits can be computed by using the hard-core of F'. The output of G will be the string h' concatenated with the above $|x'|+1$ pseudorandom bits. That is G expands its seed into a string which is at least one bit longer. The pseudorandomness of the output string is proved by noting that it is unpredictable. This is true for the h' part because it was chosen as a truly random string, and true for the other bits as guaranteed by Blum-Micali scheme. Namely, the ability to predict any of these bits would compromise the security of the hard-core of F'. The fact that the string h' is output do not help the predictor because the hard-core predicate of F' is unapproximable even when given h'. Recall that when given $F'(h',x')$ the string h' is explicitly presented.

Summarizing we get the following Lemma.

Lemma 1: Let $\tau(n)=n^2$. A sufficient condition for the existence of a pseudorandom generator is the existence of a function F of the form

$$F(h,x) = (h, F_0(h,x))$$

such that F is strongly one-way for $\tau(|x|)$ iterations.

2.2. Main Ideas

We prove the Main Theorem by transforming any regular and (strongly) one-way function into a new strongly one-way function f' for which the conditions of Lemma 1 hold.

The following are the main ideas behind this construction. Since the function f is strongly one-way, any algorithm trying to invert f can succeed only with negligible probability. Here the probability distribution on the range of f is induced by choosing a random element from the domain and applying f. However, this condition says nothing about the capability of an algorithm to invert f when the distribution on the range is substantially different. For example, there may be an algorithm which is able to invert f if we consider the distribution on the range elements induced by choosing a random element from the domain and applying f twice or more (see Appendix A). To prevent this possibility, we "randomly" redistribute, after each application of f, the elements in the range to locations in the domain. We prove the validity of our construction by showing that the probability distribution induced on the range of f by our "random" transformations (and the application of f) is close to the distribution induced by the first application of f.

The function f' we construct must be deterministic, and therefore the "random" redistribution must be deterministic (i.e. uniquely defined by the input to f'). To achieve this, we use high quality hash functions. More specifically, we use hash functions which map n-bit strings to n-bit strings, such that the locations assigned to the strings by a randomly selected hash function are uniformly distributed and n-wise independent. For properties and implementations of such functions see [CW, J, CG, Lu]. We denote this set of hash functions by $H(n)$. Elements of $H(n)$ can be described by bit strings of length n^2. In the sequel $h(\in H(n))$ refers to both the hash function and to its representation.

2.3. The Construction of f'

We view the input string to f' as containing two types of information. The first part of the input is the description of hash functions that implement the "random" redistributions and the other part is interpreted as the input for the original function f.

The following is the definition of the function f':

$$f'(h_0, \cdots, h_{t(n)-1}, i, x) = (h_0, \cdots, h_{t(n)-1}, i^+, h_i(f(x)))$$

where $x \in \{0,1\}^n$, $h_j \in H(n)$, $0 \le i \le t(n)-1$. The function $t(n)$ is a polynomial in n, and i^+ is defined as $(i+1) \bmod t(n)$.

The rest of this section is devoted to the proof of the following theorem.

Theorem 2: Let f be a regular and strongly one-way function. Then the function f' defined above is strongly one-way for $t(n)$ iterations on strings x of length n.

Our Main Theorem follows from Theorem 2 and Lemma 1 by choosing $t(n) \ge \tau(n)$.

Let $h_0, h_1, \cdots, h_{t(n)-1}$ be $t(n)$ functions from the set $H(n)$. For $r=1, \cdots, t(n)$, let g_r be the function $g_r = f\, h_{r-1} f\, h_{r-2} f \cdots h_0 f$ acting on strings of length n, let $G_r(n)$ be the set of all such functions g_r, let g be $g_{t(n)}$ and let $G(n)$ be the set of such functions g. From the above description of the function f' it is apparent that the inversion of an iterate of f' boils down to the problem of inverting f when the probability distribution on the range of f is $g_r(x)$ where $x \in_R \{0,1\}^n$. We show that, for most $g \in G(n)$, the number of preimages under g for each element in its range is close (up to a polynomial factor) to the number of preimages for the same range element under f. This implies that the same statement is true for most $g_r \in G_r(n)$ for all $r=1, \cdots, t(n)$. The proof of this result reduces to the analysis of the combinatorial game that we present in the next subsection.

2.4. The game

Consider the following game played with M balls and M cells where $t(n) \ll M \le 2^n$. Initially each cell contains a single ball. The game has $t(n)$ iterations. In each iteration, cells are mapped randomly to cells by means of an independently and randomly selected hash function $h \in_R H(n)$. This mapping induces a transfer of balls so that the balls residing (before an iteration) in cell σ are transferred to cell $h(\sigma)$. We are interested in bounding the probability that some cells contain "too many" balls when the process is finished. We show that after $t(n)$ iterations, for $t(n)$ a polynomial, the probability that there is any cell containing more than some polynomial in n balls is negligibly small (i.e. less than any polynomial in n fraction).

We first proceed to determine a bound on the probability that a specific set of n balls is mapped after $t(n)$ iterations to a single cell.

Lemma 3: The probability that a specific set of n balls is mapped after $t(n)$ iterations to the same cell is bounded above by $p(n) = \left(\frac{n \cdot t(n)}{M} \right)^{n-1}$.

Proof: Let $B = \{b_1, b_2, \cdots, b_n\}$ be a set of n balls. Notice that each execution of the game defines for every ball b_i a path through $t(n)$ cells. In particular, fixing $t(n)$ hash functions $h_0, h_1, \cdots, h_{t(n)-1}$, a path corresponding to each b_i is determined. Clearly, if two such paths intersect at some point then they coincide beyond this point. We modify these paths in the following way. The initial portion of the path for b_i that does not intersect the path of any smaller indexed ball is left unchanged. If the path for b_i intersects the path for b_j for some $j < i$ then the remainder of the path for b_i is chosen randomly and independently of the other paths from the point of the first such intersection.

Because the functions h_i are chosen totally independently of each other and because each of them has the property of mapping cells in an n-independent manner, it follows that the modified process just described is equivalent to a process in which a totally random path is selected for each ball in B. Consider the modified paths. We say that two balls b_i and b_j *join* if and only if their corresponding paths intersect. Define *merge* to be the reflexive and transitive closure of the relation join (over B). The main observation is that if $h_0, h_1, \cdots, h_{t(n)-1}$ map the balls of B to the same cell, then $b_1, b_2, \cdots, b_n$ are all in the same equivalence class with respect to the relation merge. In other words, the probability that the balls in B end up in the same cell in the original game is bounded above by the probability that the merge relation has a single equivalence class (containing all of B). Let us now consider the probability of the latter event.

If the merge relation has a single equivalence class then the join relation defines a connected graph with the n balls as vertices and the join relation as the set of edges. The "join graph" is connected if and only if it contains a spanning tree. Thus, an upper bound on the probability that the "join graph" is connected is obtained by the sum of the probabilities of each of the possible spanning trees which can be embedded in the graph. Each particular tree has probability at most $(t(n)/M)^{n-1}$ to be embedded in the graph ($t(n)/M$ is an upper bound on the probability of each edge to appear in the graph). Multiplying this probability by the (Cayley) number of different spanning trees (n^{n-2} cf. [E, Sec. 2.3]), the lemma follows. □

A straightforward upper bound on the probability that there is some set of n balls which are merged is the probability that n specific balls are merged multiplied by the number of possible distinct subsets of n balls. Unfortunately, this bound is worthless (as $\binom{M}{n} \cdot p(n) > 1$ (This phenomena is independent of the choice of the parameter n.). Instead we use the following technical lemma.

Lemma 4: Let S be a finite set, and let Π denote a partition of S. Assume we have a probability distribution on partitions of S. For every $A \subseteq S$, we define $\chi_A(\Pi)=1$ if A is contained in a single class of the partition Π and $\chi_A(\Pi)=0$ otherwise. Let n and n' be integers such that $n < n'$. Let $p(n)$ be an upper bound on the maximum over all $A \subseteq S$ such that $|A| = n$ of the probability that $\chi_A = 1$. Let $q(n')$ be an upper bound on the probability that there exists some $B \subseteq S$ such that $|B| \geq n'$ and $\chi_B = 1$. Then

$$q(n') \leq \frac{\binom{|S|}{n} \cdot p(n)}{\binom{n'}{n}}$$

Proof: For $B \subseteq S$ we define $\xi_B(\Pi)=1$ if B is exactly a single class of the partition Π and $\xi_B(\Pi)=0$ otherwise. Fix a partition Π. Observe that every B, $|B| \geq n'$, for which $\xi_B(\Pi)=1$, contributes at least $\binom{n'}{n}$ *different* subsets A of size n for which $\chi_A = 1$. Thus we get that

$$\binom{n'}{n} \cdot \sum_{B \subseteq S, |B| \geq n'} \xi_B(\Pi) \leq \sum_{A \subseteq S, |A| = n} \chi_A(\Pi)$$

Dividing both sides of this inequality by $\binom{n'}{n}$, and averaging according to the probability distribution on the partitions Π, the left hand side is an upper bound for $q(n')$, while the right hand side is bounded above by $\dfrac{\binom{|S|}{n} \cdot p(n)}{\binom{n'}{n}}$. □

Remark: Lemma 4 is useful in situations when the ratio $\frac{p(n)}{p(n')}$ is smaller than $\binom{|S|-n}{n'-n}$. Assuming that $n' \ll |S|$, this happens when $p(n)$ is greater than $|S|^{-n}$. Lemma 3 is such a case, and thus the application of Lemma 4 is useful.

Combining Lemmi 3 and 4, we get

Theorem 5: Consider the game played for $t(n)$ iterations. Then, the probability that there is $4t(n) \cdot n^2 + n$ balls which end up in the same cell is bounded above by 2^{-n}.

Proof: Let S be the set of M balls in the above game. Each game defines a partition of the balls according to their position after $t(n)$ iterations. The probability distribution on these partitions is induced by the uniform choice of the mappings h. Theorem 5 follows by using Lemma 4 with $n' = 4t(n) \cdot n^2 + n$, and the bound $p(n)$ of Lemma 3. □

2.5. Proof of Theorem 2

We now apply Theorem 5 to the analysis of the function f'. As before, let $G(n)$ be the set of functions of the form $g = f\, h_{t(n)-1} f \cdots h_0 f$. The functions $h = h_j$ are hash functions used to map the range of f to the domain of f. We let $h_0, \cdots, h_{t(n)-1}$ be randomly chosen uniformly and independently from $H(n)$, and this induces a probability distribution on $G(n)$. Denote the range of f (on strings of length n) by

$R(n) = \{z_1, z_2, \ldots, z_M\}$. Let each z_i represent a cell. Consider the function h as mapping cells to cells. We say that h maps the cell z_i to the cell z_j if $h(z_i) \in f^{-1}(z_j)$, or in other words $f(h(z_i)) = z_j$. By the regularity of the function f, we have that the size of $f^{-1}(z_i)$ (which we have denoted by $m(n)$) is equal for all $z_i \in R(n)$, and therefore the mapping induced on the cells is uniform. It is now apparent that $g \in_R G(n)$ behaves exactly as the random mappings in the game described in Section 2.4, and thus Theorem 5 can be applied. We get

Lemma 6: There is a constant c_0, such that for any constant $c > 0$ and sufficiently large n

$$Prob\left[\exists z \text{ with } |g^{-1}(z)| \geq n^{c_0} \cdot m(n)\right] \leq \frac{1}{n^c},$$

where $g \in_R G(n)$.

Let us denote by $G'(n)$ the set of functions $g \in G(n)$ such that for all z in the range of f, $|g^{-1}(z)| < n^{c_0} \cdot m(n)$. By the above lemma, $G'(n)$ contains almost all of $G(n)$. It is clear that if $g \in G'(n)$ then for all z in the range of f and for all $r = 1, \cdots, t(n)$ the function g_r defined by the first r iterations of g satisfies $|g_r^{-1}(z)| < n^{c_0} \cdot m(n)$.

Lemma 7: For any probabilistic polynomial time algorithm A, for any positive constant c and sufficiently large n and for all $r = 1, \cdots, t(n)$,

$$Prob(A(g_r, z) \in f^{-1}(z)) < n^{-c}$$

where $g_r \in_R G_r(n)$ and $z = g_r(x), x \in_R \{0,1\}^n$.

Proof: We prove the claim for $r = t(n)$ and the claim for $r = 1, \cdots, t(n)$ follows in an analogous way. Assume to the contrary that there is a probabilistic polynomial time algorithm A and a constant c_A such that $Prob(A(g, z) \in f^{-1}(z)) > n^{-c_A}$, where $g \in_R G(n)$ and $z = g(x), x \in_R \{0,1\}^n$.

By using A, we can demonstrate an algorithm A' that inverts f, contradicting the one-wayness of f. The input to A' is $z = f(x)$ where $x \in_R \{0,1\}^n$. A' chooses $g \in_R G(n)$ and outputs $A(g, z)$. We show that A' inverts f with non-negligible probability. By assumption there is a non-negligible subset $G''(n)$ of $G'(n)$ such that, for each $g \in G''(n)$, A succeeds with significant probability to compute a $y \in f^{-1}(z)$ where $z = g(x)$ and $x \in_R \{0,1\}^n$. Since $g \in G'(n)$, for all z in the range of f the probability induced by g on z differs by at most a polynomial factor in n from the probability induced by f. Thus, for $g \in G''(n)$, A succeeds with significant probability to compute a $y \in f^{-1}(z)$ where $z = f(x)$ and $x \in_R \{0,1\}^n$. This is exactly the distribution of inputs to A', and thus A' succeeds to invert f with non-negligible probability, contradicting the strong one-wayness of f. □

The meaning of Lemma 7 is that the function f is hard to invert on the distribution induced by the functions $g_r, r = 1, \ldots, t(n)$, thus proving the strong one-wayness of the function f' for $t(n)$ iterations. Theorem 2 follows.

2.6. Extensions

In the above exposition we assumed for simplicity that the function f is length preserving, i.e. $x \in \{0,1\}^n$ implies that the length of $f(x)$ is n. This condition is not essential to our proof and can be dispensed with in the following way. If f is not length preserving then it can be modified to have the following property: For every n, there is an n' such that $x \in \{0,1\}^n$ implies that the length of $f(x)$ is n'. This modification can be carried out using a padding technique that preserves the regularity of f. We can then modify our description of f' to use hash functions mapping n'-bit strings to n-bit strings. Alternatively, we can transform the above f into a length preserving and regular function f by defining $f(xy)=f(x)$, where $|x|=n$, $|y|=n'-n$.

For the applications in Section 3, and possibly for other cases, the following extension (referred to as *semi-regular*) is useful. Let $\{f_x\}_{x \in \{0,1\}^*}$ be a family of regular functions, then our construction can be still applied to the function f defined as $f(x,y)=(x,f_x(y))$ The idea is to use the construction for the application of the function f_x, while keeping x unchanged.

Another extension is a relaxation of the regularity condition. A useful notion in this context is the histogram of a function.

Definition 4: The *histogram* of the function $f: \{0,1\}^* \to \{0,1\}^*$ is a function $hist_f: N \times N \to N$ such that $hist_f(n,k)$ is the cardinality of the set

$$\{x \in \{0,1\}^n : \left\lfloor \log_2 |f^{-1}(f(x))| \right\rfloor = k\}$$

Regular functions have trivial histograms. Let f be a regular function such that for all $x \in \{0,1\}^n$, $|f^{-1}(f(x))| = m(n)$. The histogram satisfies $hist_f(n,k)=2^n$ for $k = \left\lfloor \log_2(m(n)) \right\rfloor$ and $hist_f(n,k)=0$ otherwise. Weakly regular functions have slightly less dramatic histograms.

Definition 5: The function f is *weakly regular* if there is a polynomial $p(\cdot)$ and a function $b(\cdot)$ such that the histogram of f satisfies (for all n)

i) $hist_f(n, b(n)) \geq \dfrac{2^n}{p(n)}$

ii) $\displaystyle\sum_{k=b(n)+1}^{n} hist_f(n,k) < \frac{2^n}{(n \cdot p(n))^2}$

Clearly, this definition extends the original definition of regularity. Using our techniques one can show that the existence of weakly regular strongly one-way functions implies the existence of pseudorandom generators.

Observe that if the $b(n)$-th level of the histogram contains all of the 2^n strings of length n then we can apply a similar analysis as done for the regular case. The only difference is that we have to analyze the game of subsection 2.4 not for cells of equal size, but for cells that differ in their size by a multiplicative factor of at most two. Similar arguments hold when considering the case where the $b(n)$-th level of the histogram contains at least $1/p(n)$ of the strings and the rest of strings lie below this level (i.e. $hist_f(n,k)=0$, for $k>b(n)$). Note that the "small" balls of low levels cannot cause the cells of the $b(n)$-th level to grow significantly. On the other hand, for balls bellow level $b(n)$ nothing is guaranteed. Thus, we get that in this case the function f' we construct is weakly one-way on its iterates. More precisely, it is hard to invert on its iterates for at least a $1/p(n)$ fraction of the input strings. In order to use this function for generating pseudorandom bits, we have to transform it into a strongly one-way function. This is achieved following Yao's construction [Y] by applying f' in parallel on many copies. For the present case the number of copies could be any function of n which grows faster than $c \cdot p(n) \cdot \log n$, for any constant c. This increases the number of iterations for which f' has to remain one-way by a factor equal to the number of copies used in the above transformation. That is, the number $t(n)$ of necessary iterates increases from the original requirement of $\tau(n)$ (see section 2.1) to a quantity which is greater than $c \cdot p(n) \cdot \tau(n) \cdot \log n$, for any constant c. Choosing this way the function $t(n)$ in the definition of f' in section 2.3, we get f' which is one-way for the right number of iterations.

Finally, consider the case in which there exist strings above the $b(n)$-th level. When considering the game of subsection 2.4 we want to show that, also in this case, most of the cells of the $b(n)$-th level do not grow considerably. This is guaranteed by condition (ii) in Definition 5. Consider the worst case possibility in which in every iteration the total weight of the "big" balls (those above level $b(n)$) is transferred to cells of the $b(n)$-th level. After $t(n)$ iterations this causes a concentration of "big" balls in the $b(n)$-th level having a total weight of at most $t(n) \cdot \frac{2^n}{(n \cdot p(n))^2}$. Choosing $t(n)=\frac{1}{2}p(n)n^2$ this weight will be at most $\frac{2^n}{2p(n)}$. But then one half of the weight in the $b(n)$-th level remains concentrated in balls that were not effected by the "big" balls. In other words we get that the function f' so constructed is one-way for $t(n)$ iterations on $\frac{1}{2p(n)}$ of the input strings. Applying Yao's construction , as explained above, we get a function f' which fill the criterion of Lemma 1 and then suitable for the construction of pseudorandom generators.

Further Remarks:

1) A finer analysis allows to substitute the exponent 2, in condition (ii) of Definition 5, by any constant greater than 1.

2) The entire analysis holds when defining histograms with polynomial base (instead of base 2). Namely, $hist_f(n,k)$ is the cardinality of the set

$$\{x \in \{0,1\}^n : \left\lfloor \log_{Q(n)} |f^{-1}(f(x))| \right\rfloor = k\}$$

where $Q(n)$ is a polynomial.

3. APPLICATIONS : Pseudorandom Generators Based on Particular Intractability Assumptions

In this section we apply our results in order to construct pseudorandom generators (*PRG*s) based on the assumption that one of the following computational problems is "hard on a non-negligible fraction of the instances".

3.1. PRG Based on the Intractability of the General Factoring Problem

It is known that pseudorandom generators can be constructed assuming the intractability of factoring integers of a special form [Y]. More specifically, in [Y] it is assumed that any polynomial time algorithm fails to factor a non-negligible fraction of integers that are the product of primes congruent to 3 modulo 4. With respect to such an integer N, squaring modulo N defines a permutation over the set of quadratic residues mod N, and therefore the intractability of factoring (such N's) yields the existence of a one-way permutation [R]. It was not known how to construct a one-way permutation or a pseudorandom generator assuming that factoring a non-negligible fraction of *all* the integers is intractable. In such a case modular squaring is a one-way function, but this function does not necessarily induce a permutation. Fortunately, modular squaring is a semi-regular function (see subsection 2.6), so we can apply our results.

Assumption IGF (*Intractability of the General Factoring Problem*): There exists a constant $c>0$ such that for **any** probabilistic polynomial time algorithm A, and sufficiently large k

$$Prob\left[\ A(N) \text{ does not split } N\ \right] > k^{-c},$$

where $N \in_R \{0,1\}^k$.

Corollary 8: The IGF assumption implies the existence of pseudorandom generators.

Proof: Define the following function $f(N,x)=(N,x^2 \bmod N)$. Clearly, this function is semi-regular. The one-wayness of the function follows from IGF (using Rabin's argument [R]). Using an extension of Theorem 2 (see subsection 2.6) the corollary follows. □

Subsequently, J. (Cohen) Benaloh has found a way to construct a one-way permutation based on the IGF assumption. This yields an alternative proof of Corollary 8.

3.2. PRG Based on the Intractability of Decoding Random Linear Codes

One of the most outstanding open problems in coding theory is that of decoding random linear codes. Of particular interest are random linear codes with constant information rate which can correct a constant fraction of errors. An *(n,k,d)-linear code* is an k-by-n binary matrix in which the bit-by-bit XOR of any subset of the rows has at least d ones. The Gilbert-Varshamov bound for linear codes guarantees the existence of such a code provided that $k/n < 1-H_2(d/n)$, where H_2 is the binary entropy function [McS, ch.

1, p. 34]. The same argument can be used to show (for every $\varepsilon > 0$) that if $k/n < 1 - H_2((1+\varepsilon) \cdot d/n)$, then almost all k-by-n binary matrices constitute (n,k,d)-linear codes.

We suggest the following function $f : \{0,1\}^* \rightarrow \{0,1\}^*$. Let C be an k-by-n binary matrix, $x \in \{0,1\}^k$, and $e \in E_t^n \subseteq \{0,1\}^n$ be a binary string with at most $t = \lfloor (d-1)/2 \rfloor$ ones, where d satisfies the condition of the Gilbert-Varshamov bound (see above). Clearly E_t^n can be uniformly sampled by an algorithm S running in time polynomial in n (i.e. $S : \{0,1\}^{poly(n)} \rightarrow E_t^n$). Let $r \in \{0,1\}^{poly(n)}$ be a string such that $S(r) \in E_t^n$. Then,

$$f(C,x,r) = (C, C(x) + S(r)),$$

where $C(x)$ is the codeword of x (i.e. $C(x)$ is the vector resulting by the matrix product xC). One can easily verify that f just defined is semi-regular (i.e. $f_C(x,r) = C(x) + S(r)$ is regular for all but a negligible fraction of the C's). The vector $xC + e$ $(e = S(r))$ represents a codeword perturbed by the error vector e.

Assumption IDLC (*Intractability of Decoding Random Linear Codes*): There exists a constant $c > 0$ such that for **any** probabilistic polynomial time algorithm A, and sufficiently large k

$$Prob\Big[\, A(C, C(x)+e) \neq x \,\Big] > k^{-c},$$

where C is a randomly selected k-by-n matrix, $x \in_R \{0,1\}^k$ and $e \in_R E_t^n$.

Now, either assumption IDLC is false which would be an earth-shaking result in coding theory or pseudorandom generators do exist.

Corollary 9: The IDLC assumption implies the existence of pseudorandom generators.

Proof: The one-wayness of the function f follows from IDLC. Using an extension of Theorem 2 (see subsection 2.6) the corollary follows. □

3.3. PRG Based on the Average Difficulty of Combinatorial Problems

Some combinatorial problems which are believed to be hard on the average can be used to construct a regular one-way function and hence be a basis for a pseudorandom generator. Consider, for example, the *Subset-Sum Problem*.

Input: Modulo M, $|M| = n$, and $n+1$ integers $a_0, a_1, \cdots, a_n$ of length n-bit each.

Question: Is there a subset $I \subseteq \{1, \ldots, n\}$ such that $\sum_{i \in I} a_i \equiv a_0 (\bmod M)$

Conjecture: The above problem is hard on the average, when the a_i's and M are chosen uniformly in $[2^{n-1}, 2^n - 1]$.

Under the above conjecture, the following weakly-regular function is one-way

$$f_{SS}(a_1, a_2, \cdots, a_n, M, I) = (a_1, a_2, \cdots, a_n, M, (\sum_{i \in I} a_i \bmod M))$$

ACKNOWLEDGEMENTS

We are grateful to Josh (Cohen) Benaloh, Manuel Blum, Leonid Levin, Richard Karp, Charles Rackoff, Ronny Roth and Avi Wigderson for very helpful discussions concerning this work.

The first author wishes to express special thanks to Leonid Levin and Silvio Micali for infinitely many discussions concerning pseudorandom generators.

REFERENCES

[ACGS] W. Alexi, B. Chor, O. Goldreich and C.P. Schnorr, "RSA and Rabin Functions: Certain Parts Are As Hard As the Whole", *SIAM Jour. on Computing*, Vol. 17, 1988, pp. 194-209.

[BBS] L. Blum, M. Blum and M. Shub, *A Simple Secure Unpredictable Pseudo-Random Number Generator, SIAM Jour. on Computing*, Vol. 15, 1986, pp. 364-383.

[BM] Blum, M., and Micali, S., "How to Generate Cryptographically Strong Sequences of Pseudo-Random Bits", *SIAM Jour. on Computing*, Vol. 13, 1984, pp. 850-864.

[CW] Carter, J., and M. Wegman, "Universal Classes of Hash Functions", *JCSS*, 1979, Vol. 18, pp. 143-154.

[CG] Chor, B., and O. Goldreich, "On the Power of Two-Point Sampling", to appear in *Jour. of Complexity*.

[CGG] Chor, B., O. Goldreich, and S. Goldwasser, "The Bit Security of Modular Squaring Given Partial Factorization of the Modulos", *Advances in Cryptology - Crypto 85 Proceedings*, ed. H.C. Williams, Lecture Notes in Computer Science, 218, Springer Verlag, 1985, pp. 448- 457.

[DH] W. Diffie, and M. E. Hellman, "New Directions in Cryptography", *IEEE transactions on Info. Theory*, IT-22 (Nov. 1976), pp. 644-654

[E] S. Even, *Graph Algorithms,* Computer Science Press, 1979.

[GGM] Goldreich, O., S. Goldwasser, and S. Micali, "How to Construct Random Functions", *Jour. of ACM*, Vol. 33, No. 4, 1986, pp. 792-807.

[GKL] Goldreich, O., H. Krawczyk and M. Luby, "On the Existence of Pseudorandom Generators", *Proc. 29th IEEE Symp. on Foundations of Computer Science*, 1988, pp 12-24.

[GL] Goldreich, O., and L.A. Levin, "A Hard-Core Predicate for any One-Way Function", in preparations.

[GrM] Goldreich, O., and S. Micali, "The Weakest Pseudorandom Bit Generator Implies the Strongest One", manuscript, 1984.

[GM] Goldwasser, S., and S. Micali, "Probabilistic Encryption", *JCSS*, Vol. 28, No. 2, 1984, pp. 270-299.

[ILL] Impagliazzo, R., L.A., Levin and M.G. Luby, "Pseudorandom Number Generation from any One-Way Function", preprint, 1988.

[J] A. Joffe, "On a Set of Almost Deterministic k-Independent Random Variables", *the Annals of Probability*, 1974, Vol. 2, No. 1, pp. 161-162.

[L] L.A. Levin, "One-Way Function and Pseudorandom Generators", *Combinatorica*, Vol. 7, No. 4, 1987, pp. 357-363. A preliminary version appeared in *Proc. 17th STOC*, 1985, pp. 363-365.

[L2] L.A. Levin, "Homogeneous Measures and Polynomial Time Invariants", *Proc. 29th IEEE Symp. on Foundations of Computer Science*, 1988, pp 36-41.

[Lu] M. Luby, "A Simple Parallel Algorithm for the Maximal Independent Set Problem", *SIAM J. Comput.*, Vol. 15, No. 4, Nov. 1986, pp. 1036-1054.

[LR] M. Luby and C. Rackoff, "How to Construct Pseudorandom Permutations From Pseudorandom Functions", *SIAM Jour. on Computing*, Vol. 17, 1988, pp. 373-386.

[McS] McWilliams, F.J., and N.J.A. Sloane, *The Theory of Error Correcting Codes*, North-Holland Publishing Company, 1977.

[R] M.O. Rabin, "Digitalized Signatures and Public Key Functions as Intractable as Factoring", MIT/LCS/TR-212, 1979.

[RSA] R. Rivest, A. Shamir, and L. Adleman, "A Method for Obtaining Digital Signatures and Public Key Cryptosystems", *Comm. ACM*, Vol. 21, Feb. 1978, pp 120-126

[S] A. Shamir, "On the Generation of Cryptographically Strong Pseudorandom Sequences", *ACM Transaction on Computer Systems*, Vol. 1, No. 1, February 1983, pp. 38-44.

[Y] Yao, A.C., "Theory and Applications of Trapdoor Functions", *Proc. of the 23rd IEEE Symp. on Foundation of Computer Science*, 1982, pp. 80-91.

Appendix A: One-way functions which are not one-way on their iterates

Assuming that f is a (regular) one-way function, we construct a (regular) one-way function $\bar{f}$ which is easy to invert on the distribution obtained by iterating $\bar{f}$ twice. Assume for simplicity that f is length preserving (i.e. $|f(x)| = |x|$). Let $|x| = |y|$ and let

$$\bar{f}(xy) = 0^{|y|} f(x)$$

Clearly, $\bar{f}$ is one-way. On the other hand, for every $xy \in \{0,1\}^{2n}$, $\bar{f}(\bar{f}(xy)) = 0^n f(0^n)$ and $0^n f(0^n) \in \bar{f}^{-1}(0^n f(0^n))$.

ON THE RANDOMNESS OF LEGENDRE AND JACOBI SEQUENCES

Ivan Bjerre Damgård[1]

Aarhus University, Mathematical Institute.
Ny Munkegade,
DK 8000 Aarhus C,
Denmark.

Introduction

Most of the work done in cryptography in the last few years depend on the hardness of a few specific number theoretic problems, such as factoring, discrete log, etc. Since no one has so far been able to prove that these problems are genuinely hard, it is clearly of interest to find new candidates for hard problems. In this paper, we propose such a new candidate problem, namely the problem of predicting a sequence of consecutive Legendre (Jacobi) symbols modulo a prime (composite), when the starting point and possibly also the prime is unknown. Clearly, if this problem turns out to be hard, it can be used directly to construct a cryptographically strong pseudorandom bitgenerator. Its complexity seems to be unrelated to any of the well known number theoretical problems, whence it may be able to survive the discovery of fast factoring or discrete log algorithms. Although the randomness of Legendre sequences has part of the folklore in number theory at least since the thirties, they have apparently not been considered for use in cryptography before.

We first survey some known results about the distribution of squares and nonsquares modulo a prime. These results all support the assumption that Legendre sequences look random with respect to elementary statistical tests.

We then use Levin's Isolation Theorem [BoHi] to relate the complexity of predicting Legendre sequences to the complexity of predicting Jacobi sequences. The main result of this is that if Legendre sequences are unpredictable in a very weak sense, then Jacobi sequences modulo composites with enough prime factors are strongly unpredictable, as required for cryptographic strength.

We end the paper by giving results of some emphirical tests on Legendre sequences, carried out for primes of length 25 to 400 bits. Also some possibilities for generalizing the ideas are mentioned. These ideas give significant efficiency improvements over the basic Legendre generator.

---

[1]This research was supported by the Danish Natural Science Research Council.

1. Notation

Let p be a k-bit prime, and let $a \in Z_p^*$. We then define the Legendre symbol of a modulo p, $(\frac{a}{p})$ to be 1 if a is a square modulo p, and -1 if a is a non square. For convenience, we define $(\frac{0}{p})$ to be 1.

When $n = p_1 \cdots p_r$ is a k-bit composite with prime factors $p_1, \ldots, p_r$, we define the Jacobi symbol of a modulo n to be $(\frac{a}{n}) = (\frac{a}{p_1}) \cdots (\frac{a}{p_r})$, for $0 \le a \le n-1$.

The *Legendre sequence* with starting point a and length l is the +-1 sequence

$$(\frac{a}{p}), (\frac{a+1}{p}), \cdots, (\frac{a+l}{p}).$$

Jacobi sequences are defined correspondingly.

We can now state formally our basic problem:

Problem P1.
Let L be the Legendre sequence modulo p with starting point a and length $P(k)$, for some polynomial P. Given L (but not a or p), find $(\frac{a+P(k)+1}{p})$ □

Correspondingly for Jacobi symbols:

Problem P2.
Let J be the Jacobi sequence modulo n with starting point a and length $P(k)$, for a polynomial P. Given J, find $(\frac{a+P(k)+1}{n})$ □

2. Known Results On the Distribution of Squares Modulo a Prime

The distribution of quadratic residues and non residues has been studied at least since the end of the last century. One of the first major contributions was made by Davenport [Da]:

Let S be a finite sequence of +-1's of length l. Let $p(S)$ be the number of occurrences of S in the complete Legendre sequence of p, i.e. the number of $a \in Z_p^*$ such that

$$(\frac{a}{p}), (\frac{a+1}{p}), \cdots, (\frac{a+l}{p}) = S.$$

Davenport proved that

$$p(S) = \frac{p}{2^l} + O(p^{\varepsilon}),$$

where ε is a constant between 0 and 1 which is only a function of l. In other words: the distribution of subsequences of fixed length tends to the uniform distribution exponentially in $\log_2(p)$. A uniform distribution is of course what one would expect from a really random sequence.

Perron [Pe] proved a more specific result: let $SQ(p)$ be the set of quadratic residues in Z_p^*. Then for any a, the set $a+SQ(p)$ contains almost exactly as many squares as non squares, the difference being 0 or 1, depending on whether p is 1 or 3 modulo 4. A similar result holds for $Z_p^*-SQ(p)$. This has a number of immediate consequences:

- By setting $a=1$, we get $p(1,1)\approx p(-1,1)\approx p(1,-1)\approx p(-1,-1)$, where the differences are at most 1.
- In general, pairs of symbols separated by a fixed distance are uniformly distributed.
- Define a *block* to be a run of consecutive 1's or -1's. Then by setting $a=1$ we see that half the 1's in the Legendre sequence of p are at the end of a block, and similarly for -1's. Therefore the average length of a block is 2.

Later, Burde [Bu] extended Perron's result for p's congruent to 3 modulo 4. He obtained a system of linear equations with the number of occurrences of subsequences of a fixed length as unknowns. The rank of the system is quadratic as a function of the length of the subsequences considered, and therefore the equations quickly become insufficient to determine the complete distribution. 3 is the largest length for which it can be done, and as for length 2, the distribution is uniform, apart from an "error" of order p^{-1}.

Thus the results of Perron and Burde also support the assumption that Legendre sequences will look random with respect to elementary statistical tests.

From the work of Bach [Ba], one can get very interesting estimates on the distribution of subsequences whose length is allowed to grow with the size of the prime, in contrast with Davenport's results. This is based on results from algebraic geometry by Weil. For example, for p congruent to 3 modulo 4, one can obtain that

$$p(S) \le \frac{p}{2^l} + \frac{(l-1)\sqrt{p}}{2}$$

for any S of length l. Thus, there is a limit to "how bad" the distribution of subsequences can be, for example at least

$$2^l \frac{\sqrt{p}}{\sqrt{p} + 2^{l-1}(l-1)}$$

different subsequences of length l must occur in the complete Legendre sequence. In fact, the results are much more general, and can give information also about the distribution of other character values.

But since the bounds clearly get looser as the length l increases, we are still a long way from results that would imply the impossibility of predicting Legendre sequences in polynomial time.

Many other researchers (Eliott [El], Burgess[Bur]) have looked at this problem from other angles, typically they have been concerned with finding the smallest quadratic non residue, finding the first occurrence of a given substring, etc. Thus these results do not say much about the overall distribution, which is of course what we are interested in.

The main conclusion of all this is a negative one: nothing has been found in the last 100 years or so, which immediately renders Legendre sequences useless for pseudorandom bit generation.

3. Jacobi Sequences are Harder to Predict than Legendre Sequences

Let us first define formally the *Legendre generator*:

Definition 3.1

Let Q be a polynomial. Then with security parameter value k, the Legendre generator takes as input ("seed") a randomly chosen k-bit prime p and a uniformly chosen k-bit number a. It produces as output the Legendre sequence modulo p with starting point $a \bmod p$ and length $Q(k)$, where Legendre symbols are translated into bits such that -1 corresponds to a 1-bit, while 1 corresponds to a 0-bit. This sequence will be called $L(p,a)$, and its i'th element will be denoted $L(p,a)_i$ □

Similarly, we define the *Jacobi generator*:

Definition 3.2

Let P and Q be polynomials. Then with security parameter value k, the Jacobi generator takes as input $Q(k)$ randomly chosen k-bit primes $p_1, \ldots, p_{Q(k)}$ and $Q(k)$ uniformly chosen k-bit numbers $a_1, \ldots, a_{Q(k)}$. Put $n = p_1 \cdots p_{Q(k)}$, and let a be chosen, such that a is congruent to a_i modulo p_i for $i = 1 \cdots Q(k)$ ($p_i = p_j$ for $i \neq j$ only happens with negligible probability). The generator produces as output the Jacobi sequence modulo n with starting point a and length $P(k)$, where Jacobi symbol are translated into bits as above. The sequence will be called $J(n,a)$, and its i'th element will be called $J(n,a)_i$ □

Yao [Kr] has proved that, if given a prefix of the output from a pseudorandom bit generator, it is still hard to predict the next bit, then output from the generator cannot be distinguished from truly random sequences by any feasible algorithm. Thus, informally speaking, all we have to do in order to prove the strength of our generators is to show that $P1$ or $P2$ are hard problems. This can also be stated using Levin's concept of *isolation* [BoHi]: consider some prefix of the output from the generator as a function of the seed. Then we would like the bit following the prefix to be isolated from the prefix itself.

In general, we can think of a pseudorandom bit generator as a probabilistic algorithm G which takes input x chosen from a finite set X_m, where $\{X_m\}_{m=1}^{\infty}$ is a family of finite sets, and m can be thought of as a security parameter. The output $G(x)$ is a bitstring whose i'th bit is denoted $G(x)_i$. We now have the following more formal definition of next-bit-security:

Definition 3.3

The generator G is said to be *strongly unpredictable*, if for all polynomials P and probabilistic circuits C, the following holds only for finitely many m:

there exists an i, such that

$$Prob(C(G(x)_1, \ldots, G(x)_{i-1}) = G(x)_i) \geq \frac{1}{2} + \frac{1}{P(m)},$$

where $C(G(x)_1, \ldots, G(x)_{i-1})$ denotes the output of C on input $G(x)_1, \ldots, G(x)_{i-1}$, and the probability is taken over the coinflips of C and a uniform choice of $x \in X_m$ □

Definition 3.4

The generator G is said to be *weakly unpredictable*, if the statement in Definition 3.3 holds, with the probability $\frac{1}{2} + \frac{1}{P(m)}$ replaced by $1 - \frac{1}{P(m)}$ □

Thus, for weak unpredictability, we allow algorithms that guess better as m increases, as long as the success probability does not approach 1 too rapidly.

For the next results, we need some more notation:

Definition 3.5

Let G be a pseudorandom bitgenerator as above, and let Q be a polynomial. Then G^Q is a pseudorandom bitgenerator which takes as input $x = (x_1, \ldots, x_{Q(m)})$, where $x_i \in X_m$ are cosen uniformly and independently. It produces as output the sequence whose j'th element is

$$G^Q(x)_j = G(x_1)_j \oplus \cdots \oplus G(x_{Q(m)})_j$$

We now quote from [BoHi] the following definition and theorem:

Definition 3.6

Let $\{X_m\}_{i=1}^{\infty}$ be an infinite family of finite sets, and let B be a function mapping X_m to $\{0,1\}$. Let f be a function such that $f : X_m \rightarrow \{0,1\}^{P(m)}$ for some polynomial P. Then we say that B is *(p,T)-isolated* from f if every circuit with $Q(m)$ inputs and size at most T satisfies

$$| Prob(C(f(x)) = B(x)) - \frac{1}{2} | \leq \frac{p}{2},$$

where x is chosen uniformly from X_m □

Thus, isolation measures the hardness of predicting $B(x)$ given $f(x)$.

Theorem 3.1 (Levin's Isolation Theorem)

If the functions $b_i(x_i)$ are (p,T) isolated from $f_i(x_i)$ for all $1 \leq i \leq n$, then for every $\varepsilon > 0$, the function $b_1(x_1) \oplus \cdots \oplus b_n(x_n)$ is $(p^n + \varepsilon, \varepsilon^2(1-p)^2 T)$-isolated from $f_1(x_1), \ldots, f_n(x_n)$ □

From this follows:

Theorem 3.2
Suppose that the pseudorandom bitgenerator G is weakly unpredictable. Then G^{m^2} is strongly unpredictable.

Proof.
For $x \in X_m$, let $P_i(G(x))$ denote the prefix of $G(x)$ of length $i-1$. Then weak unpredictability implies that for all i smaller than the outputlength of G, $G(x)_i$ is $(1-\frac{2}{R(m)}, T(m))$-isolated from $P_i(G(x))$ for arbitrary polynomials R and T and all sufficiently large m. Put $p(m) = 1 - \frac{2}{m}$. Using the notation of Definition 3.5, we obtain from Levin's theorem by choosing $R(m) = m$ that $G^{m^2}(x)_i$ is $(p(m)^{m^2} + \varepsilon, \varepsilon^2(1-p(m))^2 T(m))$- isolated from $P_i(G(x_1)), \ldots, P_i(G(x_{m^2}))$. From this last bit string, it is easy to compute $P_i(G^{m^2}(x))$. Moreover, $p(m)^{m^2}$ converges to 0 faster than any polynomial fraction. From these facts, we get the strong unpredictability of G^{m^2} by choosing $\varepsilon = \frac{1}{P(m)}$, where P is the polynomial from Definition 3.3 □

Alternatively, we could have used Yao's xor-Theorem, although the conclusion of that theorem as stated in [Kr] is slightly weaker than that of Levin's Theorem.

Theorem 3.2 was already known (see [Kr]) for the special case of generators constructed with an unapproximable predicate and a friendship function [BlMi].

We call attention to two points in connection with this result, which are of interest from a cryptographic point of view:

- What Theorem 3.2. proves is that the next bit of the XOR of several generators is hard to predict, even when given prefixes of the output from *each individual* generator. But in a known plaintext attack on the resulting cryptosystem, a cryptanalyst only knows the XOR of the prefixes, and therefore seems to be faced with an even harder problem. It would be very interesting to find out, whether the conditions on G needed to make G^Q strong can be relaxed using this fact.
- With essentially the same argument as for Theorem 3.2, one can prove a more general statement, which loosely speaking says that if the generator G cannot be predicted with probability better than $1 - \frac{1}{R(m)}$ for some polynomial R, then the generator $G^{mR(m)}$ is strongly unpredictable. In other words: by XOR-ing more "copies" of G, one can get away with a weaker assumption on the security of G.

It is now trivial to prove:

Corollary 3.1
If the Legendre Generator is weakly unpredictable, then the Jacobi Generator is strongly unpredictable □

Proof.
By the way we translate Legendre and Jacobi symbols into bits, it is clear that the Jacobi generator in fact outputs the xor of the output of several Legendre generators. We can therefore use Theorem 3.2 □

4. Emphirical Tests

A number of elementary statistical tests were performed on primes and sequences of various lengths. The primes were of length approximately 25, 50, 100, 200 and 400 bits, and 6 primes of each length were tested. The sequence length was fixed to $100 \cdot \log_2(p)$.

The sequences were generated using special purpose hardware (a FAP4- processor), and the tests included:

- A Chi-square test for equidistribution of subsequences of length 1 to 10. For each subsequence length a Chi-square value was computed based on the occurrences found. Representative results for the distribution of these Chi-square values can be found in Fig. 1.
- A Chi-square test on the distribution of block lengths. This produced results quite similar to those of the subsequence test.
- A test on the linear complexity of the strings. Using the Berlekamp-Massey algorithm, the linear complexity of prefixes of the Legendre sequences was computed. For a really random sequence, the linear complexity is expected to be close to half the sequence length [Ru]. Our sequences seem to fit nicely with this expectation. Fig. 2. shows a typical result.

No statistical weaknesses were found during these tests, and moving to composite numbers and Jacobi symbols produced no significant change in the results.

This can hardly be said to be surprising: as mentioned, all known results indicate, that a highly non-elementary test would be needed to detect any weakness in the Legendre generator.

Finally, let us remark that the tests mentioned here are just preliminary. Many other and more sophisticated tests with larger test material could (and should) be carried out.

5. Practical Implementation

If one wants to use special purpose hardware in implementing this system, using Gauss's Reciprocity Theorem for computing Legendre symbols hardly seems an attractive solution, at least judging from the hardware available today: most modular arithmetic chips are much better suited for exponentiation.

Computing Legendre and Jacobi symbols by exponentiation is cubic in the length of the modulus, and therefore slower asymptotically than generators based on squaring modulo a composite. In practice, the difference may not be so large, however, since it is not clear at all, that one must use primes large enough to make discrete log hard, for example.

If for example we use the Jacobi generator with 2 prime factors of size about 25 bits, this produces an effective key space of more than 90 bits, which is certainly enough to prevent exhaustive search.

If one is willing to use the same amount of hardware as would be needed for a 600 bit RSA implementation, one could compute 12 bits of the keystream in parallel, which with state of the art hardware would give a speed of about 100 Kbits pr. sec. Even use of just 2 25-bit slices would still give a speed of about 10 Kbits pr. sec.

6. Generalizations

The method we present can be generalized in several ways:

6.1. The Linear Congruence Method

First, one could consider, in stead of Legendre symbols of consecutive numbers, symbols for numbers generated by the well known Linear Congruence Generator, i.e. taking Legendre symbols for a sequence $a_1 \cdots a_n$, where

$$a_{i+1} = (m \cdot a_i + b) \bmod p ,$$

for constants m and b, and a prime p. Note, however, that since

$$a_i = m^{i-1} a_1 + m^{i-2} b + \cdots + mb + b ,$$

then by the multiplicative property of Legendre symbols,

$$\left(\frac{a'_i}{p}\right)\left(\frac{b}{p}\right) = \left(\frac{a_i}{p}\right),$$

where $a'_1 = a_1 b^{-1}$, and $a'_{i+1} = ma'_i + 1$. So if this generalization is used in a cryptosystem, there is no point in including the choice of b in the key: up to a sign, all the possible sequences can already be obtained just by using $b=1$ and varying the starting point a_1. Variation of m, on the other hand, does seem to generate new sequences compared to the basic Legendre generator. This introduces a possibility of enlarging the key-space without using a larger prime.

One should take some care, however, in choosing m, as shown by the following Lemma:

Lemma 6.1

The period of the sequence defined by $a_1 = a$ and $a_{i+1} = (ma_i + b) \bmod p$ for a prime p is

p if $m=1$

$ord(m)$ if $m \neq 1$, and $a \neq \frac{-1}{m-1}$

1 otherwise,

where $ord(m)$ denotes the order of m as element in Z_p^*.

Proof.

The $m=1$-case is trivial. For the other cases, use the recurrence

$$a_{n+k} = (m^k a_n + \frac{m^k - 1}{m-1}) \bmod p$$

and elementary number theory □

Thus, the starting point for the sequence should be chosen different from $\frac{-1}{m-1}$, and $ord(m)$ should be large. This can be ensured by choosing p such that the factorization of $p-1$ is known, computing the order of a candidate m from this, and discard low-order m's. In practice, however, it is probably better to construct p such that $p-1$ has a large prime factor q. Then an m chosen uniformly from $]0 \cdots p-1]$ will have order divisible by q with probability $1-q^{-1}$.

Finally, let us remark that the results of Perron (see Section 2) for the basic Legendre sequences are easily seen to generalize to sequences generated by the method from this section.

6.2. Using Other Character Values

Another interesting idea is to consider other characters than the quadratic one. Such character values can also be produced easily by exponentiation: If q is a divisor in $p-1$, then $a \to a^{(p-1)/q}$ is a surjective homomorphism from Z_p^* to G, the subgroup of order q. By choosing some 1-1 correspondence between elements of G and the set of complex q'th roots of unity, each element in G corresponds to one of the q possible values of the corresponding character of Z_p^*. These elements can be represented by bit strings of length approximately $\log_2(q)$. We can therefore construct a generator by computing $a^{(p-1)/q}$ for consecutive values of a, and at each point output the corresponding bit string. Clearly, there is a limit to how large q can be chosen before the generator becomes insecure (just consider $q=p-1$!). Determining the maximal useful value of q will be an interesting field for new research. This problem can be thought of as corresponding to that of finding out how many of the least significant RSA-bits are cryptographically secure (see for example [MiSc]).

7. Conclusion and Open Problems

We have presented a new pseudorandom bit generator, based on a number theoretic problem, the complexity of which may be unrelated to the well known candidate hard problems in number theory.

We have seen that to prove the cryptographic strength of the generator, it is enough to prove that Legendre Sequences are weakly unpredictable.

A number of open problems remain, however:

- Are Legendre sequences weakly unpredictable?
- Is the complexity of predicting Legendre sequences related to other number theoretic problems?
- Are other characters than the quadratic one usable for pseudo random generators?

References

[Ba] Bach: ''Realistic Analysis of Some Randomized Algorithms'', Proc. of STOC 87.

[BlMi] Blum and Micali: How to Generate Cryptographically Strong Sequences of Pseudorandom Bits'', SIAM J. of Comp., vol.13, 1984, pp.850-864.

[BoHi] Boppana and Hirschfeld: ''Pseudorandom Generators and Complexity Classes'', Manuscript, MIT, 1987.

[Bu] Burde: ''Verteilungseigenschaften von Potenzresten'', J. Reine Angev. Math 249, pp.133-172, 1971.

[Bur] Burgess: ''The Distribution of Quadratic Residues and non-Residues'', Mathematica 4 (1957), pp.106-112.

[Da] Davenport: ''On the Distribution of Quadratic Residues (mod p)'', J. London Math. Soc., 8 (1933), pp.46-52.

[El] Elliott: ''A Restricted mean value Theorem'', J. London Math. Soc.(2), 1 (1969), pp.447-460.

[Kr] Kranakis:''Primality and Cryptography'', Wiley-Teubner Series in Computer Science, 1986.

[MiSc] Micali and Schorr: ''Super-Efficient, Perfect Random Number Generators'', these proceedings.

[Pe] Perron: ''Bemerkungen uber die Verteilung der quadratischen Reste'', Math. Z. 56 (1952), pp.122-130.

[Ru] Rueppel: ''Linear Complexity and Random Sequences'', Proc. of EuroCrypt 85, pp.167-191, Springer.

EFFICIENT, PERFECT RANDOM NUMBER GENERATORS

S. Micali
Laboratory for Computer Science
MIT

C.P.Schnorr*
Fachbereich Mathematik/Informatik
Universität Frankfurt

Abstract We describe a method that transforms every perfect random number generator into one that can be accelerated by parallel evaluation. Our method of parallelization is perfect, m parallel processors speed the generation of pseudo-random bits by a factor m; these parallel processors need not to communicate. Using sufficiently many parallel processors we can generate pseudo-random bits with nearly any speed. These parallel generators enable fast retrieval of substrings of very long pseudo-random strings. Individual bits of pseudo-random strings of length 10^{20} can be accessed within a few seconds. We improve and extend the RSA-random number generator to a polynomial generator that is almost as efficient as the linear congruential generator. We question the existence of polynomial random number generators that are perfect and use a prime modulus.

*) Research performed while visiting the Department of Computer Science of the University of Chicago.
MIT - Patent Pending

1. Introduction

A *random number generator* (RNG) is an efficient algorithm that transforms short random seeds into long pseudo-random strings. A classical RNG is the linear congruential generator (LCG) that is based on the recursion $x_{i+1} := ax_i + b \pmod N$. It is well known that the LCG passes certain statistical tests, e.g. for a clever choice of the parameters a,b,N it generates well mixed numbers (see Knuth 1980). There are more elaborate statistical tests which the LCG fails. Stern (1987) shows that the sequence generated by the LCG can be inferred even if the parameters a,b,N and the seed x_0 are all unknown.

The concept of *perfect* random number generator has been introduced by Blum, Micali (1982) and Yao (1982). A RNG is *perfect* if it passes all polynomial time statistical tests, i.e. the distribution of output sequences cannot be distinguished from the uniform distribution of sequences of the same length. So far the proofs of perfectness are all based on unproven complexity assumptions. This is because we cannot prove superpolynomial complexity lower bounds.

Perfect random number generators have been established for example based on the discrete logarithm by Blum, Micali (1982), based on quadratic residuosity by Blum, Blum, Shub (1986), based on one way functions by Yao (1982), based on RSA encryption and factoring by Alexi, Chor, Goldreich and Schnorr (1984). All these RNG's are less efficient than the LCG. The RSA/RABIN-generator is the most efficient of these generators. It successively generates log n pseudo-random bits by one modular multiplication with a modulus N that is n bit long. The modulus N must be at least 512 bits long.

We extend and accelerate the RSA-generator in various ways. We give evidence for more powerful complexity assumptions that yield more efficient generators. Let $N = pq$ be product of two large random primes p and q and let d be a natural number that is relatively prime to $\varphi(N) = (p-1)(q-1)$. The number d must be small compared to log N so that the interval $[1,N^{2/d}]$ is sufficiently large. We conjecture that the following distributions are indistinguishable by efficient statistical tests (see Hypothesis 2.1):

- the distribution of $x^d \pmod N$ for random $x \in [1,N^{2/d}]$.
- the uniform distribution on $[1,N]$.

This hypothesis is closely related to the security of the RSA-scheme. Under this hypothesis the transformation

$$[1,N^{2/d}] \ni x \mapsto x^d \pmod N \in [1,N]$$

stretches short random seeds $x \in [1,N^{2/d}]$ into a pseudo-random numbers $x^d \pmod N$ in the interval $[1,N]$. We build various random number generators on this transformation. The sequential polynomial generator (SPG) generates from random seed $x \in [1,N^{2/d}]$ a sequence of numbers $x = x_1, x_2, \ldots, x_i, \ldots \in [1,N^{2/d}]$. The $n(1-2/d)$ least significant bits of the binary representation of $x_i^d \pmod N$ are the output of x_i and the $2n/d$ most significant bits form the successor x_{i+1} of x_i.

It follows from a general argument of Goldreich, Goldwasser, Micali (1986) and the above hypothesis that all these generators are perfect, i.e. the distribution of output strings is indistinguishable, by efficient statistical tests, from the uniform distribution of binary strings of the same length. The sequential generator is nearly as efficient as the LCG. Using a modulus N, that is n bit long, it outputs $n(1-2/d)$ pseudo-random bits per iteration step. The costs of an iteration step $x \mapsto x^d \pmod N$ with $x \in [1,N^{2/d}]$ corresponds to the costs of about one full multiplications modulo N. This is because the evaluation of $x^d \pmod N$ over numbers $x \le N^{2/d}$ consists almost entirely of multiplications with small numbers that do not require modular reduction.

We extend the SPG to a parallel polynomial generator (PPG). The PPG generates from random seed $x \in [1,N^{2/d}]$ a tree. The nodes of this iteration tree are pseudo-random numbers in $[1,N^{2/d}]$ with outdegree at most $d/2$. To compute the successor nodes $y(1),\ldots,y(s)$ and the output string of node y we stretch y into a pseudo-random number $y^d \pmod N$ that is n bits long. Then the successors $y(1),\ldots,y(s)$ of y are obtained by partitioning the most significant bits of $y^d \pmod N$ into $s \le d/2$ bit strings of length $\lfloor 2n/d \rfloor$. The output of node y consists of the remaining least significant bits of y^d(mod N). Any collection of subtrees of the iteration tree can be independently processed in parallel once the corresponding roots are given. In this way m parallel processors can speed the generation of pseudo-random bits by a factor m. These parallel processors need not to communicate; they are given pseudo-independent input strings and their output strings are simply concatenated. The concatenated output of all nodes of the iteration tree is pseudo-random, i.e. the parallel generator is perfect. The PPG enables fast retrieval of substrings of the pseudo-random output. To access a node of the iteration tree we follow the path from the root to this node. After retrieving a bit the subsequent bits in the output can be generated at full speed. Iteration trees of depth at most 60 are sufficient for practical purposes; they generate pseudo-random strings of length 10^{20} (for outdegree 2) such that individual bits can be retrieved within a few seconds.

The parallel generator is based on a method that has been invented by Goldreich, Goldwasser and Micali (1984) for the construction of random functions. Our contribution consists of the observation that this construction can be applied to speed every perfect random number generator by a factor m using m parallel processors. Using this principle and sufficiently many parallel processors we can generate pseudo-random bits with almost any speed. This important method of parallelization applies to all perfect RNG's but the RSA-generator is particularly suited for this method. Our method of parallelization does not apply to imperfect RNG's as the LCG since this method can further detoriate a weak generator.

The paper is organized as follows. In section 2 we formulate our basic Hypothesis which is somewhat stronger than the assumption that factoring large integers is difficult. We give support to this hypothesis and show that a weak version of it follows from the assumption that the RSA-scheme is safe. We present in section 3 sequential and parallel random number generators that are based on this hypothesis. In the open problem session we question whether there exist perfect pseudo-random number generators that use a prime modulus. This would lead to pseudo-random number generators which use a modulus that is only 224 bits long.

2. The Complexity Assumption for the Polynomial Random Generator

Let P(x) be a polynomial of degree $d \geq 2$ with integer coefficients and let N be an integer that is n bits long, i.e. $2^{n-1} \leq N < 2^n$. We denote $l = \lfloor 2n/d \rfloor$. Residue classes modulo N are identified with the corresponding integers in the interval [1,N].

The polynomial generator is based on the transformation

$$[1,M] \ni x \mapsto P(x) \bmod N \qquad (1)$$

where x ranges over a sufficiently large subinterval [1,M] of [1,N]. We would like that the outputs of (1), for random $x \in [1,M]$ and given N, M and P, be indistinguishable from random $y \in [1,N]$. The following conditions and restrictions are clearly necessary.

- the modulus N must be difficult to factor since given the factorization of N we can easily invert (1).

- The interval [1,M] must be so large that a random seed $x \in [1,M]$ cannot be easily recovered from P(x) (mod N) by guessing x. M must be sufficiently large to make P(x)/N large for almost all $x \in [1,M]$. This is because we can easily invert (1) provided that P(x)/N is small.

- P(x) must not be a square polynomial. If $P(x) = Q(x)^2$ for some polynomial Q then the Jacobi-symbol $\left(\frac{P(x)}{N}\right)$ is 1 for all x whereas $\text{prob}\left[\left(\frac{y}{N}\right) = 1\right] = \text{prob}\left[\left(\frac{y}{N}\right) = -1\right]$ for random $y \in [1,N]$. Since the Jacobi-symbol can be evaluated efficiently we can distinguish P(x) mod N from random numbers $y \in [1,N]$.

- P(x) must not be a linear transform of a square polynomial. If $P(x) = aQ(x)^2 + b$ we can, from P(x) mod N, recover $Q(x)^2$ mod N and check that $\left(\frac{Q(x)^2}{N}\right) = 1$.

We choose N,M,P(x) as to correspond to these conditions. Let N be a random number that is uniformly distributed over the set

$$S_n = \left\{ N \in \mathbb{N} \;\middle|\; \begin{array}{l} N = p\cdot q \text{ for distinct primes } p,q \\ \text{such that } 2^{n/2-1} < p,q < 2^{n/2} \end{array} \right\}$$

of integers that are products of two distinct primes which each is n/2 bits long. We choose the interval length M proportional to $2^{2n/d}$, $M = \theta(2^{2n/d})$; i.e. $1/c \le 2^{2n/d} / M \le c$ for some absolute constant $c > 0$. Then M is proportional to $N^{2/d}$ for all $N \in S_n$. The choice for the polynomials P(x) seems to be subject to only a few restrictions. We are going to study a particular class of permutation polynomials where the hypothesis below can be justified by known theory. These are the RSA-polynomials $P(x) = x^d$ with d relatively prime to $\varphi(N) = (p-1)(q-1)$.

Rivest, Shamir and Adleman (1978) have invented the RSA-cryptoscheme that is based on the multiplicative group

$$\mathbb{Z}_N^* = \{ x(\text{mod } N) \mid \gcd(x,N) = 1 \}$$

of residue classes modulo N that are relatively prime to N. The integer N is product of two odd primes, $N = p \cdot q$. The order of the group $\mathbb{Z}_N^*$ is $\varphi(N) = (p-1)(q-1)$. The transformation

$$x \mapsto x^d \pmod N \tag{2}$$

with $\gcd(\varphi(N),d) = 1$ is a permutation on the residue classes modulo N, i.e. it permutes the integers in the interval [1,N]. The inverse transformation is given by $x \mapsto x^e \pmod N$ where $e = d^{-1} \bmod \varphi(N)$. The permutation (2) with $\gcd(\varphi(N),d) = 1$ and $d \neq 1$ is an RSA-*enciphering function.* The enciphering key d does not reveal the inverse key e provided that $\varphi(N)$ is unknown. Knowledge of $\varphi(N)$ is equivalent to knowing the factorization $N = p \cdot q$. The security of the RSA-scheme relies on the assumption that RSA-enciphering $x \to x^d \pmod N$ is difficult to invert when d, N are given but $\varphi(N)$ and $e = d^{-1} \bmod \varphi(N)$ are unknown. All known methods for inverting RSA-enciphering require the factorization of N.

We are going to show that the following hypothesis is closely related to the security of the RSA-scheme. Our random number generators will rely on this hypothesis.

Hypothesis 2.1 *Let $d \geq 3$ be an odd integer and $l = \lfloor 2n/d \rfloor$. For random $N \in S_n$ such that $\gcd(d,\varphi(N)) = 1$ and for all $M = \theta(2^l)$ the following distributions on $[1,N]$ are indistinguishable by polynomial time statistical tests:*
- *the uniform distribution on $[1,N]$,*
- *$x^d \pmod N$ for random $x \in [1,M]$.*

We explain the hypothesis in more detail. The concept of a *statistical test* has been introduced by Yao (1982). A polynomial time *statistical test* is a sequence $T = (T_n)_{n \in \mathbb{N}}$ of probabilistic algorithms with a uniform polynomial time bound $n^{O(1)}$. According to Yao it is sufficient to consider statistical tests with 0,1-output. Let

$$p_n^T = \text{prob}[T_n(y,N) = 1]$$

be the probability that T_n outputs 1. The probability space is that of all integers $N \in S_n$ with $\gcd(d,\varphi(N)) = 1$, all numbers $y \in [1,N]$ and all 0-1 sequences of internal coin tosses, with uniform distribution. Let $\overline{p}_n^T(M)$ be the same probability with random numbers $y \in [1,N]$ replaced by $y = x^d \pmod N$ for random $x \in [1,M]$ and fixed d. The hypothesis means that for every polynomial time statistical test T and all $M_n = \theta(2^l)$

$$\lim_n |p_n^T - \overline{p}_n^T(M_n)|\, n^t = 0 \quad \text{for all } t > 0. \tag{3}$$

In particular the hypothesis means that any polynomial time algorithm can at most factor a negligible fraction of the integers in S_n. There are algorithms that can efficiently factor a very small fraction of the integers in S_n, e.g. Pollard's ρ-method efficiently factors all integers $N = p \cdot q$ such that either p-1 or q-1 is a product of small primes. But no algorithm is known that can factor in polynomial time a n^{-t}-fraction of the integers in S_n for some fixed $t > 0$.

We introduce some useful terminology. We say that the statistical test T *ε_n-rejects* RSA-ciphertexts $x^d \pmod N$ of random $x \in [1,M_n]$ if $|p_n^T - \overline{p}_n^T(M_n)| \geq \varepsilon_n$ for infinitely many n. If (3) holds for all polynomial time statistical tests T we call RSA-ciphertexts $x^d \pmod N$ of random messages $x \in [1,M_n]$ *pseudo-random* in [1,N]. In this case the distributions of $x^d \pmod N$ for random $x \in [1,M_n]$ and the uniform distribution on [1,N] are called *indistinguishable*.

In general two sequences of distributions $(D_n)_{n\in\mathbb{N}}$ and $(\overline{D}_n)_{n\in\mathbb{N}}$ are called *indistinguishable* if for every pol. time statistical test $(T_n)_{n\in\mathbb{N}}$, that is given random inputs with respect to D_n ($\overline{D}_n$, resp.) the probability p_n^T ($\overline{p}_n^T$, resp.) of output 1 satisfy $\lim_n |p_n^T - \overline{p}_n^T| n^t = 0$ for all $t > 0$. In case of indistinguishable distributions D_n, $\overline{D}_n$, where D_n is the uniform distribution on set C_n, random elements with respect to $\overline{D}_n$ are called *pseudo-random* in C_n. In case of pseudo-random pairs (x,y) we call x and y *pseudo-independent*. A random number generator is called *perfect* if it transforms random seeds into pseudo-random strings.

It can easily be seen that the Hypothesis 2.1 can only fail if RSA-enciphering leaks partial information on RSA-messages.

Fact 2.2 *Suppose Hypothesis 2.1 fails. Then given d and N we can distinguish between RSA-ciphertexts $x^d \pmod N$ of random messages $x \in [1,N]$ and of random messages $x \in [1,M_n]$ for some $M_n = \theta(2^l)$.*

Proof The transformation $x \mapsto x^d \pmod N$ permutes the integers in the interval [1,N].

The RSA-enciphering $x^d \pmod N$ of random messages $x \in [1,N]$ is uniformly distributed over $[1,N]$. If Hypothesis 2.1 fails the uniform distribution can be distinguished from RSA-ciphertexts $x^d \pmod N$ for random $x \in [1,M_n]$; i.e. RSA-ciphertexts $x^d \pmod N$ would leak information on whether the message x is contained in $[1,M_n]$. **QED**

Fact 2.2 does not mean that the RSA-scheme breaks down if the hypothesis fails. This is because messages in the interval $[1,2^l]$ are rather unlikely. Nevertheless the hypothesis is close to the security of the RSA-scheme. Using the following Theorem 2.3 we can relate the hypothesis to RSA-security (see Corollary 2.5).

Theorem 2.3 *Alexi, Chor, Goldreich, Schnorr (1984)*
Let d,N be integers such that $\gcd(d,\varphi(N)) = 1$. Every probabilistic algorithm AL, which given the RSA-enciphering $x^d \pmod N$ of a message x, has an ε_N-advantage in guessing the least significant bit of the message x, can be transformed (uniformly in N) into a probabilistic algorithm $\overline{AL}$ for deciphering arbitrary RSA-ciphertexts. The deciphering algorithm $\overline{AL}$, when given for input $x^d \pmod N$, d and N, terminates after at most $O(\varepsilon_N^{-8} n^3)$ elementary steps and outputs x with probability at least 1/2.

We count for *elementary steps* the $\mathbb{Z}_N$-operations (addition, multiplication, division), RSA-encryptions and calls for algorithm AL at unit cost. We say that algorithm AL has an ε_N-*advantage* in guessing the least significant bit of x if

$$\text{prob}[AL(x^d \pmod N),N) = x \pmod 2] \geq \frac{1}{2} + \varepsilon_N.$$

The probability space is the set of all $x \in [1,N]$ and all 0-1 sequences of internal coin tosses, with uniform probability.

By Theorem 2.3 the security of the RSA-scheme with parameters N, d implies that the following two distributions cannot be distinguished given only N and d:

- the uniform distribution on $[1,N]$,
- $x^d \pmod N$ for random, even $x \in [1,N]$.

Everyone who is able to distinguish these distributions can decode arbitrary RSA-ciphertexts $x^d \pmod N$ given only N and d. We will present in Corollary 2.4 a more formal version of this statement.

We say that a probabilistic algorithm AL ε_N*-rejects* the distribution D on [1,N] if

$$|p^A - \overline{p}^A| \geq \varepsilon_N$$

where p^A ($\overline{p}^A$, resp.) is the probability that AL on input $y \in [1,N]$ outputs 1. The probability space is the set of all $y \in [1,N]$, distributed according to D (with uniform distribution, resp.) and of all 0-1 sequences of internal coin tosses of algorithm AL. Using this notion we can reformulate Theorem 2.3 as follows.

Corollary 2.4 *Let d, N be integers such that* $gcd(d,\varphi(N)) = 1$. *Every probabilistic algorithm AL, that* ε_N*-rejects RSA-ciphertexts* $x^d (mod\ N)$ *of even random messages x can be transformed (uniformly in N) into a probabilistic algorithm for decoding arbitrary RSA-ciphertexts. This deciphering algorithm terminates after at most* $O(\varepsilon_N^{-8}\, n^3)$ *elementary steps (i.e.* $\mathbb{Z}_N$*-operations, RSA-encryptions and calls for AL).*

We next show that Corollary 2.4 remains valid if we replace RSA-ciphertexts of random even messages x, by RSA-ciphertexts of random messages $x \in [1,N/2]$.

Corollary 2.5 *Let d, N be odd integers such that* $gcd(d,\varphi(N)) = 1$. *Every probabilistic algorithm AL, that* ε_N*-rejects RSA-ciphertexts* $x^d (mod\ N)$ *of random messages* $x \in [1,N/2]$, *can be transformed (uniformly in N) into a probabilistic algorithm for decoding arbitrary RSA-ciphertexts. This deciphering algorithm terminates after at most* $O(\varepsilon_n^{-8}\, n^3)$ *elementary steps.*

Proof For odd N and all $x \in [1,N]$ we have

$$x \in [1,N/2] \Leftrightarrow 2x (mod\ N) \text{ is even}$$

(i.e. $x \in [1,N/2]$ iff the representative of $2x (mod\ N)$ in [1,N] is even).

We see from this equivalence that the following distributions are identical for odd N:

- $x^d (mod\ N)$ for random $x \in [1,N/2]$,
- $2^{-d} y^d (mod\ N)$ for random even $y \in [1,N]$.

Moreover we can transform in polynomial time $y^d (mod\ N)$ into $2^{-d} y^d (mod\ N)$. Thus an ε_N-rejection of RSA-encipherings $x^d (mod\ N)$ of random messages $x \in [1,N/2]$ can be transformed (uniformly in N) into an ε_N-rejection of RSA-ciphertexts $y^d (mod\ N)$ of random even $y \in [1,N]$. Corollary 2.5 follows from Corollary 2.4 by this transformation.

QED

Under the assumption that the RSA-scheme is safe Corollary 2.5 proves a slight modification of our hypothesis. The interval $[1,2^l]$ of Hypothesis 2.1 is replaced by the interval $[1,N/2]$ in this modification. This poses the question whether the length of the interval is crucial for the hypothesis to be valid. We next show that Hypothesis 2.1, with the interval $[1,2^l]$ replaced by the interval $[1,N\,2^{-\lceil \log n \rceil}]$, is valid if the RSA-scheme is safe.

Theorem 2.6 *Let d, N be odd integers such that $gcd(d,\varphi(N)) = 1$. Every probabilistic algorithm AL, that ε_N-rejects RSA-ciphertexts $x^d (mod\ N)$ of random messages $x \in [1, N2^{-k}]$, can be transformed (uniformly in N) into a probabilistic algorithm for decoding arbitrary RSA-ciphertexts. This deciphering algorithm terminates after at most $O(2^{2k}\ \varepsilon_N^{-8}\ n^3)$ elementary steps.*

Proof Under the assumption that the RSA-scheme is safe, Alexi et alii (1984) have shown that the log n least significant bits of RSA-messages x are pseudo-random when given $x^d (\text{mod } N)$, d and N. Their proof transforms every algorithm AL, that ε_N-rejects RSA-encipherings $x^d(\text{mod } N)$ of random messages x satisfying $x = 0(\text{mod } 2^k)$, (uniformly in N) into a probabilistic algorithm for deciphering arbitrary RSA-ciphertexts. This RSA-deciphering procedure terminates after at most $O(2^{2k}\ \varepsilon_N^{-8}\ n^3)$ elementary steps (i.e. $\mathbb{Z}_N$-operations, RSA-encipherings and calls for algorithm AL).

For odd N and all $x \in [1,N]$ we obviously have

$$x \in [1,N2^{-k}] \iff 2^k x(\text{mod } N) = 0(\text{mod } 2^k)$$

(i.e. $x \in [1,N\,2^{-k}]$ iff the representative of $2^k x(\text{mod } N)$ in $[1,N]$ is a multiple of 2^k). Therefore the following two distributions are identical for odd N:

- $x^d(\text{mod } N)$ for random $x \in [1,N2^{-k}]$,
- $2^{-kd} y^d(\text{mod } N)$ for random $y \in [1,N]$ satisfying $y = 0(\text{mod } 2^k)$.

Moreover we can transform in polynomial time $y^d(\text{mod } N)$ into $2^{-kd} y^d(\text{mod } N)$. Thus an ε_N-rejection of RSA-ciphertexts $x^d(\text{mod } N)$ of random messages $x \in [1,N\,2^{-k}]$ can be transformed (uniformly in N) into an ε_N-rejection of RSA-ciphertexts $y^d(\text{mod } N)$ of random messages y satisfying $y = 0(\text{mod } 2^k)$. Corollary 2.6 follows from this transformation and the above mentioned proof of Alexi et alii (1984). **QED**

Notice that the time bound for the RSA-deciphering algorithm of Corollary 2.6 is polynomially related to the time bound of algorithm AL provided that $k \le \log n$. Hence if Hypothesis 2.1 fails, with the interval $[1,2^l]$ replaced by the interval $[1,\ N2^{-\lceil \log n \rceil}]$, then RSA-ciphertexts can be deciphered in probabilistic polynomial time. Also if Hypothesis 2.1 fails, with the interval $[1,2^l]$ replaced by the interval $[1, N2^{-\lfloor\sqrt{n}\rfloor}]$, then RSA-ciphertexts can be deciphered in time $e^{O(\sqrt{n})}$. However the fastest known algorithm for RSA-deciphering, via factoring N, requires about $e^{0.693\sqrt{n \log n}}$ steps, where $0.693 \approx \log 2$. Thus if Hypothesis 2.1 fails for the interval $[1, N2^{-\lfloor\sqrt{n}\rfloor}]$, then we can speed up the presently known attacks to the RSA-scheme.

It remains the question whether the computational properties of the distribution $x^d \pmod N$ change when x ranges over very small integers x. In fact Hypothesis 2.1 does not hold for the interval $[1,N^{1/d}]$ since we have $x^d < N$ for all $x \in [1,N^{1/d}]$ and therefore RSA-ciphertexts $x^d \pmod N$ can easily be deciphered for $x \in [1,N^{1/d}]$. On the other hand the d-powers x^d are of order N^2 for almost all numbers $x \in [1,2^l]$. We conjecture that this is sufficient to make the task of deciphering $x^d \pmod N$ hard. This is justified because inverting the squaring

$$x \mapsto x^2 \pmod N$$

is known to be as hard as factoring N, and the squares x^2 are of order N^2, too.

We are going to study the question whether Hypothesis 2.1 should be extended to polynomials P(x) that are more general than RSA-polynomials $P(x) = x^d$ with $\gcd(d,\varphi(N)) = 1$. There is an obvious extension of Hypothesis 2.1 to arbitrary exponents $d \geq 2$. It seems that the condition $\gcd(d,\varphi(N)) = 1$ is not necessary for odd d. This is because no extension of the Jacobi-symbol is known for residues $x^d \pmod N$ of odd prime powers d. On the other hand we must modify the hypothesis for even d since the Jacobi-symbol gives efficient information on the quadratic residuosity. We formulate the extended hypothesis so that it can be applied in the proof of Theorem 3.1 to establish perfect RNG's. For reasons of efficiency we are particularly interested in even exponents d and in exponents that are powers of 2.

Extension to even d of Hypothesis 2.1 *For random $N \in S_n$, all $M = \theta(2^l)$, $l = \lfloor 2n/d \rfloor$, and random $x \in [1,M]$ the following holds.*

(1) $y := x^d \pmod N$ is a pseudo-random quadratic residue modulo N.

(2) Partitioning y into disjoint sections $z := \lfloor y\, 2^{-n+l} \rfloor$ and $y \pmod{2^{n-l}}$ yields pseudo-random numbers in $[1, N\, 2^{-n+l}]$ and $[1,2^{n-l}]$.

(3) $z^d \pmod N$ and $y \pmod{2^{n-l}}$ are pseudo-independent.

Article (1) of the extended hypothesis can be justified by the work of Alexi et alii (1984) for the case that N is a *Blum-integer*, i.e. N is product of two primes p and q such that p = 3(mod 4) and q = 3(mod 4). One can prove that distinguishing x^d(mod N), for random $x \in [1, N n^{-1}]$ from random quadratic residues modulo N is equivalent, by probabilistic polynomial time reductions, to factoring N. Article (2) means that neither z nor y(mod 2^{n-l}) contains efficient information on the quadratic residuosity of y. Article (3) means that the dependence of z and y(mod 2^{n-l}), via the quadratic residuosity of y, gets hidden by the transformation $z \mapsto z^d$(mod N).

Next we consider arbitrary polynomials P(x) of degree d. We are going to show that some elementary methods for distinguishing random numbers $y \in [1,N]$ and P(x) mod N for random $x \in [1,N^{2/d}]$ do not work. Theorem 2.7 is a first step in this direction. This problem clearly deserves further study.

In general we can invert the transformation

$$x \mapsto P(x) \bmod N \tag{1}$$

only if the factorization N = pq is given. Then, using Berlekamps algorithm for polynomial factorization we invert (1) modulo p and modulo q and apply the Chinese remainder construction. This can be done in probabilistic time $(nd)^{O(1)}$. Without knowing the factorization of N we do not know how to invert (1). In the particular case that P(x) divides $x^{\varphi(N)}$ we can invert (1) provided that we know the cofactor $x^{\varphi(N)}/P(x)$, but in this case we can even factor N.

Can we invert (1) for small integers x ? If $|P(x)| / N$ is small we can guess z = P(x) and factorize P(x) - z. Theorem 2.7 below shows that $|P(x)|/N$ is large for almost all $x \in [1,N^{2/d}]$ provided that P(x) has degree at most d. A degree bound is necessary since there exist polynomials of degree $N^{2/d}$ that vanish on the intervall $[1,N^{2/d}]$.

Theorem 2.7 *Let A,B,d be integers such that* $M \geq (BN)^{1/d}\, 16Ad$, *and let* $P(x) \in \mathbb{Z}[x]$ *have degree d. Then we have* $prob[|P(x)| \leq BN] \leq 1/A$ *for random* $x \in [1,M]$.

Proof Let $x_1,\dots,x_k$ be the distinct real numbers in [0,N] satisfying $P(x_i)^2 = B^2N^2$ for $i=1,\dots,k$. We have $k \leq 2d$ since $P(x)^2$ has degree 2d. We partition the real interval [0,M] into 4Ad intervals I of length M/(4Ad). A fundamental theorem in approximation theory (see e.g. Stiefel (1969), p. 236) implies that

$$\max[P(x)^2 \mid x \in I] \geq \left(\frac{M}{8Ad}\right)^{2d} 2^{-2d+1}$$

for each of these intervals I. Hence

$$\max[\,|P(x)|\; x \in I\,] > \left(\frac{M}{16Ad}\right)^{d} \geq BN.$$

This shows that every interval I, that contains an integer x satisfying $|P(x)| \leq BN$, must also contain some point x_i, $1 \leq i \leq k$. The intervals I that contain some point x_i can have at most

$$2d\left(\frac{M}{4Ad}+1\right) \leq \frac{M}{2A}+2d$$

integer points. This accounts for at most a fraction of

$$\frac{1}{2A}+\frac{2d}{M} \leq 1/A$$

of the points in [1,M]. **QED**

3. The Sequential and the Parallel Polynomial Generator

In this section we build several RNG's on polynomials P(x) of degree $d \geq 2$ that have the following generator property. The *generator property* formulates Hypothesis 2.1 for arbitrary polynomials P(x).

Definition The polynomial P(x) has the *generator property* if for random $N \in S_n$, all M proportional to $N^{2/d}$ and random $x \in [1,M]$ the number P(x) mod N is pseudo-random in [1,N].

The generator property means that P stretches random seeds $x \in [1,N^{2/d}]$ into pseudo-random numbers P(x) mod N in the interval [1,N]. By Hypothesis 2.1 RSA-polynomials $P(x) = x^d$ with $\gcd(d,\varphi(N)) = 1$ and $d \geq 3$ have the generator property.

The *sequential polynomial generator* (SPG) generates a sequence of numbers $x = x_1, x_2, \ldots, x_i, \ldots$ in $[1, N^{2/d}]$ that are represented by bit strings of length $l := \lfloor 2n/d \rfloor$. The *output* at x_i, $\mathrm{Out}(x_i) \in \{0,1\}^{n-l}$, is the bit string consisting of the $n-l$ least significant bits of the binary representation of $P(x_i) \bmod N$. The *successor* x_{i+1} of x_i is the number corresponding to the other bits of $P(x_i) \bmod N$,

$$x_{i+1} := \lceil P(x_i) \bmod N / 2^{n-l} \rceil .$$

The sequential polynomial generator can be figured by the following infinite tree

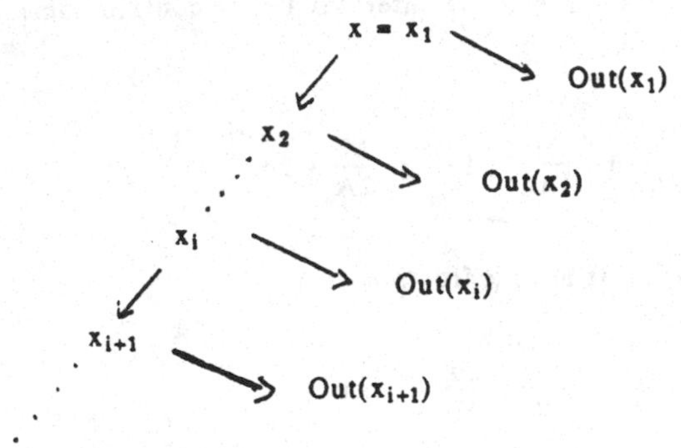

figure of the sequential polynomial generator (SPG)

Let the *k-output* of the SPG

$$\mathrm{SPG}_{k,P}(x,N) = \prod_{i=1}^{k} \mathrm{Out}(x_i)$$

be the concatenated output of the first k steps.

Notice that the most significant bits of $P(x_i) \bmod N$ are biased depending on the most significant bits of N. Even though the most significant bits of $P(x_i) \bmod N$ are not pseudo-random we can form from these bits the successor x_{i+1} of x_i. This is because the generator property and Hypothesis 2.1 imply that $P(x_i) \bmod N$ is pseudo-random if x_i is random in $[1,M]$, *for all* M proportional to 2^l.

Theorem 3.1 *Suppose that P has the generator property. Then for random* $N \in S_n$, *random*

$x \in [1, N^{2/d}]$ and polynomially bounded k (i.e. $k = k(n) = n^{O(1)}$) the k-output $SPG_{k,P}(x,N)$ of the sequential polynomial generator is pseudo-random.

Proof For random $N \in S_n$ and random $x_1 \in [1,N^{2/d}]$ the number $P(x_1)$ mod $N \in [1,N]$ is pseudo-random. It follows that the bit string $Out(x_1) \in \{0,1\}^{n-l}$ is pseudo-random and that the number $x_2 \in [1,2^l]$ is pseudo-random. We also see that the pair $(Out(x_1), x_2)$ is pseudo-random. It follows from the generator property and since x_2 is pseudo-random that

$$(Out(x_1)\, Out(x_2),\ x_3) = (SPG_{2,P}(x_1,N),\ x_3)$$

is pseudo-random, too. To prove this claim we replace in a statistical test $T = (T_n)_{n \in \mathbb{N}}$ for $z := (Out(x_1)\, Out(x_2),\ x_3)$ the pair $(Out(x_2),\ x_3)$ (the string $Out(x_1)$, resp.) by random objects generated through internal coin tosses. This transforms T into statistical tests for $P(x_1)$ mod N ($P(x_2)$mod N, resp.). If z is ε_n-rejected then either $P(x_2)$ mod N or $P(x_1)$ mod N is $(\varepsilon_n/2)$-rejected. In either case this yields a statistical test that $(\varepsilon_n/2)$-rejects $P(x_1)$ mod N.

By induction on k the same argument proves that

$$(SPG_{k,P}(x_1,N),\ x_{k+1})$$

is pseudo-random for every fixed k. The pseudo-randomness also holds if $k = k(n)$ is polynomially bounded in n, i.e. $k = n^{O(1)}$. Using the above argument we can transform a test that ε_n-rejects $(SPG_{k,P}(x_1,N),\ x_{k+1})$ into a test that (ε_n/k)-rejects $P(x_1)$ mod N. QED

It is important that the above proof also applies to polynomials $P(x) = x^d$ with even d. Instead of using the generator property of P we can use the extension to even d of Hypothesis 2.1. Speaking informally, it does not hurt that x^d(mod N) ranges over quadratic residues since the output merely contains the least significant bits of x^d(mod N) and these bits give no efficient information on the quadratic residuosity of x^d(mod N). E.g. we can use for random bit generation the polynomial $P(x) = x^8$ which yields particular efficient RNG's.

PRACTICAL SEQUENTIAL POLYNOMIAL GENERATORS: The modulus N and the number $N^{2/d}$ must be fixed in practical applications. We study the complexity conditions that N and $N^{2/d}$ must satisfy to prevent an efficient analysis of the generator output.

It must be practically impossible to factor the modulus N. For this let N be product of two random primes p and q which each is at least 256 bits long. The numbers p-1, p+1, q-1, q+1 must each have at least some prime factor which is larger than 2^{80}.

The number $N^{2/d}$ must be so large that, given $x^d \pmod N$, it is practically impossible to find $x \in [1, N^{2/d}]$ by efficient search methods. Pollard (1988) has proposed the following method to search for an input x that is product $x = uv$ of two numbers $u, v \in [1, N^{\alpha}]$:

1. Generate the set $S_1 = \{u^d (\mathrm{mod} \cdot N) \mid u \in [1, N^{\alpha}]\}$ and sort this set.
2. Generate the set $S_2 = \{x^d\, v^{-d} \pmod N \mid v \in [1, N^{\alpha}]\}$ and sort this set.
3. Test whether S_1 and S_2 have a common element. If $u^d = x^d\, v^{-d} \pmod N \in S_1 \cap S_2$ then one has found $x = uv$.

Pollard's attack performs $O(N^{\alpha})$ arithmetical steps modulo N and stores N^{α} residues modulo N. It is most efficient when x is product of two numbers in $[1, N^{1/d}]$. In order to make Pollard's attack infeasible it is sufficient that $N^{1/d}$ is at least 2^{64}.

Example 1: Let N be n = 512 bits long and let $\gcd(7, \varphi(N)) = 1$. We choose d = 7, $P(x) = x^7$. Let $Out(x_i)$ consist of the 365 least significant bits of $P(x_i)$ mod N and let x_{i+1} be the number corresponding to the 128 most significant bits of $P(x_i)$ mod N. We compute $x^7 \pmod N$ by computing x^2, x^4, $x^7 = x \cdot x^2 \cdot x^4$. Only the last multiplication requires modular reduction. The other multiplications are with small numbers. The costs of one iteration step correspond to one full modular multiplication. Thus this SPG iteratively outputs 384 pseudo-random bits at the cost of one full modular multiplication with a modulus that is 512 bits long.

Example 2: Another suitable polynomial is $P(x) = x^8$ even though this polynomial does not have the generator property. The computation of $x^8 \pmod N$ is particularly easy; we compute x^2, x^4, x^8 by successive squaring. The SPG with $P(x) = x^8$ iteratively outputs 384 bits at the cost of one full modular multiplication with a modulus N that is 512 bits long.

Efficient public key encoding and decoding. We can use the above RNG's to generate a one-time-pad for message encoding. When given the seed x_1 of the one-time-pad, encoding and decoding can be done at a speed of about n(1-2/d) bits per multiplication modulo N. A public key coding scheme as e.g. RSA can be used to encode and to decode the seed x_1.

The parallel polynomial generator. The *parallel polynomial generator* (PPG) generates from random seed $x \in [1, N^{2/d}]$ a tree with root x and outdegree at most d/2. The nodes of this *iteration tree* are pseudo-random numbers in $[1, N^{2/d}]$ that are represented by bit strings of length l.

The *successors* y(1),...,y(s) of a node y with degree s and the *output string* Out(y) of node y are defined as follows. Let $b_1, \dots, b_n$ be the bits of the binary representation of P(y) mod N, with b_1 being the most significant bit, i.e.

$$\sum_{i=1}^{n} b_i \, 2^{n-i} = P(y) \bmod N .$$

We partition the sl most significant bits into s block with l bits in each block. The corresponding numbers

$$y(j) := 1 + \sum_{i=1}^{l} b_{(j-1)l+i} \, 2^{l-i} \qquad \text{for } j = 1, \dots, s$$

are the successors of node y in the iteration tree. The *output* Out(y) at node y consists of the remaining low order bits of P(y) mod N,

$$\text{Out}(y) = b_{sl+1} \cdots b_n .$$

For convenience we denote the nodes on level k of the iteration tree as $x(j_1, \dots, j_k)$; $x(j_1, \dots, j_{k-1})$ is the direct predecessor of $x(j_1, \dots, j_k)$ and j_k ranges from 1 to s_{k-1} = "outdegree of $x(j_1, \dots, j_{k-1})$". For simplicity we let the outdegree of node $x(j_1, \dots, j_k)$ be a function depending on k only; we assume that $s_k \geq 1$.

The parallel polynomial generator can be figured by the following infinite tree

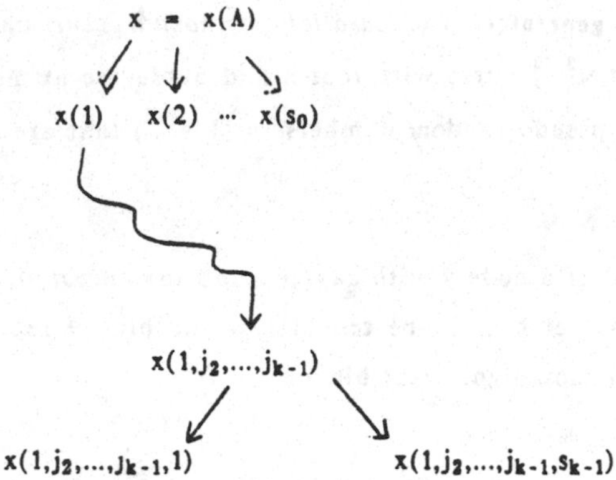

figure of the parallel polynomial generator (PPG)

We define the *k-output* $PPG_{k,P}(x,N)$ of the PPG with seed x as the concatenation of all bit strings $Out(x(j_1,...,j_i))$ on levels i with $0 \leq i \leq k$, with respect to any efficient enumeration order, as e.g. preorder traversal, postorder traversal, inorder traversal or enumeration by levels.

In the particular case that all outdegrees are one, i.e. $s_0 = s_1 = ... = s_k = 1$, the parallel and the sequential polynomial generator coincide. The argument of Goldreich, Goldwasser and Micali (1986) extends Theorem 3.1 from the SPG to arbitrary PPG's, provided that we process at most polynomially many nodes in the iteration tree. This yields the following theorem.

Theorem 3.2 *Suppose that P has the generator property. Then for random $N \in S_n$, random $x \in [1,2^l]$ the k-output $PPG_{k,P}(x,N)$ of the parallel polynomial generator is pseudo-random provided that the length of $PPG_{k,P}(x,N)$ is polynomially bounded.*

Idea of proof There is a straightforward way to extend the proof of Theorem 3.1. Suppose that the k-output $PPG_{k,P}(x,N)$ collects the outputs of $\overline{k}$ nodes. Then every statistical test that ε_n-rejects $PPG_{k,P}(x,N)$ for random $x \in [1,N^{2/d}]$ and random $N \in S_n$ can be transformed into a statistical test that $(\varepsilon_n/\overline{k})$-rejects $P(x) \bmod N$. **QED**

For the output of the PPG we can use any efficient enumeration for the nodes of the iteration tree. To support parallel evaluation we can adjust the shape of the iteration tree and the enumeration order to the number of available parallel processors. For m parallel processors we can use any iteration tree consisting of m isomorphic subtrees attached to the root; we can enumerate, in any order, the m-tuples of corresponding nodes in these subtrees. The enumeration within the subtrees can be chosen to support fast retrieval; for this we can enumerate the nodes e.g. in preorder traversal or in inorder traversal. It is an obvious but important observation that m processors can speed the pseudo-random bit generation of the PPG by a factor m. Once we are given m nodes on the same level of the iteration tree we can process the subtrees below these nodes independently by m parallel processors. These processors do not need to communicate.

Corollary 3.3 *Using m processors in parallel we can speed the pseudo-random bit generation of the parallel polynomial generator by a factor m.*

PRACTICAL PARALLEL POLYNOMIAL GENERATORS

Let N be product of two random primes so that N is 512 bits long. Let $P(x) = x^8$.

Example 3: We construct from random $x \in [1,2^{128}]$ a tree with 4 nodes per level.

1. Stretch a random seed $x \in [1,2^{128}]$ into $x^8 (\mathrm{mod}\ N)$.
2. Partition the binary representation of $x^8 (\mathrm{mod}\ N)$ into 4 bit strings $x(1),\ldots,x(4)$ of length 128. Put $k = 1$ and let $PPG_{1,P}(x,N)$ the empty string.
3. For $j = 1,\ldots,4$ let $x(j\ 1^k) \in I_{128}$ consist of the 128 most significant bits of the binary representation of $x(j\ 1^{k-1})^4 \bmod N$, and let $Out(x(j\ 1^k)) \in I_{384}$ consist of the remaining 384 least significant bits.
4. $$PPG_{k+1,P}(x,N) = PPG_{k,P}(x,N) \prod_{j=1}^{4} Out(x(j\ 1^k))$$

   $k := k + 1$, go to 3.

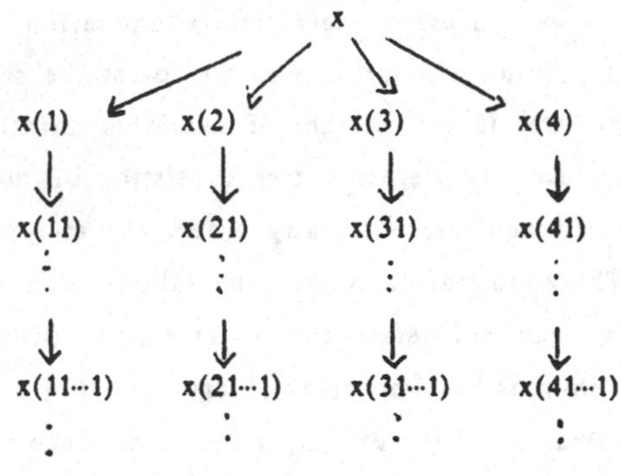

Figure of the PPG of example 3

Using 4 parallel processors this PPG iteratively generates $4 \cdot 384 = 1536$ pseudo-random bits in the time for one full modular multiplication with a modulus N that is 512 bits long. With current processors for smart cards such a full modular multiplication can be done in less than 0.2 sec. Thus 4 parallel processors can generate about 9000 pseudo-random bits per sec.

Example 4: We construct from random $x \in [1,2^{128}]$ a complete tree of outdegree 2.

1. Choose a random seed $x \in [1,2^{128}]$ for root of the tree.
2. For every node $y \in [1,2^{128}]$ of the tree compute the successors $y(1)$, $y(2)$ and the output $Out(y)$ by partitioning the binary representation B of $y^8 (\mathrm{mod}\ N)$ as

$$B = B_1 B_2 \, Out(y) \in I_{128}^2 \times I_{256} ,$$

and compute for $i = 1,2$

$$y(i) := 1 + \text{"the number with binary representation } B_i\text{"}.$$

The main interest in such a PPG comes from fast retrieval methods.

Fast retrieval for the PPG. If the PPG has a complete iteration tree one can efficiently retrieve substrings of the output. Consider example 4 with a complete iteration tree of outdegree 2. Level k of the tree has 2^k nodes and the first k levels have $2^{k+1} - 1$ nodes in total. Suppose the nodes of the tree are enumerated in preorder traversal. Each node yields 256 output bits. To retrieve node y we follow the path from the root to y. This requires processing and storage of at most k nodes and can be done at the costs of about k full modular multiplications. Once we have retrieved node y and stored the path from the root to node y, the bit string that follows Out(y) in the output can be generated using standard retrieval methods at the speed of 256 bits per modular multiplication. For most practical applications the depth k will be at most 60 which permits to generate a pseudo-random string that is $3.7 \cdot 10^{20}$ bits long. We see that retrieval of substrings is very efficient, it merely requires a preprocessing stage of a few seconds to retrieve the initial segment of the substring.

Theorem 3.4 *Every node y of depth k in the iteration tree of the PPG can be accessed and processed at the costs of O(k) modular multiplications.*

k	10	20	30	40	50	60
# nodes in the first k levels	2047	$2 \cdot 10^6$	$2.1 \cdot 10^9$	$2.2 \cdot 10^{12}$	$2.25 \cdot 10^{15}$	$2.3 \cdot 10^{18}$
# output bits	$5.2 \cdot 10^5$	$5.7 \cdot 10^8$	$5.5 \cdot 10^{11}$	$5.6 \cdot 10^{14}$	$5.8 \cdot 10^{17}$	$5.9 \cdot 10^{20}$

Table: retrieval performance of the PPG, example 4

Parallelization and fast retrieval for arbitrary perfect RNG's. It is an important observation that the above methods of parallelization and of efficient retrieval apply to every perfect RNG $(G_n)_{n \in \mathbb{N}}$. The parallel version of the generator associates an iteration tree to a random seed. For example let $G_n : I_n \to I_{3n}$ stretch a random strings in I_n into pseudo-random strings in I_{3n}. We construct from random seed $x \in I_n$ a binary iteration tree with nodes in I_n. Let x be the root of the tree. Construct the two successors y(1), y(2) and the output Out(y) of node y by partitioning $G_n(y) \in I_{3n}$ into three substrings of length n,

$$G_n(y) = y(1)\, y(2)\, \mathrm{Out}(y) .$$

Let $PG_{k,G}(x)$ be the concatenated output of all nodes with depth at most k (compare with the definition of $PPG_{k,P}(x,N)$).

Theorem 3.5 *Let $(G_n)_{n \in \mathbb{N}}$ be any perfect RNG. Then for random seed $x \in I_n$ the concatenated output $PG_{k,G}(x)$ of all nodes with depth $\le k$ is pseudo-random provided that its length is polynomially bounded in n.*

We illuminate our method of parallelization in applying it to some less efficient versions of the RSA/Rabin generator. Let N be a product of two random primes such that N is 512 bits long and $\gcd(3, \varphi(N)) = 1$.

Example 5: From random seed $x \in [1,N]$ we generate the sequence of numbers $x_1, x_2, \ldots, x_i, \ldots \in [1,N]$ as

$$x_1 = x \,, \quad x_{i+1} = x_i^3 (\text{mod } N) \,.$$

Under the assumption that the RSA-enciphering $x \mapsto x^3 (\text{mod } N)$ is safe for the *particular* N, Alexi et alii (1984) have shown that about the 16 least significant bits of x_i are pseudo-independent from x_{i+1}. This suggest the following output of x_i

$$\text{Out}(x_i) = \text{"the 16-least significant bits of } x_i\text{"}.$$

Thus for random $x_1 \in [1,N]$ and under the assumption that RSA-enciphering is safe we obtain pseudo-random bit strings $\prod_{i=1}^{100} \text{Out}(x_i)$ of length 1600. We apply a binary tree construction to the function

$$G : I_{512} \rightarrow I_{1600}$$

that stretches the binary representation of $x_1 \in [1,N]$ into $\prod_{i=1}^{100} \text{Out}(x_i)$. The binary tree has nodes in I_{512}. The successors y(1), y(2) and the output of node y are obtained by partitioning G(y) into two successor strings of length 512 and an output string $\text{Out}_G(y) \in I_{576}$. Processing a node of the binary iteration tree costs 200 modular multiplication.

Example 6: We can accelerate this generator under the reasonable assumption that the 448 least significant bits of the number x and the number $x^3 (\text{mod } N)$ are pseudo-independent for random $x \in [1,N]$. We set

$$\text{Out}(x_i) := \text{"the 448 least significant bits of } x_i\text{"} \,.$$

The assumption implies that $\prod_{i=1}^{3} \text{Out}(x_i) \in I_{1344}$ is pseudo-random for random $x_1 \in [1,N]$. We apply the binary tree construction to the function

$$G : I_{512} \rightarrow I_{1344}$$

that stretches the binary representation of $x_1 \in [1,N]$ into $\prod_{i=1}^{3} \text{Out}(x_i)$. The successors $y(1), y(2) \in I_{512}$ and the output $\text{Out}_G(y) \in I_{320}$ of node y are obtained by partitioning $G(y) \in I_{1344}$ into two strings in I_{512} and $\text{Out}_G(y) \in I_{320}$. Processing a node of the binary tree costs 6 modular multiplications.

Example 7: We can further speed up this generator under the assumption that the 448 least significant bits of random $x \in [1,N]$ and the number $x^2 \pmod N$ are pseudo-independent. (It follows from Alexi et alii (1984) that the 16 least significant bits of random $x \in [1,N]$ and the number $x^2 \pmod N$ are pseudo-independent if factoring the *particular* N is hard. Under this assumption we can replace the iteration $x_i := x_{i+1}^3 (\mathrm{mod}$ N) by $x_{i+1} := x_i^2 \pmod N$. As in Example 5 we associate with a random $x \in [1,N]$ a binary iteration tree with nodes in I_{512}. Processing a node of this tree costs about 4 modular multiplications and yields 320 pseudo-random bits for output.

It is interesting to compare the efficiency of these parallel RNG's with the parallel RNG's based on Hypothesis 2.1. For the latter RNG's in examples 1-4 the cost per node of the iteration tree is about 1 multiplication modulo N. This shows that the new perfect RNG's are more suitable for our method of parallelization and fast retrieval.

4. Open Problems: Random Number Generators Based on a Prime Modulus

In Hypothesis 2.1 we need that the modulus N is difficult to factor. This is because given the factorization of N and given $x^d \pmod N$ we can recover $x = x^{de} \pmod N$ using the inverse exponent $e = d^{-1} \pmod{\varphi(N)}$. Now suppose we are only given the least significant bits of $x^d \pmod N$. Then we cannot easily recover x even if $d^{-1} \pmod{\varphi(N)}$ is known. This poses the question whether Hypothesis 2.1 can be extended to arbitrary prime moduli p.

Problem 4.1. Let p be an arbitrary prime, $2^{n-1} < p < 2^n$, let d be relatively prime to p-1, $d \geq 3$ and let $l \geq \lfloor 2n/d \rfloor$. Is it true that for random $x \in [1, 2^l]$ and $y := x^d \pmod p$ the $n - l$ least significant bits of y are pseudo-random?

If this pseudo-randomness does not hold for all primes we ask whether it holds for random primes.

Problem 4.2. Let $d \geq 3$, $l \geq \lfloor 2n/d \rfloor$ and let p be a random prime such that $2^{n-1} < p < 2^n$ and $\gcd(d,p-1) = 1$. Is it true that for random $x \in [1, 2^l]$ and $y := x^d \pmod p$ the $n-l$ least significant bits of y are pseudo-random?

If we replace in Problem 4.2 the prime modulus p by a random composite modulus in S_n

the pseudo-randomness in question follows from Hypothesis 2.1. These problems are important since this would modify Hypothesis 2.1 so that it is no more related to the difficulty of factoring the modulus. We consider the random number generators that would follow.

The sequential generator using a prime modulus The SPG generates from a random seed $x_1 \in [1,2^l]$ a sequence of numbers $x_1, x_2, \ldots, x_i \in [1,2^l]$ that are represented by bit strings of length l. The *output* at x_i, $out(x_i) \in \{0,1\}^{n-2l}$, is the bit string consisting of the $n-2l$ least significant bits of the binary representation of $x_i^d (\bmod p)$. The *successor* x_{i+1} of x_i is the number corresponding to the next l least significant bits of $x_i^d (\bmod p)$; these are the bits in positions $n-l, \ldots, n-2l+1$ from the left.

Corollary 4.3 *If pseudo-randomness holds in problem 4.2, then the above SPG transforms for random prime p with $2^{n-l} < p < 2^n$ and every k with $k = n^{O(1)}$ a random seed $x_1 \in [0,\ldots,p-1]$ into a pseudo-random output $\prod_{i=1}^{k} out(x_i)$.*

In practical applications the number l must be so large that, given the $n-l$ least significant bits of $x^d (\bmod p)$, it is practically impossible to find $x \in [1,2^l]$. Now Pollard's attack (see section 3) does not work since the most significant bits of $x^d (\bmod p)$ are unknown. Therefore it would be sufficient to start with a random seed x_1 that is 64 bits long.

Example 8: Let p be a prime that is 224 bits long, let $\gcd(p-1,7) = 1$, $d = 7$ and $l = 64$. The output $Out(x_i)$ consists of the 96 least significant bits of $x_i^7 (\bmod p)$, the successor x_{i+1} of x_i is formed by the next 64 least significant bits of $x_i^7 (\bmod p)$. The 64 most significant bits of $x_i^7 (\bmod p)$ are not used at all. Each iteration step generates 96 pseudo-random bits roughly at the cost of one full modular multiplication with a modulus that is 224 bits long.

If we choose a 512 bit long prime modulus p and $d = 7$, $l = 64$ then we can output 384 pseudo random bits per iteration. This achieves the same performance that is obtained with a composite modulus of the same length, see example 1. However using a prime modulus that is about 224 bits long the arithmetic can be done with much smaller numbers, and thus the generator can be implemented on a cheaper chip.

Acknowledgement The second author wishes to thank the Department of Computer Science of the University of Chicago for supporting the research of this paper which was done during a stay at this department. He also wishes to thank A.K. Lenstra and A. Shamir for very inspiring discussions during this work.

References

Alexi, W., Chor, B., Goldreich, O., and Schnorr, C.P.: RSA and Rabin Functions: certain parts are as hard as the whole. Proceeding of the 25th Symposium on Foundations of Computer Science, 1984, pp. 449-457; also: Siam Journal on Comput., 17,2 (1988).

Blum, L., Blum, M. and Shub, M.: A simple unpredictable pseudo-random number generator. Siam J. on Computing (1986), pp. 364-383.

Blum, M. and Micali, S.: How to generate cryptographically strong sequences of pseudo-random bits. Proceedings of the 25th IEEE Symposium on Foundations of Computer Science, IEEE, New York (1982); also Siam J. Comput. 13 (1984), pp. 850-864.

Goldreich, O., Goldwasser, S., Micali, S.: How to Construct Random Functions. Proceedings of the 25th IEEE Symposium on Foundations of Computer Science, IEEE, New York, (1984); also Journal ACM 33,4 (1986), pp. 792-807.

Knuth, D.E.: The Art of Computer Programming. Vol. 2, second edition. Addison Wesley (1981).

Luby, M. and Rackoff, Ch.: Pseudo-random permutation generators and cryptographic composition. Proceedings of the 18th ACM Symposium on the Theory of Computing, ACM, New York (1985) pp. 356-363.

Pollard, J.: private communication (1988).

Stern, J.: Secret linear congruential generators are not cryptographically secure. Proceedings of the 28th IEEE-Symposium on Foundations of Computer Science (1987) pp. 421-426.

Stiefel, E.: Einführung in die numerische Mathematik. Teubner, Stuttgart (1969).

Yao, A.C.: Theory and applications of trapdoor functions. Proceedings of the 25th IEEE Symposium on Foundations of Computer Science, IEEE, New York (1982), pp. 80-91.

Session 6

Signatures and Authentication

Chair: E. Bach, University of Wisconsin

How To Sign Given Any Trapdoor Function

(extended abstract)

Mihir Bellare Silvio Micali*

Laboratory for Computer Science
Massachusetts Institute of Technology
Cambridge, MA 02139

Abstract

We present a digital signature scheme based on trapdoor permutations. This scheme is secure against *existential forgery under adaptive chosen message attack*. The only previous scheme with the same level of security was based on factoring.

Although the main thrust of our work is the question of reduced assumptions, we believe that the scheme itself is of some independent interest. We mention improvements on the basic scheme which lead to a memoryless and more efficient version.

1 INTRODUCTION

In 1976 Diffie and Hellman proposed that modern cryptography be based on the notions of one-way functions (functions that are easy to evaluate but hard to invert) and trapdoor functions (functions that are easy to evaluate and hard to invert without the possession of an associated secret).

In a few years, though, it became clear that basing security solely on assumptions as general as the existence of one-way or trapdoor functions was indeed a great challenge. The first provably good solutions that were found for several cryptographic problems were based on simple complexity theoretic assumptions about the computational difficulty of particular problems such as integer factorization ([GM], [Y], [BlMi]). More recently, however, it was found that pseudo-random number generation is possible if and only if certain kinds of one-way functions exist ([Y],[Le],[GKL]). Similarly we now know that secure encryption is possible if and only if trapdoor predicates exist.

Thus Diffie and Hellman's original goal was realized for two of the major cryptographic primitives. Somewhat surprisingly, in sharp contrast with the progress made

* supported in part by NSF grant DCR-84-13577 and ARO grant DAALO3-86-K-0171

in encryption and pseudo-random number generation, digital signatures, the other fundamental cryptographic primitive, was not yet based on a general assumption. The first paper to address the issues of security in a sufficiently general way and provide a signature scheme with a proof of security was that of [GMY]. Their results and the underlying notions of security were further improved in [GMR]. But both their schemes were based on factoring. Actually, [GMR] base their scheme on the existence of *claw-free pairs*, an assumption weaker than factoring but stronger than the assumption that trapdoor functions exist.

Thus, not only did we not know whether digital signatures are available based on a general complexity theoretic assumption but, even worse, digital signatures were totally linked to a single candidate hard problem. This is particularly unsatisfactory as a great many protocols make use of digital signatures and thus the computational intractability of factoring becomes a bottleneck in the assumptions of many cryptographic protocols.

The contribution of this paper is to free digital signatures from the fortunes of a specific algebraic problem by establishing a truly general signature scheme. Namely, we prove the following

Main Theorem: Secure digital signature schemes exist if *any* trapdoor permutation exists.

Thus, we show once again the feasibility of Diffie and Hellman's original proposal. To appreciate the generality of our theorem, let us clarify what "secure" and "trapdoor" mean.

By secure we mean "non existentially forgeable under an adaptive chosen message attack" as defined by [GMR]. Informally, in an adaptive chosen message attack, a polynomial time enemy (who sees the public key) can choose any message he wants and request to have it signed. After seeing the desired signature, the enemy can choose another message to be signed; and so forth for a polynomial (in the security parameter) number of times. Not to be existentially forgeable means that, after the attack, the enemy will not be able to sign any new message; that is, he will not be able to produce the signature of any string for which he had not previously requested and obtained the signature.

We believe this to be the strongest natural notion of security. In essence, in a scheme secure in this sense, signing is not only hard, but remains hard even having a "teacher" for it. This is more than we may need in practice, where an enemy may be able perhaps to see a few message-signature pairs, but is not able to ask for signatures of messages of his choice!

By trapdoor function we mean a permutation f hard to invert (without knowledge of the secret) on a polynomial fraction of the k-bit strings (when f has security parameter k). Notice that this underlying trapdoor f may by itself be insecure against an adaptive chosen message attack; that is, after being given the value of f^{-1} at a few chosen inputs, one may be able to easily invert f on all inputs. Our scheme will work with such functions as well. In fact our construction has a strengthening effect, and the resulting signing algorithm will be more secure than the trapdoor function it uses.

We are indebted to [GMR] not only as a source of ideas for the present paper, but also for their development and exposition of the notions of signatures and security on which we model our own.

2 SIGNATURE SCHEMES AND THEIR SECURITY

In a digital signature scheme, each user A publishes a "public key" while keeping secret a "secret key". User A's signature for a message m is a value depending on m and his public and secret keys such that anyone can verify the validity of A's signature using A's public key. However, it is hard to forge A's signatures without knowledge of his secret key. Below we give a more precise outline of the constituents of a signature scheme and of our notion of security against adaptive chosen message attack. We follow [GMR] for these notions.

2.1 Components of a Signature Scheme

A digital signature scheme has the following components:

- A *security parameter* k which is chosen by the user when he creates his public and secret keys and which determines overall security, the length and number of messages, and the running time of the signing algorithm.

- A *message space* which is the set of messages to which the signature algorithm may be applied. We assume all messages are binary strings, and to facilitate our exposition and proofs we assume that the message space is $\mathcal{M}_k = \{0,1\}^k$, the set of all k-bit strings, when the security parameter is k.

- A polynomial S_B called the *signature bound*. The value $S_B(k)$ represents a bound on the number of messages that can be signed when the security parameter is k.

- A probabilistic polynomial time *key generation algorithm* KG which can be used by any user to produce, on input 1^k, a pair (PK, SK) of matching public and secret keys.

- A probabilistic polynomial time *signing algorithm* $\mathcal{S}$ which given a message m and a pair (PK, SK) of matching public and secret keys, produces a signature of m with respect to PK. $\mathcal{S}$ might also have as input the signatures of all previous messages it has signed relative to PK.

- A polynomial time *verification algorithm* $\mathcal{V}$ which given S, m, and PK tests whether S is a valid signature for the message m with respect to the public key PK.

Note that the key generation algorithm must be randomized to prevent a forger from re-running it to obtain a signer's secret key. The signing algorithm need not be randomized, but ours is; a message may have many different signatures depending on the random choices of the signer.

2.2 Security against Adaptive Chosen Message Attacks

Of the various kinds of attacks that can be mounted against a signature scheme by a forger, the most general is an *adaptive chosen method attack*. Here a forger uses the signer A to obtain sample signatures of messages of his choice. He requests message signatures, with his requests depending not only on A's public key but on the signatures returned by A in response to the forger's previous requests. From the knowledge so gathered he attempts forgery.

The most general kind of forgery is the successful signing, relative to A's public key, of *any* message m. This is called an *existential forgery*. (Note that forgery of course only denotes the creation of a *new* signature; it is no forgery to obtain a valid signature from A and then claim to have "forged" it). The security we require of our scheme is that existential forgery under an adaptive chosen message attack be infeasible with very high probability. For qualifications and a more precise expression of these notions we resort to a complexity theoretic framework.

A *forger* is a probabilistic polynomial time algorithm $\mathcal{F}$ which on input a public key PK with security parameter k

- engages in a conversation with the legal signer $\mathcal{S}$, requesting and receiving signatures for messages of his choice, for a total number of messages bounded by a polynomial in k (the adaptive chosen message attack),
- then outputs S purporting to be a signature with respect to PK of a new message m (an attempt at existential forgery).

We say $\mathcal{F}$ is *successful* if the signature it creates is a genuine (i.e. $\mathcal{V}(S,m,PK) =$ `true`) forgery. We say that a signature scheme is *Q-forgable* (Q a polynomial) if there exists a forger $\mathcal{F}$ who, for infinitely many k, succeeds with probability more than $\frac{1}{Q(k)}$ on input a public key with security parameter k. The probability here is over the choice of the public key, which is chosen according to the distribution generated by KG, and over the coin tosses of $\mathcal{F}$ and $\mathcal{S}$.

The security property we are interested in consists of not being Q-forgable for any polynomial Q.

3 TRAPDOOR PERMUTATIONS

We propose here a relatively simple complexity theoretic definition of trapdoor permutations which nevertheless captures all the known candidates for trapdoor permutations.

Definition 3.1 A triplet (G, E, I) of probabalistic polynomial time algorithms is a *trapdoor permutation generator* if on input 1^k the algorithm G outputs a pair of k bit strings (x, y) such that

(1) The algorithms $E(x, \cdot)$ and $I(y, \cdot)$ define permutations of $\{0,1\}^k$ which are inverses of each other: $I(y, E(x, z)) = z$ and $E(x, I(y, z)) = z$ for all $z \in \{0,1\}^k$.

(2) For all probabilistic polynomial time (adversary) algorithms $A(.,.,.)$ and for all c and sufficiently large k,
$$Pr[E(x, A(1^k, x, z)) = z] < k^{-c}$$
when z is chosen at random from $\{0,1\}^k$ and the pair (x, y) is obtained by running the generator on input 1^k (the probability is over the random choice of z and the coin tosses of G and A).

The algorithms G, E and I are called the *generating, evaluating* and *inverting* algorithms respectively.

Definition 3.2 A function f is a *trapdoor permutation* with *security parameter* k if there is a trapdoor permutation generator (G, E, I) such that $f = E(x, \cdot)$ for a pair of strings (x, y) obtained by running G on input 1^k.

Notice that as defined above, a trapdoor permutation with security parameter k has domain all of $\{0,1\}^k$. This is not the case with known candidates such as *RSA* ([RSA]) or the trapdoor permutations of [BBS] where the domain is a subset of $\{0,1\}^k$. Also notice that the probability of inversion that we require in part (2) of the definition (k^{-c}) looks very low. Both of these, though, are not restrictions; all the known candidates can be fit into our scenario by using a cross product construction as in [Y]. This works as follows.

Given a trapdoor permutation f on a subset D of $\{0,1\}^k$ such that $|D| \geq 2^k \cdot k^{-d}$ for some d and f is hard to invert on all but a polynomial fraction of D, extend f to $\{0,1\}^k$ by defining it to be the identity function on $\{0,1\}^k - D$. This yields a permutation $\bar{f}$ on $\{0,1\}^k$. Define a function F on
$$\underbrace{\{0,1\}^k \times \ldots \times \{0,1\}^k}_{k^{d+2}}$$
by $F(x_1, \ldots, x_{k^{d+2}}) = (\bar{f}(x_1), \ldots, \bar{f}(x_{k^{d+2}}))$. F is a permutation and Yao shows that it satisfies part (2) of Definition 3.1 given our assumptions about the original f.

4 AN OVERVIEW OF THE SCHEME

We present here an overview of the scheme and a sketch of the proof of security; the succeeding sections gives a more complete description and proof. In this section, as well as in the complete scheme we describe later, we disregard efficiency completely for the sake of simplicity.

4.1 Background

In [La] Lamport suggested the following method for signing a single bit: make public f and a pair of points x^0 and x^1 and keep secret f^{-1}. The signature of a bit $b \in \{0,1\}$ is then $f^{-1}(x^b)$. The drawback of this method is that the number of bits that can be signed is limited to the number of pairs of points that are placed in the public key. Our scheme can be considered an extension of this type of scheme in that it removes the restriction on the number of bits that can be signed while using a similar basic format for signing a single bit. We do this by regenerating some of the public key information every time we sign a bit. [GMR] too uses the idea of regenerating some part of the information in the public key, but with a different, non Lamport like underlying signing method. Merkle ([M]) presents another way of extending the Lamport format; his more pragmatically oriented scheme, though, is not concerned with proofs of security.

In the scheme described below, and then in more detail in §5, we reverse the roles of functions and points in the Lamport format with respect to signing a single bit, and then sign new points as needed. (A dual and equivalent scheme consists of directly using the Lamport format but signing new *functions* instead; this was in fact the way our scheme was presented in [BeMi]).

4.2 The Signature Scheme

A user's public key in our scheme is of the form

$$PK = (f_{0,0}, f_{0,1}, \ldots, f_{k,0}, f_{k,1}, \alpha)$$

where the $f_{i,j}$ are trapdoor permutations with security parameter k and α is a random k bit string (we refer to k bit strings equivalently as *points* or *seeds*). His secret key is the trapdoor information $f_{i,j}^{-1}$. A message is signed bit by bit. The first bit b_1 is signed by sending $f_{0,b_1}^{-1}(\alpha)$ and *a signature of a new seed* α_1. The signature of the k bit string α_1 consists of sending, for each $i = 1, \ldots, k$, either $f_{i,0}^{-1}(\alpha)$ or $f_{i,1}^{-1}(\alpha)$ depending on whether the i-th bit of α_1 was a 0 or a 1.

At this point not only has the bit b_1 been signed, but the public key has been "recreated". That is, another bit can now be signed in the same manner with α_1 playing the role of α above. This process can be continued to sign a polynomial in k number of bits. The signature of a message is thus built on a chain of seeds in which each element of the chain is used to sign its successor.

4.3 Why is this Secure?

Suppose $\mathcal{F}$ is a forger (as described in §2.2). We derive a contradiction by showing that the existence of $\mathcal{F}$ implies the existence of an algorithm A which inverts the underlying trapdoor permutations with high probability.

Given a trapdoor permutation g with security parameter k and a k bit string z, the algorithm A must use the forger to find $g^{-1}(z)$. A's strategy will be to build a suitable public key and then run $\mathcal{F}$ and attempt to sign the messages requested by $\mathcal{F}$. From $\mathcal{F}$'s forged signature will come the information required to invert g.

The public key

$$PK = (f_{0,0}, f_{0,1}, \ldots, f_{k,0}, f_{k,1}, \alpha)$$

that A creates has $f_{n,c} = g$ for some n and c. All the other functions are obtained by running the generator, so A knows their inverses. In the course of signing A will use a list of seeds of the form $g(\alpha_l)$, except for some one stage at which it will use as seed the given point z. So A knows how to invert all the $f_{i,j}$ at all the seeds with the single exception of not knowing $f_{n,c}^{-1}(z)$. At this point, it is possible that A will not be able to sign a message that $\mathcal{F}$ requests. Specifically, A will not be able to sign a message m if computing the signature would require knowledge of $g^{-1}(z)$. But this is the only possible block in A's signing process, and it will happen with probability only 1/2. So A succeeds in responding to all $\mathcal{F}$'s requests with probability 1/2.

By assumption $\mathcal{F}$ will now return the signature of a message not signed previousley by A. The placement of the original function g in the public key, as well as the placement of z in the list of seeds, are unknown to $\mathcal{F}$ (more precisely, the probability distribution of real signatures and A's signatures are the same). With some sufficiently high probability, the signature of the new message will include the value of $g^{-1}(z)$ which A can output and halt.

5 THE SCHEME AND PROOF OF SECURITY

5.1 Preliminary Notation and Definitions

The i-th bit of a binary string x is denoted $(x)_i$ while its length is denoted $|x|$.

If $a = (a_1, \ldots, a_i)$ and $b = (b_1, \ldots, b_j)$ are sequences then $a*b$ denotes the sequence $(a_1, \ldots, a_i, b_1, \ldots, b_j)$. If $a = (a_1, \ldots, a_i)$ is a sequence and $j \leq i$ then $(a_1, \ldots, a_j)$ is called an initial segment of a.

We recall [GMR]'s notation and conventions for probabalistic algorithms. If A is a probabalistic algorithm then $A(x, y, \ldots)$ denotes the probability space which assigns to the string σ the probability that A, on input $x, y, \ldots$, outputs σ. We denote by $[A(x, y, \ldots)]$ the set of elements of $A(x, y, \ldots)$ which have non-zero probability. We denote by $x \leftarrow A(x, y, \ldots)$ the algorithm which assigns to x a value selected according to the probability distribution $A(x, y, \ldots)$. If S is a finite set we write $x \leftarrow S()$ for the algorithm which assigns to x a value selected from S uniformly at random. The notation

$$\boldsymbol{P}(p(x, y, \ldots) \;:\; x \leftarrow S; y \leftarrow T; \ldots)$$

denotes the probability that the predicate $p(x, y, \ldots)$ is true after the (ordered) execution of the algorithms $x \leftarrow S$, $y \leftarrow T$, etc. As an example of this notation, part (2) of Definition 3.1 would be written as

$$\boldsymbol{P}(E(x, u) = z \;:\; (x, y) \leftarrow G(1^k); z \leftarrow \{0, 1\}^k; u \leftarrow A(1^k, x, z)) < k^{-c}.$$

We let PPT denote the set of probabalistic polynomial time algorithms. We assume that a natural encoding of these algorithms as binary strings is used.

For the remainder of this section we fix a trapdoor permutation generator (G, E, I).

With some abuse of language we will often call x a function and identify it with $E(x, \cdot)$. In the scheme we now proceed to describe we will desregard efficiency completely in order to simplify the proof of security.

5.2 Building Blocks for Signing

The signing algorithm makes use of many structures. This section describes the basic building blocks that are put together to build signatures.

Let $(x_i^j, y_i^j) \in [G(1^k)]$ for $i = 0, \dots, k$ and $j = 0, 1$, and let $\vec{x} = (x_0^0, x_0^1, \dots, x_k^0, x_k^1)$, $\vec{y} = (y_0^0, y_0^1, \dots, y_k^0, y_k^1)$. Let $\alpha, \alpha' \in \{0,1\}^k$.

Definition 5.1 A *seed authenticator* $\langle \alpha'; \alpha \rangle_{\vec{x}}$ is a tuple of strings $(\alpha', \alpha, z_1, \dots, z_k)$ for which

$$E(x_i^{(\alpha)_i}, z_i) = \alpha' ,$$

for all $i = 1, \dots, k$.

Definition 5.2 A *bit authenticator* $\langle \alpha'; b \rangle_{\vec{x}}$ is a tuple of strings (α', b, z) such that $b \in \{0, 1\}$ and $E(x_0^b, z) = \alpha'$.

Definition 5.3 An *authenticator* $\langle \alpha'; c \rangle_{\vec{x}}$ is either a seed authenticator or a bit authenticator. In the authenticator $\langle \alpha'; c \rangle_{\vec{x}}$, α' is called the *root* of the authenticator, c is called the *child* of the authenticator, and $\vec{x}$ is called the *source* of the authenticator.

Given $\vec{x}$ and a tuple purporting to be an authenticator $\langle \alpha'; c \rangle_{\vec{x}}$, it is easy for anyone to check that it is indeed one. However given α', c, and $\vec{x}$ it is difficult to create an authenticator $\langle \alpha'; c \rangle_{\vec{x}}$ without the knowledge of $\vec{y}$.

Definition 5.4 A sequence $F = (F^1, \dots, F^p)$ of seed authenticators is a *spine starting at* α' if

- α' is the root of F^1.
- for $i = 1, \dots, p-1$, the root of F^{i+1} is the child of F^i.

Definition 5.5 A sequence $B = (B^1, \dots, B^q)$ of bit authenticators is *s-attached* to the spine $F = (F^1, \dots, F^p)$ if the root of B^i is equal to the child of F^{s+i-1} for $i = 1, \dots, q$. A sequence of bit authenticators $B = (B^1, \dots, B^q)$ is *attached* to the spine $F = (F^1, \dots, F^p)$ if it is s-attached for some s.

5.3 Generating Keys

The key generation algorithm KG does the following on input 1^k:

(1) Run G a total of $2k+2$ times on input 1^k to get a list of pairs (x_i^j, y_i^j) ($i = 0, \dots, k$, $j = 0, 1$).

(2) Select a random k-bit seed $\alpha \in \{0,1\}^k$.

(3) Output the public key $PK = (1^k, \vec{x}, \alpha, S_B)$ where $\vec{x} = (x_0^0, x_0^1, \dots, x_k^0, x_k^1)$ and S_B is the signature bound.

(4) Output the secret key $SK = \vec{y} = (y_0^0, y_0^1, \dots, y_k^0, y_k^1)$.

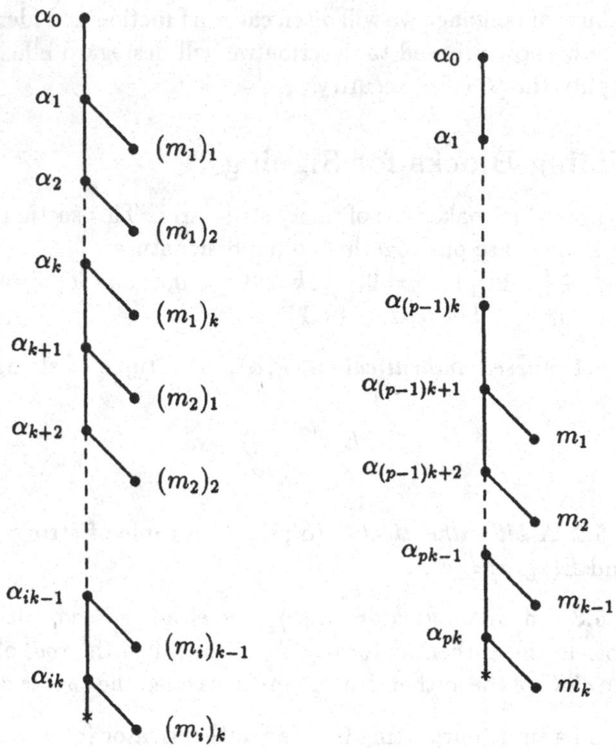

Figure 1: A signature corpus (left), and a signature of a message m (right)

5.4 What is a Signature?

Definition 5.6 A *signature* of a message $m \in \mathcal{M}_k$ with respect to a public key $PK = (1^k, \vec{x}, \alpha, S_B)$ is a triple (F, B, m) where $F = (F^1, \ldots, F^{pk})$ $(p \geq 1)$ is a spine and $B = (B^1, \ldots, B^k)$ is a sequence of bit authenticators such that

- B is $((p-1)k+1)$-attached to F.
- F starts at α.
- For all $i = 1, \ldots, k$ the child of B^i is $(m)_i$.
- The common source of all the authenticators is $\vec{x}$.

Figure 2 shows a schema of a signature for a message m with respect to a public key $(1^k, \vec{x}, \alpha_0, S_B)$; here $F^i = \langle \alpha_{i-1}; \alpha_i \rangle_{\vec{x}}$ $(i = 1, \ldots, pk)$ and $B^i = \langle \alpha_{(p-1)k+i}; (m)_i \rangle_{\vec{x}}$ $(i = 1, \ldots, k)$.

5.5 The Signing Algorithm and Signature Corpus

Let $PK = (1^k, \vec{x}, \alpha_0, S_B)$ and $SK = \vec{y}$ be a pair of public and secret keys. We presume that the signing procedure $\mathcal{S}$ is initialized with the values of PK and SK and has already signed messages $m_1, \ldots, m_{i-1}$ and kept track of the signatures $S_1 =$

$(F_1, B_1, m_1), \ldots, S_{i-1} = (F_{i-1}, B_{i-1}, m_{i-1})$ of these messages. We let F_0 be the empty sequence. To compute a signature $S_i = (F_i, B_i, m_i)$ for m_i, where $i \le S_B(k)$ and $m_i \in \mathcal{M}_k$, $\mathcal{S}$ performs the following steps:

(1) Set $l = (i-1)k$, and select k seeds $\alpha_{l+1}, \ldots, a_{l+k} \in \{0,1\}^k$ at random.

(2) Form the seed authenticators

$$F^j = \langle \alpha_{j-1}; \alpha_j \rangle_{\vec{x}},$$

for $j = l+1, \ldots, l+k$, and let F be the spine $(F^{l+1}, \ldots, F^{l+k})$.

(3) Form the bit authenticators

$$B^j = \langle \alpha_{l+j}; (m_i)_j \rangle_{\vec{x}},$$

for $j = 1, \ldots, k$, and let $B_i = (B^1, \ldots, B^k)$.

(4) Let $F_i = F_{i-1} * F$ and output $S_i = (F_i, B_i, m_i)$ as the signature of m_i.

Figure 1 shows a schema of the data structure constructed by the signing procedure as described above. This structure will be called a signature corpus below.

Definition 5.7 Let

$$(F_1, B_1, m_1), \ldots, (F_i, B_i, m_i)$$

be a sequence of the first i signatures output by our signing algorithm $\mathcal{S}$, for some $i > 0$. Let $F = F_i$ and $B = B_1 * \ldots * B_i$. We call *signature corpus* the triple $C = (F, B, (m_1, \ldots, m_i))$.

Note that a signature corpus (F, B, M) is a spine $F = (F^1, \ldots, F^p)$ to which is 1-attached the sequence of bit carrying items $B = (B^1, \ldots, B^p)$.

Definition 5.8 Let $Z = (F, B, M)$ be either a single signature or a signature corpus, relative to a public key $PK = (1^k, \vec{x}, \alpha_0, S_B)$, where $F = (F^1, \ldots, F^p)$ and $B = (B^1, \ldots, B^q)$. Then

(1) $F(Z)$ denotes F, the spine of Z, and $B(Z)$ denotes B, the sequence of bit authenticators of Z. The authenticators in F are called the seed authenticators of Z and the authenticators in B are called the bit authenticators of Z.

(2) The set of authenticators of Z is $A(Z) = \{F^1, \ldots, F^p\} \cup \{B^1, \ldots, B^q\}$.

(3) The chain of seeds of Z, denoted $P(Z)$, is the sequence of seeds which form the roots and children of the seed authenticators of F. That is, $P(Z) = (\alpha_0, \alpha_1, \ldots, \alpha_p)$, where α_i is the child of F^i for all $i = 1, \ldots, p$.

(4) The set of roots of Z, denoted $R(Z)$, is the set of roots of the seed authenticators of Z.

(5) The tuple M of messages signed by Z is denoted $M(Z)$. (If Z is the signature of a single message m, we just let $M(Z) = m$).

5.6 The Verification Algorithm

Given a public key PK and something purporting to be a signature of a message m with respect to PK, it is easy to check whether this is indeed the case. It is easy to see that checking whether a given object really has the form of definition 5.6 only requires knowledge of the public key.

5.7 Extracting Information From a Forgery

As indicated in the overview of §4.3, forgery must eventually be used to extract information about the inversion of a trapdoor function. The preliminary definitions and lemmas here are devoted to charecterizing the structure of a forgery relative to a given corpus.

Lemma 5.1 Let C be a signature corpus relative to a public key $PK = (1^k, \vec{x}, \alpha, S_B)$ and let S be a signature, relative to the same public key, of a message m not in $M(C)$. Then there is an α' in $P(C)$ such that one of the following holds:

(1) There is a pair of seed authenticators, $\langle \alpha'; h_1 \rangle_{\vec{x}}$ in $F(C)$, and $\langle \alpha'; h_2 \rangle_{\vec{x}}$ in $F(S)$, such that $h_1 \neq h_2$.

(2) α' is not in $R(C)$ (i.e. α' is the child of the last authenticator in the spine) and there is a seed authenticator $\langle \alpha'; h \rangle_{\vec{x}}$ in $F(S)$.

(3) There is a pair of bit authenticators, $\langle \alpha'; b_1 \rangle_{\vec{x}}$ in $B(C)$, and $\langle \alpha'; b_2 \rangle_{\vec{x}}$ in $B(S)$, such that $b_1 \neq b_2$.

Proof: Suppose neither (1) nor (2) holds. Since $F(S)$ and $F(C)$ both start at α, $F(S)$ must be an initial segment of $F(C)$. Thus $P(S)$ is an initial segment of $P(C)$. Since $B(S)$ is attached to $F(S)$, the roots of all the bit authenticators of S are in $P(S)$ hence in $P(C)$. So if $P(C) = (\alpha_0, \ldots, \alpha_{pk})$ then there is some i such that $\langle \alpha_{(i-1)k+j}; (m_i)_j \rangle_{\vec{x}} \in B(C)$ and $\langle \alpha_{(i-1)k+j}; (M(S))_j \rangle_{\vec{x}} \in B(S)$ for all $j = 1, \ldots, k$, where $m_i \in \mathcal{M}_k$ is the i-th message in the corpus. But $M(S)$ is not in $M(C)$, so there is some j such that $(M(S))_j \neq (m_i)_j$. Let $b_1 = (m_i)_j$, $b_2 = (M(S))_j$, and $\alpha' = \alpha_{(i-1)k+j}$. Then $\langle \alpha'; b_2 \rangle_{\vec{x}} \in B(S)$ and $\langle \alpha'; b_1 \rangle_{\vec{x}} \in B(C)$ are the desired bit authenticators which give us part (3) of the lemma. □

Let $PK = (1^k, \vec{x}, \alpha, S_B)$ be a public key, where $\vec{x} = (x_0^0, x_0^1, \ldots, x_k^0, x_k^1)$, and let C be a signature corpus relative to PK. We introduce the notion of a pair (α', x_i^j) being *unused* in C, where α' is in $P(C)$. Informally, we would like to say that (α', x_i^j) is unused if the authenticators in the corpus C do not contain $E(x_i^j, \cdot)^{-1}(\alpha')$. That is, the inversion of $E(x_i^j, \cdot)$ at α' was not required in the signing process. For technical reasons however, the formal definition that we use is rather to say that the inversion of $E(x_i^{1-j}, \cdot)$ *was* required in the signing process. Boundary conditions (being at the end of the spine) complicate things a little further.

Definition 5.9 Let PK, C be as above. We say that (α', x_i^j) is *unused* in C if α' is in $P(C)$ and one of the following holds:

(1) There is a seed authenticator $\langle\alpha'; h\rangle_{\vec{x}}$ in $A(C)$ with $(h)_i \neq j$.

(2) $i \neq 0$ and α' is not in $R(C)$. (So α' is at the tail end of the spine $F(C)$).

(3) $i = 0$ and there is a bit authenticator $\langle\alpha'; b\rangle_{\vec{x}}$ in $A(C)$ with $b \neq j$.

With PK, C as above, let S be the signature of a message m not in $M(C)$, relative to PK. We show that this signature could not have been created without inverting $E(x_i^j, \cdot)$ at α' where (α', x_i^j) was some unused pair in the corpus C.

Lemma 5.2 There is a polynomial time algorithm which takes as input PK, C, and S as described above, and outputs a triple of the form (α', x_i^j, u) such that the pair (α', x_i^j) was unused in C and $E(x_i^j, u) = \alpha'$.

Proof: Let α' be the seed of Lemma 5.1. The proof breaks down into the cases provided by Lemma 5.1, and we number the cases below accordingly. Note that given C and S it is possible for an algorithm to determine which of the cases of Lemma 5.1 applies.

(1) Since $h_1 \neq h_2$ we can find an i such that $(h_1)_i \neq (h_2)_i$. Set $j = (h_2)_i$. The authenticator $\langle\alpha'; h_2\rangle_{\vec{x}}$ provides us with the value $E(x_i^j, \cdot)^{-1}(\alpha')$, and by the first part of Definition 5.9 the pair (α', x_i^j) is unused in C.

(2) Set i to any value between 1 and k and set $j = (h)_i$. The authenticator $\langle\alpha'; h\rangle_{\vec{x}}$ provides us with the value $E(x_i^j, \cdot)^{-1}(\alpha')$, and the second part of Definition 5.9 says that (α', x_i^j) is unused in C.

(3) Set $i = 0$ and $j = b_2$. The authenticator $\langle\alpha'; b_2\rangle_{\vec{x}}$ provides us with the value $E(x_i^j, \cdot)^{-1}(\alpha')$ and the last part of Definition 5.9 says that (α', x_i^j) is unused in C. □

5.8 Proof of Security

We are finally ready to prove

Theorem 5.1 Under the assumption that (G, E, I) is a trapdoor permutation generator the above signature scheme is not even Q-forgable (see §2.2), for all polynomials Q and all sufficiently large k.

The proof of the theorem is by contradiction. Assume the existence of a polynomial Q, an infinite set $\overline{K}$, and a forger $\mathcal{F}(\cdot)$ such that for all $k \in \overline{K}$, $\mathcal{F}$ is succesful in forging with probability $\geq \frac{1}{Q(k)}$ on input a public key chosen according to the distribution induced by KG. Our goal is to construct an algorithm $A(\cdot,\cdot,\cdot) \in PPT$ which on input $1^k, x, z$ uses $\mathcal{F}$ to find $E(x, \cdot)^{-1}(z)$.

A operates as follows on input $1^k, x, z$:

(1) Let $n \leftarrow \{0, \dots, k\}()$, $c \leftarrow \{0,1\}()$, and $t \leftarrow \{0, \dots, kS_B(k)\}()$.

(2) Run G a total of $2k+1$ times on input 1^k to get (x_i^j, y_i^j) for $i = 0, \dots, k$, $j = 0, 1$, $(i,j) \neq (n,c)$. Let $x_n^c = x$, and let $\vec{x} = (x_0^0, x_0^1, \dots, x_k^0, x_k^1)$.

(3) Pick $kS_B(k)$ random k bit strings $\beta_0, \ldots, \beta_{t-1}, \beta_{t+1}, \ldots, \beta_{kS_B(k)}$, and then create the seeds

$$\alpha_l = \begin{cases} z & \text{if } l = t \\ E(x, \beta_l) & \text{otherwise.} \end{cases}$$

Let P be the sequence $(\alpha_0, \alpha_1, \ldots, \alpha_{kS_B(k)})$.

(4) Let $PK = (1^k, \vec{x}, \alpha_0, S_B)$.

(5) Invoke $\mathcal{F}$ on the public key PK, and attempt to sign the requested messages in the same manner as the signing procedure $\mathcal{S}$, but using the already generated seeds from P where $\mathcal{S}$ would pick random new seeds. The inverses of all but one of the functions in $\vec{x}$ are known, and, for that function x_n^c, the value $E(x_n^c, \cdot)^{-1}(\alpha_l) = \beta_l$ is known for all values $l \neq t$. If either $(\alpha_{t+1})_n = c$, or $n = 0$ and the sequence of requested messages has c in the t-th position, it will not be possible to sign. Output $\emptyset$ and halt in this case. If all $\mathcal{F}$'s requested messages are succesfully signed, let C be the corpus of these signatures.

(6) If $\mathcal{F}$ does not now output a signature of a message not in $M(C)$, output $\emptyset$ and halt. Otherwise, invoke the algorithm of Lemma 5.2 on input PK, C, and the signature S output by $\mathcal{F}$. This algorithm outputs a tuple (α', x_i^j, u). Now output u and halt.

We consider the distribution of A's output when its inputs are chosen at random; that is, we consider the result of executing

$$(x, y) \leftarrow G(1^k); z \leftarrow \{0,1\}^k(); u \leftarrow A(1^k, x, z).$$

Lemma 5.3 The public key PK created in step 4 has the same distribution as that induced on public keys by the key generation algorithm KG.

Proof: The functions x_i^j of step 2 were obtained by running G, as was x, so $\vec{x}$ has the right distribution. The β_l were chosen at random in step 3. Since $E(x, \cdot)$ is a permutation, the seeds α_l are also randomly distributed. Since α_0 is either one of these or the randomly chosen z, it is randomly distributed. So PK has the same distribution as generated by KG (§5.3). □

Lemma 5.4

(1) The distribution of signatures generated by the conversation between $\mathcal{F}$ and A is, at every stage in the conversation, the same as the distribution that would be generated in a conversation between $\mathcal{F}$ and the legal signer $\mathcal{S}$.

(2) With probability $\geq \frac{1}{2}$ all of $\mathcal{F}$'s requests are succesfully signed.

Proof: As noted above, the public key has the right distribution. Now the steps used by A to sign are exactly those of the signing algorithm $\mathcal{S}$, with the one exception noted in step 5 of the description of A. The signatures received by $\mathcal{F}$ upto this crucial point have the same distribution as the legal signer would have generated. Upto this point then, $\mathcal{F}$ sees no anomaly. Now at the next step A must invert either $E(x_n^0, \cdot)$ or $E(x_n^1, \cdot)$ at α_t. Since c was chosen at random, we can conclude that this stage is passed with probability $\frac{1}{2}$. Moreover, this and future signatures are still with the right distribution. Both parts of the lemma are thus verified. □

Suppose all $\mathcal{F}$'s requests are signed. By the preceding lemma, the corpus generated has the same distribution as would have been generated with the legal signer. By assumption we know $\mathcal{F}$ forges with probability $\frac{1}{Q(k)}$ on this distribution. Since the signing was accomplished with probability $\geq \frac{1}{2}$ we obtain a forgery S with probability

$$\geq \frac{1}{2Q(k)} .$$

The next step is to show that the u output by A is equal to $E(x,\cdot)^{-1}(z)$ with sufficiently high probability.

Note that $P(C)$ is an initial segment of the sequence P. If the requested messages added together to a length of more than t bits, then z is in $P(C)$. The signing process is accomplished only if inverting $E(x,\cdot) = E(x_n^c,\cdot)$ at z is avoided, so if z is in $P(C)$ then (x,z) is unused in C. We state this as a lemma.

Lemma 5.5 *If A does succeed in signing all of $\mathcal{F}$'s requests, and if z is in $P(C)$, then (z,x) is unused in C.*

Proof: If z is the last seed in the sequence $P(C)$ and $n > 0$ then we have case (2) of Definition 5.9. Otherwise, since the signing was accomplished, either (1) or (3) must hold. □

By Lemma 5.2, $u = E(x_i^j,\cdot)^{-1}(\alpha')$ for some pair (α', x_i^j) unused in C. We would like the pair to actually be (z,x), for then $u = E(x,\cdot)^{-1}(z)$. The randomization of the n and t parameters (step 1) serves to capture this event with probability at least

$$\frac{1}{(1+k)(1+kS_B(k))} .$$

We conclude that for all $k \in \overline{K}$,

$$\begin{aligned} \boldsymbol{P}(E(x,u) = z \ : \ (x,y) \leftarrow G(1^k); z \leftarrow \{0,1\}^k(); u \leftarrow A(1^k,x,z)) \\ \geq \frac{1}{2Q(k)(1+k)(1+kS_B(k))}, \end{aligned}$$

contradicting the fact that G is a trapdoor permutation generator. This completes the proof of Theorem 5.1.

6 VARIATIONS AND IMPROVEMENTS

The signatures produced by the signing algorithm of the previous section are far from compact: signatures with respect to a public key $PK = (1^k, \vec{x}, \alpha, S_B)$ could reach lengths of $O(kS_B(k))$. We describe briefly here how tree structures in the style of [GMR] could replace the linear structures of the above scheme to produce signatures of length $O(k \log S_B(k))$. The size of signatures in the modified scheme will not only be smaller but will be independent of the signatures of previous messages. The modified scheme retains the security properties of the original one.

The public key now contains $2k+1$ pairs of randomly chosen trapdoor functions of security parameter k together with, as before, a single seed. Each seed is used to sign two others, which become its right and left children in the tree; the first k pairs of the above functions are used to sign the left child, and the second k pairs to sign the right child. The tree is grown to height $k \log S_B(k)$. Each leaf can then be used as the root of a linear chain of length k which signs a single message. The proof of security needs little change for the modified scheme, and details are left to the final paper.

The assumption that messages are always of length equal to the security parameter can be removed: to sign messages of arbitrary length it suffices to first encode them with a *subsequence free encoding*. This is an encoding which guarantees that no string is a substring of the concatenation of any number of other strings, and such encodings are easy to construct.

Further, the scheme, in its tree version, can be made *memoryless* (as in the modifications of [Go] and [Gu] to the [GMR] scheme); the same ideas used by [GMR] (attributed to Levin), and extended in [Go], can be applied here. The main tool is the use of pseudo-random functions ([GGM]) whose existence is implied by our assumptions.

References

[BeMi] Bellare, M., and S. Micali, "How to Sign Given Any Trapdoor Function," *Proceedings of the 20th STOC,* ACM (1988), 32-42.

[BBS] Blum, L., M. Blum, and M. Shub, "A Simple Unpredictable Pseudo-Random Number Generator," *SIAM Journal on Computing*, Vol. 15, No. 2 (May 1986), 364-383.

[BlMi] Blum, M., and S. Micali, "How to Generate Cryptographically Strong Sequences of Pseudo-Random Bits," *SIAM Journal on Computing*, Vol. 13, No. 4 (November 1984), 850-864.

[DH] Diffie, W. and M. E. Hellman, "New Directions in Cryptography," *IEEE Trans. Info. Theory* IT-22 (November 1976), 644-654.

[Go] Goldreich, O., "Two Remarks Concerning the GMR Signature Scheme," MIT Laboratory for Computer Science Technical Report 715, (September 1986).

[GKL] Goldreich, O., M. Luby, and H. Krawczyk, "On the Existence of Pseudo-random Generators," *CRYPTO 88*.

[GGM] Goldreich, O., S. Goldwasser, and S. Micali, "How To Construct Random Functions," *Journal of the Association for Computing Machinery*, Vol. 33, No. 4 (October 1986), 792-807.

[GM] Goldwasser, S., and S. Micali, "Probabalistic Encryption," *Journal of Computer and System Sciences* 28 (April 1984), 270-299.

[GMR] Goldwasser, S., S. Micali and R. Rivest, "A Digital Signature Scheme Secure Against Adaptive Chosen-Message Attacks," *SIAM Journal on Computing*, vol. 17, No. 2, (April 1988), 281-308.

[GMY] Goldwasser, S., S. Micali, and A. Yao, "Strong Signature Schemes," *Proceedings of the 15th STOC,* ACM (1983), 431-439.

[Gu] Guillou, L., "A Zero-Knowledge Evolution of the Paradoxical GMR Signature Scheme", manuscript (February 1988).

[La] Lamport, L. "Constructing Digital Signatures from a One-Way Function," SRI Intl. CSL-98. (October 1979)

[Le] Levin, L., "One Way Functions and Pseudo Random Generators," *Proceedings of the 17th STOC,* ACM (1985), 363-365.

[M] Merkle, R., "A Digital Signature Based on a Conventional Encryption Function," *Advances in Cryptology - CRYPTO 87 (Lecture Notes in Computer Science, 293)*, Springer-Verlag, 1987.

[RSA] Rivest, R., A. Shamir, and L. Adleman, "A Method for Obtaining Digital Signatures and Public-Key Cryptosystems," *Communications of the ACM* (Feb 78), 120-26.

[Y] Yao, A. C., "Theory and Applications of Trapdoor Functions," *Proceedings of the 23rd FOCS*, IEEE (1982) 80-91.

A "Paradoxical" Indentity-Based Signature Scheme Resulting from Zero-Knowledge

Louis Claude Guillou [1] and Jean-Jacques Quisquater [2]

[1] Centre Commun d'Etudes de Télédiffusion et Télécommunications
CCETT, BP 59; F-35 512 Cesson-Sevigné Cédex, France

[2] Philips Research Laboratory Brussels
Avenue Van Becelaere, 2; B-1 170 Brussels, Belgium
E-mail: jjq@prlb2.uucp

ABSTRACT

At EUROCRYPT'88, we introduced an interactive zero-knowledge protocol (Guillou and Quisquater [13]) fitted to the authentication of tamper-resistant devices (e.g. smart cards, Guillou and Ugon [14]).

Each security device stores its secret *authentication number*, an RSA-like signature computed by an authority from the device identity. Any transaction between a tamper-resistant security device and a verifier is limited to a unique interaction: the device sends its *identity* and a random *test number*; then the verifier tells a random large *question*; and finally the device answers by a *witness number*. The transaction is successful when the test number is reconstructed from the witness number, the question and the identity according to numbers published by the authority and rules of redundancy possibly standardized.

This protocol allows a cooperation between users in such a way that a group of cooperative users looks like a new entity, having a shadowed identity the product of the individual shadowed identities, while each member reveals nothing about its secret.

In another scenario, the secret is partitioned between distinct devices sharing the same identity. A group of cooperative users looks like a unique user having a larger public exponent which is the greater common multiple of each individual exponent.

In this paper, additional features are introduced in order to provide: firstly, a mutual interactive authentication of both communicating entities and previously exchanged messages, and, secondly, a digital signature of messages, with a non-interactive zero-knowledge protocol. The problem of multiple signature is solved here in a very smart way due to the possibilities of cooperation between users.

The only secret key is the factors of the composite number chosen by the authority delivering one authentication number to each smart card. This key is not known by the user. At the user level, such a scheme may be considered as a keyless identity-based integrity scheme. This integrity has a new and important property: it cannot be misused, *i.e.* derived into a confidentiality scheme.

Keywords: cryptology, factoring, complexity, randomization, zero-knowledge interactive proofs, identity-based system, public key system, intégrity, identification, authentication, digital signature.

1 Introduction

Some problems are very asymmetric: although only inefficient methods are known for solving these problems, any proposal is easily tested in order to know whether it is a solution or not. There are two methods in order to prepare an instance of such a complex problem:

— either you prepare the instance by yourself;

— or an authority does it for you, in relation with your identity.

In the *first method*, each user picks a *trap* at random, and then deduces the text of a problem having this trap as solution. This method leads to systems where each user has his own secret key. An authority manages the system by registering the users and their public keys in a publicly available register.

> Factoring large integers is a pretty well known example of such a complex problem. The following operations are rather easy to do: selecting at random two large prime integers and computing their product. But only inefficient methods are known to factor large composite integers. Outside number theory, many other complex problems are available.

In the *second method*, each user relies upon a trusted authority, like a bank, a credit card company, a telephone operator or a transportation authority; after the signature of a contract specifying the rights and obligations of each party, the authority delivers to the new user a tamper-resistant security device, e.g. a smart card, storing a secret identity-based authentication number. This alternate method leads to a *keyless* system. Only the authority has a secret key while each card holds its own authentication number which is not a trap. Other identity-based systems have been investigated (Shamir [17], Desmedt and Quisquater [4], Quisquater [15]), but our approach is different: here we are authenticating the security device only, not its holder.

How, without revealing it, can a tamper-resistant security device convince any verifier that it knows the authentication value corresponding to its identity?

According to the zero-knowledge techniques (Goldwasser, Micali and Rackoff [8], Goldreich, Micali and Wigderson [10]), the device convinces the verifier without revealing anything on the specific value of the authentication number which remains thus an efficient identification element as long as the secret is unrevealed and as long as the (instance of the) problem remains unsolved. The *knowledge* of the authentication number makes the difference between the tamper-resistant device and the outside.

After an *interactive process*, the verifier has nothing else but an intimate conviction which cannot be transmitted to anybody else. The interactive process may be used, not only to check the identity of the device, but also to check messages endorsed by the device. This method of proof is "non-transitive".

After a *non-interactive process*, like a signature, the verifier is convinced and can convince a judge that a genuine device signed the message. A knowledge is clearly transmitted along with each signature; but while proving that the device knows its authentication number, the signature still transmits no knowledge at all on specific value of the underlying authentication number which may be used indefinitely as an identification element of the device.

The zero-knowledge techniques are very efficient in various processes aiming at protecting the integrity of data ans systems:

identification, authentication and signature.

2 The GQ authentication scheme

We found an interactive protocol aiming at verifying the presence of a secret authentication number in a tamper-resistant security device claiming its identity (Guillou and Quisquater [13]).

Each tamper-resistant security device (*e.g.* a smart card) holds its unique authentication number B related to its identity I by the following simple equation:

$$B^v \cdot J \bmod n = 1, \text{ with } J = Red(I),$$

where,

n: is a composite number;

v: is an exponent, both published by the authority and known to each verifier;

J: is the "shadowed" identity of the device, that is to say a number as large as n, including the claimed identity I, half shorter than n, completed by a redundancy (the shadow) depending on I (Guillou and Quisquater [11], Guillou, Davio and Quisquater [12]). Redundancy rules *Red* (or how constructing J from I) are published or preferably standardized.

NOTE: Let us mention that ISO is standardizing a "digital signature scheme with shadow " (see ISO-DP 9796) in the Working Group JTC1/SC20/WG2 (public-key techniques).

The authentication transaction between the verifier and the device is limited to a unique interaction, which was not the case with the previous proposals (Fiat-Shamir [5], [6]). Here is the interactive protocol described in [13]:

1. The card I transmits its identity I and a *test number* T which is the vth power in Z_n of an integer r picked at random in Z_n^*.
2. The verifier asks a *question* d which is an integer picked at random from 0 to $v-1$.

3. The card I sends a *witness number* t which is the product in Z_n of the integer r by the dth power of the authentication number B.

4. In order to verify such a witness number t, the verifier computes the product of the dth power of the shadowed identity J by the vth power of witness t, that is:

$$\begin{aligned} J^d \cdot t^v \bmod n &= J^d \cdot (r \cdot B^d)^v \bmod n \\ &= (J \cdot B^v)^d \cdot r^v \bmod n \\ &= T. \end{aligned}$$

The proof of security relies on three basic facts:

— A device knowing the authentication number can easily answer correctly any question.

— A lucky guesser has an evident winning strategy by choosing first any witness number before deducing a test number according to the guessed question.

— Knowing two correct witness numbers according to any two different questions for the same test number (anyone) reveals the authentication number.

Let us define a *cheater* as a device trying to fool the verifier, while not knowing the specific value of B.

On one hand, any cheater having guessed the question d can obviously prepare a good looking pair T and t by, firstly, picking t at random in Z_n and, secondly, deducing T by computing exactly as the verifier will do.

On the other hand, having two witnesses t' and t'' corresponding to two different questions d' and d'' for the same test number T gives a significant (and generally total) knowledge about the authentication number B (see the proof in the next section).

Any cheater is thus able to prepare in advance exactly one witness number (at least one, but not two). A lucky cheater thus fools the verifier by guessing one question amongst v possible questions. At each transaction, the verifier has $(v-1)$ chances on v to defeat a cheater. Thus, when the size of v, also named *depth of the authentication number*, is sufficient to reach directly the required level of security, there is no need to repeat the interaction.

In the GQ scheme, the size of required memory and the volume of transmitted data are reduced to *minimum minimorum.* It is well fitted to smart card authentication.

3 Security of the GQ scheme

Now let us consider more precisely the conditions on v and the factors of n in the GQ scheme. Let us consider that n has only two prime factors: p and q.

Let us consider that v is an odd integer which is an RSA-like exponent, so that: $\gcd(p-1, v) = \gcd(q-1, v) = 1$. The case where v is an even integer integer will be considered in the full paper; the exponent v may even be a power of two.

Let us consider carefully the verification formula when v is an RSA-like exponent:

$$\mathcal{F}_d(t) = J^d \cdot t^v \bmod n.$$

A collision is a set of four integers:

$$\{t', t'', d', d''\} \quad t', t'' \text{ in } Z_n^*; \quad 0 \leq d'' < d' \leq v-1$$

such that,

$$\mathcal{F}_{d'}(t') = \mathcal{F}_{d''}(t'')$$

which is,

$$J^{d'} \cdot t'^v \bmod n = J^{d''} \cdot t''^v \bmod n$$

and may be transformed in

$$J^{(d'-d'')} \cdot (t'/t'')^v \bmod n = 1.$$

According to the Bezout formula, there exists a unique pair of integers k, $0 \leq k \leq v-1$, and m, $0 \leq m \leq d' - d'' - 1$, easily computed by the Euclidean algorithm, such that:

$$m \cdot v - k \cdot (d' - d'') = \pm \gcd(v, d' - d'').$$

Let us raise the equation to the power k and substitute.

$$\begin{aligned} 1 &= J^{k \cdot (d'-d'')} \cdot (t'/t'')^{k \cdot v} \bmod n \\ &= J^{m \cdot v \pm \gcd(v,d'-d'')} \cdot (t'/t'')^{k \cdot v} \bmod n \\ &= J^{\pm \gcd(v,d'-d'')} \cdot \{J^m \cdot (t'/t'')^k\}^v \bmod n. \end{aligned}$$

Thus:

$$B^{\pm \gcd(v,d'-d'')} = J^m \cdot (t'/t'')^k \bmod n.$$

When v is prime, any collision provides B. When v is composite, generally any collision provides B as well, and in some cases, a partial knowledge of B is obtained as a power of B of a rank dividing v.

Knowing any collision in $\mathcal{F}$ is thus equivalent to knowing B or a power of B of a rank dividing v.

For a given user, J and v are fixed: the function $\mathcal{F}$ from t to $\mathcal{F}_d(t)$ is a set of permutations of Z_n indexed by d, $0 \leq d \leq v-1$.

In a way similar to what is done in the GMR scheme (Goldwasser, Micali and Rivest [9]), by composing the basic permutation $\mathcal{F}$ indexed by d, $0 < d < n/2$, a large family of permutations $\mathcal{F}$ indexed by D may be constructed. Let D be an integer written on k v–ary digits, from the most significant one $d(k-1)$ to the least significant one $d(0)$, where k is the integer such that $v^{k-1} \leq D < v^k$:

$$\mathcal{F}_D(x) = \mathcal{F}_{d(0)}(\mathcal{F}_{d(1)}(\ldots\mathcal{F}_{d(k-1)}(x)\ldots)) = J^D \cdot x^{v^k} \bmod n.$$

Knowing any collision in this composed family leads generally to knowing the solution β to the equation:

$$J \cdot \beta^{v^k} \bmod n = 1.$$

The authentication number B, such that $J \cdot B^v \bmod n = 1$, is easily deduced from β.

Collision-resistance of this set is equivalent to computing the authentication number B by inverting an RSA instance ([16]).

4 Protocols of cooperation between entities

4.1 Entities with same exponent and different identities

Let us consider two tamper-resistant security devices, each one storing its unique authentication number (B_1 or B_2) related to its identity (I_1 or I_2) by the following equations:

$$B_1^v \cdot J_1 \bmod n = 1, \text{ with } J_1 = Red(I_1),$$

$$B_2^v \cdot J_2 \bmod n = 1, \text{ with } J_2 = Red(I_2).$$

The two entities, cooperating on a *shared* Personal Computer, are negotiating an authentication transaction with a verifier according to the following protocol:

1. Entity I_1 transmits its identity I_1 and a test number T_1 which is the vth power in Z_n of an integer r_1 picked at random in Z_n^*.

   Entity I_2 transmits its identity I_2 and a test number T_2 which is the vth power in Z_n of an integer r_2 picked at random in Z_n^*.

   The Personal Computer sends to the verifier the two identities I_1 and I_2 and the *common test number* T computed from:

   $$\begin{aligned} T &= T_1 \cdot T_2 \bmod n \\ &= (r_1 \cdot r_2)^v \bmod n \\ &= r^v \bmod n \end{aligned}$$

   where r is used for the (implicit) common random number $r_1 \cdot r_2 \bmod n$.

2. The verifier asks a *question* d which is an integer picked at random from 0 to $v-1$.

3. Entity I_1 sends a witness number t_1 which is the product in Z_n of integer r_1 by the dth power of authentication number B_1.

   Entity I_2 sends a witness number t_2 which is the product in Z_n of integer r_2 by the dth power of authentication number B_2.

The Personal Computer sends to the verifier the *common witness number t:*

$$\begin{aligned} t &= t_1 \cdot t_2 \bmod n \\ &= (r_1 \cdot B_1^d) \cdot (r_2 \cdot B_2^d) \bmod n \\ &= (r_1 \cdot r_2) \cdot (B_1 \cdot B_2)^d \bmod n \\ &= r \cdot (B_1 \cdot B_2)^d \bmod n. \end{aligned}$$

4. In order to check such a witness number t, the verifier computes the product of the dth power of the shadowed identity J_1 and J_2 by the vth power of witness t, that is:

$$\begin{aligned} J_1^d \cdot J_2^d \cdot t^v \bmod n &= J_1^d \cdot J_2^d \cdot (r_1 \cdot B_1^d \cdot r_2 \cdot B_2^d)^v \bmod n \\ &= (J_1 \cdot B_1^v)^d \cdot (J_2 \cdot B_2^v)^d \cdot r^v \bmod n \\ &= T. \end{aligned}$$

This protocol of cooperation, easily extensible to any number of cooperating entities, indicates a new direction in multiple signature schemes.

4.2 Two entities with the same identity and different exponents

Let us now consider two tamper-resistant devices, each one storing its unique authentication number (B_1 and B_2) related to the same identity I by one of the following simple equations (let us consider that v_1 and v_2 are prime together):

$$B_1^{v_1} \cdot J \bmod n = 1 \text{ and } B_2^{v_2} \cdot J \bmod n = 1, \text{ with } J = Red(I).$$

The cooperation may simulate an entity having identity I with the exponent $v = v_1 \cdot v_2$,

$$B^v \cdot J \bmod n = 1,$$

with B_1 equal to $B^{v_2} \bmod n$ while B_2 is equal to $B^{v_1} \bmod n$.

The two entities, cooperating on a shared Personal Computer, are negociating an authentication transaction with a verifier according the following protocol:

1. Entity 1 transmits its identity I and a test number T_1 which is the vth power in Z_n of an integer r_1 picked at random in Z_n^*.

   Entity 2 transmits its identity I and a test number T_2 which is the vth power in Z_n of an integer r_2 picked at random in Z_n^*.

   The shared Personal Computer sends to the verifier the common identity I and the *common test number* T computed from:

$$\begin{aligned} T &= T_1^{v_2} \cdot T_2^{v_1} \bmod n \\ &= (r_1 \cdot r_2)^{v_1 \cdot v_2} \bmod n \\ &= (r_1 \cdot r_2)^v \bmod n \\ &= r^v \bmod n, \end{aligned}$$

where r is used for the (implicit) common random number $r_1 \cdot r_2 \bmod n$.

2. The verifier asks a *question* d which is an integer picked at random from 0 to $v-1$.
   The shared Personal Computer translates the question: $d_1 = d/v_2 \bmod v_1$ for the entity 1 and $d_2 = d/v_1 \bmod v_2$ for the entity 2.

3. Entity 1 sends a witness number t_1 which is the product in Z_n of integer r_1 by the d_1^{th} power of authentication number B_1.
   Entity I_2 sends a witness number t_2 which is the product in Z_n of integer r_2 by the d_2^{th} power of authentication number B_2.
   The Personal Computer sends to the verifier the *common witness number* t:
$$\begin{aligned} t &= t_1 \cdot t_2 \bmod n \\ &= r_1 \cdot r_2 \cdot B_1^{d_1} \cdot B_2^{d_2} \bmod n \\ &= r \cdot B^{d_1 \cdot v_2 + d_2 \cdot v_1} \bmod n. \end{aligned}$$

4. Let us call d' the integer $d_1 \cdot v_2 + d_2 \cdot v_1$. In order to check such a witness number t, the verifier computes the product of the dth power of the shadowed identity J by the vth power of witness t, that is:

   Is the test number T equal to $J^{d'} \cdot t^v \bmod n$?

   Proof:
$$\begin{aligned} J^{d'} \cdot t^v \bmod n &= J^{d_1 \cdot v_2 + d_2 \cdot v_1} \cdot (r_1 \cdot B_1^{d_1} \cdot r_2 \cdot B_2^{d_2})^{v_1 \cdot v_2} \bmod n \\ &= (J \cdot B_1^{v_1})^{d_1 \cdot v_2} \cdot (J \cdot B_2^{v_2})^{d_2 \cdot v_1} \cdot (r_1 \cdot r_2)^v \bmod n \\ &= T. \end{aligned}$$ □

This protocol of cooperation may easily be extended to any number of cooperating entities.

Let us remark that the protocols of cooperation solve many problems of subliminal channels in the sense of Simmons or Desmedt. One cooperating entity is then a one-way active warden (see more in the full paper).

5 Interactively authenticating both cards and messages

The authentication described in the basic method convinces the verifier that an entity knowing the authentication number is involved in the transaction.

But the interaction of simultaneous processes may be misleading: everybody knows the strategy used by the child playing chess simultaneously against two masters. The first master opens the first play, then the child reproduces this opening on the second table. The second master replies, and the child repeats this reply on the first table. While knowing nothing in chess skill, the child will not loose both plays. We must be careful in the design of a protocol, so as to avoid to give to a child the merits of a master.

Let us transpose the problem. A kitchener using a security device provided by a banker is buying oranges at a grocery, the grocer being a member of the Organization: at the same time, another member of the same Organization is negociating diamonds in a jewelry, the jeweler being unaware of any problem. When the payment operation is ready, the jeweler verifies the authenticity of the security device of the man buying diamonds. But in fact, this "security" device is connected via a full duplex radiating channel to the grocery POS terminal. And owing to this hidden synchronization, the jeweler is preparing a bill on kitchener's account number, both kitchener and jeweler being unaware of the problem. Y. Desmedt noted this problem in the rump session of CRYPTO '87.

By linking transaction purpose and buyer identity in a unique authentication process, the fraud prepared by the Organization will no more succeed. The kitchener is buying oranges, while the jeweler is selling diamonds. This message authentication must convince the verifier that the message is really sent by the entity owning the right authentication number.

Such an extension implies a hash function. Some papers (Goldreich, Goldwasser and Micali [7]) are dealing with functions *statistically undistinguishable* from really random functions *with polynomially limited resources.* Let us suppose that such a good one-way hash function h exists, while, today, no such a function is ready for standardization.

NOTE. Hash functions h may be implemented either in prover's PC or in the card. The user must control the parameters sent to the hash function. In the example, the user holds a portable device in which the card is inserted and where the hash function h is implemented.

This is a message authentication (the basic idea was already present in Fiat-Shamir [6]):

1. The user claims the message M, the identity I and the verification number V.
   At each treatment, the card picks at random an integer r in Z_n and computes a test T by raising it to the vth power in Z_n. The portable device of the user computes as the verification number V the hashing of M and T:
   $$V = h(M,T) = h(M, r^v \bmod n).$$
2. The verifier asks a question d.
   The verifier picks at random an integer d from 0 to $v-1$ and transmits it.
3. The user shows a witness t.
   The card computes as witness t the product of random elements r by the dth

power of the authentication number B:

$$t = r \cdot B^d \bmod n.$$

4. The verifier reconstructs the test number T from the question d, the identity I and the witness t. Next, the verifier reconstructs the verification number V from the message M and the test T:

$$\text{Is } V \text{ equal to } h(M, J^d \cdot t^v \bmod n) \text{ ?}$$

This is still a zero-knowledge interactive protocol.

Let us now introduce a non-interactive zero-knowledge protocol: the hash function h may be used by the prover himself to compute directly the *question* d. Some of these ideas on non-interactivity were already formulated in Fiat and Shamir [6].

6 Swapping to signatures by removing interactivity

The integrity of a transmission system is threatened in various ways:

- false information may be introduced in the system;
- a wire-tapped message may be replayed;
- the sender may be impersonated;
- false signature may be forged.

By a *signature operation*, the sender prepares a signed message.
By a *verification operation*, the receiver checks the signed message.
When the integrity is threatened, at least the receiver must protect his operation.

Each operation may be described
as an algorithm controlled by parameters such a key.
In order to protect an operation, the key at least should be kept secret.

Each signature scheme implies three fundamental operations ([12]): the *key production*, the *signature* and the *verification*. In each signature system, there are five types of partners: the *prover*, the *verifier*, the *cheater*, the trusted *authority* managing the identities of users and hot lists and the *judge* evaluating disputes and repudiations.

In an interactive authentication process, the verifier reacts in a random way. Let us use a hash function to replace the interactivity between the prover and the verifier. We are facing now a signature scheme based on a non-interactive zero-knowledge technique. Our contribution in this field is not the basic ideas ([8], [17]), but rather a first synthesis between two basic ideas.

Let us consider the security level (related to the value of v): in a proximity relation with a policeman, nobody will try to show a forged driving licence with probability $1 - 10^{-4}$ of being caught. Some people may try up 10 000 times to remotely access to a database, and in a remote control, the question must then be 20 bit long. But in a signature scheme where a simulation may be secretly forged off-line, the level of security must be raised to 60 bit long questions. Even with the most powerful computers, it is unrealistic to try 10^{18}.

Here is the *signature operation*:

1. At each signature, the card picks at random an element of Z_n, and computes as the *test* T the vth power of r in Z_n, transmitted to the PC.

2. The PC (or the card depending upon the application) hashes the message M and the test T in an integer d uniformly selected from 0 to $v-1$. This integer is transmitted to the card as the question d.

3. The card computes as the *witness* t the product in Z_n of the integer r by the dth power of the authentication number B. The consecutive computations are

$$T = r^v \bmod n; d = h(M,T); t = r \cdot B^d \bmod n.$$

The signed message consists of the message M followed by a very compact appendix including the identity I, the question d and the witness t.

The *verification operation* consists of reconstructing the test T from the witness t, the question d and the identity I, knowing n, v and the redundancy rules.

This method is still *zero-knowledge about the authentication number* included in the card. Even an enemy using a stolen card, while producing signatures, will learn nothing about the specific value of the authentication number. Ans when a card is hashing itself M and T, the property is still maintained, because the same hashing should have been done outside the card. While making forgery easy, a weak hash function should not endanger the secret.

7 The identity-based signature scheme

This signature scheme *with appendix* is a probabilistic scheme based upon an underlying signature scheme *shadow* ([12]).

The underlying signature scheme is based on user's identities. For a bank, such an identity includes an account number, a validity period and a usage code, associated with the serial number of the chip embedded in the card.

We now propose to use as hash function the collision-resistant permutations analyzed in the second paragraph and related to the underlying signature scheme with shadow. Thus the security of the hash function is homogeneous with the security of the zero-knowledge scheme.

Resulting from hashing the message M and the test T, the question d is an element of Z_n. A shortening of the question d should result in a partial collision in Z_n, which does not give the authentication number. The proof of equivalence would thus disappear. In order to accept such large questions, v is a prime between $v/2$ and n.

The resulting signature scheme is *paradoxical*:

- An enemy having received as many signatures of messages of his choice as he wants is not able to produce only one additional signature unless he has broken the underlying problem and reconstructed the authentication number of the user.

- A user trying to repudiate one signature by producing a second message with the same appendix, should reveals a collision and thus his authentication number.

This is the *signature scheme*:

1. At each signature, the card picks at random an integer r in Z_n and computes as *test* T the pth power of r in Z_n, transmitted to the PC.

2. The PC hashes the message M and the test T by computing as question d the product in Z_n of the Mth power of J by the vth power of T $(v^{k-1} \leq M \leq v^k)$.

3. The card computes as *witness* t the product in Z_n of r by the dth power of the authentication number B.

Let us summarize these successive computations (k is such that $v^{k-1} \leq M \leq v^k$):

$$T = r^v \bmod n; d = J^M \cdot T^{v^k} \bmod n; t = r \cdot B^d \bmod n.$$

NOTE. At each signature, the integer r is picked at random in Z_n. In a practical implementation in a smart card, the random generation is difficult to control. A deterministic production of r should be very useful. How to specify a secure deterministic generation of r? Such a computation should imply both the authentication number B as a secret seed and the whole message M to be signed which should include at least a time stamp.

In such an implementation, for security reasons, the whole process should be performed inside the card, like this one:

At each signature, the card receives as argument a message M to be signed.

1. From this message M and from the authentication number B, the card generates an integer r in Z_n.

2. The card raises the integer r to the vth power in Z_n to get the test number T.

3. The card computes as the *question* d the product in Z_n of the Mth power of J by the $(p^k)^{\text{th}}$ power of T $(v^{k-1} \leq M \leq v^k)$.
4. The card computes as the *witness* t the product in Z_n of r by the dth power of the authentication number B. After this sequence, the cards delivers the question d and the witness t.

The *verification operation* includes the successive reconstructions of the test T and the question d.

1. The test T is reconstructed as the product in Z_n of the dth power of the shadowed identity J by the Vth power of the witness t.
2. The question d is reconstructed as the product in Z_n of the Mth power of the shadowed identity J by the $(v^k)^{\text{th}}$ power of the test number T.

Let us summarize these computations:

$$J^d \cdot t^v \bmod n = J^d \cdot (B^d \cdot r)^v \bmod n = (J \cdot B^v)^d \cdot r^v \bmod n = r^v \bmod n = T,$$

and,

$$d = J^M \cdot T^{(v^k)} \bmod n.$$

The whole verification collapses in a simple equation:

$$\text{Is the question } d \text{ equal to } J^{M+d \cdot v^k} \cdot t^{v^{k+1}} \bmod n\ ?$$

8 Exchange authentication: *a priori* versus *a posteriori*?

Some proposals are made today to standardize authentication protocols beginning by an authentication sequence keying a pair of communicating entities. Subsequently doing the difference with the other entities on the network, this key ensures integrity of subsequent transmissions by ciphering either exchanged data or at least an imprint computed by hashing these data. This shared key must be kept secret. *A priori* authentication is mandatorily a procedure establishing a shared secret key in the pair of communicating entities. Such methods sadly confuse integrity and confidentiality while public key techniques seem powerful to provide separate solutions to the two classes of threatens against confidentiality on one hand and integrity on the other hand.

When a *priori* authentication is needed to limit misusing of gate resources by intruders, this authentication should not be used to ensure integrity of subsequent exchanges. But a second (a *posteriori*) authentication should rather be performed after the exchanges in order to check both integrity of previous exchanges and identification of communicating entities.

Operation sequencing is correct only in an a *posteriori* authentication when the authentication protocol occurs after the exchange of information. The zero-knowledge techniques are typically used after an exchange of clear information.

In a *posteriori* authentication, another subtlety appears between:

- *keyed systems*, where each user owns his secret key, like a composite number, usable for general purposes with the help of a registration authority.

- and *keyless systems*, based upon identities, where each user owns an authentication number delivered by a trusted authority for some dedicated purposes.

In a keyed system, confidentiality and integrity are both provided. The only solution (the RSA scheme) proposed today in CCITT X.509 (authentication framework) is in this category. While being useful in some circumstances, such a method is not dedicated to integrity: confidentiality is easily obtained.

In a keyless identity-based system, the communicating entities are not able to produce a common secret key: secrecy cannot be derived from the scheme.

Let us notice that in both cases, an authority (either a general multi-purpose authority or several dedicated authorities) has to play a prominent part! The keyless systems with multiple authorities fit better with the bright proposals of Chaum ([2]) on privacy protection. This is also an important point!

Integrity techniques are typically used on various remote control sysytems in such a way that no assumption has to be done on the security of networks and terminals used in the transaction. Why some assumptions should be done on the morality of potential users?

It seems to us that a *good* integrity scheme does not have to do any assumption on the integrity of the potential users.

Thus the conjunction of zero-knowledge techniques and identity-based techniques solves some political problems due to the use of cryptologic techniques on public networks. At least one signature scheme exists which cannot be misused and illegally transformed into a confidentiality scheme.

An identity-based scheme should be taken into account in X.509.

References

[1] Gilles Brassard, David Chaum and Claude Crépeau, *Minimum disclosure proofs of knowledge*, July 1987.

[2] David Chaum, *Security without identification: transaction systems to make Big Brother obsolete*, Comm. of ACM, **28**, Oct. 1985, pp. 1030–1044.

[3] Ivan Bjerre Damgård, *Collision-free hash functions and public-key signature schemes*, EUROCRYPT '87, to appear.

[4] Yvo Desmedt and Jean-Jacques Quisquater, *Public-key systems based on the difficulty of tampering*, Advances in cryptology, Proceedings of CRYPTO '86, Lectures notes in computer science, N° 263, Springer-Verlag, pp. 186–194.

[5] Amos Fiat and Adi Shamir, *How to prove yourself: practical solutions to identification and signature problems.* Springer Verlag, Lecture notes in computer science, N° 263, Advances in cryptology, Proceedings of CRYPTO '86, pp. 186–194, 1987.

[6] Amos Fiat and Adi Shamir, *Unforgeable proofs of identity*, 5th SECURICOM, Paris, 1987, pp. 147–153.

[7] Oded Goldreich, Shafi Goldwasser and Silvio Micali, *How to construct random functions*, 25th, IEEE symposium on foundations of computer science, 1984, pp. 464–479.

[8] Shafi Goldwasser, Silvio Micali and Charles Rackoff, *The knowledge of interactive proof systems*, 17th ACM symposium on theory of computing, 1985, pp. 291–304.

[9] Shafi Goldwasser, Silvio Micali and Ronald Rivest, *A paradoxical signature scheme*, 25th IEEE symposium on foundations of computer science, 1984, pp. 441–448.

[10] Oded Goldreich, Silvio Micali and Avi Wigderson, *Proofs that yields nothing but the validity of the proof*, Workshop on probabilistic algorithms, Marseille, March 1986.

[11] Louis C. Guillou and Jean-Jacques Quisquater, *Efficient digital public-key signatures with shadow*, Springer Verlag, Lecture notes in computer science, Advances in cryptology, Proceedings of CRYPTO '87, p. 223.

[12] Louis C. Guillou, Marc Davio and Jean-Jacques Quisquater, *Public-key techniques*, Cryptologia, to appear.

[13] Louis C. Guillou and Jean-Jacques Quisquater, *A practical zero-knowledge protocol fitted to security microprocessors minimizing both transmission and memory*, EUROCRYPT '88, to appear.

[14] Louis C. Guillou and Michel Ugon, *Smart card: a highly reliable and portable security device*, CRYPTO '86, Lecture notes in computer science, N° 263, Springer-Verlag, pp. 464-479.

[15] Jean-Jacques Quisquater, *Secret distribution of keys for public-key system*, Springer Verlag, Lecture notes in computer science, N° 293, Advances in cryptology, Proceedings of CRYPTO '87, pp. 203–208, 1987.

[16] Ronald Rivest, Adi Shamir and Leonard Adleman, *A method for obtaining digital signatures and public-key cryptosystems*, Comm. of ACM, **21**, Feb. 1978, pp. 120–126.

[17] Adi Shamir, *Identity-based cryptosystems and signatures schemes*, Springer Verlag, Lecture notes in computer science, N° 196, Advances in cryptology, Proceedings of CRYPTO '84, pp. 47–53, 1985.

[18] H. C. Williams, *A modification of the RSA public-key cryptosystem*, IEEE Trans. on Information Theory, **IT-26**, Nov. 1980, pp. 726–729.

A Modification of the Fiat-Shamir Scheme

Kazuo Ohta *Tatsuaki Okamoto*

NTT Communications and Information Processing Laboratories
Nippon Telegraph and Telephone Corporation
1-2356, Take, Yokosuka-shi, Kanagawa-ken, 238-03, Japan

Abstract: Fiat-Shamir's identification and signature scheme is efficient as well as provably secure, but it has a problem in that the transmitted information size and memory size cannot simultaneously be small. This paper proposes an identification and signature scheme which overcomes this problem. Our scheme is based on the difficulty of extracting the L-th roots mod n (e.g., $L = 2 \sim 10^{20}$) when the factors of n are unknown. We define some variations of no transferable information and prove that the sequential version of our scheme is a zero knowledge interactive proof system and our parallel version satisfies these variations of no transferable information under some conditions. The speed of our scheme's typical implementation is at least one order of magnitude faster than that of the RSA scheme and is relatively slow in comparison with that of the Fiat-Shamir scheme.

1. Introduction

Fiat and Shamir have proposed an identification and signature scheme which is promising because it is efficient and provably secure against any active attack [FS]. Their scheme is based on the difficulty of extracting square roots mod n when the factors of n are unknown. The Fiat-Shamir scheme consists of sequential and parallel versions. Though their sequential version is a zero knowledge interactive proof system [FFS], the iteration number must be $O(\log_2 n)$ and the communication performance is therefore low. The parallel version is more efficient than the sequential version, and it is secure because it reveals no transferable information [FFS]. There is, however, a trade-off between the transmitted information size and memory size. That is, the probability of forgery is $1/2^{kt}$, where k denotes the number of secret information integers and the overall transmitted information size is proportional to t. For example, in order to attain the security level 2^{-20}, i.e., $tk = 20$, when we reduce the information size to $t = 1$, we must store twenty ($k = 20$) secret integers. When we store only one secret integer, $k = 1$, we must send twenty ($t = 20$) times as long a message. Therefore, the efficient parameter values, $t = k = 1$, cannot be used in their scheme.

In this paper, we propose an identification and signature scheme which overcomes the above mentioned problem. Our scheme is based on the difficulty of extracting the L-th roots mod n when factors of n are unknown. In our scheme, the third design parameter L is introduced in addition to the two parameters t and k which correspond to t and k of the Fiat-Shamir scheme. Here, the security level is represented as $L^{-t \cdot k}$. Therefore, the parameter values $t = k = 1$ are applicable in our scheme if the appropriate value for L is chosen, although our scheme is relatively slow in comparison with the Fiat-Shamir scheme. Hence our scheme is suitable for smart cards, because their memory amounts are restricted.

We define new security level notions of "transferable information *with a (strict) security level ρ* " and "transferable information *with a (strict) sharp-threshold security level ρ* ." We then prove that the sequential version of our scheme is a perfect zero knowledge interactive proof system for any L. We go on to prove that our parallel version reveals no transferable information with a strict security level $1/p'$, where p and q are factors of n, $p' = (L, p-1) > 1$, $p' \geq q' = (L, q-1)$, if the factoring is difficult and an additional condition holds, where (a, b) denotes the greatest common divisor of a and b. Finally we also prove that our parallel version releases no transferable information with a strict sharp-threshold security level $1/L$ when $(L, p-1) = L$ and an additional condition holds.

Although the idea of using higher roots was implied in [FS], [GQ1] and [GQ2], its security and parameter conditions were not formally discussed.

In the following sections, we consider a typical case where $k = 1$ for the sequential version and $k = t = 1$ for the parallel version. Our results are easily extended to cases where k and t have other values.

2. Some Number-Theoretic Results

First some number-theoretic results are shown concerning the modular L-th roots.

[Lemma 1] Let p be an odd prime, L be an integer ($L \geq 2$) and $p' = (L, p-1)$. If y is the L-th residue mod p, then there are p' integers x of the L-th root mod p of y such that $x^L \equiv y \pmod{p}$.

Proof Let g be a primitive element over a finite field $GF(p)$, let α satisfy $g^\alpha \equiv x \pmod{p}$ and let β satisfy $g^\beta \equiv y \pmod{p}$. Then $x^L \equiv y \pmod{p}$ implies $(g^\alpha)^L \equiv g^\beta \pmod{p}$. Here α satisfies $\alpha L \equiv \beta \pmod{p-1}$; therefore, it has p' solutions [HW, pp.51-52]. *Q.E.D.*

[Lemma 2] Let p be an odd prime, L be an integer ($L \geq 2$) , y be the L-th residue mod p and $p' = (L, p-1) \geq 2$. If $\{x_1, \ldots, x_{p'}\}$ is the set of the L-th roots mod p of y, then any pair (x_i, x_j) satisfies $x_i^{p'} \equiv x_j^{p'} \pmod{p}$ $(1 \leq i, j \leq p')$ and there is at least one pair (x_i, x_j) such that $i \neq j$ and $x_i^{p'-1} \not\equiv x_j^{p'-1} \pmod{p}$.

Proof Let g be a primitive element over $GF(p)$ and let α_i satisfy $g^{\alpha_i} \equiv x_i \pmod{p}$ $(1 \leq i \leq p')$. Since a congruence $x_i^L \equiv x_j^L \equiv y \pmod{p}$ implies $L(\alpha_i - \alpha_j) \equiv 0 \pmod{p-1}$, $p-1$ is a divisor of $L(\alpha_i - \alpha_j)$. Here p' is the greatest common divisor of L and $p-1$. Thus, $p-1$ is a divisor of $p'(\alpha_i - \alpha_j)$. Therefore, the congruence $p'\alpha_i \equiv p'\alpha_j \pmod{p-1}$ holds, and we finally obtain $x_i^{p'} \equiv (g^{\alpha_i})^{p'} \equiv (g^{\alpha_j})^{p'} \equiv x_j^{p'} \pmod{p}$ $(1 \leq i, j \leq p')$.

Assume that any x_i satisfies $x_i^{p'-1} \equiv z \pmod{p}$ $(1 \leq i \leq p')$. Thus, there are p' integers of the $(p'-1)$-th roots mod p of z. Here, the number of the $(p'-1)$-th roots mod p of z is at most $p'-1$ according to Lemma 1 because $(p'-1, p-1) \leq p'-1$. This is a contradiction. Therefore, there is at least one pair (x_i, x_j) such that $i \neq j$ and $x_i^{p'-1} \not\equiv x_j^{p'-1} \pmod{p}$. *Q.E.D.*

We classify the L-th roots mod n of 1 in order to calculate the probability of successfully factoring n.

[Definition 1] Let L be an integer $(L \geq 2)$ and n be a composite number which is the product of two odd primes p and q. Four types of the L-th roots mod n of 1 are defined as follows:

$$\begin{aligned} &\omega \text{ is } \textit{Type1} \text{ if } \omega = [1, 1],\\ &\omega \text{ is } \textit{Type2} \text{ if } \omega = [1, \omega_q],\\ &\omega \text{ is } \textit{Type3} \text{ if } \omega = [\omega_p, 1],\\ &\omega \text{ is } \textit{Type4} \text{ if } \omega = [\omega_p, \omega_q],\end{aligned}$$

where the notation $\omega = [a, b]$ means that ω satisfies the following congruences:

$$\omega \equiv \begin{cases} a \pmod{p} \\ b \pmod{q}, \end{cases}$$

and where ω_p satisfies $1 + \omega_p + \ldots + \omega_p^{L-2} + \omega_p^{L-1} \equiv 0 \pmod{p}$ and ω_q satisfies $1 + \omega_q + \ldots + \omega_q^{L-2} + \omega_q^{L-1} \equiv 0 \pmod{q}$.

[Lemma 3] Let L be an integer $(L \geq 2)$, n be a composite number which is the product of two odd primes p and q and ω be one of the L-th roots mod n of 1. Then,

$$\begin{aligned} &\#\{\omega \mid \omega \text{ is } \textit{type1}\ \} = 1,\\ &\#\{\omega \mid \omega \text{ is } \textit{type2}\ \} = q' - 1,\\ &\#\{\omega \mid \omega \text{ is } \textit{type3}\ \} = p' - 1,\\ &\#\{\omega \mid \omega \text{ is } \textit{type4}\ \} = (p' - 1)(q' - 1)\end{aligned}$$

where $p' = (L, p-1)$, $q' = (L, q-1)$, and $\#$ denotes the number of elements of a set.

Proof The following equation with respect to ω has p' solutions in $GF(p)$ according to Lemma 1: $1 - \omega^L \equiv (1-\omega)(1 + \omega + \ldots + \omega^{L-2} + \omega^{L-1}) \equiv 0 \pmod{p}$. Thus, $\#\{\omega_p \bmod p \mid 1 + \omega_p + \ldots + \omega_p^{L-2} + \omega_p^{L-1} \equiv 0 \pmod{p)}\} = p' - 1$. Similarly, $\#\{\omega_q$ mod

$q \mid 1+\omega_q+\ldots+\omega_q^{L-2}+\omega_q^{L-1} \equiv 0 \pmod q\} = q'-1$. Moreover, $\#\{\omega \mid \#\{\omega \bmod p\} = \alpha$ and $\#\{\omega \bmod q\} = \beta\} = \alpha \cdot \beta$. Therefore, the above property is proven. *Q.E.D.*

[Theorem 1] Let L be an integer ($L \geq 2$), n be a composite number which is the product of two odd primes p and q, I be the L-th residue mod n, and AL be a probabilistic polynomial time algorithm which, given I and n, finds one of the L-th roots mod n of I with probability ($> 1/|n|^a$), where $|n|$ denotes the data length of n. If $(L, p-1) \neq 1$ or $(L, q-1) \neq 1$, then there exists a probabilistic polynomial time algorithm for factoring n using AL at most in $O(|n|^{a+2b})$ steps, where b satisfies $L = O(|n|^b)$.

Proof Choose a random integer $y \in Z_n$, where Z_n denotes $\{0, \ldots, n-1\}$, calculate $z = y^L \bmod n$ and compute x which is one of the L-th roots mod n of z by using AL. Because the distribution of x doesn't depend on which y is selected, and because y is randomly selected, $\omega = x/y \bmod n$ is uniformly distributed. If ω is *type2* or *type3*, we can calculate the factors of n by computing $(\omega - 1, n)$. Note that when ω is *type1*, $(\omega - 1, n) = n$; and when ω is *type4*, $(\omega - 1, n) = 1$. The probability of $\{\ \omega$ is *type2* or *type3* $\}$ is $\frac{p'+q'-2}{p'q'}$ according to Lemma 3. Moreover, the inequation $\frac{p'+q'-2}{p'q'} \geq \frac{1}{p'q'}$ holds because of the assumption $(L, p-1) \neq 1$ or $(L, q-1) \neq 1$. The average number of iterations for deriving x from z using AL is $|n|^a$. The average number of iterations for selecting x such that ω is *type2* or *type3* is at most $p'q' = O(|I|^{2b})$. Therefore, the total average number of iterations for the factorization of n is at most $O(|n|^{a+2b})$. *Q.E.D.*

[Definition 2] Let p be a prime and $a \in GF(p)$. An index of a over $GF(p)$, $Ind_p(a)$, is defined as follows:

$$Ind_p(a) = \min\{m \mid a^m \equiv 1 \pmod p\}.$$

[Lemma 4] Let p be a prime, $J \in GF(p)$ be the p'-th residue mod p, where $(p', p-1) = p'$ and $p' = r_1 \cdot r_2$ ($r_1, r_2 > 1$), K be one of the r_1-th roots mod p of J, $v_i (i = 1, 2, \ldots, r_2)$ be the r_2-th roots of K, v_0 be one of the p'-th roots of J, and $\omega_i = v_i/v_0 \bmod p$. If there is an integer $\delta (> 1)$ which satisfies $(\delta, r_1) = 1$ and $(\delta, r_2) = \delta$, for any K and v_0, there is at least one pair (ω_i, ω_j) such that $Ind_p(\omega_i) \neq Ind_p(\omega_j)$.

Proof Let g be a primitive element over $GF(p)$, let α satisfy $g^{r_1 \cdot r_2 \cdot \alpha} \equiv J \pmod p$. Then,

$$K = g^{r_2 \cdot \alpha + (p-1) \cdot \frac{j_1}{r_1}} \bmod p,$$

where $0 \leq j_1 \leq r_1 - 1$. Thus,

$$v_i = g^{\alpha + (p-1) \cdot \frac{j_1 + i \cdot r_1}{r_1 \cdot r_2}} \bmod p,$$

where $0 \leq i \leq r_2 - 1$. Similarly,

$$v_0 = g^{\alpha+(p-1)\cdot\frac{j_0+i_0\cdot r_1}{r_1\cdot r_2}} \bmod p.$$

Therefore,

$$\omega_i = g^{(p-1)\cdot\frac{(j_1-j_0)+(i-i_0)\cdot r_1}{r_1\cdot r_2}} \bmod p.$$

When $j_0 \neq j_1$, for any i_0, j_0 and j_1, there are two integers u $(-\delta < u < \delta)$ and v $(-r_1 < v < r_1)$ such that $(j_1-j_0)+u\cdot r_1 = v\cdot\delta$, because $(r_1,\delta) = 1$. Put $i_1 = i_0+u \bmod r_2$, then $(\omega_{i_1})^{\frac{r_1\cdot r_2}{\delta}} \equiv g^{(p-1)\cdot v} \equiv 1 \pmod p$. Thus, $Ind_p(\omega_{i_1})$ divides $\frac{r_1\cdot r_2}{\delta}$. On the other hand, there is an integer i_2 $(0 \leq i_2 \leq r_2 - 1)$ such that $(j_1 - j_0) + (i_2 - i_0)r_1 \not\equiv 0 \pmod \delta$, because $(r_1,\delta) = 1$ and $\delta \geq 2$. Thus, $Ind_p(\omega_{i_2})$ does not divide $\frac{r_1\cdot r_2}{\delta}$. Therefore, $Ind_p(\omega_{i_1}) \neq Ind_p(\omega_{i_2})$. When $j_0 = j_1$, then $\omega_{i_0} = 1$ and $\omega_i \neq 1$ $(i \neq i_0)$. Therefore, $Ind_p(\omega_{i_0}) \neq Ind_p(\omega_i)$. *Q.E.D.*

3. Sequential Version

An identification scheme is proposed here in which a prover convinces a verifier that he is a real prover. Hereafter we denote a real prover as $\overline{A}$, an invalid prover as $\widetilde{A}$, a real verifier as $\overline{B}$ and an invalid verifier as $\widetilde{B}$.

A trusted center publishes an integer L $(L \geq 2)$ and a modulus n which is the product of two secret large primes p and q. $\overline{A}$ publishes I which is calculated by $I = S^L \bmod n$ using a secret random integer $S \in Z_n$. Note that the difficulty of deriving S from I corresponds to that of breaking the RSA scheme [RSA] in the case where $(L, p-1) = 1$ and $(L, q-1) = 1$, and corresponds to the difficulty of factoring n in the case where $(L, p-1) \neq 1$ or $(L, q-1) \neq 1$ according to Theorem 1.

To generate and verify a proof of identity, the parties execute the following procedure. Repeat Steps 1 to 4 in sequence t times:

Step 1) $\overline{A}$ generates a random integer $R \in Z_n$ and sends $X = R^L \bmod n$ to $\overline{B}$.
Step 2) $\overline{B}$ sends a random integer $E \in Z_L$ to $\overline{A}$.
Step 3) $\overline{A}$ sends $Y = R \cdot S^E \bmod n$ to $\overline{B}$.
Step 4) $\overline{B}$ verifies that $Y^L \equiv X \cdot I^E \pmod n$.

Verifier $\overline{B}$ accepts prover $\overline{A}$'s proof of identity only if all the checks are successful t times. Note that there is no constraint on the relation among L, p and q.

The following theorem guarantees the security of our sequential version.

[Theorem 2] This protocol is an interactive proof system of knowledge of the S [FFS] which is perfect zero knowledge [GMW], when $t = O(|n|)$ and $L = O(1)$.

Proof (sketch) *Completeness:* To prove that $\overline{A}$'s proof always convinces $\overline{B}$, we evaluate the verification condition: $Y^L \equiv (R \cdot S^E)^L \equiv R^L(S^L)^E \equiv X \cdot I^E \pmod n$. Thus, the verifier accepts $\overline{A}$'s proof with probability 1.

Soundness: Our goal is to show that whenever $\overline{B}$ accepts $\widetilde{A}$'s proof with non-negligible probability ($> 1/|n|^a$), a probabilistic polynomial time Turing machine M can output the S', which satisfies $S'^L \equiv I \pmod{n}$, with overwhelming probability.

Let T be the truncated execution tree of $(\widetilde{A}, \overline{B})$ for input I and $\widetilde{A}$'s random tape RA. A vertex is called "heavy" if it has more than $L/2$ sons. First, we prove that at least half the vertices in at least one of the levels in T must be heavy, then that M can find a heavy vertex in T with overwhelming probability, and finally that S' can be computed from the sons of any heavy vertex when a heavy vertex is found.

Let $\alpha_i = \beta_{i+1}/\beta_i$ where β_i means the number of vertices at level i in T. If $\alpha_i < (3/4)L$ for all $1 \leq i \leq t$, then the total number of leaves in T($i.e.$, $\beta_t = \alpha_1 \cdots \alpha_{t-1} \cdot \beta_1$) is bounded by $(3/4)^{t-1}L^t$, which is a negligible fraction of the L^t possible leaves. Since we assume that this fraction is polynomial, $\alpha_i > (3/4)L$ for at least one level, which we denote i_0. Assume at least half the vertices at this level (i_0) are not heavy, then $\beta_{i_0+1} < \beta_{i_0} \cdot L - (\beta_{i_0}/2)(L/2) = (3/4)\beta_{i_0} \cdot L$, and $\alpha_{i_0} = \beta_{i_0+1}/\beta_{i_0} < (3/4)L$. Here $\alpha_{i_0} > (3/4)L$. This is a contradiction. Therefore, it is proven that at least half the vertices in at least one of the levels in T must be heavy.

In order to find a heavy vertex in T, M explores random paths in the untruncated tree by determining the degree of each vertex and restarts from the root whenever the path encounters an improperly answered query. Since a non negligible fraction ($> 1/|n|^a$) of leaves is assumed to survive the truncation, the average iteration number of the executions where the path reaches to the t-th level is $|n|^a(t \cdot L)$. Since there is at least one level in T where at least half the vertices are heavy, the average iteration number of the executions where M can find a heavy vertex is at most $2|n|^a(t \cdot L) = O(|n|^{1+a})$.

Finally, we will show how S' can be computed from the sons of any heavy vertex, when a heavy vertex is found. Let Q be the set of queries E which are properly answered by $\widetilde{A}$. Assume that all pairs of integers (E', E'') satisfy $E'' - E' > 1$ where $E', E'' \in Q$. Since $\#Q > L/2$, the largest difference between elements in Q is at least L. Here $\#Z_L = L$ and the largest difference between elements in Z_L is at most $L - 1$. This is a contradiction. Therefore, it is proven that a set Q of more than $L/2$ integers of Z_L must contain at least one pair of integers (E_1, E_2) such that $E_1 - E_2 = 1$. Since these queries were properly answered, the following verification conditions hold, where $X_i = X$; $Y_i^L \equiv I^{E_i} \cdot X \pmod{n}$ $(1 \leq i \leq 2)$. From these equations, we obtain $S' = Y_1/Y_2 \bmod n$ which satisfies the relation $S'^L \equiv (Y_1/Y_2)^L \equiv I^{E_1 - E_2} \equiv I \pmod{n}$.

Zero knowledge: Let $\widetilde{B}$ be any polynomial expected time algorithm for the verifier. The simulator $M_{\widetilde{B}}$ does the following:

repeat while $0 \leq c \leq t$

begin

choose $E' \in Z_L$ randomly and uniformly
choose $Y = R \in Z_L$ randomly and uniformly
$X = R^L / I^{E'} \pmod{n}$
$\widetilde{B}$ issues E
if $E = E'$ then halt, $c = c + 1$ and output (X, E, Y)
end

It can be demonstrated that $(\overline{A}, \widetilde{B})(n, I)$ and $M_{\widetilde{B}}(n, I)$ are identically distributed verifier's histories. For any verifier $\widetilde{B}$, the probability that $E = E'$ is at least $1/L$. Thus, the average running time of this simulator $M_{\widetilde{B}}$ is $O(t \cdot L)$, which is polynomial in $|n|$ based on our assumptions of the values of L and t. *Q.E.D.*

Remark: Evidently, we can extend the value of L to $O(|n|)$ in the above theorem.

4. Parallel Version

In this section, we consider a typical case where $k = t = 1$ and $p' = (L, p-1) > 1$ and $p' \geq q' = (L, q-1)$ for the parallel version. In this case, the difficulty of deriving S from I corresponds to that of factoring n according to Theorem 1. We define four security level notions of "transferable information with a (strict) security level ρ" and "transferable information with a (strict) sharp-threshold security level ρ," which are more rigorous than the notion "transferable information" defined by [FFS].

[Definition 3] The protocol $(\overline{A}, \overline{B})$ releases *no transferable information with a security level* ρ if:

1. It succeeds with overwhelming probability.
2. There is no coalition of $\widetilde{A}, \widetilde{B}$ with the property that, after a polynomial number of executions of $(\overline{A}, \widetilde{B})$, it is possible to execute $(\widetilde{A}, \overline{B})$ with $c \cdot \rho$ probability of success, where c is an arbitrary real constant greater than 1.

The protocol $(\overline{A}, \overline{B})$ releases *no transferable information with a strict security level* ρ if:

1. It succeeds with overwhelming probability.

2'. There is no coalition of $\widetilde{A}, \widetilde{B}$ with the property that, after a polynomial number of executions of $(\overline{A}, \widetilde{B})$, it is possible to execute $(\widetilde{A}, \overline{B})$ with $c \cdot \rho$ probability of success, where c is $(1 + 1/|n|^d)$ and d is an arbitrary constant greater than 0.

The protocol $(\overline{A}, \overline{B})$ releases *no transferable information with a sharp-threshold security level* ρ if it satisfies conditions 1 and 2 above as well as the following condition:

3. The probability of $\widetilde{A}$ cheating $\overline{B}$ is ρ.

The protocol $(\overline{A}, \overline{B})$ releases *no transferable information with a strict sharp-threshold security level* ρ if it satisfies conditions 1, 2' and 3.

It has been proven that Fiat-Shamir's parallel version of the identification scheme releases no transferable information with a sharp-threshold security level [FFS], but not with a *strict* sharp-threshold security level. The following theorem and corollary guarantee the security of our parallel version using the new notion, "transferable information with a *strict (sharp-threshold)* security level." The following results are easily extended to the situation where k and t have other values.

[Theorem 3] Let the parameters $k = t = 1$ and L satisfy $(L, p-1) = p' > 1$, $p' \geq (L, q-1) = q'$, and $L = O(1)$. When at least one of the following conditions C1, C2, C3, and C4 is satisfied, then the parallel version of our identification scheme releases no transferable information with a strict security level $1/p'$, if there is no probabilistic polynomial time algorithm of factoring.

C1. $p' = \prod_{i=1}^{N} p_i$, where p_i is a prime number, $p_i \neq p_j$ $(i \neq j)$, and $N \geq 1$.

C2. $q' = \prod_{i=1}^{M} q_i$, where q_i is a prime number, $q_i \neq q_j$ $(i \neq j)$, and $M \geq 1$.

C3. $q' = 1$.

C4. $(p', q') = 1$

Proof (sketch) Let $L = p' \cdot l_p$. To prove this theorem, we show that if $(\tilde{A}, \overline{B})$ can be executed with probability $\varepsilon = c/p' = (1 + 1/|n|^d)/p'$ after $O(|n|^e)$ executions of $(\overline{A}, \tilde{B})$, then n can be factored by a coalition of $\overline{A}, \tilde{A}, \overline{B}$ and $\tilde{B}$ at most in time $O(\|\tilde{B}\| \cdot |n|^e + \|\tilde{A}\| \cdot |n|^d)$ and with overwhelming probability, where d and e are positive constants, and $\|A\|$ and $\|B\|$ denote the time complexity of A and B.

Given any pair of unusually successful programs $\tilde{A}$ and $\tilde{B}$, we start the factorization by executing $(\overline{A}, \tilde{B})$ $O(|n|^e)$ times and relaying a transcript of the communication to $\tilde{A}$. Since $\overline{A}$ itself can be used in this part and its time complexity $\|\overline{A}\|$ is assumed to be dominated by $\|\tilde{B}\|$, these executions require $O(\|\tilde{B}\| \cdot |n|^e)$.

The possible outcomes of the executions of $(\tilde{A}, \overline{B})$ can be summarized in a large Boolean matrix H whose rows correspond to all possible choices of RA. Its columns correspond to all the possible choices L of RB, and its entries are 1 if $\overline{B}$ accepts $\tilde{A}$'s proof, and 0 if otherwise.

To factor n, the coalition tries to find at least $(l_p + 1)$ 1's along the same row in H. We call a row "heavy" if the number of 1's along it is at least $l_p + 1$. Assume that at least $1/c$ of the 1's in H are not located in heavy rows. Then the fraction of non-heavy rows in H, which we denote τ, is estimated as follows: $\tau > \frac{\varepsilon \cdot L \cdot 1/c}{l_p} = 1$. This is a contradiction. Therefore, at least $(1 - 1/c)$ of the 1's in H are located in heavy rows. We thus adopt the following strategy:

1. Probe $O(1/\varepsilon)$ random entries in H.

2. After the first 1 is found, probe $l_p O(1/\varepsilon)$ random entries along the same row. Because $\frac{1}{1-1/c} = \frac{c}{c-1} = 1 + |n|^d$, we can find a heavy row with constant probability in just $\frac{c}{c-1} \cdot \{O(\frac{1}{\varepsilon}) + l_p \cdot O(\frac{1}{\varepsilon})\} = \frac{c}{c-1} \cdot \{(1 + l_p)O(\frac{1}{\varepsilon})\} < O(|n|^d)$ probes. Again we assume that $\|\overline{B}\|$ is dominated by $\|\tilde{A}\|$, and thus the time complexity of this part of

the algorithm is at most $O(||\widetilde{A}|| \cdot |n|^d)$.

Next, we will prove that n can be factored by a coalition of $\overline{A}, \widetilde{A}, \overline{B}$ and $\widetilde{B}$ in polynomial time and with probability at least $1/p'$, when the coalition finds at least $(l_p + 1)$ 1's along the same row in H.

Let Q be the set of queries E which are properly answered by $\widetilde{A}$. Assume that all pairs of queries $(E', E")$ satisfy $E" - E' \geq p'$ where $E', E" \in Q$. Since $\#Q \geq (l_p+1)$, the largest difference between elements in Q is at least $l_p \cdot p' = L$. Here $\#Z_L = L$ and the largest difference between elements in Z_L is at most $L-1$. This is a contradiction. Therefore, it is proven that a set Q of at least $(l_p + 1)$ integers of Z_L must contain at least one pair of integers (E_1, E_2) such that $E_1 - E_2 < p'$.

Let (X, E_1, Y_1) and (X, E_2, Y_2) be the two 1's in Q, i.e., the two possible outcomes of the execution of $(\widetilde{A}, \overline{B})$, that satisfy $E_1 - E_2 < p'$. Since (X, E_1, Y_1) and (X, E_2, Y_2) satisfy the equation $(Y_1/Y_2)^L \equiv I^{E_1-E_2} \pmod{n}$, thus $Y_1/Y_2 \bmod n$ is one of the L-th roots mod n of $I^{E_1-E_2} \bmod n$, and $S^{E_1-E_2} \bmod n$ is also one of the L-th roots mod n of $I^{E_1-E_2} \bmod n$, where S is known by $\overline{A}$.

We claim that from the X's and Y's sent by $\overline{A}$ during the execution of $(\overline{A}, \widetilde{B})$ even an infinitely powerful $\widetilde{B}$ cannot determine which L-th root mod n of I $\overline{A}$ acturally uses. This can be shown as follows: let ω be one of the L-th roots mod n of 1, then $S' = \omega \cdot S$ is another L-th root mod n of I other than S. If $\overline{A}$ replaces S with S', $\overline{A}$ produces the same X, Y with the same probability distribution, shown as follows: $X \equiv R^L \equiv (R \cdot \omega^{-E})^L \pmod{n}$ and $Y \equiv S^E \cdot R \equiv S'^E(R \cdot \omega^{-E}) \pmod{n}$. Since the R's are randomly chosen, $\overline{A}$ produces the same X, Y values with the same probability distribution in both cases. Therefore, during the executions of $(\overline{A}, \widetilde{B})$ $\overline{A}$ cannot leak to $\widetilde{B}$ which L-th root mod n of $I^{E_1-E_2} \bmod n$ he can compute from the S he knows. Thus, we have proven that the L-th roots mod n of $I^{E_1-E_2} \bmod n$ which are known by $\overline{A}$ and computed by a coalition of $\widetilde{A}$, $\overline{B}$ and $\widetilde{B}$ are totally independent.

Next, we will prove that n can be factored with probability at least $1/p'$ using $S^{E_1-E_2} \bmod n$ and $Y_1/Y_2 \bmod n$, if at least one of the conditions C1, C2, C3, and C4 is satisfied, even if the value of $Y_1/Y_2 \bmod n$ is biased. Let $\omega = \frac{S^{E_1-E_2}}{Y_1/Y_2} \bmod n$ and $\omega = [\omega_p, \omega_q]$. When C1 is satisfied, the probability of successfully factoring n using ω is at least $1/p'$. This is because: if $Ind_p(\omega_p) < Ind_q(\omega_q)$, then $\omega^{Ind_p(\omega_p)} \bmod n$ is *Type2*, and if $Ind_p(\omega_p) > Ind_q(\omega_q)$, then $\omega^{Ind_q(\omega_q)} \bmod n$ is *Type3*. Therefore, when $Ind_p(\omega_p) \neq Ind_q(\omega_q)$, the probability of successfully factoring n using ω is 1. Here, we will show that the probability of $Ind_p(\omega_p) \neq Ind_q(\omega_q)$ is at least $1/p'$. Let $p' = r_1 \cdot r_2$ such that $r_1 = (p', E_1 - E_2) < p'$, $\{v_1, \cdots, v_{r_2}\} = \{S'^{E_1-E_2} \bmod p \mid S'$ is the L-th root mod n of $I\}$ and $v_0 = \frac{Y_1}{Y_2} \bmod p$, then $(r_1, r_2) = 1$ because of C1, and $r_2 \geq 2$ because of Lemma 2. Therefore, there is at least one pair $(\omega_{p,i}, \omega_{p,j})$ satisfying $Ind_p(\omega_{p,i}) \neq Ind_p(\omega_{p,j})$ according to Lemma 4, where $\omega_{p,i} = v_i/v_0 \bmod p$ $(i = 1, \cdots, r_2)$. When

C2 is satisfied, change the role of p and q in the C1 case. When C3 is satisfied and $\omega \neq 1$, then ω is *Type3* because $\omega = [\omega_p, 1]$ where $\omega_p \neq 1$. Since $\#\{S'^{E_1-E_2} \bmod p \mid S'$ is the L-th root mod n of $I\} \geq 2$, the probability of successfully factoring n using ω is at least 1/2. When C4 is satisfied and $\omega \neq 1$, then $\omega^{q'} \bmod n$ is *Type3* because $\omega^{q'} = [\omega_p, 1]$ where $\omega_p \neq 1$. Since $\#\{S'^{E_1-E_2} \bmod p \mid S'$ is the L-th root mod n of $I\} \geq 2$, the probability of successfully factoring n using ω is at least 1/2. Therefore, the probability of successfully factoring n using ω is at least $1/p'$.

Finally, We will prove that n can be factored by a coalition of $\overline{A}, \widetilde{A}, \overline{B}$ and $\widetilde{B}$ at most in time $p'\{O(\|\widetilde{B}\| \cdot |n|^e) + O(\|\widetilde{A}\| \cdot |n|^d)\} = O(\|\widetilde{B}\| \cdot |n|^e + \|\widetilde{A}\| \cdot |n|^d)$ and with overwhelming probability. When for any $f(1 \leq f < p')$, $\omega^f \bmod n$ is *Type1* or *Type4*, n cannot be factored. Then, $\overline{A}$ selects another S' randomly and calculates $I' = S'^L \bmod n$, and the coalition goes through the same procedure to factor n. The procedure is repeated until n can be factored. The average number of iterations is at most p'. *Q.E.D.*

Remark: Evidently, we can extend the value of L to $O(|n|)$ in the above theorem. However, from the practical viewpoint, it is essential that the security level is a constant value, or a non asymptotic value. Therefore, the condition of Theorem 3 for the order of L is optimal.

[Corollary] Let the parameters $k = t = 1$ and L satisfy $(L, p-1) = L$, and $L = O(1)$. If at least one of the conditions C1,C2,C3, and C4 is satisfied, then the parallel version of our identification scheme releases no transferable information with a strict ***sharp-threshold*** security level $1/L$, if there is no probabilistic polynomial time algorithm of factoring.

Proof It is proven that this protocol releases no transferable information with a strict security level $1/L$ because $(L, p-1) = L$ and because of Theorem 3. Here, $\widetilde{A}$ can cheat $\overline{B}$ with probability $1/L$ because $\widetilde{A}$ can guess RB with probability $1/L$. *Q.E.D.*

5. Applications

Signature scheme A triplet (M, E, Y) is sent as the signed message, where M is a message, h is a public pseudo-random function and $E = h(M, X) \in Z_L$, to turn the identification scheme into a signature scheme. Y here is the same as in Step 3 of the identification scheme.

N-Party authentication scheme N-party identification and signature protocol based on the Fiat-Shamir scheme was proposed by [BLY]. However, in the Fiat-Shamir scheme a large memory ($k \approx 100$) is required. Our parallel version is suited to their protocol with only one secret integer and $\log_2 L \approx 100$.

6. Efficiency

In this section, we focus on a typical implementation of our parallel version.

Secret memory size This scheme requires $|n|$ bits of secret information S, while the Fiat-Shamir scheme requires $k\ |n|$ bits of secret information. The proposed scheme is therefore more efficient than the Fiat-Shamir scheme when $k \geq 2$.

Transmission efficiency $(2|n| + |L|)$ bits are transmitted in this scheme, while $(2t|n| + kt)$ bits are transmitted in the Fiat-Shamir scheme. Note that when $t = 1$ is used in the Fiat-Shamir scheme, the k value must be large.

Processing speed The amount of processing needed for this scheme is compared with the RSA [RSA] and Fiat-Shamir schemes using the average number of modular multiplications required to generate or verify a proof of identity.

The RSA scheme requires $(3|n|)/2$ steps, the Fiat-Shamir scheme requires $t(k+2)/2$ steps, and the proposed scheme requires $(5l + 2)/2$ steps where $L = 2^l$. For example, when $tk = l = 20$, our parallel version requires 51 steps, while the Fiat-Shamir scheme requires 11 steps (where $k = 20, t = 1$) to 30 steps (where $k = 1, t = 20$), and the RSA scheme requires 768 steps where $|n| = 512$.

When a prover uses a secret integer S satisfying $I^{-1} \equiv S^L \bmod n$, a verifier checks whether $Y^L \cdot I^E \equiv X \pmod{n}$ holds. The computations of Y^L and I^E can be combined, i.e., according to the value of E, a verifier can repeatedly square the results of the intermediate calculation, or square those results multiplied by I, as appropriate. This improved calculation requires $3l/2$ steps in the verification; for example, 30 steps are required when $l = 20$.

7. Conclusion

Combining our scheme with the Fiat-Shamir scheme provides greater flexibility because three appropriate design parameters of transmitted information size, memory size and speed can be selected.

The parallel version described in Section 4 is more efficient than the Fiat-Shamir scheme from the standpoint of transmitted information size and secret information size, because it corresponds to $t = k = 1$ in their scheme. It is about one order of magnitude faster than the RSA scheme and is relatively slow in comparison with the Fiat-Shamir scheme. Our sequential and parallel versions are also shown to have the same security characteristics as the Fiat-Shamir scheme.

Finally, we conclude with an open problem relating to the security level: when $(L, p-1) = p' < L$ and at least one of conditions C1,C2,C3, and C4 is satisfied, does the parallel version of our identification scheme release no transferable information with a strict *sharp-threshold* security level $1/p'$, if there is no probabilistic polynomial time algorithm of factoring ?

Acknowledgements

The authors would like to thank Adi Shamir and Kenji Koyama for their participation in valuable discussions. In particular, we owe the improved calculation in *Section 6* to Adi Shamir.

References

[BLY] Brickell, E., Lee, P. and Yacobi, Y.: "Secure Audio Teleconference," Advances in Cryptology - Crypto'87, Lecture Notes in Computer Science 293, 1988, pp.429-433

[FS] Fiat, A. and Shamir, A.: "How to Prove Yourself: Practical Solution to Identification and Signature Problems," Advances in Cryptology - Crypto'86, Lecture Notes in Computer Science 263, 1987, pp.186-199

[FFS] Feige, U., Fiat, A. and Shamir, A.: "Zero Knowledge Proofs of Identity," Proceedings of the 19th Annual ACM Symposium on Theory of Computing, 1987, pp.210-217

[GMR] Goldwasser, S., Micali, S. and Rackoff, C.: "Knowledge Complexity of Interactive Proof Systems," Proceedings of the 17th Annual ACM Symposium on Theory of Computing, 1985, pp.291-304

[GMW] Goldreich, O., Micali, S. and Wigderson, A.: "Proofs that Yield Nothing but Their Validity and a Methdology of Cryptographic Protocol Design," Proceedings of the 27th Annual Symposium on Foundations of Computer Science, 1986, pp.174-197

[GQ1] Guillou, L.C., and Quisquater, J.J.: "A Practical Zero-Knowledge Protocol Fitted to Security Microprocessor Minimizing Both Tranamission and Memory," Eurocrypt'88 Abstracts, 1988, pp.71-75

[GQ2] Guillou, L.C., and Quisquater, J.J.: "A Paradoxical Identity-Based Signature Scheme Resulting from Zero-Knowledge," These Proceedings, 1988

[OO] Ohta, K. and Okamoto, T.: "Practical Extension of Fiat-Shamir Scheme," Electron.Lett., 24, No. 15, 1988, pp.955-956

[HW] Hardy, G.H. and Wright, E.M.: "An Introduction to the Theory of Numbers," Fifth edition, Oxford University Press, New York, 1978

[RSA] Rivest, R.L., Shamir, A. and Adleman, L.: "A Method for Obtaining Digital Signatures and Public-Key Cryptosystems," Communication of the ACM, Vol. 21, No. 2, 1978, pp.120-126

An Improvement of the Fiat-Shamir Identification and Signature Scheme

by

Silvio Micali

Lab. for Computer Science
MIT
USA

Adi Shamir

Applied Mathematics Dept.
The Weizmann Institute
Israel

Abstract

In 1986 Fiat and Shamir exhibited zero-knowledge based identification and digital signature schemes which require only 10 to 30 modular multiplications per party. In this paper we describe an improvement of this scheme which reduces the verifier's complexity to less than 2 modular multiplications and leaves the prover's complexity unchanged.

The new variant is particularly useful when a central computer has to verify in real time signed messages from thousands of remote terminals, or when the same signature has to be repeatedly verified.

1. Introduction.

Informally speaking, a digital signature is a value associated with a message which is easy to verify but difficult to forge. After having generated and verified it, the signature can be later presented to a judge since the signer cannot disown his messages. An identification scheme is a simplified signature scheme in which there are no messages disputes or judges: the proof of identity is interactive, and the verifier can either accept or reject the prover's claimed identity, with no legal or long-term consequences. To be useful and secure, the identification scheme should satisfy the following three conditions:

1) A real verifier should accept a real prover's proof of identity with overwhelming probability.

2) A real verifier should accept a cheating prover's proof of identity with negligible probability.

3) A cheating verifier should not learn anything from polynomially many interactions with a real prover that will enable him to misrepresent himself as the prover to someone else with non-negligible probability.

The best known example of a signature scheme is the RSA (Rivest, Shamir and Adleman[1978]). To use it as an identification scheme, the verifier can simply ask the prover to sign a random test message. The original scheme requires about 750 modular multiplications per party, but the verifier's complexity can be reduced to a few modular multiplications by using a low-exponent variant. A 512 bit implementation of the RSA scheme requires 10-15 seconds on IBM PC's, and several minutes on smart cards.

A faster and provably secure identification and signature scheme was proposed in Fiat and Shamir[1986]. It is based on the zero knowledge paradigm introduced in Goldwasser Micali and Rackoff[1985], and more particularly on the quadratic residuosity protocol presented by Fischer Micali and Rackoff at Eurocrypt 84. The Fiat-Shamir protocol reduces the time and communication complexities of the Fischer-Micali-Rackoff protocol by simultaneously proving the quadratic residuosity of many numbers, but by doing so it destroys the zero knowledge nature of the protocol. (The formal proof of security of the Fiat-Shamir protocol is thus based on the fact that it reveals no "transferable knowledge", which is a new measure of cryptographic strength introduced and studied in Feige, Fiat and Shamir[1987].)

In this paper we show how to substantially speed up the Fiat and Shamir scheme. There are many variants of this scheme. Our ideas speed up each single one of them. Thus below we confine ourselves to recall and speed up its simplest version.

2. The original Fiat-Shamir Scheme

Let s be a security parameter and let n be the product of two random prime numbers whose size is s. (Unlike the RSA scheme, it is not necessary to know the factorization of n in order to execute the protocol, and thus each prover can pick his own public modulus n, or use a universal modulus n published by a trusted center.)

Each prover picks a secret key consisting of k random numbers $s_1, \ldots, s_k$ in Z_n^* (the multiplicative group mod n), computes $v_j = 1/s_j^2$ (mod n) for j= 1,...,k, and publishes $v_1,\ldots,v_k$ (along with n, if it was chosen by him) in a public key directory.

The identification scheme is based on the following protocol:

1) The prover picks a random r in Z_n^*, and sends $x=r^2$ (mod n) to the verifier.

2) The verifier sends k random bits $e_1,\ldots,e_k$ to the prover.

3) The prover sends $y=r\prod_j s_j^{e_j}$ (mod n) to the verifier.

4) The verifier accepts the proof iff $x=y^2\prod_j v_j^{e_j}$ (mod n).

In practice, we would accept the probability of successful misrepresentation to be at most 1 in a million per each attempt and thus a choice of k= 20 suffices for most applications. The key size (either public or private) in 512-bit implementations is about 1.3 kilobytes, and the average number of modular multiplications per party is about 10. The communication complexity is about 1000 bits per proof, but this can be almost halfed by sending a hashed version of x to the verifier. Other optimizations and tradeoffs can be found in Fiat and Shamir[1986].

To turn this interactive identification scheme into a non interactive signature scheme, it suffices to make $e = e_1, \ldots, e_k$ to be the value of a pseudo-random function f, easy to evaluate, but hard to invert, at input (x,m), where m is the message to be signed. This pseudo random function f is universal, and its values are accessible to all the parties. The resultant signature generation protocol is:

1) Choose at random r in [0,...,n).

2) Compute $e = f(r^2(mod\ n)\ ,\ m)$ and $y = r\cdot\prod_j s_j^{e_i}$.

3) Send e and y as the signature of m.

The corresponding signature verification scheme is:

Accept the signature if *syntax error file -, between lines 234 and 234* e = f ($y^2\Pi$.

The scheme is provably secure when f is a truly random function (computed by a trusted call-up center) or when f is a strong pseudo-random function in the sense of Goldreich, Goldwasser and Micali [GGM] given to the parties in tamper-proof devices: unless factoring is easy, a cheater cannot forge the signature of a new message with non-negligible probability even after he was given polynomially many signatures of other messages and polynomially many values of f at arguments of his choice. This sketched (but formalizable) proof breaks down for technical reasons when the parties are given access to the algorithm of f (and not just to its values). However, we strongly believe that the scheme remains secure even in this case, provided that f does not interact badly with the modular multiplication operations.

Since a cheater can know in advance whether a proposed signature is valid, the value of k in practical implementations should be at least 64. This increases the key size to about 4 kilobytes, and increases the number of modular multiplications to about 32 per party. The size of a signature is 576 bits, about the same as in the RSA scheme.

3. The New Improvement

Our improvement comes about from choosing the v_i's to be the first k prime numbers (v_1=2,v_2=3,v_3=5, etc). The s_i's will then be set to be a random square root of the corresponding v_i mod n. Each prover should choose his own modulus n and use its factorization in order to extract these roots. (The factorization is now no longer needed and it can be erased.) The actual proofs of identity and signatures are generated and verified in the standard way described in the previous section.

Newly arising difficulties

Before analyzing the efficiency of this scheme, it should be noticed that we have to overcome some technical difficulties. In fact, not all of the v_i's will be quadratic residues mod n. We overcome this technical difficulty with an appropriate perturbation technique which will be described in the full version of the paper.

Gain in efficiency

The above additional difficulties are worth dealing with. Since our choice of the v_j's is universal, provers should only publish n as their public key. This reduces the size of the public key directory to 64 bytes per user, and makes it possible to use the same directory in order to verify our new signatures, as well as other signatures based on factoring, like the previous Fiat-Shamir, the RSA and the Rabin's scheme. The size of the secret key remains about 4 kilobytes, but this size is less critical since the information is stored LOCALLY rather than TRANSMITTED, and each user keeps only one such file.

The main benefit of our improvement, though, is the GREATLY reduced complexity of verification: since most of the v_j's are single-byte numbers, their product is particularly easy to compute as does not even require modular reductions! The only expensive operation left is the modular squaring of y, and thus the total complexity of verification is somewhere between 1 and 2 modular multiplications.

Security

The security of the original Fiat-Shamir scheme is based on the fact that the extraction of square roots of random vj values is as difficult as the factorization of the modulus. This proof technique is not directly applicable to the new version, since the extraction of square roots of small primes may concievably be easier than the extraction of square roots of random numbers.

For simplicitly sake, let us discuss only the security of the identification scheme (the signature scheme only needs a more complex notation). The identification scheme in question is based on on zero-knowledge proofs. Very roughly (see Feige, Fiat and Shamir for a detailed discussion) this means that the proof of identity is constituted by a proof of "knowledge of something." n our case this "something" is not a proof of quadratic residuosity (either for a particular prime or for all the primes) in the original language-theoretic sense of Goldwasser Micali and Rackoff: Since the parties execute only one round of the protocol, the prover can succeed with probability 1/2 even if all the primes are quadratic non-residues! Similarly, the protocol is not a proof of knowledge of square roots (either for a particular prime or for all the primes) as in Feige Fiat and Shamir: The knowledge tape could contain the square roots of all the 400 pairwise products of the primes, and thus a cheating prover could convince the verifier with probability 1/2 without actually knowing even one of the original roots.

A CAREFUL analysis, carefully omitted in this abstract!, shows that, in our scheme, this "something" is the square root of the product of a subset times the inverse of another subset of the first 20 primes. We thus need to argue that this piece of knowledge is not easily available to everyone, and thus distinguishes the prover from everyone else. We already know that extracting square roots modulo composite numbers is as hard as integer factorization. This implies the following fact:

> Assume there exists an algorithm A that, on input m (a s-bit long modulus) and S (a random set of quadratic residues mod m whose cardinality is k), finds a square root of the product of a subset of S and the inverse of another subset of S in time T(s). Then, there exists a factoring algorithm A' that runs essentially in time $2^k \cdot T(s)$.

The proof of this fact, though not hard, is also postponed to the final paper. One would be tempted to conclude that if the "piece of knowledge" underlying our new scheme were computable in time T(s), then one could factor in one million T(s) steps an s-bit modulus, which would imply, as we need, that T(s) is large. This is, however, a too hasty conclusion. In fact, we can assume without loss of generality that the first 20 primes are quadratic residues (since our true scheme cops with those which are not to squares mod n), but they are NOT a random subset of size 20. Thus a natural question arises: is the computational difficulty of extracting square roots of small primes any lower than for random (quadratic) residues? The answer apperas to be negative. In fact, Morrison-Brillhart type methods would be substantially sped up if square root of small primes were easier to compute! This and other details (including a formal intractability assumption) needed to transform this discussion into a proof will be given in the final paper.

Let us mention that there is a more direct way to prove the security of our scheme if one is willing to make an intractability assumption that is stronger than the one derivable by formalizing the above argument. Informally, this stronger assumption states that factoring remains difficult even when one is given the square root of a small number of small primes.

It is worth mentioning that while such an assumption is sufficient to prove the security of the new scheme, its being false DOES NOT imply that our scheme is insecure! In fact, even if a cheating verifier knows how to factor n by using the square roots of a small number of small primes, he is unlikely to get hold of these square roots since the schemes are the parallel versions of zero knowledge protocols. In other words, only the real prover is likely to benefit from such a number-theoretic breakthrough, but he already knows this factorization!

Remark

This improvement of the Fiat-Shamir scheme was discovered independently by the two authors. Additional optimization ideas will be described in the full version of this paper.

Session 7

On the Theory of Security I

Chair: R. Rivest, MIT

A Basic Theory of Public and Private Cryptosystems

by
Charles Rackoff
Dept. of Computer Science
University of Toronto

Not since the early work of [DH], [RSA], and [GM] has there been a great eal of work on the basic definition of "normal" cryptography, and on what means for a cryptosystem to be secure. By normal cryptogaphy, I mean ot protocols to accomplish sophisticated goals, but merely the situation here party *A* wishes to send a message to party *B* over a line which is eing tapped. Existing definitions of such a system, when they aren't too ague, are overly restrictive; existing definitions of security of such ystems, when given rigorously, are usually overly liberal. In this paper ll present what seem to me to be the proper definitions, give statements f the basic theorems I know about these definitions, and raise some very ındamental open questions. Most of the definitions and results appeared [R].

A cryptosystem looks like the following picture.

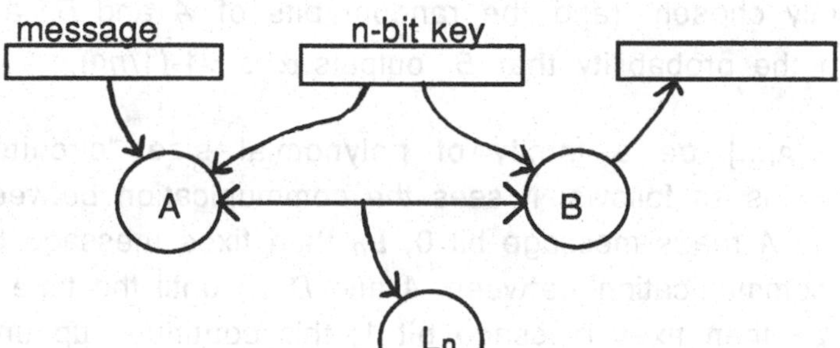

A and *B* are probabilistic interacting Turing Machines. Normally, if here is a private key, one only allows *A* to talk to *B*, but I allow *A* and *B* o talk back and forth. *n* is called the *security parameter.* The intuition ehind the (plain-text) message is that it should consist of *all* bits to be ent by the cryptosystem until the universe dies; usually this is taken to e some fixed polynomial in *n*, but I find it more pleasing to let it be ıfinite. *B* must output its guess at the ith bit of the message in time polynomial in n and i, including the computing time of A. We also need some kind of "on-line" condition; a natural (although not completely necessary) one is that A doesn't read bit $i+1$ of the message until B has output its guess at bit i. (For convenience, assume that the time at which A reads bit i of the message just depends on i and n.) A and B also read an n -bit key. The two sensitive issues are correctness and security. My definition of correctness allows a small probability of error for B. Although most definitions don't allow for this possibility, many suggested cryptosystems have in fact had this feature (since "prime" numbers that were used might not really be prime). My definition of security is intended to encompass all reasonable attacks by an eavesdropper, including chosen plain text attack. Most definitions given in the past (including those discussed in [MRS]), have the property that a secure system can be modified to be secure according to the other definition, but trivially breakable using a chosen plain text attack. The reader will note that neither the definition of correctness nor of security assumes any distribution on the message space. We will always assume (unless stated otherwise) that a cryptosystem is correct.

Correctness: For every $c,d \in \mathbf{N}$, for every sufficiently large $n \in \mathbf{N}$, for every $\alpha \in \{0,1\}^*$ of length n^c, if A tries to send a message beginning with α and the key is randomly chosen, (and the random bits of A and B are randomly chosen,) then the probability that B outputs α is $>1-(1/n^d)$.

Security: Let $L=\{L_1,L_2,...\}$ be a family of polynomial size "circuits". Actually, what L_n can do is as follows: it sees the communication between A and B up until the time A reads message bit 0; L_n then fixes message bit 0; L_n then sees the communication between A and B up until the time A reads message bit 1; L_n then fixes message bit 1; this continues up until some message bit i (i determined by L_n), is chosen randomly from $\{0,1\}$ (but not seen by L_n); L_n then sees the communication up until the time A reads message bit $i+1$; L_n then fixes message bit $i+1$; this continues until L_n chooses to output its guess at message bit i. Let p_n be the probability that L_n is successful at guessing this bit. Then for every d and sufficiently large n, $p_n<(1/2)+(1/n^d)$.

Given a definition of security, it is easy to prove many of the facts normally assumed in the folklore. Theorems 1 and 2 below are examples.

Another example is that a secure (in the sense of [GGM]) pseudo-random number generator implies the existence of a secure cryptosystem. Such a theorem, however, is only as meaningful as the definition of security is good.

In the definition of security, the listener is modeled as nonuniform "circuits" rather than as probabilistic algorithms. This isn't very important, but it makes theorems easier to prove and various things become cleaner. For example, with nonuniform circuits, it is not necessary to add probabilism to *L* since this would not affect the power of the listener. Certain other aspects of the definitions are there for cleanness and convenience. For example, if we had a system which was secure but only 3/4 correct (instead of $1-(1/n^d)$), the "majority" trick could be used to make it correct with probability exponentially close to 1, and still secure. If we had a system which was correct, but only secure if we replaced $(1/2)+(1/n^d)$ by 3/4, then the "exclusive-or" trick could be used to convert it to one which is secure (although not, as far as we know, "exponentially close to 1/2" secure), and still correct.

If the key, instead of being chosen randomly, is chosen to be 0^n, then I call the system a *public* cryptosystem. Presumably, when people talk about a secure "key exchange protocol", what they mean is a public cryptosystem which is secure for sending (say) *n* message bits; these bits can be sent as a single block (rather than one bit at a time), possibly speeding things up by a factor of *n*, but the question of the existence of such a protocol appears to be equivalent to the question of the existence of a secure public cryptosystem. Theorem 1 shows that the open question about the existence of secure public cryptography can be formulated as a question of sending only 1 message bit securely. If *A* only sends to *B*, the system is called "1-pass"; if *B* sends to *A* and then *A* sends to *B*, the system is called "2-pass"; "*j*-passes" is defined in the obvious way. A 2-pass public system is what is often called a "public key cryptosystem", where the string sent by *B* is called the "public key". Theorem 2 is part of the basis of "public key cryptography".

Theorem 1: Let (*A*,*B*) be a public cryptosystem in which the first message bit is sent securely. Then a secure public cryptosystem (for all the bits) can be obtained by *independently* running (*A*,*B*) on each of the message bits (that is, each time, *A* and *B* start over and choose new random bits).

Theorem 2: Let (A,B) be a 2-pass public cryptosystem in which the first message bit is sent securely. Then a secure (for all the bits) 2-pass public cryptosystem can be obtained by running B once to see the string ß that B would send; then run A independently on each of the message bits, where each time A uses new random bits, but the same string ß fom B.

Theorem 3 is the analogue of Theorem 1 for (private) cryptosystems.

Theorem 3: Let (A,B) be a (private) cryptosystem in which the first $n+1$ message bits are sent securely, where n is the security parameter. Then a secure (for all the bits) cryptosystem can be obtained as follows: say that α_0 is the key and that the message is $b_0b_1b_2$... ; A generates random n -bit strings $\alpha_1,\alpha_2,$... ; (A,B) is run with key α_0 on the n+1-bit message $\alpha_1 b_0$, then (A,B) is run with key α_1 on the n+1 bit message $\alpha_2 b_1$, etc. Note that if (A,B) is 1-pass, then so is the new cryptosystem. However, the new cryptosystem will be probabilistic, even if A and B are deterministic.

Open Questions: Can it help to have more than 1 pass in a (private) cryptosystem? Can it help to have more than 2 passes in a public cryptosystem? Can it help to have more than 3 passes in a public cryptosystem? For each question, either prove a negative answer, or give a convincing example where the extra passes appear to help.

It is interesting to note that there are settings, other than those discussed here, where one can either prove or give good evidence that extra interaction helps. An interesting example, of relevance to cryptography, appears in [BBR].

Theorem 5 below shows that at the moment, our ability to prove security of cryptosystems is severely limited. It is known that with a one-time pad, one can sent n message bits securely with an n-bit random key. If P=NP, which we are unable to disprove, then this is essentially the best we can do. Theorem 5 can be proven by observing that in the proof of Theorem 4, all the listener had to be able to do was "approximate counting"; this task is in the polynomial time hierarchy by a result of [St] and [Si]. Theorem 5, at least in the case $g(n)$=0, has also been observed by other people. A version of theorem 4 was first prove by Shannon [Sh].

Theorem 4: If we remove the polynomial time restriction on the listener, then there is no secure cryptosystem. In fact, a stronger result can be proved. Let g be a function such that $g(n)$ is computable in time polynomial in n, and $0 \leq g(n) \leq n$. Then there is no cryptosystem which sends $g(n)+1$ bits securely (against an unrestricted time listener) if only the first $g(n)$ bits of the key are chosen randomly (and the rest are fixed, say, to be 0).

Theorem 5: If P=NP, then for g as in Theorem 4, even if the listener is restricted to polynomial time, there is no cryptosystem which sends $g(n)+1$ securely if only the first $g(n)$ bits of the key are chosen randomly.

Probably the most important open issue in *all* of cryptography concerns the conjectures can be used to prove the existence of secure cryptosystems. The asumption P≠NP is certainly necessary, but probably not sufficient. The only natural assumptions currently in use relate to the difficulty of integer factorization or discrete log. One possible thing to search for is a "complete" cryptosystem $\mathfrak{R}$: one whose insecurity would imply the insecurity of every other system $\mathfrak{R}'$. Using ideas of Levin, such a system can be constructed by a kind of diagonalization. Such a "complete" system can also be constructed for the class of 1-pass cryptosystems, for the class of public cryptosystems, and for the class of 2-pass public cryptosystems. I will not define this notion of "complete" precisely, since in any case, it has the following problem: the time to break $\mathfrak{R}'$, given an oracle for breaking $\mathfrak{R}$, requires time only polynomial in n, but *exponential* in the size of the description of $\mathfrak{R}'$.

Open Question: Is there a cryptosystem whose security problem is "complete" in an appropriate sense?

Lastly, I'd like to point out what I have *not* talked about here. I haven't discussed the scenerio where there is a group of mutually distrusting people, each pair of which wishes to communicate in the presence of a listener. Although many soulutions to this (and more complicated) problems have been proposed, I have seen no rigorous definition of this scenerio, let alone any definition of what security would mean in such a setting. Of course, there are related subproblems which have been rigorously studied: examples are signature schemes ([GMR]) and the problems studied in this paper. But it appears to be very difficult to talk about the more complicated situation, and it can be very dangerous to think that security can necessarily be understood in terms of security in simpler situations. For example [GMT] point out that if a secure 2-pass public cryptosystem is used in the obvious way to create a "public key network of users", the result might wind up being insecure.

The difficulties involved in understanding the relatively simple situation discussed in this paper imply that one must approach the more complicated (and realistic) situations very slowly and with a great deal of care.

References

[BBR] C. H. Bennett, G. Brassard, J. Robert, "Privacy amplification by public discussion", *SIAM J. on Comput.,* 17, 1988, 210-229.

[DH] W. Diffie, M. E. Hellman, "New directions in cryptography", *IEEE Trans. Informat. Theory,* IT-22, 1976, 644-654.

[GGM] O. Goldreich, S. Goldwasser, S. Micali, "How to construct random functions", *JACM*, 33, 1986, 792-807.

[GM] S. Goldwasser, S. Micali, "Probabilistic encryption", *J. Comput. System Sci.*, 28, 1984, 270-299.

[GMR] S. Goldwasser, S. Micali, R. Rivest, "A digital signature scheme secure against adaptive chosen-message attacks", *SIAM J. on Comput.,* 17, 1988, 281-308.

[GMT] S. Goldwasser, S. Micali, P. Tong, "Why and how to establish a private code on a public network", *Proc. 23 IEEE Symp. on Foundations of Computer Science*, 1982, 134-144.

[MRS] S. Micali, C. Rackoff, B Sloan, "The notion of security for probabilistic cryptosystems", *SIAM J. on Comput.*, 17, 1988, 412-426.

[R] C. Rackoff, Class notes on Cryptography, 1985.

[RSA] R. Rivest, A. Shamir, L. Adleman, "A method for obtaining digital signatures and public-key cryptosystems", *Comm. ACM*, 21, 1978, 120-126.

[Sh] C. E. Shannon, "Communication theory of secrecy systems", *Bell Syst. Tech. J.*, vol.28, 1949, 656-715.

[Si] M. Sipser, "A complexity theoretic approach to randomness", *Proc. 15 ACM Symp. on Theory of Computing*, 1983, 330-335.

[St] L. Stockmeyer, "On approximation algorithms for #P", *SIAM J. on Comput.*, 14, 1985.

Proving Security Against Chosen Ciphertext Attacks

*Manuel Blum**
Computer Science Dept.
Univ. of Calif.
Berkely, CA

Paul Feldman
MIT Lab. for Computer Sci
Cambridge, MA

Silvio Micali†
Lab. for Computer Science
MIT
Cambridge, MA

Abstract

The relevance of zero knowledge to cryptography has become apparent in the recent years. In this paper we advance this theory by showing that interaction in *any* zero-knowledge proof can be replaced by sharing a common, short, random string. This advance finds immediate application in the construction of the *first* public-key cryptosystem secure against chosen ciphertext attack.

Our solution, though not yet practical, is of theoretical significance, since the existence of cryptosystems secure against chosen ciphertext attack has been a famous long-standing open problem in the field.

1 Introduction

Recently [GMR] have shown that it is possible to prove that some theorems are true without giving the slightest hint of why this is so. This is rigorously formalized in the somewhat paradoxical notion of a *zero-knowledge proof system*.

If secure encryption schemes exist, though, these proof systems are far from being a rare and bizar event. In fact, under this assumption, [GMW] demonstrate that any language in NP possesses zero-knowledge proof systems.

---

*Supported by NSF Grant # DCR85-13926
†Supported by NSF grant # CCR-8719689

Actually, as recently pointed out by Impagliazzo [I] and Ben-Or, Goldreich, Goldwasser, Hastad, Kilian, Micali and Rogaway [BGGHKMR], the same is true for all languages in IP; also, as pointed out by Blum [B2], any theorem at all admits a proof that conveys zero-knowledge other than betraying its own length.

Zero-knowledge proofs have proven very useful both in complexity theory and in cryptography. For instance, in complexity theory, via results of fortnow [F] and Boppana and Hastad [BH], zero-knowledge provides us an avenue to convince ourselves that certain languages are not NP-complete. In cryptography, zero-knowledge proofs have played a major role in the recently proven completeness theorem for protocols with honest majority [GMW2]. They also have inspired rigorously-analyzed identification schemes that are as efficient [FFS] and even more efficient [MS] than folklore ones.

Despite its wide applicability, zero-knowledge remains an intriguing notion: What makes zero-knowledge proofs work?

Three main features differentiate all known zero-knowledge proof systems from more traditional ones:

1. *Interaction:* The prover and the verifier talk back and forth

2. *Hidden Randomization:* The verifier tosses coins that are hidden from the prover and thus unpredictable to him.

3. *Computational Difficulty:* The prover imbeds in his proofs the computational difficulty of some other problem.

At a first glance, all of these ingredients appear to be necessary. This paper makes a first, important step in distilling what is essential in a zero-knowledge proof. We show that computational difficulty alone (for instance the hardness of distinguishing products of 2 primes from products of 3 primes) may make *inessential* the first resource (interaction) and and *eliminate the secrecy* of the second resource (randomness). That is, if the prover and the verifier share a common random string, the prover can non-interactively and yet in zero-knowledge convince the verifier of the validity of any theorem he may discover. A bit more precisely, for any constants c and d, sharing a k-bit long random string allows a prover P to prove in zero-knowledge to a poly(k)-time verifier V any k^c theorems of k^d size non-interactively; that is, without ever reading any message ¿from V.

A Conceptual Scenario: Think of P and V as two mathematicians. After having played "heads and tails" for a while, or having both witnessed the same random event, P leaves for a long trip along the world, during which he continues his mathematical investigations. whenever he discovers a theorem, he writes a postcard to v proving the validity of his new theorem in zero-knowledge. Notice that this is necessarily a non-interactive process; better said, it is a mono-directional interaction: From P to V only. in fact, even if V would like to answer or talk to P, he couldn't: P has no fixed (or predictable) address and will move away before any mail can reach him.

1.1 Our Model Versus the Old One

While the definition of zero-knowledge remains unchanged, the mechanics of the computation of the prover and verifier changes dramatically.

Notice that sharing a random string σ is a weaker requirement than being able to interact. In fact, if P and V could interact they would be able to construct a common random string by coin tossing over the phone [B1]; the converse, however, is not true.

Also notice that sharing a common random string is a requirement *even weaker* than having both parties access a random beacon in the rabin's sense (e.g. - perhaps! - the same geiger counter). In this latter case, in fact, all made coin tosses would be seen by the prover, but the future ones would still be unpredictable to him. by contrast, our model allows the prover to see in advance all the coin tosses of the verifier. That is the zero-knowledgeness of our proofs does not depend on the secrecy, or unpredictability of σ, but on the "well mixedness" of its bits! This curious property makes our result potentially applicable. For instance, all libraries in the country possess identical copies of the random tables prepared by the rand corporation. Thus, we may think of ourselves as being already in the scenario needed for non-interactive zero-knowledge proofs.

1.2 The Robustness of Our Result

As we have already said, we guarantee that all theorems proved in our proof systems are correct and zero-knowledge if the string σ is a truly random one. We may rightly ask what would happen if σ was not, in fact, truly randomly selected. fortunately, the poor randomness of σ may upset the zero-knowledgeness of our theorems, but not their correctness. That is, for almost all (poorly random) σ's, there is no wrong statement that can be accepted by the verifier. This is indeed an important property as we can never be sure of the quality of our natural sources of randomness. Unfortunately, due to the limitations of an extended abstract, we cannot further elaborate on this and similar points. We wish, however, to point out the following important corollary of our result.

1.3 Applications of our Result

A very noticeable application of non-interactive zero-knowledge is the construction of encryption schemes á la Diffie and Hellman that are secure against chosen ciphertext attacks. Whether such schemes existed has been a fundamenatal open problem ever since the appearence of complexity-based cryptography. We will discuss this application in Section 3.

1.4 What's Coming

The next section is devoted to set up our notation, recall some elementary facts from Number Theory and state the complexity assumption which sufficies to show the existence of non-interactive, zero-knowledge proofs.

In Section 3, we show the "single-theorem" case. That is, we show that if a k^4-bit string σ is randomly selected and given to both the proven and the verifier, then the first can prove, for any single string x (of length k) belonging to a NP-language L, that indeed $x \in L$; the proof will be a zero-knowledge one for whenever x is independent of σ.

In the final paper [BDFMP], we will show the "many-theorems" case. Namely, that for each fixed polynomial $Q(\cdot)$, using the *same* randomly chosen k^4-bit string, the prover can show in zero-knowledge membership in NP languages for any $Q(k)$ strings of length $Q(k)$.

The complexity assumption under which the result holds is the computational difficulty of deciding quadratic residuosity.

We would like to point out that the proof of the many-theorems result in the earlier versions of [BFM] and [DMP] contained a gap: it required, over than the stated number theoretic assumptions, a stronger property about pseudo-random generators. This stronger property is not needed in the final paper.

2 Preliminaries

2.1 Notations and Conventions

Let us quickly recall the standard notation of [GoMiRi].

We emphasize the number of inputs received by an algorithm as follows. If algorithm a receives only one input we write "$A(\cdot)$", if it receives two inputs we write "$A(\cdot,\cdot)$" and so on.

If $A(\cdot)$ is a probabilistic algorithm, then for any input x, the notation $A(x)$ refers to the probability space that assigns to the string σ the probability that A, on input x, outputs σ. If S is a probability space, then $PR_S(e)$ denotes the probability that S associates with the element e.

If $f(\cdot)$ and $g(\cdot,\ldots,\cdot)$ are probabilistic algorithms then $f(g(\cdot,\ldots,\cdot))$ is the probabilistic algorithm obtained by composing f and g (i.e. running f on g's output). For any inputs $x, y, \ldots$ the associated probability space is denoted by $f(g(x, y, \ldots))$.

If s is any probability space, then $x \leftarrow S$ denotes the algorithm which assigns to x an element randomly selected according to S. If f is a finite set, then the notation$x \leftarrow f$ denotes the algorithm which assigns to x an element selected according to the probability space whose sample space is f and uniform probability distribution on the sample points.

The notation $Pr(x \leftarrow S; y \leftarrow T; \ldots : p(x, y, ..))$ denotes the probability that the predicate $p(x, y, \ldots)$ will be true after the ordered execution of the algorithms $x \leftarrow S$, $y \leftarrow T, \ldots$

The notation $\{x \leftarrow S; y \leftarrow T; \ldots : (x, y, \ldots)\}$ denotes the probability space over $\{(x, y, \ldots)\}$ generated by the ordered execution of the algorithms $x \leftarrow S$, $y \leftarrow T, \ldots$.

Let us recall the basic definitions of [GMR]. We address the reader to the original paper for motivation, interpretation and justification of these definitions.

Let $U = \{U(x)\}$ be a family of random variables taking values in $\{0,1\}^*$, with the parameter x ranging in $\{0,1\}^*$. $U = \{U(x)\}$ is called poly-bounded family of random variables, if, for some constant $e \in \setminus$, all random variables $U(x) \in u$ assign positive probability only to strings whose length is exactly $|x|^e$.

Let $C = \{C_x\}$ be a poly-size family of boolean circuits, that is, for some constants $c, d > 0$, all C_x have one boolean output and at most $|x|^c$ gates and $|x|^d$ inputs. In the following, when we say that a random string, chosen according to $U(x)$, where $\{U(x)\}$ is a poly-bounded family of random variables, is given as input to C_x, we assume that the length of the strings that are assigned positive probability by $U(x)$ equals the number of boolean inputs of C_x.

Definition 2.1 (Indistinguishability) . *Let $L \subset \{0,1\}^*$ be a language. Two poly-bounded families of random variables $U = \{U(x)\}$ and $V = \{V(x)\}$ are indistinguishable on L if for all poly-size families of circuits $C = \{C_x\}$,*

$$\Big| Pr(A \leftarrow U(x) : C_x(a) = 1) - $$
$$Pr(a \leftarrow V(x) : C_x(a) = 1) \Big| < |x|^c$$

For all positive constants c and sufficiently large $x \in L$.

Definition 2.2 (Approximability) . *Let $L \subset \{0,1\}^*$ be a language. a family of random variables $U = \{U(x)\}$ is approximable on L if there exists a probabilistic turing machine M, running in expected polynomial time, such that the families $\{U(x)\}$ and $\{M(x)\}$ are indistinguishable on L.*

2.2 Number Theory

Let $Z_s(k)$ denote the set of integers product of $s \geq 1$ distinct primes of length k.

Let N be the set of the natural numbers, $x \in N$, $Z_x^* = \{y \mid 1 \leq y < x,\ \ gcd(x,y) = 1\}$ and $Z_x^{+1} = \{y \in Z_x^* | (y \mid x) = +1\}$, where $(y \mid x)$ is the jacobi symbol. We say that $y \in Z_x^*$ is a quadratic residue modulo x iff there is $w \in Z_x^*$ such that $w^2 \equiv y \bmod x$. If this is not the case we call w a quadratic non residue modulo x.

Define the quadratic residuosity predicate to be

$$Q_x(y) = \begin{cases} 0, & \text{if } y \text{ is a quadratic residue modulo } x; \\ 1, & \text{otherwise;} \end{cases}$$

and the languages QR and QNR as

$$QR = \{(y,x) | Q_x(y) = 0\}$$
$$QNR = \{(y,x) | y \in Z_x^{+1} \text{ and } Q_x(y) = 1\}.$$

Fact 1: Let $\sim$ be the relation so defined: $y_1 \sim y_2$ iff $Q_x(y_1y_2) = 0$. Then $\sim$ is an equivalence relation in Z_x^{+1}. Two elements are equivalents if they have the same quadratic character modulo each of the prime divisors of x. Thus, if $x \in Z_2(k)$ there are 2 equivalence classes, if $x \in Z_3(k)$ there are 4; in general if $x = p_i^{h_1} \cdots, p_n^{h_n}$ where each p_i is a prime > 2 and $p_i \neq p_i$ if $i \neq j$, then there are 2^n equivalence classes.

Fact 2: For each $y_1, y_2 \in Z_x^{+1}$ one has

$$Q_x(y_1y_2) = Q_x(y_1) \oplus Q_x(y_2).$$

Fact 3: Where "$\oplus$" denotes the *exclusive or* operator. the jacobi symbol function $x|n$ is polynomial-time computable.

We now formalize the complexity assumption that is sufficient for non-interactive zero-knowledge. Namely, that it is computationally hard to distinguish the integers product of 2 primes leftarrow the ones product of 3 primes.

2.3 A Complexity Assumption

2OR3A: for each poly-size family of circuits $\{C_k | k \in N\}$

$$\left| P_{Z_2(k)} - P_{Z_3(k)} \right| < k^{-c}$$

for all positive constants c and sufficiently large k; where

$$P_{Z_2(k)} = P_R(x \leftarrow Z_2(k) : C_k(x) = 1) \text{ and}$$
$$P_{Z_3(k)} = P_R(x \leftarrow Z_3(k) : C_k(x) = 1).$$

2OR3A is a stronger assumption than assuming that deciding quadratic residuosity is hard. (Having an oracle for $Q_n(\cdot)$, allows one to prbabilistically count the number of $\sim$ equivalence in Z_n^{+1} and thus, by fact 1, to distinguish whether $n \in Z_2(k)$ or $n \in Z_3(k)$). Thus we can freely use that quadratic residuosity is computationally hard (as formalized below) without increasing our assumption set.

Quadratic Residuosity Assumption(QRA):

For each poly-size family of circuits $\{C_k \mid k \in \mathcal{N}\}$,

$$Pr(x \leftarrow Z_2(k); y \leftarrow Z_x^{+1}: C_k(x,y) = Q_x(y))$$
$$< 1/2 + 1/k^{-O(1)}.$$

The QRA was introduced in [GM] and is now widely used in Cryptography. The current fastest algorithm to compute $Q_x(y)$ is to first factor x and then compute $Q_x(y)$, while it is well known that, given the factorization of x, $Q_x(y)$ can be computed in $O(|x|^3)$ steps. In what follows, we choose $x \in Z_2(k)$ since these integers constitute the hardest input for any known factoring algorithm.

3 Single-Theorem Non-Interactive Zero-Knowledge Proofs

To prove the existence of single-theorem Non-Interactive Zero-Knowledge Proof Systems (single-theorem non-interactive ZKPS) for all NP languages, it is enough to prove it for $3COL$ the NP-complete language of the 3-colorable graphs [GJ]. For $k > 0$, we define the language $3COL_k = \{x \in 3COL \mid |x| \leq k\}$.

Definition 3.1 . *A Single-Theorem Non-Interactive ZKPS is a pair (A,B) where A (the Prover) is a Probabilistic Turing Machine and $B(\cdot,\cdot,\cdot)$ (the Verifier) is a deterministic algorithm running in time polynomial in the length of its first input, such that:*

1. **Completeness.** *(The probability of succeeding in proving a true theorem is overwhelming.)*

   $\exists c > 0$ *such that* $\forall x \in 3COL_k$

$$Pr(\sigma \leftarrow \{0,1\}^{n^c}; y \leftarrow A(\sigma, x) : \\ B(x, y, \sigma) = 1) > 1 - n^{-O(1)}.$$

2. **Soundness.** *(The probability of succeeding in proving a false theorem is negligible.)*

   $\exists c > 0$ *such that* $\forall x \notin 3COL_k$ *and for each Probabilistic Turing Machine* A'

$$Pr(\sigma \leftarrow \{0,1\}^{n^c}; y \leftarrow A'(\sigma, x) : \\ B(x, y, \sigma) = 1) < n^{-O(1)}.$$

3. **Zero-Knowledge.** *(The proof gives no information but the validity of the theorem.)*

   $\exists c > 0$ *such that the family of random variables* $V = \{V(x)\}$ *is approximable over* $3COL$. *Where*

$$V(x) = \{\sigma \leftarrow \{0,1\}^{|x|^c}; y \leftarrow A(\sigma, x) : (\sigma, y)\},$$

Remark: Notice that, as usual, the zero-knowledge condition guarantees that the verifier's *view* can be well simulated; that is, all the verifier may see can be reconstructed with essentially the same odds. In our scenario, what the verifier sees is only the common random string and the proof, i.e., the string, received by A. Notice that in our scenario, the definition of zero-knowledge is simpler. As there is no interaction between B and A, we do not have to worry about possible cheating by the verifier to obtain a "more interesting view." That is, we can eliminate the quantification "$\forall B'$" from the original definition of [GMR].

Theorem 3.1 . *Under the QRA, there exists a Single-Theorem Non-Interactive ZKPS for 3-COL.*

This theorem will be rigorously proven in the final paper. Here we restrict ourselves to informally describe the programs P and V of a single-theorem non-interactive ZKPS (P,V) and, even more informally, to argue that they posses the desired properties.

3.1 The Proof System (P,V)

Instructions for P

1. Randomly select $n_1, n_2, n_3 \in Z_2(k)$
2. For $i = 1,2,3$ randomly select q_i such that $(q_i|n_i) = 1$ and q_i is a quadratic non-residue mod n_i.
3. Color G with colors 1,2,3.
4. For each node v of G whose color is i, label v with a randomly selected triplet $(v_1, v_2, v_3) \in Z_{n_1}^{+1} \times Z_{n_2}^{+1} \times Z_{n_3}^{+1}$ such that $Q_n(v_i) = 0$ and $Q_{n_j}(v_j) = 1$ for $j \neq i$. Call G' the so labeled G

   {**Remark 1:** WLOG (else purge σ in the "right way") let $\sigma = \sigma_1 \circ \sigma_2 \circ \sigma_3 \circ \sigma_4, \cdots$, where all triplets $(\sigma_1, \sigma_2, \sigma_3)(\sigma_4, \sigma_5, \sigma_6), \cdots$ belong to $Z_{n_i}^{+1} \times Z_{n_i}^{+1} \times Z_{n_3}^{+1}$.}

   {**Convention:** The first $8k$ triplets are assigned to the first edge of G (in the lexicographic order), the next $8k$ triplets to the second edge, and so on.}

5. For each edge (a, b) of G' (where node a has label (a_1, a_2, a_3) and node b (b_1, b_2, b_3)) and each of its $8k$ assigned triplets (z_1, z_2, z_3) compute one of the following types of *signature.*

   (**Comment:** Only one is applicable if steps 1-4 are performed correctly)}

$(\sqrt{z_1}, \sqrt{z_2}, \sqrt{z_3})$	type 0
$(\sqrt{q_1 z_1}, \sqrt{z_2}, \sqrt{z_3})$	type 1
$(\sqrt{z_1}, \sqrt{q_2 z_2}, \sqrt{z_3})$ ·	type 2
$(\sqrt{z_1}, \sqrt{z_1}, \sqrt{q_3 z_3})$	type 3
$(\sqrt{a_1 z_1}, \sqrt{a_2 z_2}, \sqrt{a_3 z_3})$	type 4
$(\sqrt{b_1 z_1}, \sqrt{b_2 z_2}, \sqrt{b_3 z_3})$	type 5
$(\sqrt{a_1 b_1 z_1}, \sqrt{a_2 b_2 z_2}, \sqrt{a_3 b_3 z_3})$	type 6
$(\sqrt{q_1 z_1}, \sqrt{q_2 z_2}, \sqrt{q_3 z_3})$	type 7

   {**Notation "by example"**: Let z_1 be a quadratic non residue mod n_1, z_2 a quadratic residue mod n_2, and z_3 is a quadratic residue mod n_2. Then the signature of the triplet (z_1, z_2, z_3) a triplet of type 1: $(\sqrt{q_1 z_1}, \sqrt{z_2}, \sqrt{z_3})$ where $\sqrt{q z_1}$ denotes a randomly selected square root of the quadratic residue $q_1 \cdot z_1$ mod n_1; and for $i = 2,3$ $\sqrt{z_i}$ denotes a randomly selected square root of z_i mod n_i}

6. Send V $n_1, n_2, n_3, q_1, q_2, q_3, G'$, and the signature of the triplets composing σ.

   {**Comment:** Note that the edges of G' are labelled with triples, not with colors!}

Instructions for V

1. Verify that n_1, n_2, and n_3 are not even and not integer powers. Verify that G' is a proper labelling of G. That is, each node v has assigned a triplet (v_1, v_2, v_3) such that $v_i \in Z_{n_i}^{+1}$ for $i = 1, 2, 3$.

2. Break σ into triplets, verify that for each edge you received a signature of some type for each of its $8k$ triplets.

3. If all the above verifications have been successfully made, accept that G is 3-colorable.

3.2 A Rough Idea of why (P,V) is a Single-Theorem Non-Interactive ZKPS

First notice that the communication is mono-directional: From P to V. Then let us convince ourselves that the statement of Remark 1 really holds without loss of generality. In our context, WLOG means with overwhelming probability.

If G has a edges, our protocol assumes σ to consist of $8 \cdot k \cdot a$ triplets in $Z_{n_1}^{+1} \times Z_{n_2}^{+1} \times Z_{n_3}^{+1}$. Such a string σ is easily obtainable from a (not too much larger) random string ρ. Consider ρ to be the concatentation of k-bit strings grouped into triplets

$$\rho = (\rho_1, \rho_2, \rho_3)(\rho_4, \rho_5, \rho_6) \cdots$$

Then obtain σ by "purging" ρ. That is, obtain σ from ρ by discarding all triplets not in $Z_{n_1}^{+1} \times Z_{n_2}^{+1} \times Z_{n_3}^{+1}$. We now argue that ρ is not much longer than σ. Let n be either n_1 or n_2 or n_3. Now a random k-bit integer (with possible leading 0's) is less than n with probability $\geq \frac{1}{2}$; a random integer less than n belongs to Z_n^* with probability $\geq \frac{1}{2}$; a random element of Z_n^* belongs to Z_n^{+1} with probability $\geq \frac{1}{2}$. Thus, we expect that at least 1 in 64 of the triplets of ρ not to be discarded.

Now let us consider the question of V's running time. V can verify in poly-time whether $n_i = x^\alpha$ (where x, α integers; $\alpha > 1$) as only values $1, \cdots, \log n_i$ should be tried for α and binary search can be performed for finding x, if it exists. All other steps of V are even easier.

Now let us give some indication that (P,V) constitute a single-theorem non-interactive ZKPS.

Completeness: Assuming that σ is already consiting of triplets in $Z_{n_1}^{+1} \times Z_{n_2}^{+1} \times Z_{n_3}^{+1}$, if P operates correctly, V will be satisfied with probability 1.

Soundness: If the verification step 1 is successfully passed, by fact 1, there must be $\geq 2 \sim$ equivalence classes in each $Z_{n_i}^{+1}$ (exactly two if P honestly chooses all the n_i's in $Z_2(k)$).

Thus, if we define two of our triplets (z_1, z_2, z_3) (w_1, w_2, w_3) to be equivalent if $z_i w_i \bmod n_i$ is a quadratic residue for $i = 1, 2, 3$, we obtain ≥ 8 equivalence classes among the triplets (exactly 8 if P is honest).

To exhibit a signature of a given type for a triplet, essentially means to put the triplet in one out of ≤ 8 possible "drawers". (there are 8 types of signatues, but they may not be mutually exclusive; thus two drawers may be equal). Moreover, it is easy to see that if two triplets are put in the same drawer, they must belong to the same equivalence class.

As σ is randomly selected, each of its triplets in $Z_{n_i}^{+1} \times Z_{n_2}^{+1} \times Z_{n_3}^{+1}$ is equally likely to belong to any of the ≥ 8 equally-numerous equivalence classes. However, since if there were > 8 classes, there would be (by fact 1) at least 16, the fact that all triplets can be fit in ≤ 8 drawers, "probabilistically proves" several facts:

1. There are exactly 8 equivalence classes among the triplets and exactly 8 distinct drawers.
2. The n_i's are product of two distinct prime powers.
3. $Q_{n_1}(q_1) = Q_{n_2}(q_2) = Q_{n_3}(q_3) = 1$
4. $Q_{n_1}(q_1) + Q_{n_2}(q_2) + Q_{n_3}(q_3) = 2$

   That is, (a_1, a_2, a_3) is a proper color (i.e., properly encodes a color: Either 1,2, or 3).
5. That (b_1, b_2, b_3) is a proper color.
6. That (a_1, a_2, a_3) and (b_1, b_2, b_3) are different colors. Else drawer 6 and drawer 0 would be the same.

Item 6 being true for all edges in G' implies that G is 3-colorable which is what was to be proven.

Zero-Knowledgeness

Let us specify the simulating machine M that, under the QRA, generates a pair (σ, proof) with the "right odds" on input G (without any coloring!)

Instructions for M

1. Randomly select $n_1, n_2, n_3, \in Z_2(k)$ together with their prime factorization.
2. Randomly select q_1, q_2, q_3 so that $Q_{n_1}(q_1) = Q_{n_1}(q_2) = Q_{n_3}(q_3) = 0$
3. For each node v of G, label v with a triplet $(v_1, v_2, v_3) \in Z_{n_1}^* \times Z_{n_2}^* \times Z_{n_3}^*$ such that $Q_{N_1}(v_1) = Q_{n_2}(v_2) = Q_{n_3}(v_3) = 0$. Call G' the so labelled graph.
4. Construct $\sigma = (\sigma_1, \sigma_2, \sigma_3)(\sigma_4, \sigma_5, \sigma_6) \cdots$, such that each triplet $(\sigma_{3j+1}, \sigma_{3j+2}, \sigma_{3j+3})$ is randomly selected so that $Q_{n_i}(\sigma_{3j+i}) = 0$ for $i = 1, 2, 3$.

   {**Remark:** Also in the simulation we only deal with already "purged strings". It is not hard to see that M could also handle generating "unpurged strings".}

5. For each edge (a, b) of G' and each of its assigned $8k$ triplets (z_1, z_2, z_3), choose an integer i at random between 0 and 7, and compute a signature of type i.

   {**Comment:** By using the prime factorization of the n_i.}

6. Output $\sigma, n_1, n_2, n_3, q_1, q_2, q_3, G'$, and the computed signatures.

We now informally argue that M is a good simulator for the view of V. Essentially, this is so because efficiently detecting that the triplets of σ are not randomly and independently drawn from the space $Z_{n_i}^{+1} \times Z_{n_2}^{+1} \times Z_{n_3}^{+1}$ is tantamount as violating the QRA (to be explained in the final paper). For the same reason, it cannot be detected efficiently that G' is an illegal labelling or that q_1, q_2, q_3 are squares mod, respectively, n_1, n_2, n_3. Given that, the distribution of the various types of signature looks "perfect".

{**Remark:** the reader is encouraged to verify that if (P,V) uses part of the used σ to show that another graph is 3-colorable, then extra knowledge would leek. For instance that there exists 3-coloring of G and H in which nodes v_1 and v_2 in H respectively have the same clolors as nodes w_1 and w_2 in G.}

4 Security Against Chosen Ciphertext Attack

One of the most beautiful gifts of complexity-based cryptography is the notion of a public-key cryptosystem. As proposed by Diffie and Hellman [DH], each user U publicizes a string P_U and keeps secret an associated string S_U. Another user, to secretely send a message m to U, computes $y = E(P_U, m)$ and sends y; upon receiving y, U retrieves m by computing $D(S_U,\ y)$; here E and D are polynomial-time algorithms chosen so that it will be infeasible, for any other user, to compute m from y.

Notice that in this set-up any other user is thought to be a "passive" adversary who tries to retrieve m by computing solely on inputs y and P_U. This is indeed a mild type of adversary and other types of attacks have been considered in the literature. It is widely believed that the strongest type of attack among all the natural ones is the *chosen-ciphertext attack.* In such an attack, someone tries to break the system by asking and receiving decryptions of ciphertexts of his choices. Rivest has shown that Rabin's scheme (whose breaking is, for a passive adversary, as hard as factoring if the messages are uniformily selected strings of a given length) is easily vulnerable to such an attack. Indeed, this is an attack feasible to any employee who works at the decoding equipment of, say, a large bank. The power by this attack is very well exemplified by an elegant scheme of Rabin [R] that is as secure as factoring (if the messages are uniformily selected strings of a given length) in the passive adversary model but is easily broken by chosen-ciphertext attack. Since observing this phenomenon, people tried to design cryptosystems invulnerable to such attacks, but in vain. A positive answer has been found [GMT] only allowing interaction, during the encryption process, between legal sender and legal receiver. However, for the standard (non-interactive) Diffie-and-Hellman model, the existence of a cryptosystem invulnerable to chosen ciphertext attack has been an open problem since 1978.

Non-interactive zero-knowledge proofs allow us to finally solve this problem. The essence of our solution (instead of its details) is informally described as follows. Instead of sending U an encryption, y, of a message m, one is required to send two strings: y and and σ, where σ is a *zero-knowledge and non-interactive proof that the sender knows the decoding of* y. The "decoding equipment" (read: the decoding function) checks that σ is convincing and, if so, outputs m, the decoding of y; Otherwise, it outputs nothing. Notice that, now, being able to use the decoding equipment provably is of no advantage! In fact, only when we feed it with ciphertexts whose decoding we can prove we know, does the decoding equipment output these decodings! In other words, the decoding equipment can only be used to output what we already know. A detailed discussion of this powerful application will appear in the final paper.

(A formal setting and the proof require some care. For instance, the decoding equipment may be used as an oracle to check whether a given string σ is a "correct proof of knowledge". Thus, in particular, one should prove that such an oracle cannot help. In the final paper we will essentially show that if one can generate a legal (y, σ) pair without having m as an input, then one can easily decrypt all messages on input y and P_U only.)

5 References

[ACGS] W. Alexi, B. Chor, O. Goldreich, and C. Schnorr *RSA/Rabin Bits Are* $1/2 + 1/_{poly}(\log N)$ *Secure*, To appear SIAM J. on Computing.

[B1] M. Blum, *Coin Flipping by Telephone*, IEEE COMPCON 1982, pp. 133-137.

[B2] M. Blum, unpublished manuscript

[BBS] M. Blum, L. Blum and M. Shub,*A simple and secure pseudo-randomnumber generator*,SIAM Journal of Computing, 1986

[BFM] Blum, De Santis, Feldman, Micali, and Persiano, *Non-Interactive Zero Knowledge and Its Applications*, in preparation.

[BGGHKMR] M. Ben-Or, O. Goldreich, S. Goldwasser, J. Hastad, J. Kilian, S. Micali, and P. Rogaway, *Everything Provable is Provable in Zero-Knowledge*, These Proceedings

[BH] R. Boppana, J. Hastad and S. Zachos, *Interactive Proofs Systems for CO-NP Imply Polynomial Time Hierarchy Collapse*, In preperation.

[BM] M. Blum and S. Micali, *How To Generate Sequences Of Cryptographically Strong Pseudo-Random Bits*, SIAM J. on Computing, Vol. 13, Nov 1984, pp. 850-864

[DH] Diffie, W., and M.E. Hellman, *New Directions in Cryptography*,IEEE Trans. on Inform. Theory,

[DMP] De Sentis, Micali, and Persiano, *Non Interactive Zero-Knowledge Proof Systems*, Proc. Crypto 87.

[F] L. Fortnow, *The Complexity of Perfect Zero-Knowledge*, Proc. 19th ann. Symp. on Theory of Computing, New York, 1987.

[FFS] Feige, Fiat and A. Shamir, *Zero-knowledge proofs of identity*, Proceedings of the 19th Annual ACM Symp. on Theory of Computing, 1987, pp. 210–217

[GM] S. Goldwasser, and S. Micali, *Probabilistic Encryption*, JCSS Vol. 28, No. 2, April 1984.
[GMR] S. Goldwasser, S. Micali and C. Rackoff, *The Knowledge Complexity of Interactive Proof-Systems*, To appear SIAM J. on Computing (manuscript available from authors).
[GoMiRi] S. Goldwasser, S. Micali, and R. Rivest, *A Digital Signature Scheme Secure Against Adaptive, Chosen Cyphertext Attack* To appear in SIAM J. on Computing (available from authors)
[GMT] S. Goldwasser, S. Micali, and P. Tong, *Why and how to establish a perivate code in a public network*, Proc. 23rd Symp. on Foundations of Computer Science, Chicago, Ill., 1982
[GMW] O. Goldreich, S. Micali and A. Wigderson, *Proofs that Yield Nothing but their Validity and a Methodology of Cryptographic Design*, Proc. of FOCS 1986.
[GMW2] O.Goldreich, S. Micali and A. Wigderson, *How to Play Any Mental Game*, Proceedings of the 19th Annual ACM Symp. on Theory of Computing, 1987, pp. 218–229.
[GS] S. Goldwasser and M. Sipser, *Private Coins versus Public Coins in Interactive Proof Systems*, Proceedings of the 18th Annual ACM Sympl on Theory of Computing, 1986, pp. 59–68.
[I] R. Impagliazzo, Personal Communication.
[MS] S. Micali and A. Shamir An improvement of the Fiat-Shamir Identification and Signature Scheme, These proceedings
[R] M. Rabin, *Digitalized signatures and public-key functions as intractable as factorization*, MIT/LCS/TR-212, Technical report MIT, 1978
[Y] A.Yao, *Theory and Application of Trapdoor Functions*, Proc. of 23rd FOCS, IEEE, Nov., 1982, pp. 80-91.

Non-Interactive Zero-Knowledge with Preprocessing

Alfredo De Santis[1,4], *Silvio Micali*[2], *Giuseppe Persiano*[3]

[1] IBM T. J. Watson Research Center, P.O. Box 218, Yorktown Heights, NY 10598

[2] Laboratory for Computer Science, MIT, Cambridge, MA 02139

[3] Aiken Comp. Lab., Harvard University, Cambridge, MA 02138

Abstract

Non-Interactive Zero-Knowledge Proof Systems have been proven to exist under a specific complexity assumption; namely, under the Quadratic Residuosity Assumption which gives rise to a specific secure probabilistic encryption scheme.

In this paper we prove that the existence of *any* secure probabilistic encryption scheme, actually any *one-way encryption scheme*, is enough for Non-Interactive Zero-Knowledge in a modified model. That is, we show that the ability to prove a randomly chosen theorem allows to subsequently prove non-interactively and in Zero-Knowledge any smaller size theorem whose proof is discovered.

A quick and dirty exposition of our results

The *one-time pad* is a well-known cipher system which achieves perfect security in the Shannon sense [Sh]. This system can be described as consisting of two stages:

1. *Preprocessing stage.* A and B interact for a while to agree on a common n-bit random string r.
2. *Communication stage.* A encrypts an n-bit plaintext by xoring it with the string r, and sends the ciphertext to B.

Notice that the communication stage is *unidirectional*: A sends the message to B, that need not to reply.

[4]Work done at IBM while on leave from Dipartimento di Informatica ed Applicazioni, Università di Salerno, 84100 Salerno, Italy.

The string r does not depend on the communication stage. When A and B agree on r, they have no idea on the message that A will send B later.

The advantage of this system is that it is the *most secure* cipher system that exists. Only B can decrypt the ciphertext, because he knows the string r. The ciphertext does not give any information to an eavesdropper, no matter how powerful he is. Indeed an eavesdropper has the same amount of information, in the Shannon sense, on the plaintext both before and after seeing the ciphertext.

This system has two drawbacks:

1. A and B have to meet beforehand in order to share the random string.
2. the length of the string r generated in the preprocessing stage *bounds* the length of the message that A can send to B in the communication stage. Indeed if the same string r is used again to send other messages, then an eavesdropper will obtain some information.

In this paper we present the notion of *Non-Interactive Zero-Knowledge Proof-System with Preprocessing* that constitutes the equivalent of one-time pad for *Non-Interactive Zero-Knowledge Proof Systems.*

Also Non-Interactive Zero-Knowledge Proof-Systems with Preprocessing have a preprocessing stage and a communication stage:

1. *Preprocessing stage.* A, the prover, chooses an n-bit theorem T_0 and proves interactively and in zero-knowledge to B, the verifier, that T_0 is true.
2. *Communication stage.* A proves to B any NP theorem of length not bigger than n^c, for some fixed positive constant $c < 1$. This proof is *unidirectional* (from A to B) and zero-knowledge (B does not get any additional knowledge of the theorem proved, but its validity).

Like in the one-time pad case, in which, when A and B interact in the preprocessing stage, they have no idea on the message that A will send later to B, also in our proof system A and B have no idea, when they interact in the preprocessing stage, on the theorem that A will later prove to B.

The advantage of this system is analogous to that of the one-time pad case. We get maximum security for the system. B does not get from the proof that receives from A any additional knowledge on the theorem but its validity. We only require that a one-way encryption scheme exists.

This system has the same two drawbacks of the one-time pad system:

1. A and B have to meet beforehand.
2. the length of the theorem T_0 *bounds* the length of the theorems that A can prove to B in the communication stage. Indeed if the same string T_0 is used again to prove non-interactively other theorems, then B will get some information.

Now we give a sketch of our protocol. To prove the existence of a Non-Interactive Zero-Knowledge Proof-System with Preprocessing for any NP language it is enough to prove it for the language of the 3-satisfiable formulae (3SAT). The following is an example of a formula in 3SAT:

$$(u_3 \vee u_2 \vee \overline{u}_1) \wedge (u_6 \vee \overline{u}_2 \vee \overline{u}_1) \wedge \ldots\ldots \wedge (u_7 \vee \overline{u}_3 \vee u_4)$$

It consists of literals ($u_1, \overline{u}_1, u_2, \overline{u}_2$...) and clauses (($u_3 \vee u_2 \vee \overline{u}_1$), ($u_6 \vee \overline{u}_2 \vee \overline{u}_1$),..., ($u_7 \vee \overline{u}_3 \vee u_4$)) Each clause contains exactly three literals. The formula is satisfiable iff there is an assignment of boolean values $\{T, F\}$ (or, equivalently, $\{1, 0\}$) to the literals such that the formula is true, that is there is at least one literal true for each clause.

Now, suppose that A wants to prove to B that a particular formula, of which he knows a satisfying assignment t, is satisfiable. A associates to each literal u an encryption of 1 if $t(u) = T$ or an encryption of 0 if $t(u) = F$. Each literal is associated with only one encryption, even if it appears in several clauses. In this way a triple of encryptions has been naturally associated to each clause.

To prove that the formula is true it is enough to show that each clause is satisfiable, i.e. that each of the triplets of encryptions associated to the clauses contains at least an encryption of 1, since this corresponds to a literal with a true value in the clause. All the interaction required for this task is squeezed in the preprocessing stage.

Here is a sketch of our protocol.

Preprocessing stage. A randomly chooses $3n$ complementary bits b_i and $\overline{b}_i = 1 - b_i$. For example, (1,0), (1,0), (0,1),.... A keeps secret these bits, and will never show them to B. These bits represent boolean values (1 stands for True, 0 for False).

Then, using a one-way encryption scheme, A computes an encryption of each bit (α_1, β_1), (α_2, β_2), (α_3, β_3),...

For each triple formed by 3 of the encryptions in the set $\{\alpha_1, \beta_1, \alpha_2, \beta_2, \alpha_3, \beta_3, \ldots\}$, A encrypts the bit that is the or of the bits encrypted by the triple. For instance, if $(\alpha_3, \alpha_2, \beta_1)$ are the encryptions of $(1, 0, 0)$ then A computes an encryption γ of the bit $1 = 1 \vee 0 \vee 0$.

Finally A concludes the preprocessing stage by sending to B all and only the encryptions he computed, and proving interactively and in zero-knowledge that these encryptions has been correctly computed. Thus B will learn that (α_i, β_i) are the encryptions of two complimentary bits, but he has no information if they are the encryptions of (1,0) or those of (0,1); moreover B will learn that also the encryptions γ's were properly computed, but he still doesn't get any information on the bits b_i and $\overline{b}_i$.

Communication stage. Now A is able to prove to B non-interactively and in zero-knowledge that any n-bit 3-satisfiable formula C is indeed satisfiable.

The written proof of A consists of an "ad hoc" association of the literals of C with the first pairs of encryptions (α_1, β_1), (α_2, β_2),.... Namely, for each variable

u_i, A considers the pair (α_i, β_i). A associates the element of the pair which is the encryption of 1 to the literal, u_i or $\overline{u}_i$, that is true under t and the remaining element of the pair to the other literal. Since each clause is formed of three literals, in this way A has associated to each clause, a triplet of encryptions. A finally looks for this triplet in the list of the triplets he gets from the preprocessing stage and shows that the associated encryption γ is an encryption of 1, by opening it.

Informally speaking this is zero-knowledge, since the only thing B will get is a zero-knowledge proof in the preprocessing stage, plus encryptions of bits. In particular the encryptions γ's, which A opens in the communication stage, are encryptions of 1, which must be certainly the case if the formula is satisfiable.

To perform our protocol, the prover A can be a probabilistic polynomial-time machine that gets an NP proof as an auxiliary input. Thus, this proof system can be used in a cryptographic scenario.

Let us proceed more formally.

1 Introduction

The notion of Non-Interactive Zero-Knowledge Proof-System has been introduced by [BlFeMi]. A Non-Interactive Zero-Knowledge Proof-System allows a prover to non-interactively and in zero-knowledge prove any number of theorems to a poly-bounded verifier, provided that the prover and the verifier share a random string. The prover, on input the common random string σ, and the theorem T writes the proof in a letter. The verifier, trusting the randomness of σ, is convinced that the theorem T is true just by reading the letter. This model is argued to be the minimal one supporting zero-knowledge proofs. [BlFeMi] and [DeMiPe] gave implementations based on the difficulty of specific computational problems. Namely the [BlFeMi] implementation relies on the difficulty of distinguishing numbers product of two primes from those product of three primes. Whereas the [DeMiPe] implementation is based on the weaker assumption of the difficulty of distinguishing a quadratic residue from a quadratic non residue.

Non-Interactive Zero-Knowledge Proof Systems are particularly useful when the prover and the verifier cannot talk each other. Indeed, though interaction is a requirement that can be met, in practice it may not be readily available. For example this is the case if the prover will leave for a 10-year trip and the mail is the only way to communicate.

In this paper we introduce the notion of Non-Interactive Zero-Knowledge Proof-System with Preprocessing. A Non-Interactive Zero-Knowledge Proof-System with Preprocessing consists of two stages:

1. *Preprocessing stage.* A chooses an n-bit string $v \in V$ and proves interactively and in zero-knowledge to B that indeed v belongs to the language V.

2. *Communication stage.* A chooses an NP theorem T of length not bigger than n^c (for some fixed positive constant c), and sends the proof that indeed T is

true to B. B, believing that $v \in V$, can check that the proof is valid without ever talking to A but gets no knowledge except the validity of the theorem.

This proof is *unidirectional* (from A to B) and zero-knowledge (B does not get any additional knowledge of the theorem proved).

A Non-Interactive Zero-Knowledge Proof-System with Preprocessing allows a prover to non-interactively and in zero-knowledge prove a theorem $\mathcal{T}$ to a poly-bounded verifier, provided that the prover and the verifier had the opportunity to meet beforehand. The prover on input the string $v \in V$ and the theorem $\mathcal{T}$ writes the proof of $\mathcal{T}$ in a letter. The verifier, being convinced in the preprocessing stage that $v \in V$, is also convinced that the theorem $\mathcal{T}$ is true, while receiving no other additional information, just by reading the letter.

This model is not as general as that considered in [BlFeMi] and [DeMiPe]. The prover and the verifier must know in advance the length of the theorem that will be later non-interactively proved. Moreover the non-interactive proof needs a string randomly chosen in a particular language V, and not just a random string. This is somewhat more difficult to obtain, even though a preprocessing step can easily handle both cases.

We give an implementation of Non-Interactive Zero-Knowledge Proof-Systems with Preprocessing based on the weakest possible assumption in Cryptography: the existence of a one-way encryption scheme. Hence making free the non-interactive proof model from the fortunes of a specifical algebraic problem.

Thus, in this paper the generality and the minimality of [BlFeMi] and [DeMiPe] is traded in exchange of the relaxation of the underlying assumption.

The proposed protocol can be used as a tool for cryptographic protocol design. It allows the squeezing, to an initial step, of all the interaction needed for zero-knowledge proofs in a multi-party protocol. It is indeed enough for a player to send the same single string to all the other players each time he needs to validate his assertion without compromising his secrets.

The reminder of the paper consists of 3 sections. In Section 2, some preliminary facts are discussed. Section 3 presents our main results: first the model is formally defined and then a protocol is given. Finally in Section 4 some open problems are presented.

2 Preliminaries

Let us quickly recall the standard notation of [GoMiRi].

We emphasize the number of inputs received by an algorithm as follows. If algorithm A receives only one input we write "$A(\cdot)$", if it receives two inputs we write "$A(\cdot,\cdot)$" and so on.

If $A(\cdot)$ is a probabilistic algorithm, then for any input x, the notation $A(x)$ refers to the probability space that assigns to the string σ the probability that A, on input

x, outputs σ. If S is a probability space, denote by $Pr_S(e)$ the probability that S associates with the element e.

If $g(\cdot)$ and $h(\cdot, \ldots, \cdot)$ are probabilistic algorithms then $g(h(\cdot, \ldots, \cdot))$ is the probabilistic algorithm obtained by composing g and h (i.e. running g on h's output). For any inputs $x, y, \ldots$ the associated probability space is denoted by $g(h(x, y, \ldots))$.

If S is any probability space, then $x \leftarrow S$ denotes the algorithm which assigns to x an element randomly selected according to S. If F is a finite set, then the notation $x \leftarrow F$ denotes the algorithm which assigns to x an element chosen with uniform probability from F. In other word, when no confusion arises, we identify a finite set with the algorithm that randomly select a point in that set.

The notation $Pr(x \leftarrow S; y \leftarrow T; ...: p(x, y, ...))$ denotes the probability that the predicate $p(x, y, ...)$ will be true after the ordered execution of the algorithms $x \leftarrow S$, $y \leftarrow T$,...

The notation $\{x \leftarrow S; y \leftarrow T; ...: (x, y, ...)\}$ denotes the probability space over $\{(x, y, ...)\}$ generated by the ordered execution of the algorithms $x \leftarrow S$, $y \leftarrow T$, ...

Let us recall the basic definitions of [GoMiRa]. We address the reader to the original paper for motivation, interpretation and justification of these definitions.

Let $U = \{U(x)\}$ be a family of random variables taking values in $\{0,1\}^*$, with the parameter x ranging in $\{0,1\}^*$. $U = \{U(x)\}$ is called poly-bounded family of random variables, if, for some constant $e \in \mathcal{N}$, all random variables $U(x) \in U$ assign positive probability only to strings whose length is exactly $|x|^e$.

Let $C = \{C_x\}$ be a poly-size family of Boolean circuits, that is, for some constants $c, d > 0$, all C_x have one Boolean output and at most $|x|^c$ gates and $|x|^d$ inputs. In the following, when we say that a random string, chosen according to $U(x)$, where $\{U(x)\}$ is a poly-bounded family of random variables, is given as input to C_x, we assume that the length of the strings that are assigned positive probability by $U(x)$ equals the number of boolean inputs of C_x.

Definition (Indistinguishability). *Let $L \subset \{0,1\}^*$ be a language. Two poly-bounded families of random variables $U = \{U(x)\}$ and $V = \{V(x)\}$ are indistinguishable on L if for all poly-size families of circuits $C = \{C_x\}$, all positive constants c and all sufficiently large $x \in L$,*

$$\left| Pr(a \leftarrow U(x) : C_x(a) = 1) - Pr(a \leftarrow V(x) : C_x(a) = 1) \right| < |x|^{-c}.$$

Definition (Approximability). *Let $L \subset \{0,1\}^*$ be a language. A family of random variables $U = \{U(x)\}$ is approximable on L if there exists a Probabilistic Turing Machine M, running in expected polynomial time, such that the families $\{U(x)\}$ and $\{M(x)\}$ are indistinguishable on L.*

The fundamental notions of security for probabilistic encryption scheme were introduced in [GoMi], see also [MiRaSl] for an extended discussion. However this probabilistic encryption scheme is a bit too powerful for our need. Indeed it is required that the encryption is easy and that also the decryption is easy provided that an extra secret key is known. In our scenario we only need that the encryption is easy and the ciphertext can be unambiguously decoded. Indeed should someone encrypts a message, he should be able later to show the message and prove that the encryption was correctly computed, by only remembering his own computation. Should he forget this computation then it will be also difficult for him to compute the message. We call such a scheme, to be formally defined below, *one-way encryption scheme.* Such a scheme has been used in the well known zero-knowledge interactive proof system for 3COL of [GoMiWi] and, more recently, also in [BGGHKMR].

A one-way encryption scheme is a probabilistic polynomial time Turing machine E that, on input x and internal coin tosses r, outputs an encryption $E(x,r)$, such that

1. *The ciphertext can be uniquely decoded.* Whatever are the coin tosses r, s and the inputs x, y, then $E(x,r) = E(y,s)$ implies $x = y$.

2. *There is no computational feasible way to distinguish the encryption of 0 from the encryption of 1.* Let $E_n(x)$ be the probability space obtained by setting $Pr(y) = 2^{-n} \cdot \mid \{r \in \{0,1\}^n : E(x,r) = y\} \mid$. Then, for any poly-size family of circuits $C = \{C_{1^n}\}$, all positive constants c and all sufficiently large n,

$$\left| Pr(a \leftarrow E_n(0) : C_{1^n}(a) = 1) - Pr(a \leftarrow E_n(1) : C_{1^n}(a) = 1) \right| < n^{-c}.$$

Using arguments similar to those of [GoMi], it is easy to show that the existence of the unapproximable predicate as defined in [BlMi] is equivalent to the existence of the one-way encryption scheme. Both the unapproximable predicate and the one-way encryption scheme exists if one makes the stronger assumption that the one-way permutation exists [Ya].

The one-way encryption scheme is an instrumental tool to commit a bit and then to decommit it. The commitment to a bit b using a security parameter n is done by choosing an n-bit random string s and sending $E(b,s)$. For the decommitment of the bit, that is the proof to a polynomial bounded machine that $E(b,s)$ is indeed an encryption of b, it suffices to exhibit s.

In the following, we denote by $E_n(b)$ the set of all possible encryptions of b, with security parameter n, i.e. $E_n(b) = \{e \mid \exists r \text{ such that } e = E(b,r) \text{ and } |r| = n\}$. Finally we denote by D_n the inverse function of E_n, that is $D_n(\alpha) = b$ if $\alpha \in E_n(b)$.

3 Non-Interactive Zero-Knowledge Proof-Systems with Preprocessing

In this section first we formally define what we mean by Non-Interactive Zero-Knowledge Proof-System with Preprocessing and then we give an implementation of it based on the assumption that one-way encryption schemes exist.

To prove the existence of Non-Interactive Zero-Knowledge Proof-System with Preprocessing for all NP languages, it is enough to prove it for the NP-complete language 3SAT [GaJo]. For $k \in \mathcal{N}$ we define the language $V_k = \{y \in V \mid |y| = k\}$.

3.1 The model

In this section we formally define what we mean by Non-Interactive Zero-Knowledge Proof-System with Preprocessing.

Definition. *Let V be a language. (A,B), where A is a Probabilistic Turing Machine and $B(\cdot,\cdot,\cdot)$ is a deterministic algorithm running in time polynomial in the length of the first input, is a Non-Interactive Zero-Knowledge Proof-System with Preprocessing if a positive constant c exists such that*

1. **(Completeness)** $\forall x \in 3SAT$, $\forall v \in V_{|x|^c}$, *for all positive constants d and all sufficiently large x,*

$$Pr\Big(y \leftarrow A(x,v)\ :\ B(x,v,y) = 1\Big) > 1 - |x|^{-d}.$$

2. **(Soundness)** $\forall x \notin 3SAT$, $\forall v \in V_{|x|^c}$, *for each Probabilistic Turing machine A', for all positive constants d and all sufficiently large x,*

$$Pr\Big(y \leftarrow A'(x,v)\ :\ B(x,v,y) = 1\Big) < |x|^{-d}.$$

3. **(Zero-Knowledge)** *The family of random variables $R = \{R(x)\}$, where*

$$R(x) = \Big\{v \leftarrow V_{|x|^c};\ y \leftarrow A(x,v) : (v,y)\Big\},$$

is approximable over $3SAT$.

Notice that in our definition we have not formalized the preprocessing stage. The preprocessing stage consists of an interactive zero-knowledge proof that a string v, randomly chosen in V, indeed belongs to V, while, in the above definition, the string $v \in V$ is seen as an additional input available to both A and B. However, we can always think of this string v as fixed in the preprocessing stage.

So, the proof goes as follows. In the preprocessing stage, A proves that $v \in V$, where V is a particular language that will be instrumental for the later non-interactive zero-knowledge proof and will be thus called V the *auxiliary language.* Later, in the communication stage, A can non-interactively prove to B any theorem of size $|v|^{1/c}$, using the fact that string $v \in V$.

In this way we have squeezed out all the interaction needed at the beginning. Notice that this is not a trivial task since A does not know which theorem he is going to prove when he chooses the string v.

Notice also that A can choose his favorite $v \in V$ (he has infinite computing power), and so in the completeness and soundness requirements we must say "for all $v \in V$" instead of saying "for an overwhelming fraction of $v \in V$".

The randomness of $v \in V$ is a requirement needed only for the zero-knowledge. It does not affect the completeness and soundness of the proof system.

Finally, notice that this model is not as general as the one proposed in [BlFeMi] and [DeMiPe]. In our model the size of the theorem that can be proved is determined by the length of the shared string v. In the [BlFeMi] and [DeMiPe]'s model the prover can prove any theorem of size polynomial in the length of the shared string. On the other hand in this model we still have to make some specific computational assumptions, while in our weaker model we only make the natural assumption that one way functions exist. Thus the auxiliary language V and the bound on the size of the theorem are the price we pay to relax the underlying assumption.

3.2 Our protocol

In this section we describe first informally and then formally our protocol for the Non-Interactive Zero-Knowledge Proof-System with Preprocessing.

Our protocol is based on the following observation. Suppose we associate to each literal u an encryption of 1 if $t(u) = T$ or an encryption of 0 if $t(u) = F$. In this way a triple of encryptions has been naturally associated to each clause c_j. Thus to show that the clause c_j is satisfiable it is enough to show that the associated triple of encryption contains at least an encryption of 1, since this correspond to a literal with a true value in c_j.

The language V is designed to simplify all the future work of proving the needed relationships between these encryptions. To better understanding the language V it is useful to regard the string $v \in V$ as "made" by two parts. The first part consists of $3n$ pairs (α_i, β_i) that are the encryptions of two complementary bits b_i and $\overline{b}_i$. The second part is a table in which, given any three $\nu_1 \in E_n(d_1), \nu_2 \in E_n(d_2), \nu_3 \in E_n(d_3)$, of the $6n$ encryptions $\alpha_1, \alpha_2, ..., \alpha_{3n}, \beta_1, ..., \beta_{3n}$, considered in the first part, we obtain an encryption γ of the bit $d_1 \vee d_2 \vee d_3$.

Thus, given such a string $v \in V$ the non-interactive proof of A consists of an "ad hoc" association of the literals of C with the encryptions contained in the first part of $v \in V$. Namely, for each variable u_i, A considers the pair (α_i, β_i) in $v \in V$. A associates the element of the pair that is the encryption of 1 to the literal, u_i or $\overline{u}_i$, that is true under t and the remaining element of the pair to the other literal. Since each clause c_j is formed of three literals, in this way A has associated to each clause, a tern of encryptions. A finally looks for this tern in the second part of v and shows that the associated γ is an encryption of 1, by revealing the random bits used for its encryption.

Let us now proceed more formally. We first define the language V we use, and then we describe our protocol.

3.2.1 The language V

In this section we describe the auxiliary language $V = \{V_n\}_{n\in\mathcal{N}}$ that we use in our protocol.
Let E be a one-way encryption scheme.

A string $v \in V_n$ is the concatenation of two substrings $v_1 \circ v_2$.

1. Let $\alpha_1, \alpha_2, \ldots, \alpha_{3n}$ be encryptions, computed using E, of $3n$ bits $b_1, b_2, \ldots, b_{3n}$.
   The first substring v_1, is formed by concatenating the α_i, $1 \le i \le 3n$, each one followed by an encryption β_i of $\overline{b}_i = 1 - b_i$.
   That is $v_1 = v_{1,1} \circ \ldots \circ v_{1,3n}$, where $v_{1,i} = \alpha_i \circ \beta_i$, $\alpha_i \in E_n(b_i)$ and $\beta_i \in E_n(\overline{b}_i)$ for $1 \le i \le 3n$.

2. For the second substring v_2, consider the set

$$S = \bigcup_{1 \le h \le i \le j \le 3n} \{\alpha_h, \beta_h\} \times \{\alpha_i, \beta_i\} \times \{\alpha_j, \beta_j\}$$

   Notice that $|S| = O(n^3)$.

   Let $z_1, z_2, \ldots, z_{|S|}$ be an arbitrary but fixed ordering (e.g. the lexicographic one) of the elements of S. Then $v_2 = v_{2,1} \circ \ldots \circ v_{2,|S|}$, where $v_{2,h} = z_h \circ \gamma_h$ and $\gamma_h \in E_n(D_n(\nu_{1,h}) \vee D_n(\nu_{2,h}) \vee D_n(\nu_{3,h}))$ if $z_h = (\nu_{1,h}, \nu_{2,h}, \nu_{3,h})$.

Notice that $V = \{V_n\}_{n\in\mathcal{N}}$ is an NP language. If you "guess" all the bits b_i as well as the coin tosses used for the encryptions, then you can verify in time polynomial in $|v|$ that indeed $v \in V$.

3.2.2 The implementation

In this section we exhibit a Non-Interactive Zero-Knowledge Proof-System with Preprocessing. We use the language described in Section 3.2.1 as the auxiliary language V.

Theorem 1. *If one-way encryption schemes exist, then there exists a Non-Interactive Zero-Knowledge Proof-System (A, B) with Preprocessing.*

To prove the theorem we start by formally describing the protocol.

A's protocol.

When we say A "writes τ", we mean that A appends τ followed by a special symbol, such as #, to the string $Proof$ that will be sent to B.

Preprocessing stage.
A randomly chooses a v in V. A sends v to B and proves interactively and in zero-knowledge to B that indeed v belongs to the language V.

Communication stage.

Let $v = \alpha_1 \circ \beta_1 \circ \alpha_2 \circ \beta_2 \ldots \circ \alpha_{3n} \circ \beta_{3n} \circ z_1 \circ \gamma_1 \ldots \circ z_{|S|} \circ \gamma_{|S|}$

Let $C = \{c_1, \ldots c_n\}$ be a collection of clauses over the set of variables $U = \{u_1, \ldots, u_k\}$ and $t : U \rightarrow \{T, F\}$ a truth assignment satisfying C.

1. A sets $Proof=$ empty string.

2. A repeats step 2.1 for $i = 1, \ldots, k$.

   2.1 If $t(u_i) = T$ and $D_n(\alpha_i) = 1$ then A writes (α_i, β_i).
   If $t(u_i) = T$ and $D_n(\alpha_i) = 0$ then A writes (β_i, α_i).
   If $t(u_i) = F$ and $D_n(\alpha_i) = 1$ then A writes (β_i, α_i).
   If $t(u_i) = F$ and $D_n(\alpha_i) = 0$ then A writes (α_i, β_i).

3. A repeats step 3.1 for $i = 1, \ldots n$.

   3.1 Let z_h be the tern in v_2 associated to c_i. A looks for z_h in v_2 and shows that the associated γ_h is an encryption of 1, by revealing the random n-bit seed used for its computation.

Now, we show that the protocol meets the Completeness and the Soundness requirements.

Suppose that the formula $C = \{c_1, \ldots, c_n\}$ is satisfied by the truth assignment t. If A follows the specification of the protocol then the γ associated to each formula will be certainly an encryption of 1. Therefore B always accepts.

On the other hand, suppose that C is not satisfied by any truth assignment. Since each pair (α_i, β_i) is formed by the encryption of two different bits, in any way A associates such a pair to the literal, there will be at least a γ that is the encryption of 0. Therefore B always rejects C.

Now we show that the protocol meets the Zero-Knowledge requirement.

We exhibit a Probabilistic Turing Machine M, running in expected polynomial time that approximates the family of random variables $R = \{R(x)\}$, where

$$R(x) = \left\{ v \leftarrow V_{|x|^c};\ y \leftarrow A(v, x) : (v, y) \right\}$$

for a certain constant c.

The basic idea is to make M to perform the same protocol as A, having in input a string $v' \in \mathcal{V}$, where $\mathcal{V}$ is a particular language defined in the following, instead of having a string in V. The language $\mathcal{V}$ is similar to V, with only one exception: the first part of each string of $\mathcal{V}(n)$ is formed of $6n$ encryptions of 1.

On input the collection of clauses C, M first outputs a string $v' \in \mathcal{V}$. M, then randomly associates the strings α_i, β_i to the literals $u_i, \overline{u}_i$, for each variable u_i, $i = 1, \ldots k$. At this point M has associated to each clause a tern of values that are encryptions of 1 and so also the associated γ will be an encryption of 1. M simply shows the n random bits used to compute such a γ.

M's output is different from a real proof only for what concerns the common string. In fact the second part of the string used by M is formed by the $3n$ encryptions of 1 while the one used by A is constituted by the encryption of n random bits and their complement. If E is a one-way encryption scheme then the two distributions are indistinguishable.

In the following we formally describe M's program.

1. M sets v=empty string.

2. M repeats steps 2.1-2.2 for $i = 1, \ldots, 3n$.

   2.1 M randomly chooses two n-bit strings s_i, s'_i.

   2.2 M appends $(\alpha_i, \beta_i) = (E(1, s_i), E(1, s'_i))$ to the string v.

3. M sets $S = \bigcup_{1 \leq h \leq i \leq j \leq 3n} S_h \times S_i \times S_j$ where $S_l = \{\alpha_l, \beta_l\}$, $l = 1, \ldots, 3n$.

4. M repeats steps 4.1-4.2 for each $z_h = (\alpha, \beta, \delta) \in S$.

   4.1 M chooses a random n-bit seed r_h and computes $\gamma_h = E(1, r_h)$.

   4.2 M appends $z_h \circ \gamma_h$ to v.

5. M repeats steps 5.1-5.2 for $i = 1, \ldots, k$.

   5.1 M tosses a fair coin.

   5.2 If HEAD then M writes (α_i, β_i) else M writes (β_i, α_i).

6. Repeat step 6.1 for $i = 1, \ldots, n$.

   6.1 Let z_h be the tern in S associated to c_i. M writes r_h.

Comment. In the protocol (A, B), above described, there is no need for the prover to have infinite computing power. Indeed it is enough for him to be a probabilistic polynomial-time machine that gets an NP proof as an auxiliary input. Namely, in the case $x \in 3SAT$ and $v \in V$, the prover needs only to known a satisfying assignment for x and the bits b_i along with the coin tosses used for the encryptions in the string v.

4 Open Problems

Our results can be extended in two directions.

The first extension concerns the length of the theorem that can be proved. In fact, in our model the length of the theorem that can be proved is determined by the length of the shared string. Rephrasing this in term of our example, suppose that, before A leaves, A and B agree on a n^c-bit long string $v \in V$, where c is the suitable constant. What happens if A finds the proof of a theorem of size, say, $2n$? Certainly, A will not be able to prove this theorem to B in Zero-knowledge, using our protocol.

One would like to have a protocol that allows to prove any polynomial number of theorems of length polynomial in the length of the shared string v.

The second extension concerns the auxiliary language V. Namely, can we replace the string $v \in V$ with a random string σ?

References

[BGGHKMR] M. Ben-Or, O. Goldreich, S. Goldwasser, J. Hastad, J. Kilian, S. Micali, and P. Rogaway, *Everything Provable is Provable in Zero-Knowledge*, CRYPTO 88.

[BlFeMi] M. Blum, P. Feldman, and S. Micali, *Non-Interactive Zero-Knowledge Proof Systems and Applications*, Proceedings of the 20th Annual ACM Symposium on Theory of Computing, Chicago, Illinois, 1988, pp. 103–112.

[BlMi] M. Blum and S. Micali, *How to generate cryptographically strong sequences of pseudo random bits*, SIAM Journal on Computing, vol. 13, n. 4, November 1984, pp. 850–864.

[DeMiPe] A. De Santis, S. Micali, and G. Persiano, *Non-Interactive Zero-Knowledge Proof Systems*, in "Advances in Cryptology - CRYPTO 87 Proceedings", pp. 52–72, vol. 293 of "Lecture Notes in Computer Science", Springer Verlag.

[GoMi] S. Goldwasser and S. Micali, *Probabilistic Encryption*, Journal of Computer and System Science, vol. 28, n. 2, 1984, pp. 270–299.

[GoMiRi] S. Goldwasser, S. Micali and R. Rivest, *A Digital Signature Scheme Secure Against Adaptive Chosen-Message Attack*, SIAM Journal on Computing, vol. 17, n. 2, April 1988, pp. 281–308.

[GaJo] M. Garey and D. Johnson, *Computers and Intractability: a Guide to the Theory of NP-Completeness*, W. H. Freeman & Co., New York, 1979.

[GoMiRa] S. Goldwasser, S. Micali, and C. Rackoff, *The Knowledge Complexity of Interactive Proof-Systems,* Proceedings of the 17th Annual ACM Symposium on Theory of Computing, Providence, RI, May 1985, pp. 291–304.

[GoMiWi] O. Goldreich, S. Micali, and A. Wigderson, *Proofs that yield nothing but their validity and a methodology of cryptographic protocol design*, Proceedings of the 27th Annual Symposium on Foundations of Computer Science, October 1986, pp. 174–187.

[MiRaSl] S. Micali, C. Rackoff, and B. Sloan, *The Notion of Security for Probabilistic Cryptosystems*, SIAM Journal on Computing, vol. 17, n. 2, April 1988, pp. 412–426.

[Sh] C. E. Shannon, *Communication Theory of Secrecy Systems*, Bell System Tech. J., vol. 28, Oct. 1949, pp. 656-715.

[Ya] A. C. Yao, *Theory and Applications of Trapdoor Functions*, Proceedings of the 23rd IEEE Symposium on Foundation of Computer Science, 1982, pp. 80–91.

Session 8

On the Theory of Security II

Chair: R. Rivest, MIT

The Noisy Oracle Problem

U. Feige, A. Shamir, M. Tennenholtz
Department of Applied Mathematics
The Weizmann Institute of Science
Rehovot 76100, Israel

Abstract

We describe a model in which a computationally bounded verifier consults with a computationally unbounded oracle, in the presence of malicious faults on the communication lines. We require a fairness condition which in essence says that some of the oracle's messages arrive uncorrupted. We show that a deterministic polynomial time verifier can test membership in any language in P-space, but cannot test membership in languages not in P-space, even if he is allowed to toss random coins in private. We discuss the zero knowledge aspects of our model, and demonstrate zero knowledge tests of membership for any language in P-space.

1 Introduction

The original GMR [7] model of interactive proof systems (IPS) is based on a powerful prover P who wants to convince a polynomial verifier V that a certain assertion Q is true. The model considers two separate scenarios: P can be trustworthy (and then V should accept true assertions), or P can be a cheater (and then V should reject false assertions). By introducing interaction into the proof process, GMR hoped to achieve two goals:

- To increase the class of provable assertions beyond NP.
- To provide zero knowledge proofs.

Recently, Ben-or, Goldwasser, Killian and Wigderson (BGKW [1]) have introduced a general model of multi-prover protocols, and studied a variant which makes it possible to provide perfect zero knowledge proofs, but does not seem to increase the power of the proof system. Motivated and inspired by their work, we consider in this paper a different variant which does extend the class of provable assertions.

In our new Noisy Oracle model, a probabilistic polynomial time verifier V tries to decide whether an input assertion Q is true or false. He is aided by a trusted and infinitely powerful prover P, but their communication is disrupted by an adversary

A who can block or modify their messages. If A does not exist, V can just believe the one-bit advice provided by P. If A can totally block the communication, P cannot help V. In our model we consider the interesting case in which A is computationally unbounded but an imperfect jammer in the sense that a non-negligible fraction of the messages exchanged by P and V reach their destination unaltered, in spite of the malicious interference. The problem from V's point of view is that he does not know which messages are authentic, and the only assumption we allow him to make is that occasionally he gets good advice. In particular, if he asks the same question sufficiently many times, at least one of the answers will be correct with overwhelming probability, but he is still left with the problem of deciding which one it is.

An alternative model which has essentially the same properties as the Noisy Oracle model is the following Multi-Oracle model: V interacts with several infinitely powerful oracles $P_1,...,P_n$, and has to decide whether the common input Q is true or false. To make the model non-trivial, we assume that at least one of the oracles is trustworthy and at least one of the oracles is a cheater, but V does not know who is who. The powerful cheaters can break cryptosystems, forge signatures, and test all the possible outcomes of their actions, while the limited verifier has to rely on the (unknown) trustworthy prover to refute incorrect claims.

As a motivating example to the Multi-Oracle model, consider a court of law in which two lawyers (oracles) present their conflicting views of the same case in front of a judge (verifier). Both lawyers have spent much time learning all details of the case. On the other hand, the judge does not have much time to spend. He must use the lawyer's knowledge in order to extract the facts correctly in a short time, and then reach a decision based on the facts he knows. One of the lawyers wants the judge to learn the facts correctly, because then his client is bound to win the case. The other lawyer naturally wants to fool the judge, as otherwise his client will lose the case. The judge does not know which is the truthful lawyer and which is the cheating lawyer. Using our Multi-Oracle model we characterize the facts that the judge can extract efficiently despite the presence of a cheating lawyer.

In the Multi-Oracle presentation of our model, we restrict the cheating oracles by limiting their numbers. In this the model resembles models for distributed computing in the presence of faults (and in particular Byzantine agreement HM [9]). But there is a great difference: we are interested in proving mathematical facts, which the verifier could not prove by himself because of limited computing power. In the Byzantine agreement model, the emphasis is on reaching a consensus about the environment (e. g. the value of a global coin), rather than testing the correctness of mathematical assertions.

In section 2 we present our model. In section 3 we consider verifiers bounded to polynomial time. Our main result is that in the main variant of the Noisy Oracle model, the testable assertions are exactly those which lie in the complexity class P-space. This is in contrast to what is believed to be the case with IPS, where proving even co-NP-complete assertions will lead to the collapse of the polynomial hierarchy (see GS [8]). In section 4 we define the zero-knowledge aspects of the model, and demonstrate zero-knowledge protocols for all statements in P-space. Section 5 suggests directions for further research, and in particular discusses some recent results on multi oracle protocols with space bounded verifiers.

2 The Model

A computationally bounded verifier V interacts with a set of oracles, where the size of the set (denoted by n) is fixed, depending on the length of the input. t of the oracles are potential cheaters, where $\frac{n}{2} \leq t < n$. (The case where $t < \frac{n}{2}$ is uninteresting, as V can take a majority vote to determine the correct answer). In the multi-oracle model V receives a set of up to n answers, t of which may be incorrect, but at least one answer must be correct. In the noisy oracle model V receives one answer, where the probability of it being correct is $\frac{n-t}{n}$. Thus in the noisy oracle model, we can only demand that the verifier be convinced with overwhelming probability, as V is not guaranteed to ever receive a correct answer.

In the multi oracle model we may consider different network configurations. If we have a star shaped (or similar) network, with the verifier at the center and the oracles at the leaves, then the oracles become addressable. The verifier can ask each oracle a different question, and can ignore certain oracles once it has determined that they are cheaters. On a broadcast network, the oracles are anonymous, and the verifier cannot associate between messages and their senders. In the configuration of the network we include the ability or inability of one oracle to perform interactions with other oracles, or to perform eavesdropping on lines not his own.

An *interactive test* is a protocol in which the verifier manages to test an assertion, even if the adversary (cheaters) displays his worst case behavior. The assertion is one concerned with the common input, such as: "the following instance x is a legitimate member of language L". In our model, a language L is said to be testable if the following condition holds:

For every input x, and for any adversary A,
if $x \in L$ then the verifier outputs "$x \in L$" ,
and if $x \notin L$ the verifier outputs "$x \notin L$" .
(In some cases we should add — "with overwhelming probability").

3 Polynomial Time Verifiers

First we demonstrate a protocol in which the polynomial time verifier tests an NP-complete assertion. n is assumed to be equal to the length of the input. Note that in this section we do not care about the zero knowledge aspects of the protocols.

NP-protocol:

- V sends the problem instance and asks for a witness.

- P (a truthful prover) sends a correct witness if such exists, and null otherwise. C (cheater) sends any message.
- V tests if at least one of the messages was a valid witness to the statement. If so, he concludes that the instance was a yes instance. If not, he concludes it was a no instance.

The correctness of this protocol follows from the fact that $t < n$. V knows that at least one answer is correct, so if he receives no witness, he can conclude that no witness exists. Note that the same protocol is applicable to co-NP statements, if P sends a counter witness whenever it exists.

Next we demonstrate a protocol for a P-space complete problem. In order to simplify the initial demonstration, we start with a simple version of the model: $n = 2, t = 1$, addressable model. That is: we have a star configuration with a verifier interacting separately with two oracles. One of the oracles may be a cheater, but the verifier does not know which one. We shall demonstrate the protocol on any P-space complete game (as can be found in GJ [5]). In such a game we ask whether the first player to move ("white"), or the other player ("black") has a winning strategy. The players alternate in the moves they make, and the game is guaranteed to end after a polynomial number of moves.

Simple P-space protocol:

1. V sends the instance (initial position) of the game to the two oracles, and asks them who wins. If their answers agree V stops.
2. V sends the current position to the oracle who claims that white wins (denote him by W) and asks for a winning move for white.
3. If W does not reply with a move, V concludes that he was cheating, and so black is the one who wins. So W replies with a move w for white.
4. V computes the new position generated by applying w to the previous position, sends it to B (the oracle who claims that black wins), and asks for a winning move for black.
5. If B does not reply with a move, V concludes that white wins. So B replies with b. V applies b to the current position and returns to step 2.

After not more than a polynomial number of moves, either one of the two oracles is caught not following the protocol, or the game reaches its natural end. In either case V concludes correctly who wins in the initial position, as he knows that the trustful oracle must have chosen the wining side, and must have made the optimal moves.

The above protocol may easily be generalized to a star network with one prover and $n - 1$ cheaters. The verifier asks each oracle for the outcome of the game instance. If all agree, there is no problem. If V receives conflicting answers, he chooses two oracles who gave conflicting answers, and lets them play one against the other. The loser of the game is marked as a cheating oracle, and is discarded from the network. Now V remains with a network of $n-1$ oracles, at least one of which is truthful. By induction on the size of the network, we see that V determines the correct outcome of the game.

The above protocols cannot be directly applied to the broadcast model or to the noisy oracle model, where V receives unordered sets of answers. The problem is that V does not know which answer corresponds to which oracle. For each position, half the cheaters can actually claim the correct answer, but then try to discredit it with bad moves. If V tries to discard some of the moves, he might discard the alternative suggested by the real prover. If V does not discard moves, this results in an exponential number of variations of the ongoing game that V has to keep track of. Nevertheless, a good protocol can be constructed.

THEOREM 1: Let n be polynomial in the length of the input, with $t < n$. Any statement in P-space can be tested even in the broadcast model (where V receives unordered sets of answers).

We shall introduce terminology and a few short lemmas simplifying the proof of the above theorem.

Let G be a two player (W and B) game, in which the player to move looses if he has no legal move. The *outcome* of a position of G is W (B respectively) if W (B) has a winning strategy. An *n-hyper-position* HP of G is a $k_1 < n$ by $k_2 < n$ matrix in which each entry is a position in G, and in all positions the same player is to move. A *hyper-move* is made by choosing a column in HP, transposing it to a row, and for each entry in the row — performing a legal move of G. An *HP-transposition* is constructed by considering $k_3 < n$ H-moves (from now on, H stands for *hyper*) as rows of a new H-position HP^t. Note that if W is about to move in the entries of HP then B is about to move in the entries of HP^t, and vice versa. A *n-hyper-game* is played by n *hyper-players*. For each H-position each H-player may choose to make one H-move, resulting in an HP-transposition. The H-game ends with one of the two *outcomes* W or B in one of the two ways:

1. Agreement: All n H-players choose to make an H-move in the same H-position. The outcome of the H-game is W if W has the move in the entries of HP, and B otherwise.

2. Resignation: No player makes an H-move. The outcome of the H-game is W if B has the move in the entries of HP, and B otherwise.

Note that a 2-H-game can be viewed as just an ordinary two player game.

HP is *row-dominated* (*column-dominated* respectively) if it has a row (column) in which all entries are positions of G in which the previous (next) player wins. Note that HP cannot be both row-dominated and column-dominated simultaneously.

Lemma: If HP is column-dominated then there exists an H-move such that HP^t is row-dominated.

Proof: Pick the dominating column in HP. In all its entries the next player wins. For each entry perform the winning moves in G and transpose the column. This is the desired H-move. QED.

An H-move as described above shall be called *optimal*.

Lemma: If HP contains a dominating row then HP^t contains a dominating column.

Proof: Assume row j is dominating HP. Column j will necessarily dominate HP^t. QED.

An *optimal* H-player makes an optimal H-move whenever one exists, and no H-move otherwise.

Lemma: If HP is dominated, and if at least one of the H-players is optimal, then if the H-game ends, it ends with a outcome equal to the outcome of any of the positions in the entries of the row/column dominating HP.

Proof: The existence of the optimal H-player ensures that row/column domination will alternate in HP-transpositions, that resignation will not occur in column-dominated H-positions, and that agreement will not occur in row-dominated H-positions. So the H-game may not end with a outcome different than stated in the lemma. QED.

Now we return to the proof of theorem 1.

Proof (theorem 1): It is sufficient to consider any game G complete in P-space, because reductions into P-space complete problems can be done in polynomial time. V can monitor an n-H-game based on G, where the oracles act as H-players:

1. The initial HP consists of only one entry: the initial position of G. This implies that the initial HP is dominated.
2. V sends the current HP as a challenge to all oracles.
3. V receives a set of H-moves, at most one from each oracle.
4. V constructs HP^t from the legal H-moves, makes it the current HP and returns to step 2 (unless the H-game ended).

The truthful prover plays the role of the optimal H-player. By the above lemmas the outcome of the H-game is identical to the outcome of G. The size of each n-HP and the number of HP-transpositions are bounded by a polynomial in the length of the input. So V can perform the protocol in polynomial time. QED.

Corollary: Any assertion in P-space can be tested with overwhelming probability in a system where the probability of a correct response is non-negligible. (The Noisy Oracle model).

Proof: Let n be the length of the input, let p be a polynomial, and let $\frac{1}{p(n)}$ be the probability of a correct response. Then by repeating each question $n \cdot p(n)$ times V receives a set of answers which with overwhelming probability includes at least one correct answer. Because the whole protocol is limited to a polynomial number of steps, the probability that each of the steps contains at least one correct answer remains overwhelming. Thus the verifier can transform his original noisy-oracle setting to a new multi-oracle setting, but with modified parameters: $n \cdot p(n)$ oracles and one true prover. The proof follows, as V monitors an $n \cdot p(n)$-H-game. QED.

We have seen that V can test any P-space assertion even in the most difficult scenario — the noisy oracle model with low probability of correct answers, and no coin tossing allowed. We shall now show that even in a more favorable scenario, the multi-oracle broadcast model with only one cheating oracle and with a secret source of randomness, V cannot test assertions not in P-space.

THEOREM 2: In a broadcast network, if in every round the adversary sees the message of the prover before deciding on his own message, a probabilistic polynomial time verifier cannot test assertions not in P-space.

Proof (sketch): Suppose a language L is testable in the broadcast model. We demonstrate a P-space algorithm for testing whether an instance x of length n belongs to L. V is assumed to truthfully follow a certain algorithm, and to output either "$x \in L$" or "$x \notin L$" after a number of steps bounded by some polynomial $p(n)$. This implies bounds of $p(n)$ on the number of coin tosses V makes, on the number of messages he expects to receive and on the maximal message length. We construct a game tree of depth $3p(n)$, composed of three types of nodes: O_1, O_2, V. The edges leading out of nodes of O_1 (O_2, V respectively) correspond to *all* possible messages (whether they make sense or not) of oracles trying to prove $x \in L$ (oracles trying to prove $x \notin L$, verifier respectively). A string S is said to *agree* with the path from the root to a leaf, if V's algorithm, given S on his random tape and assuming O_1 and O_2 send the messages implied by their edges along the path, would indeed cause V to send the messages implied by his edges. The *value* R of a leaf is the number of length-$p(n)$ random strings which agree with the corresponding path, and which cause V to output "$x \in L$". The value of an inner node of the tree is defined recursively: An O_1-node maximizes over its sons (nodes of O_2). An O_2-node minimizes over its sons (nodes of V). A V-node sums over its sons (nodes of O_1). We shall prove that the value of the root is greater than $2^{p(n)-1}$ iff $x \in L$.

Assume $x \in L$. We want to show that this implies $R > 2^{p(n)-1}$. Consider an adversary who takes the role of O_2 in the game tree constructed, and chooses his messages optimally so as to minimize R. Because we assume that L is testable, there exists a strategy for the real prover which causes V to output "$x \in L$" with overwhelming probability. In particular, the *optimal* strategy of taking the role of O_1 and choosing messages so as to maximize R must work. Note that there exists no better strategy, because we assumed the adversary may choose his messages in each round after viewing the prover's messages. This optimal strategy convinces V with probability $\frac{R}{2^{p(n)}}$, and so $R > 2^{p(n)-1}$.

Assume $x \notin L$. We want to show that this implies $R < 2^{p(n)-1}$. Consider an adversary who takes the role of O_1 in the game tree constructed, and chooses his messages optimally so as to maximize R. Because we assume that L is testable, there exists a strategy for the real prover which causes V to output "$x \notin L$" with overwhelming probability. In particular, the ***better than optimal*** strategy of taking the role of O_2 and choosing messages so as to minimize R must work. This strategy is better than optimal because the prover chooses his messages *after* viewing the adversary's messages. This better than optimal strategy convinces V with probability $\frac{2^{p(n)}-R}{2^{p(n)}}$, and so $R < 2^{p(n)-1}$.

Finally, the value of the root can be computed by a Depth First Search traversal algorithm in polynomial space. This is a consequence of the simplicity we maintain in the construction of the game tree. We do not try to consider only "sensible" messages of the oracles, as a P-space algorithm cannot judge what messages are considered sensible by computationally unbounded and possibly cheating oracles. Furthermore, the edges leading out of V nodes are not restricted only to those corresponding to messages V really sends, as this will imply extensive bookkeeping. The only information our algorithm needs to save is both a message number and an R-counter for each node along the current path considered in the tree, and this requires $O(p^2(n))$ space. QED.

The assumption that the adversary sees the prover's message before deciding on his own was crucial in the above proof. In particular, if the situation is known to be reversed, the verifier can correctly test any assertion, as the prover just sends a message whose exclusive-or with the adversary's messages gives the correct answer. In addressable models, we assume no oracle knows which messages were sent by other oracles, and so again the proof of theorem 2 does not hold. A major open question is whether addressable models allow probabilistic polynomial time verifiers to test membership in languages not known to be in P-space.

4 Knowledge complexity

GMR introduced the notion of knowledge complexity of a language. We extend their definitions so as to apply to our model with a probabilistic polynomial time verifier.

One may view the interaction between the verifier and the noisy oracle as an interaction between two parties, where one of the parties wants to test the value of a predicate through the other party. The adversary (noise) models possible trouble. Allowing for this possibility, the oracle agrees to help the verifier filter out incorrect answers, by demonstrating their incorrectness. In *zero-knowledge* protocols the oracle does not agree to allow a cheating verifier's claim of alleged imperfect communication to cause the oracle to reveal additional information. For example, the oracle may not agree to give the verifier specific witnesses which demonstrate the correctness of the claims made by the oracle, or which demonstrate the incorrectness of claims made by the adversary. We model the above setting by stating that there is no adversary along the communication lines. A cheating verifier interacts with a cautious oracle, and *claims* that he receives conflicting messages. The corresponding definition for the *multi oracle* model allows polynomial time cheating verifier V^* to manage the behavior of the cheating oracles as best suits him. A protocol is in zero knowledge if V^* can not increase his knowledge beyond the one bit he receives anyway (the truth value). Note that if we do not limit the cheaters to polynomial time, nothing prevents them from revealing information to the verifier.

As an example motivating the zero knowledge concept, we return to our two lawyers making their claims in front of the judge. The defending lawyer claims that the defendant has a perfect alibi for the night of the murder, while the prosecution claims the opposite. The defense is in a delicate position because the alibi is rather embarassing: the defendant has spent the night of the murder with the judge's wife! How can the defense convince the judge that the defendant is not guilty (of murder), without revealing the actual alibi? The answer is simple — do it in zero knowledge.

We assume the reader is familiar with the definitions of zero-knowledge. We just sketch our definition.

Definition: In the noisy oracle model, a protocol is said to be *zero knowledge* if the following condition holds: There exists a probabilistic polynomial time algorithm M (which may run V^* as a subroutine), where M is given the truth value of the assertion involved, such that for any non-uniform polynomial time algorithm V^* which manages the behavior of the verifier and of the cheaters, M's output is indistinguishable from V^*'s view of the communication with the real prover.

We shall consider the model in which n is equal to the length of the input, $t = n - 1$ and the oracles are not addressable. The adversary A manages the behavior of all the cheaters. In proving that a protocol is in zero-knowledge, V^* manages the behavior of V and of A, but is limited to polynomial time. P denotes the truthful prover.

The trivial protocol for NP languages, (sending a witness whenever one exists), reveals no information in the case where $x \notin L$ (co-NP statements), as P sends no message, and M has nothing to simulate. But the protocol as a whole is not zero knowledge, as a polynomial time simulator M is not guaranteed to produce witnesses for NP statements in cases where $x \in L$. So in order to construct zero knowledge protocols in our model, we shall apply techniques used in IPS. We shall demonstrate that care should be taken when doing so.

Following the footsteps of GMW [6], we assume that safe encryption functions exist. We use a protocol proposed by Manuel Blum, and sketch its basic structure in order to make this paper self contained:

1. V asks for an encrypted random permutation on the graph G.

2. P sends the encryption, and A adds whatever messages he wants.

3. For each message he receives, V requests either it's full decryption and the permutation used, or partial decryption revealing a Hamiltonian cycle.

4. P sends the requested information, and A adds whatever messages he wants.

5. V checks that at least in one of the received messages the protocol was followed correctly.

In order to diminish the chance of cheating, this basic structure may be iterated n times (serial version). Alternatively V may ask at step 1 for n encrypted random permutations of G (parallel version). We shall demonstrate that both approaches are not recommended.

1. Serial version:

   Here the adversary has a good chance of cheating in a broadcast network. Suppose G does not have a Hamiltonian cycle. The adversary may convince V to the contrary: when asked for an encrypted permutation of the graph, the adversary sends $\frac{n}{4}$ such encryptions, $\frac{n}{4}$ encryptions of Hamiltonian cycles and $\frac{n}{2}$ null answers. (We assume the prover returns a null answer as well). When V asks for decryptions, A has 0.5 chance to succeed for each message. Because there are $O(n)$ messages, he has exponentially high probability of cheating at least once. Even if the protocol is iterated a polynomial number of times, the adversary has high probability of succeeding in all rounds. In the broadcast model, V does not know if there existed one oracle which succeeded in all iterations, in which case V should accept, or whether each oracle failed at least once, in which case V should reject.

2. Parallel version:

   Here the adversary has a negligible chance of cheating, but now the simulator (M) has an impossible task: he himself needs exponential time in order to simulate a run (assuming he does not know any Hamiltonian cycle).

Since both the serial version and the parallel version are incorrect, we shall use a combined version. V asks for $\log n$ encryptions of different random permutations of G, all in parallel. This basic protocol is iterated n times. In each iteration, the adversary has $\frac{1}{n}$ probability of cheating with any single message, or constant probability of cheating with at least one message. So A has negligible chance of cheating at least once in every iteration of the whole protocol. On the other hand, M can simulate each iteration in $O(n)$ trials, and the whole protocol in $O(n^2)$ steps.

The zero knowledge protocol presented above suits Co-NP statements as well. If each iteration contains at least one good message, the verifier concludes that there exists a witness. Otherwise, he concludes that no witness exists. This is true of protocols in our model in general: they either demonstrate that an assertion is correct, or that it is not correct. In no case does the verifier remain in state of doubt.

Finally, we consider the P-space complete game discussed in Theorem 1. The H-game constructed in the proof of this theorem can be played even if the entries to the H-positions are encrypted. The powerful provers play by breaking the corresponding cryptosystems, choosing an H-move and encrypting it. Polynomial time V has the task of blindly constructing encrypted H-positions from the encrypted H-moves he receives. When the encrypted H-game ends (it must end in a polynomial number of rounds), the verifier tests in zero knowledge a statement in NP — that the decryption of the whole history of the H-game would give the desired outcome.

Theorem 3: Under the assumption that encryption functions exist, any statement in P-space can be tested in zero-knowledge (in all models).

A proof can be constructed from the discussion above.

The above theorem can be extended to any protocol in which all V does is shift messages back and forth. For all such protocols we can easily construct zero-knowledge versions. On the other hand, in protocols in which V takes actions which depend upon the messages he receives and upon his coin tosses, the above technique does not give a zero-knowledge protocol, since V does not know which action to take if the messages he receives are encrypted.

5 Further Research

In this paper we presented a new model, which has many variants: Addressable versus broadcast communication, time bounded versus space bounded verifiers, etc. There are many open questions, of which we want to point out one in particular: Is there any language not known to be in P-space which is testable by a probabilistic polynomial time verifier? As can be derived from theorem 2, the best model for looking for such a language is the addressable model with private coins.

It may be interesting to mix our model with the multi-oracle model of BGKW [1]. We may assume that at least two oracles are trustworthy, that oracles cannot communicate among themselves, and that the good oracles share a read only common random string. The results of BGKW transform to this model, giving perfect zero knowledge proofs for every language in NP. Furthermore, this automatically gives perfect zero knowledge proofs of all languages in co-NP, as the model is closed under complement. Can these results be pushed further up in the polynomial hierarchy?

A different line of research is to study the structure of our models. In what models are public coins as powerful as private coins? In what ways does the number of rounds of a protocol correspond to the complexity level of the tested language in the polynomial hierarchy?

Interactive protocols with space bounded verifiers received much attention recently (see Condon and Ladner [2], Dwork and Stockmeyer [3], Kilian [10] and others). The fact that if some information is hidden from the players, the outcome of very complex games can be tested in very little space, was demonstrated by Reif [12] and by Peterson and Reif [11]. In a paper now in preperation (FS [4]), the notion of probabilistic space bounded verifiers in addressable multi oracle systems is defined. The following surprising results are proved:

Theorem 4 (FS [4]): In our multi oracle model, a log-space verifier can test any P-space assertion in polynomial time.

Theorem 5 (FS [4]): In our multi oracle model, the set of elementary recursive languages (that is, languages recognized by a Turing machine in time $2^{\cdot^{\cdot^{\cdot^{2^n}}}}$, with a fixed number of exponentiations) is strictly contained in the set of languages testable by log-space verifiers.

Theorem 6 (FS [4]): In the multi prover model of BGKW [1], the set of elementary recursive languages is strictly contained in the set of languages testable by log-space verifiers. Furthermore, if protocols are not requested to end with probability 1, constant space verifiers can recognize any partial recursive language.

References

[1] M. Ben-or, S. Goldwasser, J. Killian, A. Wigderson, *Multi Prover Interactive Proofs: How to Remove Intractability* Proc. of 20th STOC 1988, pp. 113-131.

[2] A. Condon, R. Ladner, *Probabilistic Game Automata* Proc. Structure in Complexity, 1986, pp. 144-162.

[3] C. Dwork, L. Stockmeyer, *Zero-Knowledge with Finite State Verifiers* these proceedings.

[4] U. Feige, A. Shamir, *Multi Oracle Interactive Protocols with Space Bounded Verifiers* in preperation.

[5] M. Garey, D. Johnson, *Computers and intractability, A guide to theory of NP-Completeness* Freeman 1979.

[6] O. Goldreich, S. Micali, A. Wigderson, *Proofs that Yield Nothing But their Validity and a Methodology of Cryptographic Protocol Design* Proc. 27th FOCS, 1986, pp. 174-187.

[7] S. Goldwasser, S. Micali, C. Rackoff, *The Knowledge Complexity of Interactive Proofs* Proc. of 17th STOC, 1985, pp. 291-304.

[8] S.Goldwasser, M.Sipser , *Arthur Merlin Games versus Interactive proof systems* 18th STOC, 1986, pp. 59-68.

[9] J. Y. Halpern, Y. Moses, *Knowledge and common knowledge in a distributed environment* Proceedings of the Fourth PODC, 1985, pp. 224-236.

[10] J. Kilian, *Zero-Knowledge with Log-Space Verifiers* to be presented at FOCS 1988.

[11] G. L. Peterson and J. H. Reif, *Multiple-Person Alternation* Proceedings of 20th FOCS, 1979, pp. 348-363.

[12] J. H. Reif, *The complexity of two-player games of incomplete information* JCSS 29, 1984, pp 274-301.

On Generating Solved Instances of Computational Problems

Martín Abadi*
Eric Allender†
Andrei Broder*
Joan Feigenbaum‡
Lane A. Hemachandra§

Abstract: We consider the efficient generation of solved instances of computational problems. In particular, we consider *invulnerable generators*. Let S be a subset of $\{0,1\}^*$ and M be a Turing Machine that accepts S; an accepting computation w of M on input x is called a "witness" that $x \in S$. Informally, a program is an *α-invulnerable generator* if, on input 1^n, it produces instance-witness pairs $\langle x, w\rangle$, with $|x| = n$, according to a distribution under which any polynomial-time adversary who is given x fails to find a witness that $x \in S$, with probability at least α, for infinitely many lengths n.

The question of which sets have invulnerable generators is intrinsically appealing theoretically, and the results can be applied to the generation of test data for heuristic algorithms and to the theory of zero-knowledge proof systems. The existence of invulnerable generators is closely related to the existence of cryptographically secure one-way functions. We prove three theorems about invulnerability. The first addresses the question of which sets in NP have invulnerable generators, if indeed any NP sets do. The second addresses the question of how invulnerable these generators are.

Theorem (Completeness): If any set in NP has an α-invulnerable generator, then SAT has one.

Theorem (Amplification): If $S \in$ NP has a β-invulnerable generator, for some constant $\beta \in (0,1)$, then S has an α-invulnerable generator, for every constant $\alpha \in (0,1)$.

---

*DEC Systems Research Center, Palo Alto, CA 94301.
†Rutgers University, New Brunswick, NJ 08903. Research supported in part by NSF grant CCR-8810467.
‡AT&T Bell Laboratories, Murray Hill, NJ 07974.
§Columbia University, New York, NY 10027. Research supported by NSF grant CCR-8809174 and a Hewlett-Packard Corporation equipment grant.

Our third theorem on invulnerability shows that one cannot, using techniques that relativize, resolve the question of whether the assumption that P $\neq$ NP alone suffices to prove the existence of invulnerable generators. Clearly there are relativized worlds in which invulnerable generators exist; in all of these worlds, P $\neq$ NP. The more subtle question, which we resolve in our third theorem, is whether there are also relativized worlds in which P $\neq$ NP and invulnerable generators do not exist.

Theorem (Relativization): There is an oracle relative to which P $\neq$ NP but there are no invulnerable generators.

1 Introduction

Sanchis and Fulk have studied the complexity of constructing test instances of hard problems, and the connections between such construction and the structure of complexity classes [20,21]. In this paper, we consider the efficient generation of solved instances of computational problems. For example, if $S = \{x\colon \exists w.p(x, w)\}$ is a set in NP, we may wish to generate instance-witness pairs $\langle x, w\rangle$ according to a specified distribution. The relationship between the complexity of generating pairs $\langle x, w\rangle$ and the complexity of finding w given x is intrinsically interesting theoretically, and it is also important to the testing of heuristic algorithms for hard problems and the proposed applications of zero-knowledge proof systems.

Specifically, we ask is whether it is possible to generate what we call an *invulnerable* distribution of instance-witness pairs. For example, is it possible to generate pairs $\langle f, a\rangle$, where f is a boolean formula and a is a satisfying assignment, give the secret a to one user A, publish the formula f, and remain reasonably confident that a polynomial-time adversary would be unable to find a satisfying assignment a' for f and thus to impersonate A? Feige, Fiat, and Shamir proposed this use of "zero-knowledge proofs of identity" as a security mechanism; the specific scheme they suggest is based on the Quadratic Residuosity Problem (QRP, [6]). Zero-knowledge proofs of identity may still be useful even if the QRP turns out to be easier than is widely assumed; furthermore, even if the QRP is hard, it may be possible to base a scheme on another problem and achieve more security. Thus, it is important to have a complexity-theoretic framework in which to consider whether a scheme for generating instance-witness pairs produces a secure distribution.

When Goldwasser, Micali, and Rackoff first introduced zero-knowledge proof systems, they postulated an all-powerful prover ([10]). Since then, they and others (e.g., [3], [5]) have considered a model in which prover and verifier have the same computational resources, and the prover's only advantage is that he happens to know the witness w for a particular instance x of the hard problem at hand, perhaps because he constructed x and w simultaneously. This model, together with the proof that all sets in NP have zero-knowledge proof systems ([4], [11]), forms the basis for the "compilation" of multi-party protocols into "validated" protocols ([7], [11]). Thus, it is important to realize that the model is meaningful only if there is a way for an efficient program to generate harder instances than the verifier can solve.

Many NP-Complete sets have obvious, simple generation schemes. For example, Hamiltonian graphs on n vertices can be generated by choosing a random circuit and then adding each other possible edge independently with probability 1/2. The probability of generating a particular graph is proportional to the number of Hamiltonian circuits it has. However, the following examples show that some natural methods of generating solved instances are not secure. The first method succumbs to a very simple algorithm; the second can be cracked by a sophisticated technique.

Example: 3SAT. A 3SAT instance is a set of variables $U = \{u_1, u_2, \ldots, u_n\}$ and a set of clauses $C = \{c_1, c_2, \ldots, c_m\}$, where each clause consists of three literals. The question is whether there exists a truth assignment that satisfies C. (See [8] for definitions.)

A "natural" way to generate solved 3SAT instances is as follows. Choose a truth assignment t uniformly from the 2^n possibilities. For each i between 1 and m, choose three distinct variables uniformly at random; of the eight sets of literals that correspond to these variables, seven are true under t. Choose clause c_i from those seven, uniformly at random. This scheme produces each set C of m clauses satisfied by t with equal probability.

A polynomial-time adversary can reconstruct t with high probability, if the number of clauses m is large enough. The basic observation is that if $t(u_i) = \text{TRUE}$ then

$$\frac{\Pr(u_i \in c_j)}{\Pr(\overline{u_i} \in c_j)} = \frac{4}{3}, \qquad \text{for every } i \text{ and } j.$$

Therefore, if $m \geq kn \ln n$ for a suitable constant k, then with probability $1 - o(1)$ for every i simultaneously, the literal u_i appears in C more often than the literal $\overline{u_i}$ if and only if $t(u_i) = \text{TRUE}$.

One can try to improve this generation scheme by choosing the literals in each clause so that at least one is FALSE and at least one is TRUE. Then the expected number of occurences of $\overline{u_i}$ is equal to the expected number of occurences of u_i, for all i. However, the improved scheme can be cracked easily if $m \geq kn^2 \ln n$ by observing statistics about pairs of variables. ∎

Example: Subset Sum. A Subset Sum instance consists of a finite set $A = \{a_1, a_2, \ldots, a_n\}$ of positive integers and a positive integer M. The question is whether there exists a set $A' \subset A$ that has sum equal to M. The difficulty of the Subset Sum problem is the justification of knapsack-type public key cryptosystems.

One can generate solved Subset Sum instances as follows. Choose a vector $e = (e_1, \ldots, e_n)$ of zeroes and ones, uniformly at random. Fix a positive integer B. Choose each $a_i \in A$ uniformly at random from $\{1, 2, \ldots, B\}$. Let $M = \sum_{1 \leq i \leq n} a_i e_i$.

This generation scheme can be cracked with an algorithm due to Lagarias and Odlyzko ([16]). If B is sufficiently large, then every instance is almost certainly solvable by their ingenious application of the LLL basis-reduction algorithm. ∎

In Section 3 below, we define precisely what it means for a generation scheme to be invulnerable. We then prove a Completeness Theorem that states that, if any set in NP has an invulnerable generator, SAT has one. In particular, under the Quadratic Residuosity Assumption, the Discrete Logarithm Assumption, or the Factoring Assumption, one can generate a hard distribution of SAT.[1] This is not surprising. What is more interesting is that, even if all of these assumptions turn out to be false, one can still generate a hard distribution of SAT, provided one can generate a hard distribution of anything in NP. Our construction of an invulnerable generator for SAT incorporates whatever invulnerability is present in any possible generator for an NP set and does not assume it knows where the invulnerability comes from (as it would be assuming if it built hard instances by multiplying distinct primes, as in [6], etc.). Section 3 also contains an Amplification Theorem, which shows how to enhance the invulnerability of any generable distribution, and a Relativization Theorem — the existence of invulnerable generators clearly implies that P $\neq$ NP, but the converse cannot be proven by techniques that relativize.

In Section 4, we discuss briefly the general question of which sets can be generated

[1] Various forms of these assumptions are ubiquitous in the cryptographic literature (see, e.g., [1], [2], [9], [23]), and we don't need precise statements of them for this informal discussion. For our purposes, it suffices to note that it is possible to generate instances of these number-theoretic problems in randomized polynomial time and that it is widely assumed that, for each of the three problems, for any constant fraction, each polynomial-time algorithm fails to solve that constant fraction of the instances of length n, for all sufficiently large n.

according to which distributions, consider several related works, and propose directions for future research. Section 2 contains terminology and notation that is used extensively in the rest of the paper. We have deferred full proofs until the final version of the paper in order to save space; whenever possible, we give sketches that convey some of the essential points.

2 Terminology, Notation, and Conventions

We call a program that flips coins and terminates in worst-case polynomial time on all inputs a ***randomized polynomial-time*** program. Let $\{M_i\}$ denote a standard enumeration of the randomized polynomial-time programs. Let $\{N_j\}$ denote a standard enumeration of polynomial-time nondeterministic programs; thus, each NP set is recognized by at least one program in our enumeration. We use $L(N_j)$ to denote the set (or language) recognized by N_j.

Let N be a nondeterministic polynomial-time program and S be $L(N)$. We call each accepting path of N on input x a ***witness*** that the ***instance*** x is in S. We assume without loss of generality that, for any fixed program N, the length n of an instance determines the length m of a witness and that the function $n \mapsto n + m$ is one-to-one. We let ω_x^N denote the set of witnesses that $x \in S$.

We use PF to denote the class of polynomial-time computable functions; a function $f \in$ PF need not have range $\{0, 1\}$, and thus PF is a proper superset of the functions that compute membership of strings in sets in P.

We let S_n denote the elements of S that have length n. The symbol Λ denotes the default output of a program; it may be used to indicate that the desired output does not exist or that the program failed to find it. All of the generation programs that we consider take as input the length n, written in unary, run in polynomial time, and produce elements of S_n; thus we have, by definition, restricted attention to efficient generation.

3 Invulnerable Generators

In this section, we provide a complexity-theoretic framework in which to consider the generation of hard, solved instances. We define precisely what it means for a distribution of instance-witness pairs to be "secure against polynomial-time adversaries." Our first theorem addresses the question of which sets in NP have invulnerable generators, if indeed any such sets have them. Theorem 2 addresses the question of exactly how invulnerable these generators are. Finally, Theorem 3 addresses the question of what complexity-theoretic assumptions are needed to prove the existence of invulnerable generators.

Definition: The $(i,j)^{\text{th}}$ *generation scheme*, which we denote $G_{i,j}$, is a program that, on input 1^n, first simulates M_i on input 1^n and obtains an output string y. If y is of the form $\langle x, w\rangle$, where $|x| = n$ and w is an accepting computation of N_j on input x, then $G_{i,j}$ outputs $\langle x, w\rangle$; otherwise, it outputs Λ.

Consider the following game, played between a generation scheme $G_{i,j}$ and an adversary f in PF. The input to the game is a string 1^n; the first move is a run of $G_{i,j}$ on input 1^n. If $G_{i,j}$ outputs a pair $\langle x, w\rangle$, then the second move is for f to output $f(x)$; if $G_{i,j}$ outputs Λ, then the game ends after the first move. The function f wins the game if the generator outputs Λ, or if the generator outputs $\langle x, w\rangle$ and $f(x)$ is an accepting computation w' of N_j on input x; otherwise, the generator wins. Note that w' need not equal w; for example, in the identification scheme of Section 1, the adversary f can compromise the security of user A if he computes any satisfying assignment for A's public formula — he need not discover the private assignment that A was given during key-distribution.

Definition: A generation scheme is *α-invulnerable*, where α is a constant in $[0,1]$, if, for all $f \in$ PF, there are infinitely many lengths n for which the probability that f wins on input 1^n is at most $1-\alpha$. This probability is computed over runs of the game on input 1^n.

Definition: A set S in NP is *α-invulnerable*, where α is a constant in $[0,1]$, if there is a pair (i,j) for which $G_{i,j}$ is *α-invulnerable* and $S = L(N_j)$.

Notice that invulnerable generators are closely related to cryptographically secure one-way functions. Let g be a length-preserving function in PF, and assume that any polynomial-time program fails to invert at least a constant fraction of g's outputs, on infinitely many lengths (where "invert" means "find some element of the preimage"). Then the image of g has an invulnerable generation scheme: on input 1^n, generate a random w of length n and let x equal $g(w)$. Similarly, an invulnerable generation scheme $G_{i,j}$ gives rise to a cryptographically secure one-way function. The program M_i can be viewed as a mapping from coin-toss sequences to pairs $\langle x, w\rangle$. Let g be the function that takes a coin-toss sequence to the first component x of the pair output by M_i. Then g must be hard for any polynomial-time adversary to invert on infinitely many lengths; if it weren't the adversary could discover a coin-toss sequence that gives rise to $\langle x, w\rangle$, and the scheme $G_{i,j}$ would be vulnerable. The same remarks apply if we require in both cases that adversaries fail on all sufficiently high lengths instead of just infinitely many lengths.

We do not claim that a generation scheme that is invulnerable according to our definition is necessarily useful in practice. For example, the key-distributor in [6] would certainly like to know more than that there *exist* infinitely many lengths on which a particular polynomially bounded adversary can be thwarted with high probability; he would also like to know that such lengths are of practical size and to have a procedure for finding them. Our definition of invulnerability does, however, provide a good place to start a complexity-theoretic investigation.

Theorem 1 (Completeness): If any NP set is α-invulnerable, for some positive α, then SAT is also α-invulnerable.

Proof (sketch): The full proof proceeds in three stages. First, we construct a "universal generation scheme" G_U that simulates all possible generation schemes, capturing a constant fraction of whatever invulnerability is present in any of them. Next we construct a generator for SAT that applies Cook's reduction to the set S_U generated by G_U in a way that preserves invulnerability. Finally, we show that the lost fraction of invulnerability can be recaptured.

For the universal generator G_U, we need one program M_U, whose running time is bounded by a specific polynomial, to simulate infinitely many programs, whose individual running times may be arbitrarily high degree polynomials. We overcome that obstacle with the following lemma; it guarantees that we need only consider generators $\{G_k\}$ in which the program M runs in quadratic time.

Lemma: If $G_{i,j}$ is α-invulnerable, then there is an α-invulnerable generation scheme $G_{i',j'}$ in which $M_{i'}$ runs in quadratic time.

We cannot use a "generic reduction" such as the one used in Cook's proof of the NP-Completeness of SAT in order to construct a universal generator. Such a reduction would not necessarily be length-consistent (i.e., map instances of the same length to instances of the same length). Furthermore, even if our generic reduction mapped instances of length n to instances of length n^k, it may not preserve invulnerability: informally, if the "hard instances" output by a particular generator G_m represent a constant fraction α of the probability mass at length n, their images do not necessarily represent a constant fraction of the probability mass at length n^k, simply because there are so many more instances of length n^k.

We use a nonstandard pairing function to overcome this difficulty. It partitions the positive integers into "columns" as follows: column m, consists of all integers of the form $2^{m-1} + k \cdot 2^m$, where $k \geq 0$. Each input length n falls into exactly one column — the one whose index is one more than that of the least significant "1"-bit in the binary representation of n. On input 1^n, G_U first finds m, the index of the column containing n, then chooses an integer l uniformly from the interval $[n - 2^m, n)$. Next, G_U simulates G_m on input 1^l to obtain $\langle x, w \rangle$, pads x, and outputs $\langle x10^{n-l-1}, w \rangle$.

Lemma: If G_m is α-invulnerable, then G_U is $(\alpha/2^m)$-invulnerable.

Informally, to show that, for all f in PF, there are infinitely many lengths n on which f fails to "crack" the output of G_U with probability at least $\alpha/2^m$, we show that any such f corresponds to a function f' that fails to crack the output of G_m on infinitely many lengths n' with probability at least α. The loss of a factor of 2^m occurs because the "hard length" n' (for f' and G_m) corresponds to the hard length n (for f and G_U) such that $n' \in [n - 2^m, n)$; thus G_U only chooses to simulate G_m on input $1^{n'}$ with probability 2^{-m}. (Note that $\alpha/2^m$ really *is* a constant, because m is just the (fixed) index of a generator in our enumeration $\{G_k\}$.)

To construct an $(\alpha/2^m)$-invulnerable generator G_{SAT} for SAT, we use the fact that the program M_U in generator G_U runs in cubic time. We modify Cook's reduction so that, when applied to NP machines that run in cubic time, it takes instances of length k and produces instances of length exactly k^4. This modified Cook's reduction r also induces a mapping from witnesses of membership in S_U to satisfying assignments of elements of SAT. Thus G_{SAT} behaves as follows on input 1^n. If n is not a perfect fourth power, it outputs Λ. Otherwise, it simulates G_U on input 1^k, where $k^4 = n$, obtains a pair $\langle x, w \rangle$, and outputs $r(\langle x, w \rangle)$. We prove in the full paper that G_{SAT} is at least as invulnerable as G_U.

Theorem 2, below, guarantees that, if SAT has an $(\alpha/2^m)$-invulnerable generator, then it also has an α-invulnerable generator. ▮

Corollary: Under the Quadratic Residuosity Assumption, the Discrete Logarithm Assumption, or the Factoring Assumption, there is an α-invulnerable generator for SAT, for some $\alpha \in (0, 1)$.

Remark 1: For cryptographic purposes, one would really want more than that "there exists an infinite set of hard lengths" for cryptographic purposes. Note that the proof of Theorem 1 gives some hope because, if some G_m defeats an adversary on $t(n)$ lengths between 1 and n, then G_U defeats the corresponding adversary on $\Omega(t(n))$ lengths between 1 and n. (This would not have been true had we used a standard pairing function that stretches both of its arguments quadratically.)

Theorem 2 (Amplification): If an NP set S is β-invulnerable, for some positive β, then S is also α-invulnerable, for all $\alpha \in (0,1)$.

Proof (sketch): It suffices to show that α-invulnerability implies $2\alpha/(1+\alpha)$-invulnerability, because the limit of the sequence defined by $\alpha_0 = \alpha$, $\alpha_i = 2\alpha_{i-1}/(1+\alpha_{i-1})$ is 1.

Intuitively, we will show how to increase the level of invulnerability in the most natural way: generate instances, try to crack them, and throw out the cracked ones. Suppose that $G_{i,j}$ is α-invulnerable and that $S = L(N_j)$. If $G_{i,j}$ is $(\alpha + (1-\alpha)/2)$-invulnerable, then we are done, because $(\alpha + (1-\alpha)/2) > (2\alpha/(1+\alpha))$; so suppose that it isn't. Then, by definition, there is some $f \in$ PF that wins against $G_{i,j}$ on all but finitely many inputs 1^n with probability greater than $(1-\alpha)/2$.

Consider the generator $G_{i',j}$ that works as follows on input 1^n: first it runs M_i on input 1^n, just as $G_{i,j}$ does. If M_i outputs $\langle x, w\rangle$, then $G_{i',j}$ computes $f(x)$ and checks whether it is an accepting computation of N_j on input x. If it is, then $G_{i',j}$ runs M_i again on input 1^n; otherwise, $G_{i',j}$ outputs $\langle x, w\rangle$. If f wins a sufficiently large number of successive runs of the game, then $G_{i',j}$ outputs Λ.

Clearly, $G_{i',j}$ generates the same set as $G_{i,j}$, namely $L(N_j)$. In the full paper, we show that $G_{i',j}$ is $(2\alpha/(1+\alpha))$-invulnerable and derive a good enough bound on the number of runs of the game between f and $G_{i,j}$ that $G_{i',j}$ has to simulate. ∎

Remark 2: For simplicity, we have modeled the adversary as a deterministic polynomial-time function. Clearly, in practice one would have to guard against randomized polynomial-time adversaries. Theorems 1 and 2 as stated hold even if we quantify over all randomized polynomial-time functions in the definition of invulnerability. We give details in the full paper.

Is it possible, in Theorem 1, to weaken the hypothesis that at least one set in NP is α-invulnerable? There are clearly oracles relative to which invulnerable generators exist. Indeed a random oracle will do ([19]). In all of these relativized worlds, P $\neq$ NP. Is the assumption that P $\neq$ NP sufficient to prove that invulnerable generators exist? Our next theorem shows that such a proof would not relativize.

Theorem 3 (Relativization): There is an oracle B such that $P^B \neq NP^B$, and invulnerable generators do not exist relative to B.

Proof (sketch): Let B = QBF$\oplus K$, where $\oplus$ is disjoint union, and K is an extremely sparse set of strings of maximum Kolmogorov complexity. Specifically, K contains one string of each length n_i, where the sequence $n_1, n_2, \ldots$ is defined by: $n_1 = 2$, n_i is triply exponential in n_{i-1}, for $i > 1$; if $x \in K$ and $|x| = n$, then x has Kolmogorov complexity n. The inclusion of QBF gives machines with access to B the full power of PSPACE.

It is straightforward to prove that $\mathrm{P}^{\mathrm{B}} \neq \mathrm{NP}^{\mathrm{B}}$ using the techniques in [13].

To show that no invulnerable generators exist relative to B, let $G_{i,j}$ be a generation scheme that has access to the oracle, and assume that it is α-invulnerable, for some constant α in $(0,1)$. We derive a contradiction by producing an adversary f in PF^{B} that can crack a higher fraction than $1-\alpha$ of all of the instances of any length. Here is an informal description of f and why it works:

The generator $G_{i,j}$ involves a randomized polynomial-time program M_i and a nondeterministic polynomial-time program N_j, both of which can query B at any step. Let n^{k_1} and n^{k_2} be bounds on the running times of M_i and N_j, and let k be an integer greater than $\max(k_1, k_2)$. When trying to crack an instance x of length n, f first constructs the set K' consisting of all elements of K that have length less than $\log(n^{ck})$, where c is a suitably chosen constant. Because K is so sparse, there is at most one string r in $K \setminus K'$ about which $G_{i,j}$ may have queried B in generating an x of length n.

Assume that N is the integer closest to n for which there is a string in K of length N. The difficult case is when $\log(N^{ck}) \leq n \leq 2^{N/ck}$; otherwise, f can construct a witness that $x \in L(N_j)$ by using queries to $\mathrm{B}' = \mathrm{QBF} \oplus K'$. So assume, for example, that $n = 2^{N/ck}$.

The cracker f first uses B$'$ to determine whether there is a coin-toss sequence s that would cause $G_{i,j}$ to output x on input 1^n if $G_{i,j}$ were using B$'$. If such an s exists, then f can use PSPACE to construct one and in turn to construct a witness; this construction may or may not involve the discovery of the random string $r \in K \setminus K'$. If such an s does not exist, then f is not able to construct a witness. We show, however, that the only time there is no such s (and hence the only time f fails) is when $G_{i,j}$ actually queried B about the membership of r in K. We complete the proof with a counting argument that shows that, if this happens with any constant probability α, then r cannot have maximum Kolmogorov complexity. ∎

4 Discussion, Related Work, and Open Problems

Let S be an NP set and fix a specific machine N that accepts S. Recall that ω_x^N is the set of accepting paths of N on input x. We say that S is *canonically generable* if there is a randomized polynomial-time program that, on input 1^n, generates pairs $\langle x, w\rangle$, where $|x| = n$, such that the probability accorded x is proportional to $|\omega_x^N|$. The straightforward generation schemes given in Section 1 for Hamiltonian graphs, 3SAT formulas, and Subset Sum instances are all canonical, with respect to the usual types of witnesses for these sets.

We call these generators canonical mainly because, in a sense, all generators for NP sets are canonical. If $G_{i,j}$ is a generation scheme for S, then a coin-toss sequence that causes M_i to output $\langle x, w\rangle$ is a witness that $x \in S$, and the probability accorded a particularly x is clearly proportional to the number of coin-toss sequences that cause it to be output.

The straightforward canonical generation scheme for Hamiltonian graphs has this general form: generate w uniformly and then pick x uniformly from the set of all instances such that w is a witness that x is in S. In fact, many sets in NP (e.g., SAT, graphs with perfect matchings, graphs with cliques of size $|V(G)|/2$) have canonical generation schemes of this form with respect to the usual types of witnesses. This leads naturally to the question of whether every set in NP has such a canonical generation scheme with respect to every type of witness. The answer to this question is no, unless the construction problem for NP sets can always be solved in polynomial time. (The construction problem is: given an instance x, find a witness if x is a yes-instance, and say that there is no witness if x is a no-instance.)

An interesting area for further research is the relationship between generable (i.e., canonical) distributions and the "hard-on-average" distributions studied by Levin et al. ([17], see also [12], [15], and, more recently, [22]). Levin's randomized NP (denoted RNP) is a class of pairs (D, μ), where D is any decision problem in NP and μ is any probability function on $\{0,1\}^*$ (interpreted as instances of D) for which the cumulative distribution function $\mu^*(x) = \Sigma_{z<x}\mu(z)$ is polynomial-time computable. In [22], Venkatesan and Levin extend the definition to construction problems in NP; the distributions they allow are still those with polynomial-time computable μ^*.

Venkatesan and Levin exhibit a construction problem that is RNP-hard, i.e., if there is an algorithm that can solve it in expected polynomial time, then all RNP-construction problems can be solved in expected polynomial time. The distribution of instances that they consider is easy to generate; however, it assigns positive probability to no-instances.

This suggests some natural questions. Is there an RNP-hard distribution (of instances of a construction problem) that assigns positive probability only to yes-instances? Can that distribution be generated efficiently if one insists on generating witnesses along with the instances? Are the requirements that a distribution be efficiently generable and that it have an efficiently computable μ^* mutually exclusive? For example, our canonical generation scheme for Hamiltonian graphs produces a distribution that probably does not have a polynomial-time computable μ^*: if it did, then the #P-Complete problem of computing the number of Hamiltonian cycles in a graph would be solvable in polynomial time.

Finally, we would like to mention that generation of solved instances has also been considered by Rardin, Tovey, and Pilcher [18]; their goal is the construction of test instances for heuristic algorithms.

5 Acknowledgements

We thank Mike Foster, Steve Mahaney, Steven Rudich, and Mihalis Yannakakis for helpful discussions. We are particularly grateful to Laura Sanchis.

References

[1] M. Blum and S. Micali. "How to Generate Cryptographically Strong Sequences of Pseudo-random Bits," SIAM J. on Comput. (13), 1984, 850–864.

[2] R. Boppana and R. Hirschfeld. "Pseudorandom Generators and Complexity Classes," to appear in Advances in Computer Research, Silvio Micali (ed.), JAI Press (pub.), 1987.

[3] G. Brassard and C. Crépeau. "Non-transitive Transfer of Confidence: A *Perfect* Zero-Knowledge Interactive Protocol for SAT and Beyond," Proceedings of the 27th FOCS, IEEE, 1986, 188–195.

[4] G. Brassard and C. Crépeau. "Zero-Knowledge Simulation of Boolean Circuits," Advances in Cryptology — CRYPTO86 Proceedings, Andrew Odlyzko (ed.), Springer-Verlag (pub.), 1987, 223–233.

[5] G. Brassard, D. Chaum, and C. Crépeau. "Minimum Disclosure Proofs of Knowledge," to appear.

[6] U. Feige, A. Fiat, and A. Shamir. "Zero Knowledge Proofs of Identity," Proceedings of the 19th STOC, ACM, 1987, 210–217.

[7] Z. Galil, S. Haber, and M. Yung. "Cryptographic Computation: Secure Fault-Tolerant Protocols and the Public-Key Model," Advances in Crytology — CRYPTO87 Proceedings, Carl Pomerance (ed.), Springer-Verlag (pub.), 1988, 135–155.

[8] M. Garey and D. Johnson. *Computers and Intractability: A Guide to the Theory of* NP-*Completeness*, Freeman, San Francisco, 1979.

[9] S. Goldwasser and S. Micali. "Probabilistic Encryption," JCSS (28), 1984, 270-299.

[10] S. Goldwasser, S. Micali, and C. Rackoff. "The Knowledge Complexity of Interactive Proof Systems," to appear in SIAM J. on Comput.

[11] O. Goldreich, S. Micali, and A. Wigderson. "Proofs that Yield Nothing but their Validity and a Method of Cryptographic Protocol Design," Proceedings of the 27th FOCS, IEEE, 1986, 174–187.

[12] Y. Gurevich. "Complete and Incomplete Randomized NP Problems," Proceedings of the 28th FOCS, IEEE, 1987, 111–117.

[13] J. Hartmanis. "Generalized Kolmogorov Complexity and the Structure of Feasible Computations," Proceedings of the 24th FOCS, IEEE, 1983, 439–445.

[14] M. Jerrum, L. Valiant, and V. Vazirani. "Random Generation of Combinatorial Structures from a Uniform Distribution," TCS (43), 1986, 169–188.

[15] D. Johnson. "The NP-Completeness Column, An Ongoing Guide," JOA (5), 1984, 284–299.

[16] J. Lagarias and A. Oldlyzko. "Solving Low-Density Subset Sum Problems," JACM (32), 1985, 229–246.

[17] L. Levin. "Average Case Complete Problems," SIAM J. on Comput. (15), 1986, 285–286.

[18] R. Rardin, C. Tovey, and M. Pilcher. "Polynomial Constructability and Traveling Salesman Problems of Intermediate Complexity," ONR-URI Computational Combinatorics Report CC-88-2, Purdue University, November, 1988.

[19] S. Rudich, private communication.

[20] L. Sanchis and M. Fulk. "Efficient Language Instance Generation", University of Rochester Computer Science Department TR 235, 1988.

[21] L. Sanchis. "Test Instance Construction for NP-hard Problems," University of Rochester Computer Science Department TR 206, 1987.

[22] R. Venkatesan and L. Levin. "Random Instances of a Graph Coloring Problem are Hard," Proceedings of the 20th STOC, ACM, 1988, 217–222.

[23] A. C. Yao. "Theory and Applications of Trapdoor Functions," Proceedings of the 23rd FOCS, IEEE, 1982, 80–91.

Bounds and Constructions for Authentication - Secrecy Codes with Splitting

Marijke De Soete
Seminar of Geometry and Combinatorics
State University of Ghent, Krijgslaan 281, B–9000 Ghent Belgium.

It is the aim to deal with codes having unconditional security, which means that the security is independent of the computing power. Analogously to the theory of unconditional secrecy due to Shannon [12], Simmons developed a theory of unconditional authentication [10]. In this paper we give some new bounds and constructions for authentication/secrecy codes with splitting.

Consider a transmitter who wants to communicate a source to a remote receiver by sending messages through an imperfect communication channel. Then there are two fundamentally different ways in which the receiver can be deceived. The channel may be noisy so that the symbols in the transmitted message can be received in error, or the channel may be under control of an opponent who can either deliberately modify legitimate messages or else introduce fraudulent ones. Simmons [10] showed that both problems could be modeled in complete generality by replacing the classical noisy communications channel of coding theory with a game - theoretic noiseless channel in which an intelligent opponent, who knows the system and can observe the channel, plays so as to optimize his chances of deceiving the receiver. To provide some degree of immunity to deception (of the receiver), the transmitter also introduces redundancy in this case, but does so in such a way that, for any message the transmitter may send, the altered messages that the opponent would introduce using his optimal strategy are spread randomly. Authentication theory is concerned with devising and analizing schemes (codes) to achieve this "spreading".
In the mathematical model there are three participants: a *transmitter*, a *receiver* and an *opponent*. The transmitter wants to communicate some information to the receiver. The opponent wanting to deceive the receiver, can either impersonate the receiver, making him accept a fraudulent message as authentic, or, modify a message which has been sent by the transmitter.
Let S denote the set of k source states, M the set of v messages and E the set of b encoding rules.

A *source state* $s \in S$ is the information that the transmitter wishes to communicate to the receiver. The transmitter and receiver will have secretly chosen an *encoding rule* $e \in E$ beforehand. An encoding rule e will be used to determine the message $e(s)$ to be sent to communicate any source state s. In a model with *splitting*, several messages can be used to determine a particular source state. However, in order for a receiver to be able to uniquely determine the source state from the message sent, there can be at most one source state which is encoded by any given message $m \in M$, for a given encoding rule $e \in E$ (this means: $e(s) \neq e(s')$ if $s \neq s'$).

An opponent will play *impersonation* or *substitution.* When the opponent plays impersonation, he sends a message to the receiver, attempting to have the receiver accept the message as authentic. When the opponent plays substitution, he waits until a message m has been sent, and then replaces m with another message m', so that the receiver is misled as to the state of source. More generally, an opponent can observe i (≥ 0) distinct messages being sent over the channel knowing that the same key is used to transmit them, but ignoring this key. If we consider the code as a secrecy system, then we make the assumption that the opponent can only observe the messages being sent. Our goal is that the opponent be unable to determine any information regarding the i source states from the i messages he has observed.

We shall use the following notations. Given an encoding rule e, we define $M(e) = \{e(s)|s \in S\}$, i.e. the set of messages permitted by encoding rule e, and let $|M(e)| = k(e)$. For a set of distinct messages $M' \subset M$ and an encoding rule e, define $f_e(M') = \{s \in S|e(s) \in M'\}$, i.e. the set of source states which will be encoded under encoding rule e by a message in M'. Define also $E(M') = \{e \in E|M' \subseteq M(e)\}$, i.e. the set of encoding rules under which all the messages in M' are permitted.

The following scenario for authentication is investigated. After the observation of i messages $M' \subset M$, the opponent sends a message m' to the receiver, $m' \notin M'$, hoping to have it accepted as authentic. This is called a *spoofing attack of order i* [6], with the special cases $i = 0$ and $i = 1$ corresponding respectively to the impersonation and substitution game. The last games have been studied extensively by several authors (see [2], [5], [10], [13]).

For any i, there will be a probability on the set of i source states which occur. We ignore the order in which the i source states occur, and assume that no source state occurs more than once. Also, we assume that any set of i source states has a non-zero probability of occuring. Given a set of i source states, we define $p(S)$ to be the probability that the source states in S occur.

Given the probability distributions on the source states described above, the receiver and transmitter will choose a probability distribution for E, called an *encoding strategy.* If splitting occurs, then they will also determine a *splitting strategy* to determine $m \in M$, given $s \in S$ and $e \in E$ (this corresponds to non-deterministic encoding). The transmitter/receiver will determine these strategies to minimize the chance that an opponent can deceive them.

Once the transmitter/receiver have chosen encoding and splitting strategies, we can define for each $i \geq 0$ a probability denoted P_{d_i}, which is the probability that the opponent can deceive the transmitter/receiver with a spoofing attack of order i. We denote by $AC(k, v, b)$ an authentication system with k source states, v messages and b encoding rules.

1 Secrecy

Considering the secrecy of a code, we desire no information be conveyed by the observation of the messsages. A code has *perfect L-fold secrecy* (Stinson [14]) if, for every set M_1 of at most L messages observed in the channel, and for every set S_1 of at most $|M_1|$ source states, we have $p(S_1/M_1) = p(S_1)$. This means that observing a set of at most L messages in the channel does not help the opponent to determine the L source states. On the other hand, a code is said to be *Cartesian* ([2], [13]) if any message uniquely determines the source state, independent of the particular encoding rule being used .

2 Bounds on P_{d_i} and b

Bounds on P_{d_0} and P_{d_1} for authentication codes with splitting depending on the entropies of the various probability distributions can be found in [2], [9], [10], [13] and [14]. The most important bounds are given by:

$$P_{d_0} \geq 2^{H(MES)-H(E)-H(M)} = 2^{H(M|ES)+H(S)-H(M)}$$

and for a substitution with secrecy

$$P_{d_1} \geq 2^{-H(E/M)} = 2^{H(M)-H(E)-H(S)+H(M/ES)}.$$

The following bounds for an impersonation, resp. a substitution game are proven in [4]:

$$P_{d_0} \geq min_{e\in E}\frac{k(e)}{v} \text{ (see also [9] [10]),}$$

$$P_{d_1} \geq min_{e\in E}\frac{k(e) - max_{s\in S}|e(s)|}{v - max_{s\in S}|e(s)|}.$$

For codes without splitting this results in the known bounds $P_{d_0} \geq k/v$ and $P_{d_1} \geq (k-1)/(v-1)$ ([6], [13], [14]).
These bounds can also be generalized for a spoofing attack of order i [4] to

$$P_{d_i} \geq min_{e\in E}\frac{k(e) - i \cdot max_{s\in S}|e(s)|}{v - i \cdot max_{s\in S}|e(s)|}.$$

An authentication system which achieves equality $\forall i,\ 0 \leq i \leq L$, is called *L-fold secure against spoofing* (this is a generalization of the definition for codes without splitting, see [6], [14]).

The number of keys is basically influenced by the following two aspects: (i) the distribution on the source states and (ii) the secrecy of the code. In [4] we obtain the following bound:
If a code achieves perfect L-fold secrecy and is $(L-1)$-fold secure against spoofing, then

$$b \geq \frac{v \cdot (v - max_{s\in S}|e(s)|) \cdots (v - (L-1) \cdot max_{s\in S}|e(s)|)}{L!}.$$

Analogously as for codes without splitting [14], we define an *optimal L-code* to be a code which achieves perfect L-fold secrecy, which is $(L-1)$-fold secure against spoofing and which meets equality in the foregoing formula.

3 Constructions for authentication codes with arbitrary source distribution

3.1 Authentication codes derived from partial geometries

A (finite) *partial geometry* (PG) is an incidence structure $\mathcal{G} = (P, B, I)$ in which P and B are disjoint (nonempty) sets of objects called *points* and *lines* resp., and for which I is a symmetric point-line incidence relation satisfying the following axioms:

1. Each point is incident with $1 + t$ lines ($t \geq 1$) and two distinct points are incident with at most one line.

2. Each line is incident with $1 + s$ points ($s \geq 1$) and two distinct lines are incident with at most one point.

3. If x is a point and L a line not incident with x, then there are exactly α, ($\alpha \geq 1$) points $x_1, x_2, \ldots, x_\alpha$ and α lines $L_1, L_2, \ldots, L_\alpha$ such that $x \; I \; L_i \; I \; x_i \; I L$, $i = 1, 2, \ldots, \alpha$.

Partial geometries were introduced by R. C. Bose. The partial geometries with $\alpha = 1$ are the *generalized quadrangles* (GQ).

There holds $|P| = (s+1)(st+\alpha)/\alpha$, $|B| = (t+1)(st+\alpha)/\alpha$, $\alpha(s+t+1-\alpha)|st(s+1)(t+1)$ and $(s+1-2\alpha)t \leq (s-1)(s+1-\alpha)^2$ (and dually). We remark that the dual incidence structure $G' = (P', B', I')$, $P' = B$, $B' = P$, $I' = I$, is a partial geometry with parameters $t' = s$, $s' = t$ and $\alpha' = \alpha$.
Further information about PG and GQ can be found in [7].

1. From a generalized quadrangle of order (s, t), $s, t > 1$, we can define the following two authentication codes without splitting [3].

   - *A GQ of order (s,t) defines a cartesian $AC(t+1, (t+1)s, ts^2)$ which is 0-fold secure against spoofing and for which $P_{d_1} = 1/s$.*

   - *If the GQ contains a regular point, the foregoing code can be improved to an $AC(t+1, (t+1)s, (t+1)s^2)$ which is 0-fold secure against spoofing, which has perfect 1-fold secrecy, and for which $P_{d_1} = 1/s$.*

2. A PG with parameters $s, t \geq 1$, $\alpha > 1$ defines an $AC(t+1, (t+1)s, (t+1)st(s+1-\alpha))$ code which has 0-fold security against spoofing and which has perfect 1-fold secrecy [4].

3. A *spread* of a PG $\mathcal{G}$ is a set $\mathcal{R}$ of lines of $\mathcal{G}$ such that each point of $\mathcal{G}$ is incident with a unique line of $\mathcal{R}$. Hence there holds $|\mathcal{R}| = (st + \alpha)/\alpha$.
Let $\mathcal{G}$ be a PG with parameters $s, t > 1$, $\alpha \geq 1$, containing a spread $\mathcal{R}$. Then we can define the following authentication codes.

   - *For* $\alpha > 1$, $\mathcal{G}$ *defines an optimal* 1*-code with splitting* [4].
   - *For* $\alpha = 1$, $\mathcal{G}$ *defines an optimal* 1*-code without splitting* [3].

3.2 Authentication codes derived from designs

Consider an *affine resolvable BIB-design*. This is a 2-(v, k, λ) design $\mathcal{D} = (P, B, I)$ for which there exists a partition of $B = B_1 \cup B_2 \ldots B_r$ of the block set, $|B_i| = n$, such that each point occurs exactly once in the blocks of any set B_i, $1 \leq i \leq r$ and any two blocks of different sets have exactly μ, $\mu > 0$, points in common [1]. There holds $|B| = rn$, $|P| = kn$, $\lambda = r(k-1)/(nk-1)$ and $k = \mu n$.

In [4] we construct the following authentication code with splitting:

An affine resolvable design $\mathcal{D}$ *defines an* $AC(n, kn, (r-1)n^2)$ *which is* 0*-fold secure against spoofing, which has* 1*-fold secrecy, and for which* $P_{d_1} = \lambda/(r-1)$.

References

[1] Beth T., Jungnickel D., Lenz H., *Design Theory*. Wissenschaftsverlag Bibliografisches Institut Mannheim, 1985.

[2] Brickell E. F., *A few results in message authentication*. Proc. of the 15th Southeastern Conf. on Combinatorics, Graph theory and Computing, Boca Raton LA (1984), 141–154.

[3] De Soete M., *Some Constructions for Authentication / Secrecy codes*, Proceedings of Eurocrypt'88, Davos, L.N.C.S., to appear.

[4] De Soete M., *New Bounds and Constructions for Authentication / Secrecy Codes with Splitting.* In preparation.

[5] Gilbert E. N., MacWilliams F. J., Sloane N. J. A., *Codes which detect deception.* Bell Sys. Techn. J., Vol.53–3 (1974), 405–424.

[6] Massey J. L., *Cryptography - A Selective Survey.* Proc. of 1985 Int. Tirrenia Workshop on Digital Communications, Tirrenia, Italy, 1985, Digital Communications, ed. E. Biglieri and G. Prati, Elsevier Science Publ. 1986, 3–25.

[7] Payne S. E., Thas J. A., *Finite generalized quadrangles.* Research Notes in Math. #110, Pitman Publ. Inc. 1984.

[8] Shrikhande S. S., *Affine resolvable balanced incomplete block designs: a survey.* Aequat. Math. 14 (1976), 251–269.

[9] Simmons G. J., *Message Authentication: A Game on Hypergraphs.* Proc. of the 15th Southeastern Conf. on Combinatorics, Graph Theory and Computing, Baton Rouge LA Mar 5–8 1984, Cong. Num. 45 (1984), 161–192.

[10] Simmons G. J., *Authentication theory / Coding theory.* Proc. of Crypto'84, Santa Barbara, CA, Aug.19–22, 1984, Advances in Cryptology, ed. R. Blakley, Lect. Notes Comp. Science 196, Springer 1985, 411–432.

[11] Simmons G. J., *A natural taxonomy for digital information authentication schemes.* Proc. of Crypto '87, Santa Barbara, CA, Aug 16–20, 1987, to appear in Advances in Cryptology, ed. C. Pomerance, Springer Verlag, Berlin.

[12] Shannon C. E., *Communication Theory of Secrecy Systems.* Bell Technical Journal, Vol.28 (1949), 656–715.

[13] Stinson D. R., *Some Constructions and Bounds for Authentication Codes.* J. Cryptology 1 (1988), 37–51.

[14] Stinson D. R., *A construction for authentication / secrecy codes from certain combinatorial designs.* Crypto '87, Santa Barbara, CA, Aug 16–20, 1987, to appear in J. Cryptology.

Session 9

Protocols

Chair: G. Brassard, University of Montreal

Untraceable Electronic Cash †
(Extended Abstract)

David Chaum [1] *Amos Fiat* [2] *Moni Naor* [3]

[1] Center for Mathematics and Computer Science
Kruislaan 413, 1098 SJ Amsterdam, The Netherlands

[2] Tel-Aviv University
Tel-Aviv, Israel

[3] IBM Almaden Research Center
650 Harry Road, San Jose, CA 95120

Introduction

The use of credit cards today is an act of faith on the part of all concerned. Each party is vulnerable to fraud by the others, and the cardholder in particular has no protection against surveillance.

Paper cash is considered to have a significant advantage over credit cards with respect to privacy, although the serial numbers on cash make it traceable in principle. Chaum has introduced unconditionally untraceable electronic money([C85] and [C88]). But what is to prevent anyone from making several copies of an electronic coin and using them at different shops? On-line clearing is one possible solution though a rather expensive one. Paper banknotes don't present this problem, since making exact copies of them is thought to be infeasible. Nor do credit cards, because their unique identity lets the bank take legal action to regain overdrawn balances, and the bank can add cards to a blacklist.

Generating an electronic cash should be difficult for anyone, unless it is done in cooperation with the bank. The RSA digital signature scheme can be used to realize untraceable electronic money as proposed in [C85 and C88]. This money might be of the form $\left(x, f(x)^{1/3} \pmod{n}\right)$ where n is some composite whose factorization is known only to the bank and f is a suitable one-way function. The protocol for issuing and spending such money can be summarized as follows:

1. Alice chooses a random x and r, and supplies the bank with $B = r^3 f(x) \pmod{n}$.

† Work done while the second and third authors were at the University of California at Berkeley. The work of the second author was supported by a Weizmann Postdoctoral Fellowship and by NSF Grants DCR 84-11954 and DCR 85-13926. The work of the third author was supported by NSF Grants DCR 85-13926 and CCR 88-13632.

2. The bank returns the third root of B modulo n: $r \cdot f(x)^{1/3} \pmod{n}$ and withdraws one dollar from her account.
3. Alice extracts $C = f(x)^{1/3} \bmod n$ from B.
4. To pay Bob one dollar, Alice gives him the pair $(x, f(x)^{1/3} \pmod{n})$.
5. Bob immediately calls the bank, verifying that this electronic coin has not already been deposited.

Everyone can easily verify that the coin has the right structure and has been signed by the bank, yet the bank cannot link this specific coin to Alice's account.

Among other advantages, the new approach presented here removes the requirement that the shopkeeper must contact the bank during every transaction. If Alice uses a coin only once, her privacy is protected unconditionally. But if Alice reuses a coin, the bank can trace it to her account and can *prove* that she has used it twice.

Our work is motivated by that on minimum disclosure ([C86], [BC86a], [BC86b] and [BCC]) and on zero-knowledge ([GMR], [GMW86a] and [GMW86b]). Our scheme protects Alice's privacy unconditionally as is possible with the former, rather than computationally as in the latter. Using these very general results – which seem to be infeasible in practice – the security of the protocols presented here could be reduced to, say factoring (or any onw-way permutation if Alice's privacy is only computationally secure). Instead, We use the cut-and-choose methodology (first introduced in [R77]) directly, yielding quite practical constructions.

The next section presents our basic scheme, which guarantees untraceability, yet allows the bank to trace a "repeat spender". We then show how to modify the protocol so that the bank can supply incontestable proof that Alice has reused her money. Finally, we give a more efficient variant and briefly discuss further work.

1. Untraceable Coins

The bank initially publishes an RSA modulus n whose factorization is kept secret and for which $\phi(n)$ has no small odd factors. The bank also sets some security parameter k.

Let f and g be two-argument collision-free functions; that is, for any particular such function, it is infeasible to find two inputs that map to the same point. We require that f be "similar to a random oracle". For unconditional untraceability we also require g to have the property that fixing the first argument gives a one-to-one (or c to 1) map from the second argument *onto* the range.

Alice has a bank account numbered u and the bank keeps a counter v associated with it. Let $\oplus$ denote bitwise exclusive or and $\|$ denote concatenation.

To get an electronic coin, Alice conducts the following protocol with the bank:

1. Alice chooses a_i, c_i, d_i and r_i, $1 \leq i \leq k$, independently and uniformly at random from the residues (mod n).
2. Alice forms and sends to the bank k *blinded candidates* (called B for mnemonic purposes)

$$B_i = r_i^3 \cdot f(x_i, y_i) \bmod n \qquad \text{for} \quad 1 \leq i \leq k,$$

where

$$x_i = g(a_i, c_i) \qquad y_i = g(a_i \oplus (u \| (v+i)), d_i).$$

3. The bank chooses a random subset of $k/2$ blinded candidate indices $R = \{i_j\}$, $1 \leq i_j \leq k$ for $1 \leq j \leq k/2$ and transmits it to Alice.
4. Alice displays the r_i, a_i, c_i and d_i values for all i in R, and the bank checks them. Note that $u \| (v+i)$ is known to the bank. To simplify notation we will assume that $R = \{k/2+1, k/2+2, \ldots, k\}$.
5. The bank gives Alice

$$\prod_{i \notin R} B_i^{1/3} = \prod_{1 \leq i \leq k/2} B_i^{1/3} \bmod n$$

and charges her account one dollar. The bank also increments Alice's counter v by k.
6. Alice can then easily extract the electronic coin

$$C = \prod_{1 \leq i \leq k/2} f(x_i, y_i)^{1/3} \bmod n.$$

Alice reindexes the candidates in C to be lexicographic on their representation: $f(x_1, y_1) < f(x_2, y_2) < \cdots < f(x_{k/2}, y_{k/2})$. Alice also increments her copy of the counter v by k.

Note: For any fixed ϵ, if fewer than $(1-\epsilon)$ of the k blinded candidates B_i's have the proper form $(r^3 f(g(a_i, c_i), g(a_i \oplus (u \| (v+i)), d_i)))$, then Alice is caught with probability $1 - \exp(-c\epsilon k)$ for some constant c.

To pay Bob one dollar, Alice and Bob proceed as follows:

1. Alice sends C to Bob.
2. Bob chooses a random binary string $z_1, z_2, \ldots, z_{k/2}$.
3. Alice responds as follows, for all $1 \leq i \leq k/2$:
   a. If $z_i = 1$, then Alice sends Bob a_i, c_i and y_i.
   b. If $z_i = 0$, then Alice sends Bob x_i, $a_i \oplus (u \| (v+i))$ and d_i.
4. Bob verifies that C is of the proper form and that Alice's responses fit C.
5. Bob later sends C and Alice's responses to the bank, which verifies their correctness and credits his account.

The bank must store C, the binary string $z_1, \ldots, z_k$ and the values a_i (for $z_i = 1$) and $a_i \oplus (u \| v)$ (for $z_i = 0$).

If Alice uses the same coin C twice, then she has a high probability of being traced: with high probability, two different shopkeepers will send complementary binary values for at least one bit z_i for which B_i was of the proper form. The bank can easily search its records to ensure that C has not been used before. If Alice uses C twice, then, with high probability, the bank has both a_i and $a_i \oplus (u \| (v+i))$ with high probability. Thus, the bank can isolate u and trace the payment to Alice's account.

A possible problem with this scheme is a collusion between Alice and a second shopkeeper Charlie. After the transaction with Bob, Alice describes the transaction to Charlie, and both Bob and Charlie send the bank the same information; the bank knows that with very high probability one of them is lying, but has no way of telling which one, and cannot trace the coin to Alice's account

By fixing Bob's challenge to Alice, however, such a coalition can be kept from defrauding the bank. Every shopkeeper has a fixed query string, and every two strings have Hamming distance at least ck for some constant c. To prevent Alice from reusing the same coin at the same shop part of the challenge should still be random, or the shopkeeper should maintain his own list.

The scheme we describe above requires Alice to hold several coin denominations and use them to pay the exact amount. Section 3 presents a more efficient way to handle exact amounts.

2. Proving Multiple Spending

The scheme we describe above has the unfortunate property that the bank can frame Alice as a multiple spender. This means that these schemes cannot have any legal significance. To prevent a frame-up we assume that Alice has a digital signature scheme and a certified copy of her public key. Because we use digital signatures, Alice is protected against frame-up only computationally, not unconditionally. Yet, Alice's privacy remains unconditionally protected.

Rather than use the same account number u for all coins given to Alice, u will vary from coin to coin and from one blinded candidate to the next. We describe only the modifications to the basic scheme of section one.

Alice chooses two random integers z_i' and z_i'' for every i; u_i could then be chosen of the form "Alice's Account Number" $\|\ z'\ \|\ z''$. Along with the blinded candidates (the B_i values) Alice supplies the bank with a digital signature on

$$g(z_1', z_1'') \| g(z_2', z_2'') \| \cdots \| g(z_k', z_k'').$$

During the cut-and-choose, the bank verifies that each of the $k/2$ B_i's it examines generate an appropriate u_i. The bank has legal proof that Alice reused the coin whenever it can present the preimage of at least $k/2+1$ of the $g(z_i', z_i'')$.

Of course Alice has no hope if the bank can break the signature scheme she has chosen. Assuming the bank cannot forge her signature, then even if the bank can break g, its bijective property mentioned earlier ensures, with high probability, that she can prove g was broken by showing her (z_i', z_i'') for any broken $g(z_i', z_i'')$. This is a proof, since the assumption is that only the bank and not Alice can break g.

3. Untraceable Checks

The following scheme emulates the concept of guaranteed checks (similar to that of EuroChecks), but ensures untraceability. Alice requests a set of checks, whereby she can use each check for any single amount up to its limit and can later request a refund for the difference (limit minus actual sum). The bank will not know where the money was spent, nor the individual transaction amounts.

Alice can generate several checks in one interaction with the bank. The checks are similar to the basic version described in section one, but the first j factors are used to encode the purchase sum and the next $k-j$ factors are used to prevent Alice from using any check more than once.

The bank publishes two different RSA moduli, n and n', which are used for two different kinds of digital signature.

Alice's u can be used either as in section one or in section two. As before, let v be Alice's personal counter.

Alice sends the bank t pairs of major and minor candidates. For every major candidate Alice chooses b, c, d, and a at random; a major candidate M_i is of the form $f(x,y)$ where $x = g(a\|b,c)$ and $y = g(a \oplus (u\|(v+i)), d)$. Each minor candidate is of the form $g(b,e)$ where e is chosen at random. Alice generates several major candidates $M_1, M_2, \ldots, M_t$ and their related minor candidates $m_1, m_2, \ldots, m_t$.

Alice blinds the major and minor terms before submitting them to the bank. Blinded major candidates are of the form $B(M_i) = r^{3^k} \cdot M_i \bmod n$, where r is chosen at random; blinded minor candidates are of the form $B(m_i) = r^{3^k} \cdot m_i \bmod n'$. If the bank provides some $3^i th$ root of a blinded major(minor) term, $i \leq k$, then, as before, Alice can extract the appropriate root of the major(minor) term itself.

Alice sends the blinded M_i's and m_i's to the bank. Much as in section two, the bank performs a cut-and-choose operation, verifying that 1/2 of the pairs have the proper form. Then the bank performs a random permutation of the rest, grouping them into ordered sets of size k. Let one such set be denoted for simplicity

$B(M_1), B(M_2), \ldots, B(M_k)$. The bank extracts the following roots:

$$F_i = B(M_i)^{1/3^i} \pmod n \quad \text{for} \quad 1 \leq i \leq k,$$
$$D_i = B(m_i)^{1/3^i} \pmod{n'} \quad \text{for} \quad 1 \leq i \leq j.$$

The bank now returns the product of the k roots of blinded major candidates $(\prod_{i=1}^{k} F_i)$; the appropriate roots of the j blinded minor candidates are returned individually. Alice extracts the check

$$C = \prod_{i=1}^{k} M_i^{1/3^i}$$

and $E_1, E_2, \ldots, E_j$, where $E_i = m_i^{1/3^i}$.

The bank now increments Alice's counter v by t, Alice does likewise to her local copy.

To make a purchase with such a check Alice encodes the purchase sum by regarding the first j of the M_i locations as denominations $1, 2, \ldots, 2^{j-1}$. If the *ith* denomination is a term in the purchase sum, then Alice reveals to the shopkeeper the appropriate y_i and the preimages of the x_i; if the *ith* denomination is not a term in the purchase sum, then Alice reveals x_i and y_i. Thus, later presenting E_i and the internal structure of the matching m_i term to the bank for a refund is safe exactly when the denomination is not spent.

Note: Given a root of the form $x^{1/3^i}$, it is trivial to compute roots of the form $x^{1/3^j}$ for $j \leq i$. Thus, Alice could use the denomination 2^j, not use the denomination 2^i, $j < i$, and present the bank with the value

$$E_i^{i-j} = (m_i^{1/3^i})^{i-j} = m_i^{1/3^j} = g(b,e)^{1/3^j} \pmod{n'},$$

claiming that this is a signed minor term for an unused 2^j denomination. The bank has no trace of the appropriate b value and would grant the refund. Fortunately, this would not be in Alice's interest, since she would get a smaller refund than she is entitled to.

The last $k - j$ major terms prevent Alice from using the check more than once, Even if the purchase amount is exactly the same. As in section one, the shopkeeper could present a random challenge or every shopkeeper has a probe sequence for these $k - j$ terms chosen from a code with large Hamming distance.

Alice does however, have a good chance of successfully cheating the bank with respect to the refund. All she needs is two unrelated major and minor terms. Still, this type of cheating is far less dangerous than having an open check that can be used over and over again. The bank could penalize Alice whenever it detects an attempt at cheating, negating Alice's expected profit from cheating attempts. A variation would allocate two major terms per denomination, making the probability of cheating much smaller.

4. Blacklisting Withdrawals

It may be desirable that if Alice uses a coin twice then the bank can blacklist all of the coins Alice has withdrawn. Obviously, this means that all her coins must be related in some manner. The idea is to encrypt some redundancy in Alice's "random" choices; this redundancy can be recognized only when Alice spends a coin more than once. Alice's privacy is thus protected only computationally, *not* unconditionally.

Consider the basic scheme: Alice sends the bank k blinded candidates of the form $r^3 f(g(a,c), g(a \oplus (u \| (v+i)), d))$ where a, c and d are chosen at random by Alice, v is Alice's counter, i is the candidate serial number and u is Alice's account number. We modify the protocol so that Alice generates b electronic coins simultaneously.

Alice sends the bank bk blinded candidates as a matrix

$$\begin{pmatrix} B_{11} & B_{12} & \dots & B_{1k} \\ B_{21} & B_{22} & \dots & B_{2k} \\ \vdots & \vdots & \ddots & \vdots \\ B_{b1} & B_{b2} & \dots & B_{bk} \end{pmatrix}.$$

The bank ask's to see $k/2$ *columns* in their entirety. Each B_{ij} should be of the form

$$r_{ij}^3 f(g(a_{ij}, c_{ij}), g(a_{ij} \oplus (u \| k_j \| (v + ki + j)), d_{ij})).$$

Alice chooses r_{ij}, a_{ij} at random per blinded term and chooses k_j at random per column.

Let $\{h_l\}$ be a family of one-way functions. Each c_{ij} is of the form $h_{l_j}(c'_{ij}) \| c'_{ij}$; each d_{ij} is of the form $h_{l_j}(d'_{ij}) \| d'_{ij}$. Alice chooses c'_{ij} and d'_{ij} at random per blinded term.

The bank can easily verify that each of the $k/2$ columns it sees is of the proper form. For notational simplicity, we assume that the bank asks to see columns $k/2+1, \dots, k$. The bank then supplies Alice with b products

$$P_i = \prod_{1 \le j \le k/2} B_{ij}^{1/3} \pmod{n}$$

and charges her account b dollars.

Alice can then easily extract

$$C_i = \prod_{1 \le j \le k/2} f(g(a_{ij}, c_{ij}), g(a_{ij} \oplus (u \| l_j \| (v + ki + j)), d_{i,j}))^{1/3} \pmod{n}, for 1 \le i \le b.$$

Alice also arranges the factors into lexicographic sequence.

These coins are used exactly as in the basic scheme, except that the shopkeeper has the set of blacklisted indices L. If the merchant sends $e_j = 1$, then Alice must reveal the appropriate a, c and y. The shopkeeper computes $f(g(a,c),y)$ and checks that $c = c''\|c'$ does not satisfy $c'' = h_l(c')$ for all $l \in L$. Similarly, if the shopkeeper sends $e_j = 0$, then Alice must reveal the appropriate $x, a \oplus (u\|h_j\|(v + ki + j))$ and $d = d''\|d'$. Again, the shopkeeper checks that $d'' \neq h_l(d')$ for all $l \in L$.

If Alice uses *any* coin more than once then the bank adds the appropriate revealed k_j's to the blacklist supplied to the merchants.

5. Further Work

In forthcoming work, Chaum and Impagliazzo investigate formal requirements for the function f and den Boer has proposed suitable g's whose security is reducible to factoring or to discrete log. A good deal of progress has been made towards establishing the overall security of similar protocols [CE87]. Formal proofs for the protocols of this paper, however, remain an open challenge.

Acknowledgements

It is a pleasure to thank Russell Impagliazzo for inspiring this work and Eugene van Heigst for making several helpful comments.

References

[BC86a] Brassard, G. and C. Crépeau, Zero-knowledge simulation of Boolean circuits, presented at Crypto '86, 1986.

[BC86b] Brassard, G. and C. Crépeau, Zero-knowledge simulation of Boolean circuits, Proc. of 27*th* Symp. on Foundations of Computer Science, 1986.

[BCC] Brassard, G., Chaum D., and C. Crépeau, Minimum Disclosure Proofs of Knowledge. Center for Mathematics and Computer Science, Report PM-R8710, December 1987.

[C85] Chaum, D., Security without identification: transaction systems to make big brother obsolete, Comm. ACM 28, 10 (October 1985).

[C86] Chaum, D., Demonstrating that a Public Predicate can be Satisfied Without Revealing Any Information About How, presented at Crypto '86, 1986.

[C88] Chaum, D., Privacy Protected Payments: Unconditional Payer And/Or Payee Untracability, Smartcard 2000, North Holland, 1988.

[CE] Chaum, D. and J.H. Evertse, Showing credentials without identification: signatures transferred between unconditionally unlinkable pseudonyms, Proceedings of Crypto '86, Springer-Verlag, 1987.

[CDG] Chaum, D., I. Damgaard, and J. v.d. Graaf, Multiparty computations ensuring the privacy of each party's input and the correctness of the result, Proceedings of Crypto 87, Springer-Verlag, 1988.

[GMR] Goldwasser, S., S. Micali, and C. Rackoff, The Knowledge Complexity of Interactive Proof Systems, Proc. of 17*th* ACM symp. on Theory of Computation, 1985.

[GMW86a] Goldreich, O., S. Micali, and A. Wigderson, How to prove all NP-statements in zero-knowledge, and a methodology of cryptographic protocol design, presented at Crypto '86, 1986.

[GMW86b] Goldreich, O., S. Micali, and A. Wigderson, Proofs that yield nothing but their validity and a methodology of cryptographic protocol design, Proc. of 27*th* Symp. on Foundations of Computer Science, 1986.

[R77] Rabin, M.O., Digitalized signatures, in Foundations of Secure Computation, Academic Press, NY, 1978.

[RSA] Rivest, R.L., A. Shamir, and L. Adleman, A Method for Obtaining Digital Signatures and Public-Key Cryptosystems, CACM 21, 2 (February 1978).

PAYMENT SYSTEMS AND CREDENTIAL MECHANISMS WITH PROVABLE SECURITY AGAINST ABUSE BY INDIVIDUALS

(Extended Abstract)

Ivan Bjerre Damgård[1]

Aarhus University, Mathematical Institute,
Ny Munkegade,
DK 8000 Aarhus C,
Denmark.

Summary

Payment systems and credential mechanisms are protocols allowing individuals to conduct a wide range of financial and social activities while preventing even infinitely powerful and cooperating organizations from monitoring these activities. These concepts were invented and first studied by David Chaum.

Clearly, such systems must also be secure against abuse by individuals (prevent them from showing credentials that have not been issued to them, etc.). In this work, we present constructions for which we can prove, that no individual can cheat successfully, unless he possesses an algorithm that contradicts a single plausible intractability assumption. This can be done while maintaining the unconditional security against abuse by organizations.

Our construction will work using any general two-party computation protocol with unconditional privacy for one party, and any signature scheme secure against adaptive chosen message attacks (these concepts are explained in more detail later). From the signature scheme by Bellare and Micali [BeMi] and the multiparty computation protocol by Chaum, Damgård and van de Graaf [ChDaGr], it will be clear that both requirements can be met if pairs of claw free functions and trapdoor one-way permutations exist. This, in turn, is satisfied, for example if factoring Blum integers is a hard problem.

For credential mechanisms, we obtain an additional advantage over one earlier proposals [ChEv], where a center trusted by the organizations (but not by individuals) was needed. This center possessed a "master" secret allowing it to issue all types of credentials supported by the system. Moreover, the center had to be on-line permanently. In our construction, only an off-line center is needed, which only has to be trusted as far as validating the identity of each individual is concerned. Only organizations authorized to issue a given type of credential have the ability to compute them.

---

[1]This research was supported by the Danish Natural Science Research Council.

1. Related Work

In earlier work, Chaum [Ch] and Chaum and Evertse [ChEv] have proposed ways to implement payment systems and credential mechanisms, and have established their security against infinitely powerful organizations by information theoretic arguments. While these constructions were quite practical, they left some open questions with regard to the security against abuse by individuals: For payment systems, this security depended on the assumption that RSA used for signatures, along with some redundancy scheme or one way function, is secure against a chosen message attack. So far, no one has been able to reduce this to some widely accepted intractability assumption, and in fact many proposed redundancy schemes have subsequently been broken. For credential mechanisms, the security could only be proved in a restricted, formal model, where potentially bad interaction between RSA and the one way function used was abstracted away. Moreover, an assumption about a very powerful center was needed, as outlined in the summary above (note, however, that Chaum [Ch2] has later modified the construction to do without this last assumption).

Chaum [Ch3] has also designed a credential mechanism which has provable security, but is based on the specific homomorphic properties of RSA. This protocol is much slower than [ChEv], although not completely unresonable in practice.

By contrast, our work is of mainly theoretical interest: while the protocols constructed are probably not practical in the forseeable future, the main purpose of our work is to establish the existence of credential mechanisms and payment systems with respect to as weak an intractability assumption as possible.

In independent work, Chaum [Ch2] has designed a protocol construction with some properties quite similar to ours, in terms of feasibility, the intractability assumption needed, and the problems that can be solved by the protocols. Chaum's solution uses interactive proofs, and not multiparty computations. Compared to our work, the process of creating a credential is simplified, while the process of showing one is slightly more complicated.

2. Basic Results

A pair of functions (f_0, f_1) is called *claw free* if

- $\mathrm{Im}(f_0) = \mathrm{Im}(f_1)$.
- Both functions are t to 1 mappings for some constant t.
- Both f_0 and f_1 are easy to compute, but it is hard to find *a claw*, i.e. r, s, such that $f_0(r) = f_1(s)$.

It is well known that claw free pairs of permutations exist, for example if factoring a Blum-integer is hard. A Blum-integer is an integer $n = pq$, where p and q are primes congruent to 3 modulo 4. As an easy example, consider

$$f_0(x) = x^2 \bmod n, \text{ and } f_1(x) = (a \cdot x)^2 \bmod n,$$

where a has Jacobi symbol -1. It is elementary to prove that these functions permute the set of quadratic residues modulo n, and that knowledge of a claw immediately implies knowledge of the factors of n. If the functions are easy to invert, given some extra information, they are called *trapdoor*. The example above clearly has this property: knowledge of the factors of n suffices to extract square roots modulo n. Although more details would be required for a formal definition, the above will do for this abstract. Details on claw-free functions and their cryptographic applications can be found in [Da].

Our protocols are based on the following two results:

Theorem 1. [BeMi]
If one-way trapdoor permutations exist, then there exists a signature scheme which is secure against an adaptive chosen message attack □

Here, "secure" means that an enemy will not be able to produce even a single message m and a valid signature for it, if he has not seen a signature for m produced by the real signer. This signature system will be called ''The BM signature scheme'' in the following.

Theorem 2. [ChDaGr]
If clawfree pairs of functions exist, and trapdoor one-way permutations exist, then there exists a protocol allowing parties A and B to carry out any (probabilistic) computation with private input, such that the secrets of A are unconditionally protected, and the secrets of B are protected, if A cannot invert the one way trapdoor permutation used □

This protocol can easily be generalized such that also the output is kept secret to one party. Note also that the unconditional protection of one party is essential to the "unconditional untraceability", that we want from the systems constructed in the following. Therefore other general computation protocols [Ya], [GoMiWi] cannot be used.

One of the main ideas in this protocol is that, using a pair (f_0, f_1) of claw free functions, it is possible for participant A to *commit* to a choice of a bit, without giving away any Shannon information about her choice: having chosen $b \in \{0,1\}$, A chooses uniformly $x \in domain(f_b)$, and computes the *commitment*, $f_b(x)$. If she chooses to do so, A can later *open* the commitment by revealing x, this will convince everybody about her original choice.

Since both functions are t to 1 mappings, even an infinitely powerful receiver will not be able to compute anything about b from the commitment; and by the claw freeness, a polynomially bounded A will not be able to open a commitment in more than one way. Note, however, that any method for establishing such commitments can be used by the protocol, and that the existence of pairs of claw free functions is not a necessary condition for the existence of bit commitment schemes.

In the protocols considered in this paper, we have two kinds of participants: individuals with limited (polynomially bounded) computing power, and organizations, which may have unlimited computing power, but are not required to use it in the protocols. Given organization O and individual A, consider the following interaction:

1) O chooses an instance of the BM signature scheme, and sends the public key to A.
2) A chooses some message m.
3) A and O use the protocol from Theorem 2 to compute O's signature on m. The protocol is set up, such that A is unconditionally protected and enters m as private output, while O enters the secret key to the signature scheme. Also, the signature is private output for A.
4) Steps 2) and 3) are repeated a number of times, polynomial in the security parameter.

Let us remark that the security parameter is simply an integer that measures the work that has to be done in the protocol, and the cryptographic security.

Theorem 3.
After the above interaction, the following hold:

i) O has no information in the Shannon sense about the m's chosen by A.
ii) A is not able to compute O's signature on any message with non negligible probability, unless it has been chosen in step 2) at some point.

Proof (sketch).
i) is clear from Theorem 2.
Assume ii) is false. Then the following procedure will break the BM scheme under an adaptive chosen message attack, contradicting Theorem 1:
We simply run A's algorithm, and each time A has executed step 2), we use the chosen message attack to obtain a valid signature on the m that was chosen. With this information, we can simulate A's interaction with O in step 3) without knowing the secret key. By the minimum-knowledge property of the computation protocol, the messages sent in the simulated interaction have a distribution which is polynomially indistinguishable from those sent in a conversation with the real O. In particular, this means that A's probability of outputting a new, signed message is essentially the same in the simulation as in the actual interaction with O □

3. Payment Systems

In a payment system, we have one special participant called the bank (B). In addition, we have a set of individuals, and a set of organizations.

Each individual can do a special interaction with B called a *withdrawal* (one can think of this as the individual withdrawing money from his account). If B is willing to participate, then after completion of the withdrawal, the individual can compute one element in a set of numbers called EC. A number in EC is called an *electronic coin* (ec). Each individual can submit the ec's he possesses to organizations as payment. The organization will then, possibly by interacting with B, decide whether to accept the payment. The purpose of a payment system is to ensure that:

1) Each ec can be submitted and accepted as payment exactly once.

2) At some point of time, consider the set Ω of successful withdrawals. Let Δ be the set of *ec* 's accepted by organizations. Assuming that 1) holds, Δ must correspond in a natural way to a subset of Ω, i.e. there is an injective map, $f : \Delta \rightarrow \Omega$, such that when $f(\delta) = \omega$, then ω is exactly the withdrawal which enabled that individual to later transmit δ to some organization. We now require, that at each point of time, no matter which strategy the organizations (including the bank) follow, and no matter how much computing power they have, the probability distribution on f they can compute will be the uniform distribution over all injective mappings from Δ to Ω.

Based on Theorem 3., a payment system is easily designed: Assume individual A has an account in bank B. The bank chooses an instance of the BM-signature scheme, and we fix the rule that any number signed with this instance is an *ec*.

When A wishes to conduct a withdrawal, he chooses a random number R and gets the bank's signature on it by doing the computation protocol from Theorem 2 with the bank. Since R is entered as private input from A, B gets no information on the numbers signed. After this, the bank deducts the corresponding amount from A's account. When A wants to spend his money, say in shop S, he gives R and the signature to S. S will send this to B, who will check if R has been submitted before, and whether the signature is valid. The bank then puts money on the account of S and informs S about acceptance of the payment.

It follows easily from Theorem 3 that A will not be able to spend money without receiving it from the bank first, and that the bank will not be able to trace any number it receives, back to a particular individual, i.e. condition 2) above is satisfied, and condition 1) holds relative to our intractability assumption.

In contrast with the credential mechanism to be outlined later, this system needs an on-line participant, namely the bank. This seems to be an inherent property in systems were numbers are worth money, and you want to prevent individuals from using a number more than once.

4. Credential Mechanisms

For this, we need the concept of unconditionally secure *bit commitments*, as explained in Section 2.

What we are looking for is a method allowing organization O to transmit personal information about individual A, say, to some other organization. Typically, this information takes the form of a *credential*, i.e. a message saying that a given individual satisfies some "predicate": he can drive a car, passed an exam, etc. At the same time, we want to prevent organizations from building complete records on the behavior of an individual, i.e. find out which credentials he possesses, who he shows them to, etc. Following the ideas of [ChEv], we will let each individual represent himself by different *pseudonyms* with different organizations. Assume that some unique bit string $ID(A)$ (name, address, etc.) is attached to each individual A. Then a pseudonym in our case will be a set of unconditionally secure bit commitments to the bits in $ID(A)$. A will compute one such set for each organization, he interacts with.

In order for a credential mechanism to be useful, it has to satisfy 2 basic properties:

1) No individual can show a credential to anyone, unless it has been properly issued to him.

2) The credential mechanism reveals no Shannon information about which pseudonyms apply to the same individual.

Property 2) must be stated a little more precisely before it can be formally proved, but it will do for the informal reasoning in this abstract.

A more complete and formal definition of the concept of a credential mechanism can be found in [ChEv].

To set up our construction, we need one special organization, called Z, which will be used to validate once and for all each individual in the system. Z starts by choosing its own instance of the BM scheme and sending the public key to all participants.

The following protocol is executed for each individual A:

a) A sends $ID(A)$ to Z, and Z checks this against A. Z also makes sure that A has not entered the system before.

b) The following steps c)-d) are executed for each organization O, that A wants to interact with later:

c) A chooses a random bitstring R_O, which must contain as many bits as is needed as random input to the computation of A's pseudonym with O.

d) A and Z do a computation protocol, where Z signs a bitstring which is the concatenation of $ID(A)$, $ID(O)$ and R_O. A is unconditionally protected, and enters R_O as private input, while $ID(A)$ and $ID(O)$ are public. Z enters its secret key to the signature scheme as private input. The resulting signature is given to A as private output.

After this, A can compute his pseudonym with O, $PS_O(A)$, based on R_O. When he starts interacting with O, he must first convince O that he knows Z's signature on a string which is the concatenation of $ID(A)$, $ID(O)$ and a string R_O, and also that this string has the property that computing a pseudonym for $ID(A)$ based on R_O leads to the pseudonym $PS_O(A)$ that A wants to use with O. Using the general computation protocol with no private input from O, this can be done while revealing no information to O about $ID(A)$ or R_O.

Lemma 1

The above ensures that each individual is represented by at most 1 pseudonym with each organization, and that different individuals have different pseudonyms with the same organization.

Proof.

Assume the first statement is false, and let A be an individual with 2 pseudonyms representing him with O. Since $PS_O(A)$ is uniquely determined by $(ID(A), R_O)$, this means that A must have Z's signature on at least two strings of the form $(ID(A), ID(O), R_O)$, $(ID(A), ID(O), R'_O)$. But since Z only signs 1 string starting with

$ID(A), ID(O)$, this contradicts Theorem 3. If the second statement is false, this trivially implies that some conspiracy of individuals has been able to find a claw for the pair of functions used in computing commitments. But this contradicts the basic assumption on claw freeness □

Now, for each type of credential, an instance of the BM scheme is chosen, and each organization authorized to issue that type is assumed to have a copy of the secret key. We then fix the rule that a given type of credential applies to A, if he possesses a signature in the corresponding signature scheme on $ID(A)$.

O can now issue a credential to A by doing a computation protocol with him, where O signs $ID(A)$. During this protocol, it is checked by using the commitments in $PS_O(A)$ that A really enters the correct ID-string as private input.

A can show this credential to O' by convincing O' that he knows a signature in the relevant signature scheme on the string committed to in $PS_{O'}(A)$. As before, this can be done while revealing no information about $ID(A)$ or the signature.

Theorem 4.
The credential mechanism outlined above satisfies properties 1) and 2).

Proof.
1): By correctness of the computation protocol, it is clear that A cannot show a credential, unless he really knows the relevant signature on $ID(A)$. By Lemma 1, he cannot pretend being someone else, and in that way fool an organization into computing this signature for him. Thus, if he did not receive the credential from an organization, the only possibility is that he computed the signature himself, which contradicts Theorem 3.
2): follows easily from the fact that all individuals are unconditionally protected in all interactions with organizations □

It might be argued that this system, like any system that identifies people by numbers, does not protect against different *physical* persons sharing the same *digital* identity (see for example [De]). A solution to this would of course have to deal with the problem of checking the physical identity of a person. Numerous solutions using tamper resistant devices, photos, hand-written signatures and the like can be developed. Note that such a solution does not have to violate condition 2) above (the untraceability), because the identity check does not have to be executed by the organizations themselves, but could be done e.g. by an independent tamper resistant device.

Acknowledgement

The author would like to thank David Chaum for many inspiring and helpful discussions on credential mechanisms in general, and on this work in particular.

References

[BeMi] Bellare and Micali: "How to Sign Given any Trapdoor Function", these proceedings.

[Ch] Chaum: "Privacy Protected Payments", Preprint., available from author.

[Ch2] Chaum: Private communication.

[Ch3] Chaum: "Elections with Unconditionally Secret Ballots and Disruption Equivalent to Breaking RSA", to appear in Proc. of EuroCrypt 88.

[ChDaGr] Chaum, Damgård and van de Graaf: "Multiparty Computations Ensuring Privacy of each Party's Input and Correctness of the Result", Proc. of Crypto 87, Springer.

[ChEv] Chaum and Evertse: "A Secure and Privacy Protecting Protocol for Transmitting Personal Information Between Organizations", Proc. of Crypto 86, Springer.

[Da] Damgård: "The Application of Claw Free Functions in Cryptography; Unconditional Protection in Cryptographic Protocols", phd.-thesis, Aarhus University, 1988.

[De] Desmedt: "Special Uses and Abuses of the Fiat-Shamir Passport Protocol", Proc. of Crypto 87, Springer.

[GoMiRi] Goldwasser, Micali and Rivest: "A Paradoxical Solution to the Signature Problem", Proc. of FOCS 84, pp.441-448.

[GoMiWi] Goldreich, Micali and Wigderson: "How to Play Any Mental Game", proc. of FOCS 87.

[Ya] Yao: "How to Generate an Exchange Secrets", proc. of FOCS 86.

A Universal Problem in Secure and Verifiable Distributed Computation

Ming-Deh A. Huang* Shang-Hua Teng†

Department of Computer Science
University of Southern California
Los Angeles, California 90089

Abstract

A notion of **reduction** among multi-party distributed computing problems is introduced and formally defined. Here the reduction from one multi-party distributed computing problem to another means, roughly speaking, a secure and verifiable protocol for the first problem can be constructed solely from a secure and verifiable protocol of the second. A **universal** or **complete** multi-party distributed computing problem is defined to be one to which the whole class of multiparty problems is reducible. One is interested in finding a simple and natural multi-party problem which is universal. The *distributed sum problem*, of summing secret inputs from N parties, is shown to be such a universal problem. The reduction yields an efficient systematic method for the automatic generation of secure and verifiable protocols for all multi-party distributed computing problems. Incorporating the result from [14], it also yields an alternative proof to the completeness theorem of [9] that assuming honest majority and the existence of a trap-door function, for all multi-party problems, there is a secure and verifiable protocol.

---

*Supported in part by National Science Foundation through grant CCR-8701541 & USC Faculty Research Award.

†Supported in part by National Science Foundation through grant CCR-87-13489. Currently visiting Computer Science Department, Carnegie Mellon University, Pittsburg, PA 15213-3890

1 Introduction

We are concerned with the problem of computing correctly and securely in a distributed environment. This problem, raised by Goldreich, Micali, and Wigderson, was called the *multi-party protocol problem* [9]. Informally, the multi-party protocol problem can be stated as: given a description of a game with incomplete information of any number of players, produce a protocol for playing the game that leaks no partial information, provided that the majority of the players is honest. Such protocols are called *secure and verifiable protocol* they simultaneously guarantee correctness of the corresponding games and privacy of all players.

In [9], Goldreich, Micali, and Wigderson presented the first solution to the multi-party protocol problem and derived a completeness theorem for the class of distributed protocol problems with honest majority, namely, if any trap-door function exists, then for all games, there is secure and verifiable protocol provided that more than half of the players are honest. Ben-Or, Goldwasser and Wigderson [2], Chaum, Crepeau and Damgdra [5] independently prove a completeness result for multi-party protocol problem in a non-cryptographic setting.

In this paper, the relationship among the multi-party problems is studied. We formalize the notion of **reduction** among multi-party problems. Roughly speaking, a multi-party problem $\mathcal{P}$ is **reducible** to a set $\mathcal{S}$ of multi-party problems if a secure and verifiable protocol for $\mathcal{P}$ can be constructed solely from the combination of secure and verifiable protocols for problems in $\mathcal{S}$. From the notion of reduction, the concept of **universal set** and **universal multi-party problem** is defined. A set $\mathcal{S}$ of multi-party problems is a *universal set* if all multi-party problems are reducible to $\mathcal{S}$. In other words, secure and verifiable protocols for a universal set can be used as fundamental building block for constructing secure and verifiable protocols for all multi-party problems. A multi-party problem $\mathcal{P}$ is **universal** if itself forms a universal set.

We are interested in finding a simple and natural multi-party problem that is universal for the whole class of multi-party problems. The *distributed sum problem*, of summing secret inputs from N parties, is shown to be such a universal problem. Besides being a universal problem, the distributed sum problem itself is also an important problem. For example, the well-known election problem [6,14,13,4,7,18] is the distributed sum problem when the secret inputs are restricted to 0 and 1.

We prove that, assuming honest majority, designing a secure and verifiable protocol for any N-player multi-party problem is reducible to the design of secure and verifiable problem for distributed sum problem over N players. This reduction demonstrates that the distributed sum problem is universal, and gives an efficient systematic method for the automatic generation of secure and verifiable protocol for all multi-party problems.

Incorporating the result from [14][1], it yields an alternative proof to the completeness theorem of Goldreich, Micali, and Wigderson [9].

2 Preliminary

The computation model used for multi-party problems is a complete synchronous network of N nodes. Each node (node i) has a probabilistic Turing machine ($\mathcal{U}_i$), called a *user*, with its own private read-only input tape, write-only output tape, and work tape. There is a common read-only tape, a common write-only tape, and a global clock shared by all machines.

Various models can be defined according to the different means of communication among the machines [9,2,6].

- **Private Channel Model**: There are $\frac{N(N-1)}{2}$ perfectly secure private communication tapes. The ith machine communicates with the jth machine, and vice verse, via tape $i \leftrightarrow j$. No other machines can read the message on the $i \leftrightarrow j$ tape.

- **Common Tape Model**: There is only one communication tape. Each machine can read the message from the tape and write message on the tape.

- **Bulletin Board Model**: There are N publicly readable tape, $\mathcal{BB}_1, \ldots, \mathcal{BB}_N$, called *Bulletin Boards*, where $\mathcal{BB}_i$ is writable only by the ith machine.

Throughout this paper, the bulletin board model is assumed. Note that using digital signatures [16] to authenticate the sender, protocols designed on the bulletin board model can be implemented on the common tape model. Also, using Byzantine agreement [15], all machines can agree on what message machine i has sent to machine j at certain time. Hence, protocols designed on the bulletin board model can be implemented on the private channel model.

A distributed protocol $\mathcal{DP}$ consists of a set of probabilistic algorithms $\{\mathcal{A}_i : 1 \leq i \leq N\}$ to be run on a distributed system of N parties $\mathcal{U}_1, \ldots, \mathcal{U}_N$. The algorithm $\mathcal{A}_i$ runs on $\mathcal{U}_i$. The initial content of the shared input tape is the *common input*, and the initial content of the private input tape of $\mathcal{U}_i$ is the *secret input* to $\mathcal{U}_i$. The common input typically consists of the agreed upon *verifiability and security parameters* denoted by V_N and K_N respectively. The final content on the shared output tape is the *public output* of $\mathcal{D}$, and the secret outputs of $\mathcal{U}_i$ appear on the private output tape of $\mathcal{U}_i$.

[1] It was proven in [14] that there is an optimally secure and verifiable protocol for the distributed sum problem.

Let $\mathcal{DP} = \{\mathcal{A}_i : 1 \leq i \leq N\}$ be a distributed protocol of N parties $\mathcal{U}_1, \ldots, \mathcal{U}_N$. A party $\mathcal{U}_i$ is *honest* if it runs its preassigned algorithm $\mathcal{A}_i$ faithfully and only runs $\mathcal{A}_i$, and is called *dishonest* otherwise. We allow the possibility of sharing information among the dishonest parties. A dishonest party can be either *passive* or *malicious* in the sense of [9]. We also allow each party to become dishonest in a dynamic fashion during the execution of the protocol.

A *conspiracy* $\mathcal{C}$ among s dishonest parties is a set of probabilistic polynomial time algorithms $\{\mathcal{C}_i : 1 \leq i \leq N\}$ and a dishonest parties $\mathcal{U}_a$, where $\mathcal{C}_i = \mathcal{A}_i$ if $\mathcal{U}_i$ is honest. The common input of $\mathcal{C}$ and the secret input to the honest $\mathcal{U}_i$ are the same as those in $\mathcal{DP}$. The output of $\mathcal{C}$ is defined to be the private output of $\mathcal{U}_a$, and is either one or zero.

For the ease of understanding, we restrict our consideration to a special subclass of multi-party problems, *distributed transformation problem.* The result achieved for this subclass can be generalized to the general multi-party problems [9,2,5].

The N-party *distributed transformation problem* is stated as: given a $2N$-ary formula[2] $\mathcal{CF}(x_1, \ldots, x_N, y_1, \ldots, y_N)$, design a protocol $\mathcal{P}$ such that on each *tuple of secret inputs* $(s_1, \ldots, s_N)$, the application of the protocol outputs a tuple of secret outputs $(z_1, \ldots, z_N)$, such that:

- **Verifiable Correctness:** $\mathcal{CF}(s_1, \ldots, s_N, z_1, \ldots, z_N) = 1$.
- **Privacy**: No subset of less than $\lceil N/2 \rceil$ parties can extract any more information about s_i's and z_i's from execution of $\mathcal{P}$ than it is already contained in the formula $\mathcal{CF}(s_1, \ldots, s_N, z_1, \ldots, z_N) = 1$ and their shared secret inputs.

where a *tuple of secret inputs* $(s_1, \ldots, s_N)$ means that s_i is the secret input of $\mathcal{U}_i$, and a *tuple of secret outputs* $(z_1, \ldots, z_N)$ means that z_i is the secret output of $\mathcal{U}_i$.

The distributed transformation problem can be interpreted as: at the beginning of the execution, the ith party owns a private database $\mathcal{DB}_i$, the application of the protocol transforms the ith database securely into a new database $\mathcal{DB}_i'$ which satisfies the predefined properties without revealing any more information about $\mathcal{DB}_i$'s and $\mathcal{DB}_i'$'s.

The Turing machine game, defined by Goldreich, Micali, and Wigderson [9], is a subclass of the distributed transformation problem defined above. Informally, the Turing machine game can be described as: N parties, respectively owning secret inputs $s_1, \ldots, s_N$, are to *correctly* run a given Turing machine $\mathcal{M}$ on $s_1, \ldots, s_N$ while keeping the maximum possible privacy of all parties. Clearly, the Turing machine game with Turing machine $\mathcal{M}$

[2]It is usually assumed that the formula $\mathcal{CF}$ of a distributed transformation problem is random polynomial time computable in the sense that we can construct a *random* algorithm $\mathcal{A}_{\mathcal{CF}}$ which on each tuple of inputs $(s_1, \ldots, s_N)$ outputs a tuple of output, in random polynomial time, a tuple $(z_1, \ldots, z_N)$ such that $\mathcal{CF}(s_1, \ldots, s_N, z_1, \ldots, z_N) = 1$.

is a distributed transformation problem with formula $\mathcal{CF}_{\mathcal{M}}$:

$$\mathcal{CF}_{\mathcal{M}}(x_1,\ldots,x_N,y_1,\ldots,y_N) = 1 \text{ } if \text{ } y_1 = y_2 = \ldots = y_N = \mathcal{M}(x_1,\ldots,x_N)$$

A distributed protocol $\mathcal{DP}$ is a **s-secure** protocol for a distributed transformation problem with formula $\mathcal{CF}$, if the following condition is satisfied.

For all conspiracy $\mathcal{C}$ among a set of s dishonest users, for all pairs of $2N$-ary tuples $(x_1,\ldots,x_N,u_1,\ldots,u_N)$ and $(y_1,\ldots,y_N,v_1,\ldots,v_N)$ with

$$\mathcal{CF}(x_1,\ldots,x_N,u_1,\ldots,u_N) = \mathcal{CF}(y_1,\ldots,y_N,v_1,\ldots,v_N) = 1$$

and $x_i = y_i$ if $\mathcal{U}_i$ is dishonest, for all $k \in \mathcal{N}$,

$$prob\{\mathcal{C}(x_1,\ldots,x_N,u_1,\ldots,u_N) = 1\} - prob\{\mathcal{C}(y_1,\ldots,y_N,v_1,\ldots,v_N) = 1\} \leq \frac{1}{(N+z+K_N)^k}$$

where z is the input size which equals to the maximum binary-length of x_i and y_i.

Informally, the above condition says that $(x_1,\ldots,x_N)$ and $(y_1,\ldots,y_N)$ are *polynomial time indistinguishable* to the dishonest users.

A distributed protocol $\mathcal{DP}$ is **s-verifiable** for a distributed transformation problem with formula $\mathcal{CF}$ if for all inputs $S = (s_1,\ldots,s_N)$, the probability that $\mathcal{CF}(S,Z) = 1$, where $Z = (z_1,\ldots,z_N)$ is the output of $\mathcal{DP}$ on $s_1,\ldots,s_N$, is at least $1 - \frac{1}{(N+z+V_N)^k}$ for all $k \in \mathcal{N}$, provided no more than s users are dishonest.

A distributed protocol $\mathcal{DP}$ is an **optimally secure and verifiable** protocol for a distributed transformation problem iff it is s-secure and s-verifiable for all $1 \leq s \leq N$.

3 Complete Sets and Universal Problems

Throughout the development of computational complexity theory, an important notion has been the *reduction* among a class $\mathcal{CP}$ of problems. Informally, reduction from one problem to another shows that the first problem is essentially no harder than the second. The notion of reduction introduces a partial order among problems in $\mathcal{CP}$. A problem $\mathcal{P}$ is **complete** or **universal** for the whole class of problems if all problems in $\mathcal{CP}$ are reducible to $\mathcal{P}$. More generally, a subset $\mathcal{S} \subseteq \mathcal{CP}$ is a *complete set* if all problems in $\mathcal{CP}$ are reducible to $\mathcal{S}$.

The completeness of a problem $\mathcal{P}$ is often used as a strong evidence that $\mathcal{P}$ is intractable up to certain computation power. For example, if a problem $\mathcal{P}$ is complete for the class of recursive functions (NP, P) under recursive reduction (polynomial-time reduction, NC-reduction, respectively), then $\mathcal{P}$ is undecidable (unlikely in P, unlike in NC, respectively). However, in the case where $\mathcal{P}$ admits an efficient solution, a constructive proof of completeness provides a systematic method for solving all problems in $\mathcal{CP}$. In this case, we also call $\mathcal{P}$ a universal problem for the class $\mathcal{CP}$.

Informally, the reduction from a multi-party problem $\mathcal{P}$ to another multi-party problem $\mathcal{P}'$ means that $\mathcal{P}$ can be solved by alternating applications of local computation by individual parties, and a secure and verifiable protocol for $\mathcal{P}'$. More specifically, each distributed protocol can be decomposed into a sequence of local transformation where each user computes locally and securely; and distributed transformation where all users work together to transform a tuple of secret inputs to a tuple of secret outputs satisfying some predefined conditions. Let $Program(S)$ be the set of all distributed programs consisting of alternating local transformation and distributed transformation protocols from S. Informally, a set of protocols is *complete* iff for all multi-party problems P, there is a distributed program from $Program(S)$ that is secure and verifiable for P. A multi-party problem Q is *universal* iff each secure and verifiable protocol for Q by itself forms a complete set.

3.1 Local Transformation vs Distributed Transformation

The class of distributed transformation problem can be partitioned into two subclasses according to the input–output dependency. Let us first see some examples:

Problem 3.1 *There are N users. User i has a secret value s_i. User i wants to compute the largest perfect square which is smaller than s_i. In other words, problem 3.1 is a distributed transformation problem with formula CF:*

$$\mathcal{CF}(x_1,\ldots,x_N,y_1,\ldots,y_N) = 1 \textit{iff } y_i = \max\{z^2 \mid z^2 \leq x_i\}$$

Problem 3.2 *There are N users. User i has a secret value s_i. User i wants to compute $\sum_{j=1}^{N} s_j^i$.*

In problem 3.1, each user can locally compute its secret output from its secret input; while in problem 3.2, the secret output of each user depends on the secret inputs of all other users. Hence, each user, by itself, can not obtain the correct secret output. In order to perform the computation, each user has to communicate with other users.

In general, a distributed transformation problem $\mathcal{P}$ with formula $\mathcal{CF}$ is *locally computable* if there are N functions $f_1,\ldots,f_N \in \mathcal{RPU}$, such that for all

$$\mathcal{CF}(s_1,\ldots,s_N,z_1,\ldots,z_N) = 1 \Longleftrightarrow z_i = f_i(s_i),$$

where $\mathcal{RPU}$ stands for the class of probabilistic polynomial time computable unary functions.

The computation of a locally computable distributed transformation problem is called *local transformation*, and the computation of a distributed transformation problem which involves inter-user communication is called a *distributed transformation.*

Using the probabilistic public-key cryptosystem of Goldwasser and Micali [11] and two party zero knowledge proof protocols from [12,10,3], or using the verifiable secret sharing (VSS) [2,5,17], each party can prove to all other parties the correctness of its local computation without leaking any information about its secrets. Such a scheme can be found in [1,2,5,14]. Thus, it is assumed that the local transformation of each party in all distributed protocols is performed securely with verifiable correctness.

3.2 Reducibility

Two operators are defined on the set of distributed protocols to formalize the concept of reduction from one multi-party problem to another.

Definition 3.1 (Composition) *Let $\mathcal{DP}_1$ and $\mathcal{DP}_2$ be two N-party distributed protocols, and $F = (f_1, \ldots, f_N) \in \mathcal{RPU}^N$. The F-composition of $\mathcal{DP}_1$ and $\mathcal{DP}_2$ forms a new N-party distributed protocol, denoted by $\mathcal{DP}_2 \odot_F \mathcal{DP}_1$, which is composed of the following three steps: (1) apply $\mathcal{DP}_1$ on a tuple of secret input $(s_1, \ldots, s_N)$ to compute a tuple of secret outputs $(u_1, \ldots, u_N)$; (2) each party $\mathcal{U}_i$ performs a local transformation to compute $f_i(u_i)$; (3) apply $\mathcal{DP}_2$ on $(f_1(u_1), \ldots, f_N(u_N))$ to compute the final tuple of secret outputs $(z_1, \ldots, z_N)$.*

Definition 3.2 (Combination) *Let $\mathcal{DP}_1, \ldots, \mathcal{DP}_k$ be k N-party distributed protocols, the combination of these k protocols defines a new N-party distributed protocol, denoted by $\biguplus_{i=1}^{k} \mathcal{DP}_i$, which is specified as: on a tuple of secret inputs $((s_{1,1}, , , s_{k,1}), \ldots, (s_{1,N}, \ldots, s_{k,N}))$, for $i = 1$ to k, apply $\mathcal{DP}_i$ on S_i to compute a tuple of secret outputs $(z_{i,1} \ldots, z_{i,N})$. Then the final tuple of secret output is $((z_{1,1}, , , z_{k,1}), \ldots, (z_{1,N}, \ldots, z_{k,N}))$.*

Let $\mathcal{IDP}_N$ denote the identity distributed protocol whose application on any tuple of secret inputs $(s_1, \ldots, s_N)$ outputs the tuple of secret outputs $(s_1, \ldots, s_N)$.

Definition 3.3 (Protocol Circuit) *An N-party protocol circuit C is a labeled directed acyclic simple graph with a unique sink r_C in which each vertex v is labeled by an ordered pair $(F_v, \mathcal{DP}_v)$, where $F_v \in \mathcal{RPU}^N$ and $\mathcal{DP}_v$ is a N-party distributed protocol. The* value *of each vertex v in a protocol circuit is a N-party distributed protocol which is defined inductively:*

- *If v is a leaf vertex with label $(F_v, \mathcal{DP}_v)$, then $value(v) = \mathcal{DP}_v \odot_{F_v} \mathcal{IDP}_N$.*

- *If v is an internal vertex, labeled by $(F_v, \mathcal{DP}_v)$ and with children $w_1, \ldots, w_k$, then:*

$$value(v) = \mathcal{DP}_v \odot_{F_v} (\biguplus_{i=1}^{k} value(w_i))$$

The distributed protocol defined a protocol circuit C, denoted by protocol(C), is value(r_C).

We can evaluate a protocol circuit C on a tuple of secret inputs $(s_1, \ldots, s_N)$ according to the definition of composition and combination. Note that the evaluation is composed of an alternating applications of local transformation within each party and some distributed protocols associated with the vertices in C. This yields a general paradigm for solving multi-party problems.

Definition 3.4 (Reduction) *A set $\mathcal{S}$ of multi-party problems is* **reducible** *to another set $\mathcal{T}$ of multi-party problems iff for all $\mathcal{P} \in \mathcal{S}$, there exist protocol circuit C, with protocol labels only from the set of protocols which are secure and verifiable for problems in $\mathcal{S}$, that defines a secure and verifiable protocol for $\mathcal{P}$.*

Let $F = (f_1, \ldots, f_N)$ be an N-tuple of random polynomial computable unary functions. Let $\mathcal{DP}_1, \ldots, \mathcal{DP}_k$ be k $2N$-ary formulas. The *F-composition* of $\mathcal{CF}_1$ and $\mathcal{CF}_2$, denoted by $\mathcal{CF}_2 \odot_F \mathcal{CF}_1$, is defined as: for all N-tuples, $X = (x_1, \ldots, x_N)$, $Y = (y_1, \ldots, y_N)$,

$$\mathcal{CF}_2 \odot_F \mathcal{CF}_1(X, Y) = 1, \ iff \ \exists U = (u_1, \ldots, u_N), \mathcal{CF}_1(X, F(U)) = \mathcal{CF}_2(F(U), Y) = 1 \quad (1)$$

Where $F(U) = (f_1(u_1), ..., f_N(u_N))$.

The *combination* of $\mathcal{CF}_1, \ldots, \mathcal{CF}_k$, denoted by $\uplus_{i=1}^{k} \mathcal{CF}_i$, is defined as $1 \leq i \leq k$, $1 \leq j \leq N$, for all $S_i = (s_{i,1}, ..., s_{i,N})$, $Z_i = (z_{i,1}, ..., z_{i,N})$, $U_j = (s_{1,j}, , , s_{k,j})$, $V_j = (z_{1,j}, , , z_{k,j})$,

$$\uplus_{i=1}^{k} \mathcal{CF}_i\{(U_1, V_1), ..., (U_N, V_N)\} = 1 \ iff \ \prod_{i=1}^{k} \mathcal{CF}_i(S_i, Z_i) = 1 \quad (2)$$

If for all vertex v is a protocol circuit, $\mathcal{DP}_v$ is a distributed protocol for a distributed transformation problem with formula $\mathcal{CF}_v$, then C defines a formula, denoted by $formula(C)$, by Relation (1), and Relation (2) in a natural way. We can prove the following lemma.

Lemma 3.1 *If for all v, $\mathcal{DP}_v$ is a s-verifiable distributed protocol for the distributed transformation problem with formula $\mathcal{CF}_v$, then protocol(C) is a s-verifiable distributed protocol for the distributed transformation problem with formula $formula(C)$.*

4 The Distributed Sum Problem

Formally, the distributed sum problem is a distributed transformation problem with formula

$$\mathcal{CF}(x_1,\ldots,x_N,y_1,\ldots,y_N)=1\ iff\ y_1=y_2=\ldots=y_N=\sum_{i=1}^{N}x_i.$$

where the problem domain is a subset of $\mathcal{Z}$, the set of integers. It will be shown in next section that the distributed sum problem is universal over all multi-party problems.

An optimally secure and verifiable protocol is presented in [14] based on the efficient construction of *perfectly secure patterns*.

Lemma 4.1 ([14]) *There is an optimally secure and verifiable protocol for the distributed sum problem.*

Other secure and verifiable protocols for the distributed sum problem are also implied in [2,5,9,8].

An important variance of the distributed sum problem, denoted by $\mathcal{DSP}_i^N$, is the one that after the computation, only the ith party correctly computes the sum, and all other parties can extract no information about the sum. In other words, $\mathcal{DSP}_i^N$ is a distributed transformation problem with formula:

$$\mathcal{CF}(x_1,\ldots,x_N,y_1,\ldots,y_N)=1\ iff\ y_i=\sum_{j=1}^{N}x_i$$

Lemma 4.2 *the distributed sum problem and $\{\mathcal{DSP}_i^N \mid 1\le i\le N\}$ are reducible between each other.*

[PROOF] It can be easily shown that the distributed sum problem is reducible to $\{\mathcal{DSP}_i^N \mid 1\le i\le N\}$. We now show that $\{\mathcal{DSP}_i^N \mid 1\le i\le N\}$ is reducible to the distributed sum problem $\mathcal{DSP}^N$. Let $(s_1,\ldots,s_N)$ be a tuple of secret inputs, let $s=\sum_{i=1}^{N}s_i$, the application of $\mathcal{DSP}_i^N$ on $(s_1,\ldots,s_N)$ can be done by:

1. The ith party $\mathcal{U}_i$ randomly chooses $w_1,w_2\in\mathcal{Z}$, such that $w_1+w_2=s_i$.

2. Apply a protocol for the distributed sum problem on $(s_1,..,s_{i-1},w_1,..,s_N)$ to produce $(y,...,y)$, where $y=w_1-s_i+\sum_{j=1}^{N}s_j=s-w_2$.

3. $\mathcal{U}_i$ locally compute w_2+y to get s.

Note that y contains no information about s, therefore the above protocol is s-secure, if the protocol for the distributed sum problem is s-secure. □

5 Universality of the Distributed Sum Problem

A natural universal problem for multi-party distributed computation is sought. And the simpler the universal problem, the better. In this section, the very simple the distributed sum problem is proven to be a universal multi-party problem. Moreover, the proof is constructive.

Theorem 5.1 (Main Theorem) *The distributed sum problem is a universal multi-party problem.*

Corollary 5.1 *For all multi-party problem, there is a secure and verifiable protocol assuming honest majority.*

5.1 Distributed Boolean Circuit Problem

The proof of Theorem 5.1 consists of a sequence of reductions. The first step is to reduce the general distributed transformation problem to a special distributed transformation problem, the *distributed boolean circuit problem.*

The distributed boolean circuit problem is proposed by the following observation.

For each formula $\mathcal{CF}(x_1,\ldots,x_N,Y_1,\ldots,Y_N)$, we can construct a *probabilistic* algorithm $\mathcal{A}_{\mathcal{CF}}$ which on each tuple of inputs $(s_1,\ldots,s_N)$ outputs a tuple $(z_1,\ldots,z_N)$ such that $\mathcal{CF}(s_1,\ldots,s_N,z_1,\ldots,z_N)=1$. In turn, we can construct a Boolean circuit[3], $\mathcal{C}_{\mathcal{CF}}$ to implement $\mathcal{A}_{\mathcal{CF}}$ such that the size of $\mathcal{C}_{CF}$ is polynomially bounded by the time complexity of $\mathcal{A}_{\mathcal{CF}}$. In the context of secure distributed computation, $\mathcal{N}$-parties, each holding some secret inputs to $\mathcal{C}_{\mathcal{CF}}$, want to evaluate $\mathcal{C}_{\mathcal{CF}}$ to correctly compute their corresponding secret output, i.e. $\mathcal{U}_i$ holding secret input s_i is to securely and correctly compute the value of z_i. In circuit $\mathcal{C}_{CF}$, s_i corresponds to a subset of input Boolean variables, and z_i to a subset of output Boolean variables. The distributed boolean circuit problem is defined formally as:

Definition 5.1 (Distributed Boolean circuit problem) *Given a Boolean circuit $\mathcal{C}$ of m input variables $x_1,\ldots,x_m$, and n output variables $b_1,\ldots,b_n$. Each party $\mathcal{U}_i$ owns a nonempty subset of input variables X_i such that $\cup_{i=1}^N X_i=\{x_1,\ldots,x_m\}$ and $X_i\cap X_j=\phi$, for all $1\le i\ne j\le N$. The distributed Boolean circuit problem with circuit $\mathcal{C}$ is a distributed transformation problem with formula $\mathcal{CF}_\mathcal{C}$:*

$$\mathcal{CF}_\mathcal{C}(X_1,\ldots,X_N,Y_1,\ldots,Y_N)=\bigwedge_{j=1}^{n}\{\mathcal{F}_j(X_1,\ldots,X_N)=(\sum_{i=1}^{N}y_{i,j}) \bmod 2\}$$

[3]A Boolean circuit is a labeled directed acyclic graph in which the leaves are labeled by distinct Boolean variables, and the internal nodes are labeled from the set of Boolean operators. Each node v in the Boolean circuit is associated with a Boolean formula which is defined in a natural way.

Where $\mathcal{F}_i$ is the boolean formula defined by the Boolean circuit C on output variable z_i, $Y_i = \{y_{i,1}, ..., y_{i,n}\}$.

Lemma 5.1 *The distributed transformation problem is reducible to the distributed Boolean circuit problem and the distributed sum problem.*

[**PROOF**] Given a distributed transformation problem with formula $\mathcal{CF}$, first we construct a Boolean circuit $C_{\mathcal{CF}}$ of output Boolean variables $b_1, \ldots, b_n$, then we apply the secure and verifiable protocol for the distributed Boolean circuit problem on $C_{\mathcal{CF}}$. Suppose b_j is a bit in the secret output z_i of $\mathcal{U}_i$, we apply the secure and verifiable protocol for $\mathcal{DSP}_i$ on $(y_{1,j}, ..., y_{N,j})$ to transform b_j securely and correctly to $\mathcal{U}_i$. The verifiability and the security of the above reduction can be easily verified. □

5.2 Two Primitives for the Distributed Boolean Circuit Problem

We will construct a distributed protocol which evaluates a Boolean circuit sequentially gate after gate in such a way that after evaluating one gate b, each party $\mathcal{U}_i$ obtains a fraction of information y_i about the value of b defined on the secret inputs $S_1, \ldots, S_N$, and $\sum_{i=1}^{N} y_i \bmod 2 = value(b)$. Moreover, no proper subset less than $N/2$ parties can extract any information about $value(b)$ and S_i's more than those contained in their secret inputs. Note that $\wedge, \oplus_2$ are **complete** in zero-one Boolean Algebra in the sense that all boolean operators can be respented by those two operators. Therefore, for all Boolean circuits C of size n, there is an equivalent Boolean circuit C' built up by $\oplus_2$ and $\wedge$ only, whose size is polynomial in n, computes the same function as C does. This reduces the distributed Boolean circuit problem to the following set of problems.

- **Distributed $\oplus_2$-problem:** a distributed transformation problem with formula:

$$\mathcal{CF}((x_1, y_1), ..., (x_N, y_N), z_1, \ldots, z_N) = 1 \; iff \sum_{i=1}^{N} z_i \bmod 2 = (\sum_{i=1}^{N} x_i \bmod 2) \oplus_2 (\sum_{i=1}^{N} y_i \bmod 2)$$

- **Distributed $\wedge$-problem:** a distributed transformation problem with formula:

$$\mathcal{CF}((x_1, y_1), ..., (x_N, y_N), z_1, \ldots, z_N) = 1 \; iff \sum_{i=1}^{N} z_i \bmod 2 = (\sum_{i=1}^{N} x_i \bmod 2) \wedge (\sum_{i=1}^{N} y_i \bmod 2)$$

We can prove the following lemma:

Lemma 5.2 *The distributed Boolean circuit problem is reducible to the distributed $\oplus_2$-problem, distributed $\wedge$-problem, and the distributed sum problem.*

Motivated by finding reduction from the distributed $\wedge$-problem and the distributed $\oplus_2$-problem to the distributed sum problem, we introduce the following set of equivalent distributed transformation problems.

Let $\mathcal{Z}[x]$ denote the set of polynomials in x whose coefficients are in $\mathcal{Z}$. Let $\mathcal{Z}[x]^N$ stand for the set of integral polynomials of degree N. We define a function $\mathcal{PARITY}$: $\mathcal{Z}[x] \rightarrow \{0,1\}$ as: for all $a[x] = \Sigma_{i=1}^{N} a_i x^i \in \mathcal{Z}[x]$,

$$\mathcal{PARITY}(a[x]) = \sum_{i=1}^{N} a_i \bmod 2$$

• **$\oplus_2$-simulation-Problem** *is a distributed transformation problem with formula:*

$$\mathcal{CF}(F,H) = 1 \; iff \; \mathcal{PARITY}\left(\sum_{i=1}^{N} h_i[x]\right) = \mathcal{PARITY}\left(\sum_{i=1}^{N} f_i[x]\right) \oplus_2 \mathcal{PARITY}\left(\sum_{i=1}^{N} g_i[x]\right) \quad (3)$$

• **$\wedge$-simulation-problem** *is a distributed transformation problem with formula:*

$$\mathcal{CF}(F,H) = 1 \; iff \; \mathcal{PARITY}\left(\sum_{i=1}^{N} h_i[x]\right) = \mathcal{PARITY}\left(\sum_{i=1}^{N} f_i[x]\right) \wedge \mathcal{PARITY}\left(\sum_{i=1}^{N} g_i[x]\right) \quad (4)$$

Where $F = ((f_1[x], g_1[x]), ..., (f_N[x], f_N[x]))$, and $f_i[x], g_i[x], h_i[x] \in \mathcal{Z}[x]^N$.

We observe that $\mathcal{PARITY}$ defines a homomorphism from $\mathcal{Z}[x]$ to $\mathcal{F}_2 = \{0, 1, \oplus_2, \wedge\}$. Therefore, the solution to the above two problems can be applied for N parties to perform secure $\wedge$ and $\oplus_2$ operations. This reduces the distributed $\oplus_2$-problem and the distributed $\wedge$-problem to the $\oplus_2$-simulation problem and the $\wedge$-simulation problem.

Observer that, for all $i : 1 \leq i \leq N$, letting $h_i[x] = f_i[x] + g_i[x]$,

$$\mathcal{PARITY}\left(\sum_{i=1}^{N} h_i[x]\right) = \mathcal{PARITY}\left(\sum_{i=1}^{N} f_i[x]\right) \oplus_2 \mathcal{PARITY}\left(\sum_{i=1}^{N} g_i[x]\right)$$

Hence, the $\oplus_2$-simulation problem can be solved solely by local computation. Consequently, we have:

Lemma 5.3 *The distributed Boolean circuit problem, hence the distributed transformation problem, is reducible to the $\wedge$-simulation problem, and the distributed sum problem.*

5.3 Reducing the $\wedge$-Simulation Problem to the Distributed Sum Problem

In this section, we complete the proof of the main theorem by showing that the $\wedge$-simulation problem is reducible to the distributed sum problem.

Let $f[x] = \sum_{j=1}^{N} f_j[x]$, $g[x] = \sum_{j=1}^{N} g_j[x]$, and $h[x] = f[x]g[x]$. Then $h[x]$ is a polynomial of degree no more than $2N$, and

$$\mathcal{PARITY}(h[x]) = \mathcal{PARITY}(f[x]) \wedge \mathcal{PARITY}(g[x])$$

Lemma 5.4 *Given a secure and verifiable protocol for the distributed sum problem, there is a secure and verifiable protocol to transform the tuple of secret inputs*

$$((f_1, g_1), \cdots, (f_i, g_i), \cdots, (f_N, g_N))$$

to the tuple of secret outputs

$$((h[2], h[2N+2], h[4N+2]), \cdots, (h[2i], h[2N+2i], h[4N+2]), \cdots, (h[2N], h[4N], h[4N+2])).$$

[PROOF] Protocol 1:

1. for all i, locally, $\mathcal{U}_i$ computes $f_i[2j]$ and $g_i[2j]$, $1 \leq j \leq 2N+1$
2. for $i = 1$ to N, apply a protocol $\mathcal{DSP}_i^N$ on tuples of secret inputs $(f_1[2i], \cdots, f_N[2i])$ and $(f_1[2N+2i], \cdots, f_N[2N+2i])$, $(g_1[2i], \cdots, g_N[2i])$ and $(g_1[2N+2i], \cdots, g_N[2N+2i])$ to transfer $f[2i] = \sum_{j=1}^N f_j[2i]$, $f[2N+2i] = \sum_{j=1}^N f_j[2N+2i]$, $g[2i] = \sum_{j=1}^N g_j[2i]$, $g[2N+2i] = \sum_{j=1}^N g_j[2N+2i]$ securely and correctly to $\mathcal{U}_i$.
3. apply a protocol for the distributed sum problem on the tuple of secret inputs $(f_1[4N+2], \cdots, f_N[4N+2])$ and $(g_1[4N+2], \cdots, g_N[4N+2])$ to transfer $f[4N+2] = \sum_{j=1}^N f_j[4N+2]$, and $g[4N+2] = \sum_{j=1}^N g_j[4N+2]$ securely and correctly to all parties.
4. Each party $\mathcal{U}_i$ computes $h[2i]$, $h[2N+2i]$ and $h[4N+2]$ locally. □

By interpolation law, we have:

$$h[x] = \sum_{k=1}^{2N+1} \frac{\prod_{j\neq k}(x-2j)}{\prod_{j\neq k}(2k-2j)} h[2k] = \frac{1}{A}\sum_{k=1}^{2N+1} H_k[x]$$

Where $A = \frac{1}{\prod_{1\leq i<j\leq 2N+1}(2i-2j)}$, and for all $k : 1 \leq k \leq 2N+1$,

$$H_k[x] = \prod_{j\neq k}(x-2j) \prod_{1\leq a<b\leq 2N+1, a\neq k, b\neq k} (2a-2b)h[2k]$$

Let $H_k[x] = \sum_{j=0}^{2N} H_{k,j}x^j$, let $c_j = \sum_{k=1}^{2N+1} H_{k,j}$. Then $d_j = c_j/A \in \mathcal{Z}$.

The following procedure forms a reduction for the $\wedge$-simulation-problem to $\{\mathcal{DSP}_i \mid 1 \leq i \leq N\}$ and the distributed sum problem.

Reduction Protocol:

1. apply Protocol 1.

2. By interpolation law, each party $\mathcal{U}_i$ computes polynomial $H_i[x]$, $H_{N+i}[x]$ and $H_{2N+1}[x]$ locally. Then for all $j : 0 \leq j \leq 2N$, $\mathcal{U}_1$ computes $c_{i,j} = H_{1,j} + H_{N+1,j} + H_{2N+1,j}$, and all other parties $\mathcal{U}_i$ computes $c_{i,j} = H_{i,j} + H_{N+i,j}$ locally.

3. for $i = 1$ to N, apply a protocol for $\mathcal{DSP}_i^N$ on tuples of secret inputs $(c_{1,i}, ..., c_{N,i})$ and $(c_{1,N+i}, ..., c_{N,N+i})$ to transfer $c_i = \sum_{k=1}^{2N+1} H_{k,i}$, and $c_{N+i} = \sum_{k=1}^{2N+1} H_{k,N+i}$, securely and correctly to $\mathcal{U}_i$.

4. apply a protocol for the distributed sum problem on the tuple of secret inputs $(c_{1,0}, ..., c_{1,0})$ to transfer $c_0 = \sum_{k=1}^{2N+1} H_{k,0}$, securely and correctly all parties.

5. *Local Transformation*: each party $\mathcal{U}_i$ computes $d_i = \frac{c_i}{A}$, $d_{N+i} = \frac{c_{N+i}}{A}$ and $d_0 = \frac{c_0}{A}$ locally. Then for all $j : 1 \leq j \leq N$, $\mathcal{U}_1$ randomly generate a polynomial $h_1[x] \in \mathcal{Z}[x]^N$ locally such that $\mathcal{PARITY}(h_1) = (d_1 + d_{N+1} + d_0) \bmod 2$, and all other parties $\mathcal{U}_i$ randomly generate a polynomial $h_i[x] \in \mathcal{Z}[x]^N$ locally such that $\mathcal{PARITY}(h_i) = (d_i + d_{N+i}) \bmod 2$.

By interpolation law, we have:

$$\mathcal{PARITY}\left(\sum_{j=1}^{N} h_j[x]\right) = \mathcal{PARITY}(h[x]) = \mathcal{PARITY}(f[x]) \wedge \mathcal{PARITY}(g[x])$$

The verifiability of the above reduction can be easily verified. We now show that the above reduction is secure under the assumption of honest majority.

We first make an observation based on the classical information theory.

Lemma 5.5 (Composition Lemma) *If $a_1, \ldots, a_k$ be k ($K > 1$) random numbers in $\mathcal{Z}$, then for all $X \subset \{a_1, \ldots, a_k\}$, X implies no information about $\Sigma_{i=1}^k a_i$.*

It follows from the Composition Lemma (Lemma 5.5), any proper subset of $\{f_1, \ldots, f_N\}$ or $\{g_1, \ldots, g_N\}$ contains no information about $\mathcal{PARITY}(f[x])$ and $\mathcal{PARITY}(g[x])$. So, after running the protocol, the only additional information obtained by the ith party is the values of $f[2i], f[2N+2i], f[4N+2]$, $g[2i], g[2N+2i], g[4N+2]$, hence $h[2i], h[2N+2i], h[4N+2]$. It follows that for any subset $S \subset \{1, ..., N\}$, the information on the parity of f, g and h that can be obtained by S is all that is implied by the interpolated values held by the party in S. Treating the coefficients of f (g, h) as variables, each value $f[2j]$ ($g[2j]$, $h[2j]$ respectively) determines a linear equation in the coefficients of f (g, h respectively). So the values held by the users in S determines a linear system L. Consequently, the security of our scheme relies on the parity of the solutions in the solution space of L.

The following Lemma can be proved via linear algebraic analysis (the proof will appear in the full paper).

Lemma 5.6 *For all* $p[x] = \sum_{i=0}^{N} p_i x^i \in \mathcal{Z}[x]$, *for all* $X \subset \{2,4,6...,4N+2\}$, *if* $|X| \leq N$, *then:*

$$\frac{|SOLU_0|}{|SOLU_1|} = 1$$

Where for $j \in \{0,1\}$

$$SOLU_j = \{(c_0, c_1, \ldots, c_N) \mid \sum_{i=0}^{N} c_i = j mod2 \ \& \ \forall x \in X, p[x] = \sum_{i=0}^{N} c_i x^i\}$$

Therefore, it follows from the above lemmas that any subset of s dishonest users can extract no more information about $\mathcal{PARITY}(f_i)[x]$, $\mathcal{PARITY}(g_i)[x]$, $\mathcal{PARITY}(h_i[x])$, $\mathcal{PARITY}(f[x])$, $\mathcal{PARITY}(g[x])$, and $\mathcal{PARITY}(h[x])$ provided all applications of the protocols for the distributed sum problem and its variance are s-secure, and $s \leq \lceil \frac{N}{2} \rceil$.

Lemma 5.7 *The* $\wedge$*-simulation problem is reducible to the distributed sum problem.*

Remark 5.1 *It is assumed in Lemma 5.6 that the domain of the distributed sum problem is* $\mathcal{Z}$, *the set of integers. This assumption is not realistic in the sense that there is no bound on the size of integers. The following are some results when the size of the integers is bounded.*

Let S_N be a security parameter agreed upon all users, let $d_N = 2^{S_N}$ and $D_N = \{-d_N, ..., d_N\}$, and for all $i : N-1 \leq i \leq 0$, $d_i = \Theta(N^2 d_{i+1})$ and $D_i = \{-d_i, ..., d_i\}$. The following Lemma can be proved via linear algebraic analysis.

Lemma 5.8 *For a random polynomial* $p[x] = \sum_{i=0}^{N} p_i x^i \in \mathcal{Z}[x]$ *such that* $p_i \in \mathcal{D}_i$, *for all* $X \subset \{2,4,6...,4N+2\}$, *if* $|X| \leq N$, *then with probability at least* $1 - poly(N)(\frac{1}{2})^{S_N}$:

$$\frac{|SOLU_0|}{|SOLU_1|} = 1 \pm poly(N) \cdot (\frac{1}{2})^{S_N}$$

Where $poly(N)$ *mean a polynomial in* N.

Assuming the coefficients of the secret polynomials f and g are bounded as in the above lemma, it follows from the above lemma that, with very high probability, no subset of s dishonest users can extract any more information about $\mathcal{PARITY}(f[x])$, $\mathcal{PARITY}(g[x])$ and $\mathcal{PARITY}(h[x])$, where $s \leq \lceil \frac{N}{2} \rceil$.

Remark 5.2 *Note that, if s, the number of dishonest users, is greater than* $\lfloor \frac{N}{2} \rfloor$, *they can compute* $f[x]$, $g[x]$ *by interpolation law, thus* $\mathcal{PARITY}(h[x])$. *Hence, our scheme is not secure against the dishonest majority.*

Acknowledgement: We would like to thank Len Adleman and Gary Miller for helpful discussion.

References

[1] J. Benaloh. Secret sharing homomorphisms: keeping shares of a secret secret. In *CRYPTO*, 1986.

[2] Michael Ben-or, Shafi Goldwasser, and Avi Wigderson. Completeness theorems for non-cryptographic fault-tolerant distributed computation. In *Proceedings of the 20th Annual ACM Symposium on Theory of Computing*, pages 1–10, ACM, May 1988.

[3] Gilles Brassard and Claude Crepeau. Non-transitive transfer of confidence: a perfect zero-knowledge interactive protocol for sat and beyond. In *27th Annual Symposium on Foundations of Computer Science*, pages 188–195, IEEE, October 1986.

[4] D. L. Chaum. Untraceable electronic mail, return addresses, and digital pseudonyms. *CACM*, 24():84–88, Feb. 1981.

[5] David Chaum, Claude Crepeau, and Ivan Damgdra. Multi-party unconditionally secure protocols. In *Proceedings of the 20th Annual ACM Symposium on Theory of Computing*, pages 11–19, ACM, May 1988.

[6] J. Cohen and M. Fisher. A robust and verifiable cryptographically secure election scheme. In *FOCS25*, pages 372–382, IEEE, October 1985.

[7] R. A. Demillo, N. A. Lynch, and M. J. Merritt. Cryptographic protocols. In *Proceedings of the 14h Annual ACM Symposium on Theory of Computing*, pages 383–400, ACM, May 1982.

[8] Z. Galil, S. Haber, and M. Yung. Cryptographic computation: secure fault-tolerant protocols and the public-key model. In *CRYPTO*, 1987.

[9] O. Goldreich, S. Micali, and A. Wigderson. How to play any mental game. In *Proceedings of the 19th Annual ACM Symposium on Theory of Computing*, pages 218–229, ACM, 1987.

[10] O. Goldreich, S. Micali, and A. Wigderson. Proofs that yield nothing but their validity and a methodology of cryptographic protocol design. In *27th Annual Symposium on Foundations of Computer Science*, pages 174–187, IEEE, 1986.

[11] S. Goldwasser and S. Micali. Probabilistic encryption. *JCSS*, 28(2):270–299, 1984.

[12] S. Goldwasser, S. Micali, and C. Rackoff. The knowledge of complexity of interactive proof-systems. In *Proceedings of the 17th Annual ACM Symposium on Theory of Computing*, pages 291–304, ACM, May 1985.

[13] Ming-Deh A. Huang and Shang-Hua Teng. Election schemes of optimal security and verifiability. manuscript, usc. 1988.

[14] Ming-Deh A. Huang and Shang-Hua Teng. Secure and verifiable schemes for election and general distributed computing problems. In *7th Annual Symposium on Principles of Distributed Computing*, ACM, 1988.

[15] M. Pease, R. Shostak, and L. Lamport. Reaching agreement in the presence of faults. *JACM*, 27:228–234, 1980.

[16] R. Rivest, A. Shamir, and L. Adleman. A method for obtaining digitial signatures and public–key cryptosystems. *CACM*, 21(2):120–126, 1978.

[17] A. Shamir. How to share a secret. *CACM*, 22(11):612–613, 1979.

[18] A. Yao. Protocols for secure computations. In *23th Annual Symposium on Foundations of Computer Science*, pages 160–164, IEEE, 1982.

[19] A. Yao. Theory and application of trapdoor functions. In *23th Annual Symposium on Foundations of Computer Science*, pages 80–91, IEEE, 1982.

Session 10

Security Concerns

Chair: G. Brassard, University of Montreal

An Abstract Theory of Computer Viruses

Leonard M. Adleman*

Department of Computer Science
University of Southern California

1 Introduction

In recent years the detection of computer viruses has become common place. It appears that for the most part these viruses have been 'benign' or only mildly destructive. However, whether or not computer viruses have the potential to cause major and prolonged disruptions of computing environments is an open question.

Such basic questions as:

1. How hard is it to detect programs infected by computer viruses?
2. Can infected programs be 'disinfected'?
3. What forms of protection exist?
4. How destructive can computer viruses be?

have been at most partially addressed [Co1][Co2][1]. Indeed a generally accepted definition of computer virus has yet to emerge.

For these reasons, a rigorous study of computer viruses seems appropriate.

*Research supported by NSF through grant CCR 8519296

[1]It appears that F. Cohen is the first researcher in an academic setting to consider the practical and theoretical aspects of computer viruses. The formalism presented here differs considerably from that explored by Cohen [Co1][Co2].

2 Basic Definitions

For the purpose of motivating the definitions which follow, consider this (fabricated) 'case study':

A text editor becomes infected with a computer virus. Each time the text editor is used, it performs the text editing tasks as it did prior to infection, but it also searches the files for a program and infects it. When run, each of these newly infected programs performs its 'intended' tasks as before, but also searches the files for a program and infects it. This process continues. As these infected programs pass between systems, as when they are sold, or given to others, new opportunities for spreading the virus are created. Finally, after Jan. 1, 1990, the infected programs cease acting as before. Now, each time such a program is run, it deletes all files.

Such a computer virus can easily be created using a program scheme (in an *ad hoc* language) similar to that found in [Co1]:

```
{main:=
     call injure;
     ...
     call submain;
     ...
     call infect;
}

{injure:=
     if condition then whatever damage is to be done and halt
}

{infect:=
     if condition then infect files
}
```

where for the 'case study virus':

```
{main:=
     call injure;
     call submain;
     call infect;
```

```
}

{injure:=
    if date ≥ Jan. 1, 1990 then
        while files ≠ ∅:
            file = get-random-file;
            delete file;
        halt;
}

{infect:=
    if true then
    file = get-random-executable-file;
    rename main routine submain;
    prepend self to file;
}
```

By modifying the scheme above, a wide variety of viruses can be created. Even 'helpful' viruses may be created. For example the following minor variant of Cohen's [Co1] compression virus which saves storage space:

```
{main:=
    call injure;
    decompress compressed part of program;
    call submain;
    call infect;
}

{injure:=
    if false then halt
}

{infect:=
    if executable-files ≠ ∅ then
    file = get-random-executable-file;
    rename main routine submain;
    compress file;
    prepend self to file;
}
```

With the 'case study virus' and all of those which could be created by the scheme above, it appears that the following properties are relevant:

1. For every program, there is an 'infected' form of that program. That is, it is possible to think of the virus as a map from programs to ('infected') programs.

2. Each infected program on each input (where here by input is meant all 'accessible' information: e.g. the user's input, the system's clock, files containing data or programs) makes one of three choices:

   Injure:
   Ignore the 'intended' task and compute some other function. Note that in the case study, which inputs result in injury (i.e. those where the system clock indicates that the date is Jan. 1, 1990 or later), and what kind of injury occurs (file deletion) are the same whether the infected program is a text editor or a compiler or something else. Thus which inputs result in injury and what form the injury takes is independent of which infected program is running and is actually dependent solely on the virus itself.

   Infect:
   Perform the 'intended' task and if it halts, infect programs. Notice in particular that the clock, the user/program communications and all other 'accessible' information other than programs, are handled just as they would have been had the uninfected version of the program been run. Further, notice that whether the infected program is a text editor or a compiler or something else, when it infects a program the resulting infected program is the same. Thus the infected form of a program is independent of which infected program produces the infection.

   Imitate:
   Neither injure nor infect. Perform the 'intended' task without modification. This may be thought of as a special case of 'Infect', where the number of programs getting infected is zero. (In the case study, imitation only occurs when no programs are accessible for infection).

A formal definition of computer virus is presented next.

Notation 1

1. S denotes the set of all finite sequences of natural numbers.

2. *e denotes a computable injective function from $S \times S$ onto N with computable inverse.*

3. *For all $s, t \in S$, $< s, t >$ denotes $e(s, t)$.*

4. *For all partial $f : N \rightarrow N$, for all $s, t \in S$, $f(s, t)$ denotes $f(< s, t >)$.*

5. *e' denotes a computable injective function from $N \times N$ onto N with computable inverse such that for all $i, j \in N$, $e'(i, j) \geq i$.*

6. *For all $i, j \in N$, $< i, j >$ denotes $e'(i, j)$.*

7. *For all partial $f : N \rightarrow N$, for all $i, j \in N$, $f(i, j)$ denotes $f(< i, j >)$.*

8. *For all partial $f : N \rightarrow N$, for all $n \in N$, write $f(n) \downarrow$ iff $f(n)$ is defined.*

9. *For all partial $f : N \rightarrow N$, for all $n \in N$, write $f(n) \uparrow$ iff $f(n)$ is undefined.*

Definition 1 *For all partial $f, g : N \rightarrow N$, for all $s, t \in S$, $f(s, t) = g(s, t)$ iff either:*

1. *$f(s, t) \uparrow$ & $g(s, t) \uparrow$ or*
2. *$f(s, t) \downarrow$ & $g(s, t) \downarrow$ & $f(s, t) = g(s, t)$.*

Definition 2 *For all $z, z' \in N$, for all $p, p', q = q_1, q_2, ..., q_z, q' = q'_1, q'_2, ..., q'_{z'} \in S$, for all partial functions $h : N \rightarrow N$, $< p, q > \overset{h}{\sim} < p', q' >$ iff:*

1. *$z = z'$ and*
2. *$p = p'$ and*
3. *there exists an i, with $1 \leq i \leq z$ such that $q_i \neq q'_i$ and*
4. *for $i = 1, 2, ..., z$, either*
   (a) *$q_i = q'_i$ or*
   (b) *$h(q_i) \downarrow$ and $h(q_i) = q'_i$.*

Definition 3 *For all partial $f, g, h : N \rightarrow N$, for all $s, t \in S$, $f(s, t) \overset{h}{\sim} g(s, t)$ iff $f(s, t) \downarrow$ & $g(s, t) \downarrow$ & $f(s, t) \overset{h}{\sim} g(s, t)$.*

Definition 4 *For all partial $f, g, h : N \to N$ for all $s, t \in S$, $f(s,t) \stackrel{h}{\cong} g(s,t)$ iff $f(s,t) = g(s,t)$ or $f(s,t) \stackrel{h}{\sim} g(s,t)$.*

Definition 5 *For all Gödel numberings of the partial recursive functions $\{\phi_i\}$, a total recursive function v is a* virus *with respect to $\{\phi_i\}$ iff for all $d, p \in S$, either:*

1. Injure:

$$(\forall i, j \in N)[\phi_{v(i)}(d,p) = \phi_{v(j)}(d,p)]$$

2. Infect or Imitate:

$$(\forall j \in N)[\phi_j(d,p) \stackrel{v}{\cong} \phi_{v(j)}(d,p)]$$

Remark 1 *The choice of symbols d, p above is intended to suggest the decomposition of all 'accessible' information into 'data' (information not susceptible to infection) and 'programs' (information susceptible to infection).*

3 Types of Viruses

In this section the set of viruses is decomposed into the disjoint union of four principal types. The nature of so called 'Trojan horses' is considered.

Definition 6 *For all Gödel numberings of the partial recursive functions $\{\phi_i\}$, for all viruses v with respect to $\{\phi_i\}$, for all $i, j \in N$:*

i is pathogenic *with respect to v and j iff*

$$i = v(j) \;\&$$
$$(\exists d, p \in S)[\phi_j(d,p) \stackrel{v}{\not\cong} \phi_i(d,p)]$$

i is **contagious** *with respect to v and j iff*

$$i = v(j) \ \& \\ (\exists d, p \in S)[\phi_j(d,p) \overset{v}{\sim} \phi_i(d,p)]$$

i is **benignant** *with respect to v and j iff*

$i = v(j)$ &
i is not pathogenic with respect to j &
i is not contagious with respect to j

i is a **Trojan horse** *with respect to v and j iff*

$i = v(j)$ &
i is pathogenic with respect to j &
i is not contagious with respect to j

i is a **carrier** *with respect to v and j iff*

$i = v(j)$ &
i is not pathogenic with respect to j &
i is contagious with respect to j

i is **virulent** *with respect to v and j iff*

$i = v(j)$ &
i is pathogenic with respect to j &
i is contagious with respect to j

When there exists a unique j such that $i = v(j)$ (e.g. when v is injective) then if i is pathogenic (contagious, benignant, a Trojan horse, a carrier, virulent) with respect to v and j, the reference to j will be dropped and i will be said to be pathogenic (contagious, benignant, a Trojan horse, a carrier, virulent) with respect to v.

Hence, if with respect to some virus an infected program is benignant, then it computes the same function as its uninfected predecessor. If it is a Trojan horse then it is incapable of infecting other programs. It can only imitate or injure, and under the right conditions it will do the latter. If it is a carrier, it is incapable of causing injury but under the right conditions it will infect other programs.

Definition 7 *For all Gödel numberings of the partial recursive functions* $\{\phi_i\}$, *for all viruses v with respect to* $\{\phi_i\}$:

v is benign *iff both:*

$$(\forall j \in N)[v(j) \textit{ is not pathogenic with respect to } v \textit{ and } j]$$
$$(\forall j \in N)[v(j) \textit{ is not contagious with respect to } v \textit{ and } j]$$

v is Epeian [2] *iff both:*

$$(\exists j \in N)[v(j) \textit{ is pathogenic with respect to } v \textit{ and } j]$$
$$(\forall j \in N)[v(j) \textit{ is not contagious with respect to } v \textit{ and } j]$$

v is disseminating *iff both:*

$$(\forall j \in N)[v(j) \textit{ is not pathogenic with respect to } v \textit{ and } j]$$
$$(\exists j \in N)[v(j) \textit{ is contagious with respect to } v \textit{ and } j]$$

v is malicious *iff both:*

$$(\exists j \in N)[v(j) \textit{ is pathogenic with respect to } v \textit{ and } j]$$
$$(\exists j \in N)[v(j) \textit{ is contagious with respect to } v \textit{ and } j]$$

The next theorem records some simple facts about types of viruses.

Theorem 1 *For all Gödel numberings of the partial recursive functions* $\{\phi_i\}$ *for all viruses v with respect to* $\{\phi_i\}$:

1. $(\exists j \in N)[v(j)$ *is benignant with respect to v and j*$]$
2. *v is benign iff*

$$(\forall j \in N)$$
$$[v(j) \textit{ is benignant with respect to } v \textit{ and } j]$$

2

Now shift your theme, and sing that wooden horse
Epeios built, inspired by Athena -
the ambuscade Odysseus filled with fighters
and sent to take the inner town of troy

The Odyssey of Homer, 8.492-495.
translation by Robert Fitzgerald
Doubleday & Co., NY, 1961

3. if v is Epeian then

$$(\forall j \in N)$$
$$[[v(j) \textit{ is benignant with respect to } v \textit{ and } j] \textit{ or }$$
$$[v(j) \textit{ is a Trojan horse with respect to } v \textit{ and } j]]$$

4. if v is disseminating then

$$(\forall j \in N)$$
$$[[v(j) \textit{ is benignant with respect to } v \textit{ and } j] \textit{ or }$$
$$[v(j) \textit{ is a carrier with respect to } v \textit{ and } j]]$$

Proof

Part 1 follows immediately from the recursion theorem.

All other parts follow immediately from the definitions.

□

Thus, all programs infected by a benign virus are benignant with respect to their uninfected predecessors. They function just as if they had never been infected. Viruses in this class appear to be the least threatening. This class includes many 'degenerate' viruses such as the identity function and 'padding' functions.

Programs infected by an Epeian virus can only be benignant or Trojan horses with respect to their uninfected predecessors. Further the latter option must sometimes occur. Epeian viruses will not be able to spread themselves; however, an infected program may imitate the 'intended' task of its uninfected predecessor until some 'trigger' causes it to do damage. Among the Epeian viruses are the 'degenerate' class of constant functions, which never imitate-or-infect but only injure.

Programs infected by a disseminating viruses can only be benignant or carriers with respect to their uninfected predecessors. Further the latter option must sometimes occur. Thus programs infected with such viruses are never pathogenic. However, it is worth noting that disseminating viruses may modify the size of programs or their complexity characteristics, and by this means become detectable or cause harm (or benefit as in the case of the compression virus). In fact, size and complexity may be important properties when considering viruses. An extension of the current theory to account for size and complexity seems appropriate (see §*further research*).

Malicious viruses can both spread and produce injuries. They appear to be the most threatening kind of virus. The 'case study virus' in §*basic definitions* is malicious.

Remark 2 *It may be appropriate to view contagiousness as a necessary property of computer viruses. With this perspective, it would be reasonable to define the set of viruses as the union of the set of disseminating viruses and the set malicious viruses, and to exclude benign and Epeian viruses altogether.*

4 Detecting The Set Of Viruses

The question of detecting viruses is addressed in the next theorem:

Theorem 2 *For all Gödel numberings of the partial recursive functions* $\{\phi_i\}$*:*

$$V = \{i|\phi_i \textit{ is a virus}\} \textit{ is } \Pi_2 - \textit{complete}$$

Proof

Let $T = \{i|\phi_i \text{ is a total}\}$. It is well known (§13 and §14 [Ro]) that T is Π_2 – complete.

To establish that $T \leq_1 V$, let $j \in V$ (for example let j be an index for the identity function) and consider the function $g : N \to N$ such that for all $i, y \in N$:

$$g(i,y) = \begin{cases} \phi_j(y) & \text{if } \phi_i(y)\downarrow \\ \uparrow & \text{otherwise} \end{cases}$$

Then g is a partial recursive function. Let k be an index for g, and let $f : N \to N$, be such that:

$$(\forall i \in N)[f(i) = s(k,1,i)]$$

where s is as in the $s - m - n$ theorem [Ro].

Then f is a total recursive function and:

$$(\forall i, y \in N)[\phi_{f(i)}(y) = \phi_{s(k,1,i)}(y) = \phi_k(i, y) = g(i, y) = \begin{cases} \phi_j(y) & \text{if } \phi_i(y)\downarrow \\ \uparrow & \text{otherwise} \end{cases}]$$

It follows that:

$$i \in T \Leftrightarrow f(i) \in V$$

Thus $T \leq_m V$. It follows, as in §7.2 [Ro], that $T \leq_1 V$ as desired.

To establish that $V \in \Pi_2$, consider the following formula for V which arises directly from the definition of virus:

$$
\begin{array}{ll}
(\forall j)(\exists k, t) & [H(i, j, k, t)] \\
\& & \\
(\forall < d, p >) & [(\forall j_1, k_1, t_1) \\
& [H(i, j_1, k_1, t_1) \Rightarrow \\
& (\forall < e, q >, t_2)[\neg H(k_1, < d, p >, < e, p >, t_2)]] \\
& \text{or} \\
& (\forall j_1, k_1, t_1, j_2, k_2, t_2) \\
& [[H(i, j_1, k_1, t_1) \ \& \ H(i, j_2, k_2, t_2)] \Rightarrow \\
& (\exists < e, q >, t_3, t_4) \\
& [H(k_1, < d, p >, < e, q >, t_3) \ \& \\
& H(k_2, < d, p >, < e, q >, t_4)]] \\
& \text{or} \\
& (\forall j_1, k_1, t_1, < e, q >, t_2) \\
& [[H(i, j_1, k_1, t_1) \ \& \ H(j_1, < d, p >, < e, q >, t_2)] \Rightarrow \\
& (\exists < e', q' >, t_3, t_4) \\
& [H(k_1, < d, p >, < e', q' >, t_3) \ \& \\
& L(i, < e, q >, < e', q' >, t_4)] \\
& \& \\
& [H(i, j_1, k_1, t_1) \ \& \ H(k_1, < d, p >, < e, q >, t_2)] \Rightarrow \\
& (\exists < e', q' >, t_3, t_4) \\
& [H(j, < d, p >, < e', q' >, t_3) \ \& \\
& L(i, < e', q' >, < e, q >, t_4)]]]
\end{array}
$$

Where H is a 'step counting' predicate for $\{\phi_i\}$ such that:

$$
\begin{array}{ll}
(\forall i, j, k) & \\
\text{if } \phi_i(j) = k \text{ then} & (\exists t)[H(i, j, k, t)] \\
\text{if } \phi_i(j) \neq k \text{ then} & (\forall t)[\neg H(i, j, k, t)]
\end{array}
$$

And where L is a predicate for $\{\phi_i\}$ such that:

$$(\forall i, <e,q>, <e',q'>, t)$$
$$\text{if } <e,q> \overset{\phi_i}{\sim} <e',q'> \text{ then } \quad (\exists t)[L(i, <e,q>, <e',q'>, t)]$$
$$\text{if } <e,q> \overset{\phi_i}{\not\sim} <e',q'> \text{ then } \quad (\forall t)[\neg L(i, <e,q>, <e',q'>, t)]$$

Since for all acceptable Gödel numberings of the partial recursive functions $\{\phi_i\}$ it is easily seen that there exist recursive predicates H and L as above, it follows that $V \in \Pi_2$.

□

Thus detecting viruses is quite intractable, and it seems unlikely that protection systems predicated on virus detection will be successful.

5 Isolation As A Protection Strategy

As noted in [Co1] isolating a computing environment from its surroundings is a powerful method of protecting it from viruses. For example, if no new programs can be introduced, no old programs can be updated, and no communication can occur, then it seems viruses are no threat.

Unfortunatly, such isolation is unrealistic in many computing environments. The next theorems explore the possibility of protecting computing environments with less severe forms of isolation.

Definition 8 *For all Gödel numberings of the partial recursive functions $\{\phi_i\}$, for all viruses v with respect to $\{\phi_i\}$, let:*

The infected set of v

$$I_v = \{i \in N | (\exists j \in N)[i = v(j)]\}$$

Definition 9 *For all Gödel numberings of the partial recursive functions $\{\phi_i\}$, for all viruses v with respect to $\{\phi_i\}$, v is* absolutely isolable *iff I_v is decidable.*

Clearly if a virus is absolutely isolable, then (at least in theory) it can be neutralized. Whenever a program becomes infected, it is detected and removed. The following is a simple fact about absolutely isolable viruses:

Theorem 3 *For all Gödel numberings of the partial recursive functions* $\{\phi_i\}$, *for all viruses* v *with respect to* $\{\phi_i\}$ *if for all* $i \in N$, $v(i) \geq i$ *then* v *is absolutely isolable.*

Proof trivial.

□

Thus the case study virus, as implemented using the scheme in §*basic definitions* would be absolutely isolable. In fact, what little experience with viruses there is to date seems to suggest that in practice people who produce viruses begin by producing ones with the increasing property necessary for theorem 3 to apply. Unfortunately, not all viruses have this property. For example, with any reasonable compression scheme, the compression virus of §*basic definitions* would not have this property. Nonetheless, the compression virus is absolutely isolable. Given a program with the proper syntax, it is in the infected set if and only if decompressing the compressed part results in a legitimate program.

Is every virus absolutely isolable?

Regretably, the next theorem shows that the answer is no.

Theorem 4 *For all Gödel numberings of the partial recursive functions* $\{\phi_i\}$, *there exists a total recursive function* v *such that:*

1. v *is a malicious virus with respect to* $\{\phi_i\}$
2. I_v *is* Σ_1*-complete.*

Proof

Let f be a total recursive function such that:

$$Rg(f) = K = \{i | \phi_i(i) \downarrow\}$$

Let $j_1 : N \rightarrow N$ be a $1-1$ total recursive function such that for all $i, x \in N$:

$$\phi_i = \phi_{j_1(i,x)} \tag{1}$$

Such a function, known as a padding function, exists by Proposition 3.4.5 [MY].

Let $j_2 : N \to N$ be such that:

$$(\forall i, x \in N)[j_2(i, x) = j_1(i, y)]$$

where y is the least natural number such that, for all $i', x' \in N$ with $< i', x' >< < i, x >$, $j_2(i', x') < j_1(i, y)$.

Then j_2 is a monotonically increasing total recursive function and by (1), it follows that:

$$(\forall i, x \in N)[\phi_i = \phi_{j_2(i,x)}] \tag{2}$$

Let $j' : N \to N$ be such that for all $i \in N$:

$$j'(i) = \begin{cases} y + 1 & \text{if } i = j_2(1, y) \\ 0 & \text{otherwise} \end{cases}$$

Then since j_2 is monotonically increasing, it follows that j' is a total recursive function.

Consider the function $b_1 : N \to N$ such that for all $d, p \in S$ and $i, k \in N$:

$$\phi_{b_1(i,k)}(d, p) = \begin{cases} 0 & \text{if } d \text{ is even} \\ < e, [\phi_k(q)] > & \text{if } d \text{ is odd } \& \ \phi_i(d, p) =< e, [q] > \text{ and } \phi_k(q) \downarrow \\ \uparrow & \text{if } d \text{ is odd } \& \ \phi_i(d, p) =< e, [q] > \text{ and } \phi_k(q) \uparrow \\ \phi_i(d, p) & \text{otherwise} \end{cases}$$

where for all $q \in N$, $[q]$ denotes the one element sequence in S consisting only of q.

Then by standard arguments, b_1 is a total recursive function and:

$$(\forall i, x, k \in N)[\phi_{b_1(i,k)} = \phi_{b_1(j_2(i,x),k)}]. \tag{3}$$

Let $b_2 : N \to N$ be such that for all $i, k \in N$:

$$b_2(i, k) = \begin{cases} j_2(b_1(i, k), f(0)) & \text{if } j'(i) = 0 \\ j_2(b_1(1, k), f(y)) & \text{if } j'(i) = y + 1 \end{cases}$$

Then b_2 is a total recursive function and it follows from (2) and (3) that:

$$(\forall i, k \in N)[\phi_{b_2(i,k)} = \phi_{b_1(i,k)}]. \tag{4}$$

Applying the s-m-n theorem there exists a total recursive function g such that for all $i, k \in N$:

$$\phi_{g(k)}(i) = b_2(i, k)$$

By the recursion theorem, there exists an $h \in N$ such that for all $i \in N$:

$$\phi_h(i) = b_2(i, h)$$

Let $v = \phi_h$. Then v is a total recursive function since b_2 is.

Let $d, p \in S$, then using that fact that $v = \phi_h$ is a total recursive function and applying (4) gives:

$$\begin{aligned}
\phi_{v(i)}(d,p) &= \phi_{b_2(i,h)}(d,p) \\
&= \phi_{b_1(i,h)}(d,p) \\
&= \begin{cases} 0 & \text{if } d \text{ is even} \\ < e, [\phi_h(q)] > & \text{if } d \text{ is odd } \& \ \phi_i(d,p) =< e, [q] > \ \& \ \phi_h(q) \downarrow \\ \uparrow & \text{if } d \text{ is odd } \& \ \phi_i(d,p) =< e, [q] > \ \& \ \phi_h(q) \uparrow \\ \phi_i(d,p) & \text{otherwise} \end{cases} \\
&= \begin{cases} 0 & \text{if } d \text{ is even} \\ < e, [v(q)] > & \text{if } d \text{ is odd } \& \ \phi_i(d,p) =< e, [q] > \\ \phi_i(d,p) & \text{otherwise} \end{cases}
\end{aligned}$$

1 of the theorem now follows directly from the definition of malicious virus.

Since, for all total recursive functions m, $Rg(m)$ is recursively enumerable, it follows that $I_v = Rg(v) \in \Sigma_1$.

Let $c : N \to N$ be such that for all $x \in N$, $c(x) = j_2(b_1(1,h), x)$. Since j_2 is $1-1$ so is c. Then $x \in K$ implies the existence of a $y \in N$ such that $f(y) = x$. Let $i = j_2(1, y)$, then:

$$c(x) = j_2(b_1(1,h), x) = j_2(b_1(1,h), f(y)) = b_2(i,h) = v(i) \in I_v$$

On the other hand, assume $x \notin K$ and $c(x) \in I_v$. Then there exists an $i \in N$ such that:

$$j_2(b_1(1,h),x) = c(x) = v(i) = b_2(i,h) = \begin{cases} j_2(b_1(i,h), f(0)) & \text{if } j'(i) = 0 \\ j_2(b_1(1,h), f(y)) & \text{if } j'(i) = y+1 \end{cases}$$

Since j_2 is $1-1$, it follows that $x = f(y) \in K$. $\Rightarrow\Leftarrow$. Hence, $K \leq_1 I_v$ and 2 of the theorem holds.

□

Thus, for the viruses described in the previous theorem, protection cannot be based upon deciding whether a particular program is infected or not. Paradoxically, despite this, it is often possible to defend against such viruses. How such a defense could be mounted will be described below; however, a few definitions are in order first.

Definition 10 *For all Gödel numberings of the partial recursive functions $\{\phi_i\}$, for all viruses v with respect to $\{\phi_i\}$, let:*

The germ set of v

$$G_v = \{i | i \in N \ \& \ (\exists j \in N)[\phi_i = \phi_{v(j)}]\}$$

Thus the germs of a virus are functionally the same as infected programs, but are syntactically different. They can infect programs, but cannot result from infection. They may start 'epidemics', but are never propagated with them.

Definition 11 *For all Gödel numberings of the partial recursive functions $\{\phi_i\}$, for all viruses v with respect to $\{\phi_i\}$, v is* **isolable within its germ set** *iff there exists an $S \subseteq N$ such that:*

1. $I_v \subseteq S \subseteq G_v$.

2. *S is decidable.*

Notice that if a virus is isolable within its germ set by a decidable set S, then not allowing programs in the set S to be written to storage or to be communicated will stop the virus from infecting. Further, the isolation of some uninfected germs by this process appears to be an added benefit.

Returning now to the viruses described in the previous theorem: assume that the function b_1 above had the property that for all i, k, $b_1(i,k) >< i,k >$. The proof of the previous theorem could easily have been modified to assure this. Further, in Godel numberings derived in the usually fashion from natural programming languages, a b_1 constructed in a straightforward manner would have this property. Consider the set

$$S = \{j_2(b_1(i,h), y) | i, y \in Z_{>0}\}$$

By the monotonically increasing property of j_2 and the property of b_1 which is being assumed, S is decidable. On the other hand if $a \in I_v$ then there exists i such that

$$a = v(i) = b_2(i,h) = \begin{cases} j_2(b_1(i,h), f(0)) & \text{if } j'(i) = 0 \\ j_2(b_1(1,h), f(y)) & \text{if } j'(i) = y+1 \end{cases}$$

And it follows that $a \in S$. On the other hand if $a \in S$ then there exist an y, i such that

$$a = j_2(b_1(i,h), y)$$

By (2) and (4):

$$\phi_a = \phi_{j_2(b_1(i,h),y)} = \phi_{b_1(i,h)} = \phi_{b_2(i,h)} = \phi_{v(i)}$$

And hence $a \in G_v$ as desired.

Thus viruses like the ones in theorem 4 demonstrate that decidability of I_v is sufficient but not necessary for neutralization. Apparently, more work needs to be done before a clear idea of the value of isolation will emerge. Are all viruses isolable within their germ set? The answer is no (proof omitted). Are all disseminating viruses isolable within their germ set? The answer is not known. Are there notions of isolation which provide significant protection at a reasonable cost?

6 Further Research

The study of computer viruses is embryonic. Since so little is known, virtually any idea seems worth exploring.

Listed below are a few avenues for further investigation.

1. *Complexity theoretic and program size theoretic aspects of computer viruses.*

   Introduce complexity theory and program size theory into the study of computer viruses. As noted earlier, even disseminating viruses may affect the complexity characteristics and size of infected programs and as a result become detectable or harmful.

   Complexity theory and program size considerations can be introduced at a abstract level (see for example [MY]) or a concrete level.

   For example, viruses in the 'real world' would probably have the property that the running time of an infected program, at least while imitating or infecting, would be at most polynomial (linear) in the running time of its uninfected precursor. Does this class of 'polynomial (linear) viruses' pose a less serious threat? Do NP-completeness considerations, or cryptographic considerations come into play?

2. *Protection Mechanisms*

   In this paper one form of protection mechanism, isolation, was briefly considered. In addition to considering isolation in greater depth, numerous other possibilities exist. For example:

   Quarinteening
   Is there value in taking a new program and running it in a safe environment for a while before introducing it into an environment were it could spread or do harm? For example, putting the new program on an isolated machine with dummy infectable programs and with a variety of settings of the system clock might evoke behavior indicative of infection. In particular would this be helpful with the class of polynomial viruses or linear viruses?

   Disinfecting
   Under what circumstances can an infected program be disinfected? Certainly when a virus is absolutely isolable there exists a procedure which when given an infected program will return a program which 'infects to' the original one. How general is this phenomena?

   Certificates
   Can some programs be given a 'clean bill of health'? For example, if it is know that a certain virus is about, would it be possible for a vendor to 'prove' that his program was not in the germ set? Would it be possible to prove that the software was not in the germ set of a large class of viruses?

Operating System Modification

Could modifications to the operating system provide some protection. For example, assume that the (secure) operating system required that the user 'initiate' all new programs by designating the files which the program is given the privilege to read and write. Then, for example, a simple program (e.g. a game) could be given only the privilege to read and write files it creates. If the program was uninfected it might perform satisfactorily under this constraint. If however the program was infected, this constraint might severely limit the damage due to the virus. (This example arose during joint work with K. Kompella).

3. *Other Models Of Computer Viruses.*

The notion of computer viruses presented here is not the only one possible. It was selected because it seemed to be an adequate place to begin an investigation. More general, and more restrictive notions are possible. Indeed it seems possible that no definition will conform to everyone's intuitions about 'computer viruses'.

More 'machine dependent' approaches could be considered. Approaches which take into account the communications channels over which viruses pass seem particularly important.

One interesting generalization of the current notion is inspired by [Co1], where viruses are assumed to be capable of evolving. The 'Mutating Viruses' (μ-viruses) partially defined next are an attempt to capture this property.

Definition 12 *For all* $z, z' \in N$, *for all* $p, p', q = q_1, q_2, ..., q_z, q' = q'_1, q'_2, ..., q'_{z'} \in S$, *for all sets* H *of partial functions from* N *to* N, $<p, q> \overset{H}{\sim} <p', q'>$ *iff:*

(a) $z = z'$ *and*

(b) $p = p'$ *and*

(c) *there exists an* i, *with* $1 \leq i \leq z$ *such that* $q_i \neq q'_i$ *and*

(d) *for* $i = 1, 2, ..., z$, *either*

i. $q_i = q'_i$ *or*

ii. *there exists an* $h \in H$ *such that* $h(q_i) \downarrow$ *and* $h(q_i) = q'_i$.

Definition 13 *For all sets of partial functions H from N to N, for all partial $f, g : N \to N$, for all $s, t \in S$, $f(s,t) \overset{H}{\sim} g(s,t)$ iff $f(s,t) \downarrow$ & $g(s,t) \downarrow$ & $f(s,t) \overset{H}{\sim} g(s,t)$.*

Definition 14 *For all sets of partial functions H from N to N, for all partial $f, g : N \to N$, for all $s, t \in S$, $f(s,t) \overset{H}{\cong} g(s,t)$ iff $f(s,t) = g(s,t)$ or $f(s,t) \overset{H}{\sim} g(s,t)$.*

Definition 15 *For all Gödel numberings of the partial recursive functions $\{\phi_i\}$, a set M of total recursive functions is a* mutating virus, *μ-virus, with respect to $\{\phi_i\}$ iff both:*

(a) for all $m \in M$, for all $d, p \in S$ either:

i. Injure:

$$(\forall i, j \in N)[\phi_{m(i)}(d,p) = \phi_{m(j)}(d,p)]$$

ii. Infect or Imitate:

$$(\forall j \in N)[\phi_j(d,p) \overset{M}{\cong} \phi_{m(j)}(d,p)]$$

Some computer viruses which have recently caused problems (e.g. the so called 'Scores virus' [Up] which attacked Macintosh computers) are μ-viruses and not just viruses. Hence this generalization of the notion of virus may be of more than theoretical interest.

This is only a partial definition because some notion of 'connectivity' is needed. That is, the union of two μ-viruses, neither of which 'evolves' into the other should not be a μ-virus. Many definitions of 'connectivity' can be defined, but further study will be required to choose those which are most appropriate. Once an appropriate choice is made, an important question will be whether the set of infected indices of a μ-virus can be harder to detect than those of a virus.

4. *Computer Organisms.*

This issue has evolved during joint work with K. Kompella.

There appear to be programs which can reproduce or reproduce and injure but which are not viruses (e.g. programs which just make copies of themselves but never 'infect'). These 'computer organisms' may be a serious security problem.

It may be appropriate to study 'computer organisms' and treat 'computer viruses' as special case.

7 Acknowledgments

I would like to thank Dean Jacobs, and Gary Miller for contributing their ideas to this paper.

I would also like to thank two of my students: Fred Cohen and Kireeti Kompella. Cohen brought the threat of computer viruses to my (and everyone's) attention. Kompella has spent many hours reviewing this work and has made numerous suggestions which have improved it.

References

[Co1] Cohen F. Computer Viruses. Ph.D. dissertation, University of Southern California, Jan. 1986.

[Co2] Cohen F. Computer Viruses - Theory and Experiments. Computers and Security 6 (1987) 22-35. North-Holland.

[MY] Machtey M, Young P. An introduction to the general theory of algorithms. North-Holland, NY 1978.

[Ro] Rogers, H Jr. Theory of Recursive Functions and Effective Computability. McGraw-Hill Book Co., NY 1967.

[Up] Upchurch, H. The Scores Virus, unpublished manuscript , 1988.

Abuses in Cryptography and How to Fight Them

(Extended Abstract)

Yvo Desmedt

Dept. EE & CS, Univ. of Wisconsin – Milwaukee
P.O. Box 784, WI 53201 Milwaukee, U.S.A.

Abstract. *The following seems quite familiar: "Alice and Bob want to flip a coin by telephone. (They have just divorced, live in different* countries, *want to decide who will have the children during the next holiday.)..." So they use [Blu82]'s (or an improved) protocol.* However, *Alice and Bob's* divorce has been set up to cover up their spying activities. *When they use [Blu82]'s protocol, they don't care if the "coin-flip" is random, but they want to* abuse the protocol *to send secret information to each other. The counter-espionage service, however, doesn't know that the divorce and the use of the [Blu82]'s protocol are just cover-ups.*

In this paper, we demonstrate how several modern crypto-systems can be abused. We generalize [Sim83b]'s subliminal channel and [DGB87]'s abuse of the [FFS87, FS86] identification systems and demonstrate how one can prevent abuses *of crypto-systems.*

1 Introduction

[Sim83b] introduced the notion of subliminal channel. His example is related to two prisoners who are communicating authenticated messages in full view of a warden, who is able to read the messages. The subliminal consists in hiding a message *through* the authentication scheme such that the warden *cannot detect its use nor read the hidden part.* At Crypto'87, [DGB87] discussed a similar scenario by demonstrating that the [FFS87,FS86] identification systems can be abused for sending secret messages in an undetectable way. Claiming that he is identifying himself, [DGB87] enables, for example, a mafia Godfather to communicate under the F.B.I.'s very nose without having to worry that it would be detected. In this paper we will *generalize* these undetectable abuses and subliminal channels. We prefer to use the term *abuse* in the general context and reserve the word subliminal in the special context that it is an abuse of an authentication or signature system. We will briefly demonstrate that many modern crypto-systems can be abused (see Section 2). Abuses (in particular subliminal channels) are not covert channels in the strict way, as will briefly be discussed in Section 2.2.

The *main purpose of this paper* is to make *abuse-free crypto-systems* (including protocols). In Section 2.3, we will propose the main tool, while in Section 3 we will give general solutions to solve the abuse problem. Specific applications, such as coin flipping over the telephone, and subliminal-free authentication- and signature systems will be focused on in more detail in Section 4.

2 Abuses

2.1 An introduction

The problem of fraud, such as eavesdropping and modification of messages, is well known. It can be said that modern cryptography studies the methods used to protect data against several types of fraud. A crypto-system protects data against a subset of frauds. For example, [BG84] protects the information against an eavesdropper who would use ciphertext-only and known-plaintext, but not chosen ciphertext attacks.

Let us now discuss what an abuse is. In order for our definition to make sense, we need a warden, as in [Sim83b]. If A uses a crypto-system or is a party of a protocol, we say that A can *abuse* the system if she is able to use it for a *different purpose* than for which it is intended. An abuse is *undetectable* if it is impossible for the warden to detect (in polynomial time) that A uses the abuse. It is trivial to make detectable abuses. So we will *only discuss undetectable abuses and will call them briefly*: "abuses". A *particular* abuse is that A is able to send (encrypted) information to other parties involved besides the warden.

A formal and more general definition of an abuse can be given (see [Des]), but this formal definition is complex and therefore not covered here. Informally, if A is supposed to use a *crypto-system* C (or is a party of a cryptographic protocol), but uses a different special system (C'), we say that A can *abuse* the system if the *numbers* that she sends:

- do not allow the warden to distinguish (in polynomial time) between normal use and special use,
- allow a participant (e.g. B) to distinguish with high enough probability.

It is trivial to understand that an abuse can consist *in replacing the random* which is used in a crypto-system *by pseudo-random*, or even by the output of a one-time pad. *The user who abuses a system can find the use of his abuse more important than endangering the security of the system*, in particular *his secret*. Abusing systems as: [Blu82], and zero-knowledge is trivial and it is remarkable that this aspect of crypto-systems has never been studied before beside in a narrow context as authentication [Sim83b] and identification [DGB87]. ***Remark that the goal of an abuse can be considered as the opposite of the goal of zero-knowledge.***

2.2 ABUSES VERSUS COVERT CHANNELS

It is important to remark that, strictly speaking, abuses are not covert channels. According to [Lam73, p. 614]:

> *Covert* channels, i.e. those not intended for information transfer at all, such as the service program's effect on the system load.

A more general definition can be found in [Dep83, p. 110].

It is very important to observe that the Lampson definition implies that *abuses are* not *covert channels.* Indeed, messages are transmitted in crypto-systems so they are intended for information transfer. For example, a zero-knowledge protocol intends a very small information transfer.

In ***this paper we will only discuss abuses and not covert channels.*** Leaking information through methods such as time jitter, crosstalk and amplitude modulation, as discussed in [Sim88, p. 626], are covert channels, and thus not a topic of this paper.

Whether our solutions against abuses can be extended to covert channels is a new, open question. The author admits that the difference between abuses and covert channels is debatable, and that one could claim that abuses are very special covert channels. In this paper we consider them to be different. What makes abuses so unique is that the hidden information is a "number", and that one can hide it by using a ***crypto-system.*** Formalizing the definition of covert channels could imply that the new open problem can be solved.

2.3 ABUSE-FREENESS

We will say that a warden is passive if he is just observing the communication between the participants. In the narrow context of authentication, which was studied in [Sim83b], the warden was also passive. Our main solution against abuses is based on an *active* warden W who does not only listen to catch up subliminal senders, but also *interacts* in the communication in a special way to better enforce the subliminal-freeness. In other words, he participates actively in the communication between *all* participants and he can modify the "numbers" that are communicated. The only trust in the active warden consists in believing he will not help to set-up an abuse. One could compare the active wardens with Simmons' idea used to exclude the use of analog covert channels [Sim83a, p. 65]. The main difference is that the active warden is digital.

Informally, we said that a system can be abused if another system exists such that the warden cannot distinguish between normal execution and the execution of the other system, while a participant can. So, abuse-freeness can be considered as the logical negation of the existence of an abuse. It means: if a participant is able to distinguish between execution of the normal system and a different one, then the warden can distinguish it also, as well as for all possible different systems, which are different from the normal system. Hereto the warden will modify the numbers that are transmitted. An exact and more formal definition is given in [Des], but not included here because it is too lengthy and complicated.

Evidently, the action of the warden may not endanger the security of the system.

2.4 Practical aspects of the warden

One could wonder if it is possible to construct abuse-free crypto-systems where the warden is passive. In Section 3.3 we will briefly discuss this topic.

In our solution, one assumes that there is *one* (active) warden. In some situations, as in the verification of treaties [Ada88,Sim88], it is in reality sufficient and achievable to have only one active warden. However, in some circumstances having only one warden is insecure or impractical. [Des] discusses these situations and proposes better models. Let us briefly overview them.

Goutier [DGB87] remarked that *the subliminal sender can also send information to others who are eavesdropping on the communications.* Indeed, passive and active eavesdroppers can be subliminal receivers. *In this paper, we assume that this isn't the case.* It is not hard to generalize our results in order to solve the problem of eavesdroppers who are subliminal receivers by using *two wardens* to protect a communication link, one at each end of the line. These two wardens could trust or distrust each other. Some of our protocols can easily be adapted to it. Other problems are:

- that we assume that the warden himself will not try to abuse the system; implying that we trust the warden will not try to send hidden information,
- speed and number of interactions between warden and participants,
- that it is not excluded that participants have been able to hide a covert or physical channel with small capacity. The warden is unaware of this extra channel. The existence of this hidden channel could imply that the system, which was originally abuse-free, is no more due to this extra information.

We call this last problem the *collapse problem.* These topics and how to solve them are discussed in [Des].

We now discuss how the idea of an active and censoring warden can actually be used. The main techniques that we use are: commitment, zero-knowledge [GMR], and the one-time pad crypto-system.

We will start by discussing the more general cases first. Proofs will not be given because they require a formal definition.

3 Abuse-free systems in a general context

Zero-knowledge allows A to restrict leaks of information *if A wants.* We demonstrate that a warden can *enforce* A not to leak information, even if A tries her hardest. Hereto, we first discuss in Section 3.1, in general terms, how to generate a public key in an abuse-free way. We then discuss how to make interactive and non-interactive zero-knowledge abuse-free.

3.1 HOW TO GENERATE A PUBLIC KEY IN AN ABUSE-FREE WAY

Motivation

Publishing a public key can be abused. To illustrate, suppose that A publishes a public key $n = pq$, where p and q are primes of 100 digits. If A is able and/or allowed to give B a 100 digit number *e.g.*, p, it is trivial to understand that, *by publishing n, A is able to leak 100 digits of extra information to B* (for improvements see [RS85]). Another method for leaking information is to choose p and q such that the least significant bits of n have a special form not required by the specifications.

So the process of publishing a key is abuse-free if the key is guaranteed to be random beside the specifications *e.g.*, a product of two large different primes both congruent to 3 modulo 4.

A solution could be that the warden chooses the public key of A. However, this allows the warden to become Big Brother. We exclude this solution.

A solution

To generate the public key, A normally chooses some random number, R and verifies if R satisfies conditions, C and if so calculates public key, $P = \text{GEN}(R)$, where GEN is a publicly known algorithm. However it is also possible that R is not suited (does not satisfy C) *e.g.*, p and q that are composite numbers are unsuited for RSA. In the last case, we require that A must be able to convince the active warden, W that this R is unsuited to make P. Roughly speaking, to obtain an abuse-free public key, A will use her own generated random (R) exored with random (R') generated by W, to make the public key P. The following protocol makes it clear that no cheating is possible.

First W and A agree on a commitment algorithm (or circuit) E, such that the commitment can be verified in random polynomial time. They also agree on algorithm TESTC to verify that R satisfies conditions C, and on algorithm TESTNOC to verify that R does not. We then have:

Step 1 A chooses a (random) binary string $R = (r_1, r_2, \ldots, r_l)$ of l bits, an appropriate k, and A sends $M = E_k(R)$ to W as a commitment for R.

Step 2 W chooses a truly random binary string $R' = (r'_1, r'_2, \ldots, r'_l)$ of l bits and sends it to A.

Step 3 A calculates $S = (r_1 \oplus r'_1, r_2 \oplus r'_2, \ldots, r_l \oplus r'_l)$ ($\oplus$ is exclusive or). Shortly we denote $S = R \oplus R'$. *If* S satisfies conditions C (case 1) *then* A calculates $P = \text{GEN}(S)$ and *sends* P *to* W. A then proves to W that there exists $R = (r_1, r_2, \ldots, r_l)$ and k such that:

$$M = E_k(R) \quad \wedge \quad \text{TESTC}(R \oplus R') \quad \wedge \quad P = \text{GEN}(R \oplus R').$$

This proof has to be zero-knowledge. *Else* (case 2) A convinces W that S does not satisfy conditions C. In this case, A can even reveal R and k to convince W.

Step 4 W verifies A's proof. *If* this proof fails, W stops protocol, *else* one continues. In the case that a P was delivered by A (case 1), W publishes A's public key P and protocol halts. Else (in case 2) the protocol restarts from the beginning (Step 1).

Important remarks

Security (privacy of the secret key) and abuse-freeness of this protocol are proven in [Des]. The security is based on the assumption that the commitment algorithm is hard to invert. Indeed, if the commitment algorithm could be broken, the warden will know A's secret key. The abuse-freeness is unconditional.

In most cases the zero-knowledge proof which is given in Step 3 is impractical and too slow, certainly when it has to be based on [GMW86]. But because public keys are only generated occasionally, this is of less importance. If we would use [BC86], the unconditionality of the abuse-freeness disappears.

3.2 ABUSE-FREE INTERACTIVE ZERO-KNOWLEDGE

Many practical zero-knowledge protocols can be made abuse-free. We will mainly give a general result by demonstrating how to make the [GMW86] zero-knowledge proof of 3-colourability abuse-free, and indicate how one can make the [BC86] zero-knowledge proof for SAT abuse-free. Let us start with [GMW86].

We use the same notation as in [GMW86, pp.176–177], but we number the edges as $E = \{0, 1, \ldots, m-1\}$ and we call the prover, A and the verifier, B. One agrees that the protocol will always end after l iterations, l in function of the (security) parameters. The main problem is that A will reveal $(\pi(\phi(u)), r_u)$ and $(\pi(\phi(v)), r_v)$, which could be abused. A similar approach as in Section 3.1 could be followed but would be very impractical. The following protocols avoid this. *It is organized*

such that all numbers that are sent cannot be abused. The warden will influence all numbers that are transmitted from A to B and vice-versa.

Step 1 A chooses a (random) permutation $\pi \in Sym(\{1,2,3\})$ and (random) r_v, r'_v and k_v (large enough) and computes the commitments $R_v = f(\pi(\phi(v)), r_v)$ and $K_v = f(k_v, r'_v)$ (for all $v \in V$), and sends R_1, R_2, ..., R_n and $K_1, K_2, \ldots, K_n$ to W.

Step 2 W chooses a truly random $\pi' \in_R Sym(\{1,2,3\})$ and truly random k'_1, k'_2, ..., k'_n and sends them to A.

Step 3 A calculates $\pi'' = \pi'\pi$ and $r''_v = k_v \oplus k'_v$ and $R'_v = f(\pi''(\phi(v)), r''_v)$ (for all $v \in V$) and sends $R'_1, R'_2, \ldots, R'_n$ to W.

Step 4 W chooses truly random s_v and calculates $R''_v = f(R'_v, s_v)$ (for all $v \in V$) and sends $R''_1, R''_2, \ldots, R''_n$ to B.

Step 5 B selects (at random) an edge $e \in E$ and a (random) t and sends $S = f(e,t)$ to W.

Step 6 W chooses a truly random $e' \in_R E$ and sends it B.

Step 7 B reveals e and t to W.

Step 8 W verifies if $e \in E$ and checks whether $S = f(e,t)$. *If* both conditions are satisfied, *then* W calculates $e'' = e + e' \bmod m$ (edges where hereto specially numbered) and sends e'' to A. *Else* W stops protocol.

Step 9 Let (u, v) correspond with e'', where $u, v \in V$ and $u < v$. A reveals $(\pi''(\phi(u)), r''_u, r'_u, r_u)$ and $(\pi''(\phi(v)), r''_v, r'_v, r_v)$ to W. If $e'' \notin E$ (W cheats), then A stops.

Step 10 W uses the information revealed by A to check R'_u. Then W calculates k_u ($k_u = r''_u \oplus k'_u$) and verifies K_u. He then calculates $\pi(\phi(u))$ (starting from $\pi''(\phi(u))$ and π') and verifies R_u. He does exactly the same to verify R'_v, K_v and R_v. W then checks if $\pi(\phi(u)) \neq \pi(\phi(v))$ and $\pi(\phi(u)), \pi(\phi(v)) \in \{1,2,3\}$. *If* either condition is violated, W rejects and stops protocol, *else* W reveals $(\pi''(\phi(u)), r''_u, s_u)$ and $(\pi''(\phi(v)), r''_v, s_v)$ to B.

Step 11 B checks if $R''_u = f(f(\pi''(\phi(u)), r''_u), s_u)$ and similar for R''_v. He also checks if $\pi''(\phi(u)) \neq \pi''(\phi(v))$ and $\pi''(\phi(u)), \pi''(\phi(v)) \in \{1,2,3\}$. *If* either condition is violated, B rejects and stops protocol, *else* one continues with the next iteration if the number of iterations is less than l (else stops and B accepts).

As in [GMW86], ϕ is never released. Their main theorems remain valid, so abuse-free zero-knowledge protocols exist for all NP languages. The abuse-freeness of the above protocol is not unconditional.

If W is forced to stop the protocol (see Step 8 and Step 10), one could correctly remark that A or B has succeeded in leaking one bit of information. However, this is not an abuse according to our definition because the warden can detect it, too. In practice it means that one is able to leak one bit of information (the fact that W was forced to stop the protocol), *however*, the risk to be caught is too high to attempt it. *The same remark is valid for most protocols that we will discuss further in this paper. We will not repeat this remark.*

Let us now explain how to make [BC86] zero-knowledge proof abuse-free. We will use the same notations as in [BC86]. Here, we only demonstrate how A can prove, in an abuse-free way, to B that $b_1 = b_2$ without revealing them. It is then trivial (see [Des]) to extend the results to make the [BC86] zero-knowledge proofs abuse-free. We assume that B has published abuse-free $y \in QR_n$ and n, such that $n = pq$ where p and q are both primes congruent to 3 modulo 4 and convinced A by using a [GMW86] type abuse-free zero-knowledge protocol that y and n satisfy the conditions. Remark that the requirement that y, n and the last proof have to be abuse-freeness can be relaxed. This means, for example, that if A would know the factorization of n, it would *not* help him to abuse the following protocol. The abuse-free protocol to prove that $b_1 = b_2$ is as follows:

Step 1 A chooses a (random) w_1, such that $\gcd(w_1, n) = 1$ and calculates $z_1 = \pm w_1^2 y^{b_1} \bmod n$. A calculates z_2 in a similar way. A then calculates w as in [BC86] (if $b_1 = 1$ and $b_2 = 1$ then $w = w_1 w_2 y \bmod n$, else $w = w_1 w_2 \bmod n$). A sends z_1, z_2 and w to W.

Step 2 W verifies if $z_1 z_2 = \pm w^2 \bmod n$ or if $z_1 z_2 = \pm w^2 y \bmod n$. He also verifies if the Jacobi symbols $(z_1 \mid n) = 1$ and $(z_2 \mid n) = 1$. If either condition is violated W stops protocol, else he chooses truly random ψ_1 and ψ_2 coprime with n and sends $z_1' = \pm \psi_1^2 z_1 \bmod n$, $z_2' = \pm \psi_2^2 z_2 \bmod n$ and $\omega = \psi_1 \psi_2 w \bmod n$ to B.

Step 3 B verifies if $z = z_1' z_2' = \pm \omega^2 \bmod n$, then $b_1 = b_2$, else $b_1 \neq b_2$.

The abuse-freeness is unconditional and the protocol is practical. Remark that in the original [BC86] proof, n didn't have to have the special form we request, and the $\pm$ were not used in the protocol. Without these modifications it would have been impossible to make the protocol abuse-free without increasing the overhead enormously. Purdy made the observation to the author that the test of the Jacobi symbol can be eliminated by choosing $p \equiv 3 \bmod 8$ and $q \equiv 7 \bmod 8$. However, it must then be replaced by a test for gcd, which is almost as involved.

3.3 ABUSE-FREE NON-INTERACTIVE ZERO-KNOWLEDGE

Non-interactive zero-knowledge protocols were introduced by [BFM88]. Let us briefly discuss, from our point of view, the main ideas used in it. Prover and verifier share a common random string (the rand tables). The verifier does not need to toss secret coins, *however*, the prover tosses secret coins. When A proves a theorem, we say that she feeds here private random coin tosses into the proof mechanism.

The main problem from our viewpoint is that nothing guarantees that the prover will indeed toss coins and will not proceed differently in order to abuse. We now sketch how one can make abuse-free non-interactive zero-knowledge (more details are in [Des]).

In the set-up process each individual makes a secret abuse-free seed, in a similar way as in Section 3.1. This means that the individual, A, chooses a number, R, and that she sends the warden W a commitment to R. The warden, W, chooses a random number R' and sends it to A. We call $S = R \oplus R'$ (bit by bit exclusive or) the abuse-free seed. Let us now discuss how to proceed when A wants to prove a theorem τ in non-interactive zero-knowledge. A uses a commonly agreed upon pseudo-noise generator which starts from the abuse-free seed. The output of that generator is fed to the proof mechanism. To understand the idea, it is important to observe that by knowing the seed, the proof mechanism is a deterministic process. A now also generates a non-interactive proof to demonstrate that she used the seed S in the correct way and that indeed $S = R \oplus R'$, where the R' had been chosen earlier by the warden and that she committed herself to R. R itself will never be released! So the warden receives *two* non-interactive zero-knowledge proofs: the first for theorem τ and the second to prove that A "decently behaved" when she was proving the first theorem. The warden verifies both theorems and will censor the second one. If both are correct, the warden publishes (or sends) the first proof.

The abuse-freeness is *not* unconditional. A problem of the above solution is that it suffers from the collapse problem. Indeed, suppose that there exists for a few days a covert channel with small capacity between the prover and a verifier, which cannot be controlled by the warden, W. If the prover sends to that verifier the seed S (secret previously unknown), then the zero-knowledge disappears from a practical point of view (it is still theoretically zero-knowledge). The prover can then later abuse the non-interactive protocol to send the complete proof to the verifier and the warden will believe falsely that the verifier will not learn more than the fact to be convinced that the prover knows a proof. The protocols discussed earlier in this paper didn't suffer from this collapse problem.

The warden in this scenario is less active than in previous ones. His only action is verification and censoring, we therefore call him: *censoring warden*. The idea of censoring warden opens the question if it is possible to reduce the warden's role to a passive one keeping the abuse-freeness. *If it would be possible to generate* true *randomness and to prove in some zero-knowledge way that indeed the numbers are truly random, then the above open problem could be solved.* In this context, one could think to use [GMW86, p. 182] ideas to prove *pseudo*randomness, however in many of the systems discussed here the abuse-freeness is unconditional. So the question is if one can benefit from both unconditional abuse-freeness and a passive warden. Making the warden less active does not necessarily imply that the system becomes more practical. Indeed, the above solution is, for the moment, completely impractical because the prover has to perform a tremendous amount of work.

4 Abuse-free crypto-systems: in narrow contexts

We will briefly discuss particular abuse-free crypto-systems which are more useful for daily life applications.

4.1 ABUSE-FREE PRIVACY

It is possible to make probabilistic public-key encryption systems as [BG84] abuse-free. This may seem meaningless. However, it could be that a warden allows A to send m encrypted bits but no more. This can be achieved by making it abuse-free, regardless of the fact that [BG84] expands the data.

4.2 ABUSE-FREE AUTHENTICATION AND SIGNATURES

Based on [GMR88], zero-knowledge and the idea of an active warden, a "practical" abuse-free (public-key) authentication system and a less practical abuse-free signature system were presented by the author in [Des88]. The author, [Des88] observed also that in the case that zero-knowledge is combined with [GMR88] for authentication purposes, the authentication tree can be dropped without endangering the proven secure aspect of the scheme. We now discuss a more practical abuse-free signature system.

We briefly discuss here how one can make an abuse-free signature system based on [FFS87,FS86]. Let $(n, I_1, \ldots, I_k)$ be an abuse-free public key, such that $n = pq$, where both p and q are congruent to 3 modulo 4 and $I_i = s_i^2 \bmod n$. We assume that n has indeed this form and that the warden has been convinced of this for once and for all. When A wants to sign the message m:

Step 1 W chooses truly random $r_i \in \mathrm{QR}_n$, and random k_i and sends $E_{k_i}(r_i)$ $(1 \le i \le t)$ as commitments to A.

Step 2 A picks (random) $\rho_i \in \mathrm{QR}_n$ and computes $x_i = \rho_i^2 \bmod n$ and sends these x_i to W.

Step 3 W reveals (r_i, k_i) to A.

Step 4 A verifies commitment. If satisfied, A computes $x_i' = r_i^2 x_i$ and computes e_{ij} starting from x_i', in a similar way as in [FS86].

Step 5 One continues in a similar way as in [FS86] with the x_i'. W verifies if A

has used the r_i. *A has to prove (using zero-knowledge) to W that* all *the numbers that she has sent, except the* e_{ij}, are quadratic residues *mod* n. Then W publishes the signature.

The problem that f is non-random remains, which implies that problems arise to prove the security of the signature system, similarly as in [FS86]. Two aspects of the above protocol can be improved by modifying it. The above protocol is not unconditionally abuse-free. Indeed if the commitment function E can be broken, then A can abuse it. This evil can be overcome by changing the protocol so that A is committing herself, instead of the warden, similarly as it was in all previous protocols. In Step 5, it was mentioned that A has to prove that all numbers (except e_{ij}) are quadratic residues, this includes all ρ_i. One can drop the zero-knowledge proof for ρ_i if $\rho_i \in Z_n^{+1} = \{y \in Z_n^* \mid (y \mid n) = 1\}$, *and* if the warden can choose randomly, with uniform distribution, in polynomial time y's with Jacobi symbols $(y \mid n) = 1$. Details of the modified algorithm are described in [Des].

4.3 Abuse-free coin flipping over the telephone

We now apply our tools to make a modified version of Blum's protocol abuse-free.

We will base our solution on this assumption: it is hard to determine if a number is a quadratic residue mod n. If Alice (A) and Bob (B) want to flip a coin, then the following protocol is abuse-free:

Step 1 A (with W) generates an abuse-free public key n, which is the product of *two* large distinct primes both congruent to 3 modulo 4, using the protocol of Section 3.1.

Step 2 A generates a (random) X, such that $\gcd(X,n) = 1$ and sends $Y = \pm X^2 \bmod n$ to W.

Step 3 W checks if the Jacobi symbol $(Y \mid n) = 1$. *If* it is *not* 1, *then* W stops protocol (or asks another Y), *else* W generates truly random X' and ± 1, such that $\gcd(X', n) = 1$ and sends $Y'' = \pm Y X'^2 \bmod n$ to B. *If* the warden's ± 1 is 1, *then* $g' = 1$, *else* $g' = 0$.

Step 4 B guesses if Y'' is a quadratic residue mod n. *If* he thinks it is, *then* he sends $g = 1$ to W, *else* he sends $g = 0$.

Step 5 W sends $g'' = g \oplus g'$ to A.

Step 6 A calculates the outcome of the protocol $g'' \oplus q$, where $q = 1$ *when* Y is quadratic residue (mod n), *otherwise* $q = 0$. A then reveals X to W.

Step 7 W verifies Y. *If* it is correct, *then* W reveals $X'' = X \cdot X' \bmod n$ to B, *else* W stops protocol.

Step 8 B verifies Y''. B must still be convinced that n is of the appropriate form. Hereto:

Step 9 A (with W) proves to B that n is of the appropriate form by using an abuse-free zero-knowledge protocol.

Step 10 B verifies this abuse-free zero-knowledge protocol. If satisfied, B calculates $g \oplus q''$ as the outcome of the protocol, where $q'' = 1$ if Y'' turns out to be a quadratic residue, else $q'' = 0$. Remark that A and B have the same outcome, in other words $g \oplus q'' = g'' \oplus q$.

Notice that A and B are not able to abuse one bit, not even the outcome bit. Even if W collaborates with A (or similarly with B), A cannot benefit from this collaboration to influence the outcome of the protocol in her favor.

The above protocol is not unconditionally abuse-free. The impossibility that A could abuse the coin-flip is unconditional, but B could do it if it were easy for him to determine whether a number is a quadratic residue. This means that there is a small collapse problem. Indeed, if a hidden channel not under control by the warden, exists for a while between A and B, A could use it to send the factorization of n to B. Later (when the hidden channel is no longer), when A would choose her $\pm$ in Step 2 in a way that B could predict, B is able to calculate, from Y'', what the warden's choice for g' was. This allows him to choose his g in such a way that $g'' = g \oplus g'$ contains the subliminal information for A. The collapse problem is a direct consequence of the fact that the protocol is not unconditionally abuse-free.

The above protocol is very practical beside the fact that in Step 1 one makes an abuse-free public key n. The question, if it is allowed to drop this condition as long as A proves to W that n is of the appropriate form, can be derived from the discussion about the small collapse problem, higher up.

[Blu82] already suggested authenticating all communications. It is possible to come up with a system such that A and B can flip a coin in an abuse-free way and authenticate the coin-flip in an abuse-free way by using our ideas of Section 4.2.

5 Theoretical and practical consequences: conclusion

Zero-knowledge allows A to restrict leaks of information *if A wants*. This paper demonstrates that a warden can *enforce* A not to leak information, even if A tries her hardest. Solutions were presented in a general context and practical solutions were presented in particular contexts. In some of these practical protocols, the warden's role is minimal and it is mainly a multiplication.

Applications of abuse-free cryptography are in the area of international communications as: international bank transfers, authentication of international messages, and treaty verification [Sim83a,Sim88] (in the context of authentication without privacy). If coin-flipping and similar protocols had been used on a large international scale, abuses would have formed a threat for (national) security. The above protocols prevent this danger in such a way that users of the system do not have to trust the warden's integrity. Objections against the use of (public key) cryptography (as in [PK79, p. 344]), in particular against authentication, grounded on the fear that terrorists would be able to communicate encrypted information, now vanishes when abuse-free systems are used. So, it also promotes the commercial use of cryptology.

One can wonder if the above solutions are applicable to covert-channel-free computation and computer-security in general.

Acknowledgements: The author thanks the anonymous referees for encouraging the author to make the definitions more precise and to discuss different models for wardens, which are both included in [Des], and to emphasize the difference between covert channels and subliminal channels. George Purdy called the author's attention to the collapse problem. Russell Impagliazzo suggested the name: censoring warden. The author wishes to thank all those who showed interest for this work, in particular, Manuel Blum, Gilles Brassard, David Chaum, Shimon Even, and Jennifer Seberry.

6 References

[Ada88] J. A. Adam. Ways to verify the U.S.-Soviet arms pact. *IEEE Spectrum*, pp. 30–34, February 1988.

[BC86] G. Brassard and C. Crepeau. Non-transitive transfer of confidence: a perfect zero-knowledge interactive protocol for SAT and beyond. In *27th Annual Symp. on Foundations of Computer Science (FOCS)*, pp. 188–195, IEEE Computer Society Press, October 27–29 1986. Toronto, Ontario, Canada.

[BFM88] M. Blum, P. Feldman, and S. Micali. Non-interactive zero-knowledge and its applications. In *Proceedings of the twentieth ACM Symp. Theory of Computing, STOC*, pp. 103 - 112, May 2–4, 1988.

[BG84] M. Blum and S. Goldwasser. An efficient probabilistic public-key encryption scheme which hides all partial information. In *Advances in Cryptology. Proc. of Crypto '84 (Lecture Notes in Computer Science 196)*, pp. 289–299, Springer–Verlag, New York, 1985. Santa Barbara, August 1984.

[Blu82] M. Blum. Coin flipping by telephone - a protocol for solving impossible problems. In *digest of papers COMPCON82*, pp. 133–137, IEEE Computer Society, February 1982.

[Dep83] *Department of Defense Trusted Computer System Evaluation Criteria.* U.S. Department of Defense, August 15 1983. Also known as the Orange Book.

[Des] Y. Desmedt. Abuse-free cryptosystems: particularly subliminal-free authentication and signature. In preparation, available from author when finished.

[Des88] Y. Desmedt. Subliminal-free authentication and signature. May 1988. Presented at Eurocrypt'88, Davos, Switzerland, to appear in: Advances in Cryptology. Proc. of Eurocrypt 88 (Lecture Notes in Computer Science), Springer–Verlag.

[DGB87] Y. Desmedt, C. Goutier, and S. Bengio. Special uses and abuses of the Fiat–Shamir passport protocol. In C. Pomerance, editor, *Advances in Cryptology, Proc. of Crypto'87 (Lecture Notes in Computer Science 293)*, pp. 21–39, Springer–Verlag, 1988. Santa Barbara, California, U.S.A., August 16–20.

[FFS87] U. Feige, A. Fiat, and A. Shamir. Zero knowledge proofs of identity. In *Proceedings of the Nineteenth ACM Symp. Theory of Computing, STOC*, pp. 210 - 217, May 25–27, 1987.

[FS86] A. Fiat and A. Shamir. How to prove yourself: Practical solutions to identification and signature problems. In A. Odlyzko, editor, *Advances in Cryptology, Proc. of Crypto'86 (Lecture Notes in Computer Science 263)*, pp. 186–194, Springer–Verlag, 1987. Santa Barbara, California, U. S. A., August 11–15.

[GMR] S. Goldwasser, S. Micali, and C. Rackoff. The knowledge complexity of interactive proof systems. to appear in Siam J. Comput., vol. 18, No. 1, January 1989.

[GMR88] S. Goldwasser, S. Micali, and R. Rivest. A digital signature scheme secure against adaptive chosen-message attacks. *Siam J. Comput.*, 17(2), pp. 281–308, April 1988.

[GMW86] O. Goldreich, S. Micali, and A. Wigderson. How to prove all NP statements in zero-knowledge and a methodolgy of cryptographic protocol design. In A. Odlyzko, editor, *Advances in Cryptology, Proc. of Crypto'86 (Lecture Notes in Computer Science 263)*, pp. 171–185, Springer–Verlag, 1987. Santa Barbara, California, U. S. A., August 11–15.

[Lam73] B. W. Lampson. A note on the confinement problem. *Comm. ACM*, 16(10), pp. 613–615, October 1973.

[PK79] G. J. Popek and C. S. Kline. Encryption and secure computer networks. *ACM Computing Surveys*, 11(4), pp. 335–356, December 1979.

[RS85] R. L. Rivest and A. Shamir. Efficient factoring based on partial information. In F. Pichler, editor, *Advances in Cryptology. Proc. of Eurocrypt 85 (Lecture Notes in Computer Science 209)*, pp. 31–34, Springer–Verlag, Berlin, 1986.

[Sim83a] G. J. Simmons. Verification of treaty compliance–revisited. In *Proc. of the 1983 IEEE Symposium on Security and Privacy*, pp. 61–66, IEEE Computer Society Press, April 25–27 1983. Oakland, California.

[Sim83b] G. J. Simmons. The prisoners' problem and the subliminal channel. In D. Chaum, editor, *Advances in Cryptology. Proc. of Crypto 83*, pp. 51–67, Plenum Press N.Y., 1984. Santa Barbara, California, August 1983.

[Sim88] G. J. Simmons. How to insure that data acquired to verify treaty compliance are trustworthy. *Proc. IEEE*, 76(5), pp. 621–627, May 1988.

HOW TO (REALLY) SHARE A SECRET[1]

Gustavus J. Simmons
Sandia National Laboratories
Albuquerque, New Mexico 87185

Introduction

In information based systems, the integrity of the information (from unauthorized scrutiny or disclosure, manipulation or alteration, forgery, false dating, etc.) is commonly provided for by requiring operation(s) on the information that one or more of the participants, who know some private piece(s) of information not known to all of the other participants, can carry out but which (probably) can't be carried out by anyone who doesn't know the private information. Encryption/decryption in a single key cryptoalgorithm is a paradigm of such an operation, with the key being the private (secret) piece of information. Although it is implicit, it is almost never stated explicitly that in a single-key cryptographic communications link, the transmitter and the receiver must unconditionally trust each other since either can do anything that the other can.

Even if it can't be assumed that all of the elements in a system are trustworthy, so long as there exists at least one identified unconditionally trustworthy element (individual or device), it is generally possible to devise protocols to transfer trust from this element to other elements of unknown trustworthiness to make it possible for users to trust the integrity of the information in the system even though they may not trust all of the elements. A paradigm for such a protocol is the cryptographic key distribution system described in ANSI X9.17 which makes it possible for

1. This work performed at Sandia National Laboratories supported by the U. S. Department of Energy under contract no. DE-AC04-76DP00789.

users who have had no previous contact, nor any reason to trust each other, to trust a common cryptographic session key because they each unconditionally trust the key distribution centers (KDC).

The more common (and hence the more realistic) situation is that there are no identified unconditionally trustworthy elements in a system. Instead, the most that can be assumed is that while any specific element may be suspect, i.e., possibly subject to either deliberate or inadvertent compromise, and hence untrustworthy insofar as the faithful execution of the part of the protocol entrusted to it, that there are some (unidentified) elements in the system which are trustworthy. Under these circumstances there is apparently only one way to improve the confidence one can have in the integrity of the system over the confidence one has in the integrity of the individual elements, and that is by introducing some form of redundancy. To protect against random failures of devices, this is commonly achieved by parallel or by series-parallel operation of redundant elements or by even more complex logical interconnections. In the case of individuals, though, since the failure may be both deliberate and clandestine, redundancy typically takes the form of requiring the concurrence of two or more knowledgeable persons to carry out an action. A paradigm for this would be the well-known two-man control rule for access to, or the control of, nuclear weapons. The k-out-of-ℓ shared secret or threshold schemes first discussed by Blakley [10] and Shamir [33], and subsequently by numerous other authors [see the bibliography], are a natural generalization of this concept. In fact, shared secret schemes exist that are adequate to the task of insuring shared capability if all that is needed is a simple k-out-of-ℓ participation for the reconstruction of a secret piece of information essential to the system functioning. Ideally, any collusion of k-1 or fewer of the holders of information -- even if they pool their private pieces of information in an effort to cheat the system -- should have no better chance of success than an outsider who knows no private information at all. Schemes in which this latter condition holds have been characterized as "perfect" by Stinson [33,34]. We merely remark that several perfect k-out-of-ℓ shared secret or threshold schemes have been described in the literature. Many of these schemes are also unconditionally secure in the sense that the security they provide is independent of the computing time or power that an opponent may bring to bear on subverting the system, or, put in another way, even with infinite computing power would-be cheaters can do no better than guess (with a uniform probability distribution on the choices available to them) at the secret. If the secret is a function (such as one of the coordinates, or the largest coordinate, or the norm of the coordinates, etc.) of a (secret) point in some n-dimensional vector space over a finite field GF(q), then by choosing q large enough we can make the system be as secure as we wish for an arbitrary $k < \ell$. These are "plain vanilla" shared secret schemes for which several implementations have been devised [see the references flagged with an * in the bibliography]. Consequently, there is no difficulty in providing (and implementing) simple shared secret schemes for arbitrary choices of k and ℓ and for any desired level of security.

Real-world applications, however, require rather considerably more in the way of capabilities in shared secret schemes than a simple k-out-of-ℓ concurrence for an action to be initiated. In this paper we will do two things: first enumerate and briefly describe eight of these extended capabilities and then (in compliance with the unanimous recommendation of the reviewers) describe in detail how to realize only one class of these extensions in order to keep the length of this paper within reasonable bounds.

Capabilities Required for Various "Real" Applications of Shared Secret Schemes

The new capabilities (over and above the simple k-out-of-ℓ shared secret schemes) are:

- Compartmented[2] k_i-out-of-ℓ_i shared secret schemes in which the private information is partitioned in such a way that reconstruction of the secret requires a specified level of concurrence by the participants in some specified number (perhaps all) of the compartments (k_i concurrence is required of the members of the ith compartment).
- Multilevel[2] k_i-out-of-ℓ_i shared secret schemes in which the private information is partitioned into two or more levels (classes) in such a way that concurrence of the specified number of participants at any one of the levels will permit the secret to be reconstructed (k_i concurrence by the members of the ith or higher levels is required).
- Extrinsic as opposed to intrinsic shared secret schemes, i.e., schemes in which the value of a private piece of information to the reconstruction of the secret depends only on its functional relationship to other pieces of private information, and not on its information content (in an information theoretic sense).
- Prepositioned shared secret schemes in which the holders of the private pieces of information are unable to recover the secret information, even if they all collude to do so, until such time as the scheme is activated by communicating additional information.
- Prepositioned shared secret schemes in which the same collection of private pieces of information can be

2. We have adopted standard security terminology in which information is classified into levels (classifications) and into compartments (need to know) to describe the two types of partitioning of the private pieces of information in a shared secret scheme.

used to reveal different secrets depending on the choice of the activating information.

- Proof of correctness of the reconstructed secret information to a confidence of $\approx 1-P_d$, where P_d is the probability of guessing the secret.
- Tolerance of erroneous inputs of some number, s, of the private pieces of information, i.e., the correct secret information will be calculated even though s of the inputs are in error, where s is a design parameter.
- A cryptographically secure mnemonic technique to make it possible for the participants to recover a private piece of information that they can't remember using a piece that they can.

It is easy to conceive of situations in which it might be desirable that some action require a preselected level of concurrence by two or more parties in order for the action to be executed. For example, a treaty might require that two out of a Russian control team and two out of a U. S. team agree that the controlled action is to be taken before it could be initiated. What is different about such a compartmented scheme from the simple k-out-of-ℓ schemes, is that no matter how many of the participants of one nationality (compartment or part) concur, the action is to be inhibited unless the preselected number of the other nationality also concur. Clearly, there is nothing special about partitioning the private information into only two parts (compartments). The specific application will determine how many parts are needed to effect the type of concurrence desired.

In *Animal Farm*, George Orwell's animals have a slogan "All animals are equal, but some animals are more equal than others" which is certainly descriptive of the apportionment of authority in most organizations. While it is not true, for example, that two members of the Joint Chiefs of Staff equal one President, it is easy to conceive of circumstances in which the President might wish to delegate authority to the Joint Chiefs to initiate some action with the proviso that "If two of you agree that the circumstances warrant, then this is what you should do...." On the other hand, there are also plausible scenarios in which the concurrence of larger numbers of persons with lesser authority (and responsibility) could act in the stead of smaller numbers of higher authority. For example, it might well be the case that any senior officer of a bank can authorize an electronic funds transfer up to some specified limit, but that in the absence of a senior officer, any two senior tellers could do so, etc. The point is that authority in the real world is typically different for different classes (levels) -- like it or not -- and that consequently control schemes for information, i.e., shared secret schemes, need to reflect this class structure. We describe such schemes as multilevel k_i-out-of-ℓ_i schemes, where realistically the number of levels is small and the values of the k_i are determined by the requirements of the application. The notion of a hierarchy of shared secret schemes was already anticipated in Shamir's paper, but in a form (intrinsic) that as we shall see has very serious deficiencies for real-world applications.

In a multilevel system, the persons holding the private pieces of information are grouped into classes (levels) such that the private information one class has is more (or less) valuable in recovering the secret than that which another class has. In all of the perfect shared secret schemes that we know of, the private pieces of information are not used to directly reconstruct the "secret" itself but instead are used to reconstruct an algebraic variety (a line, a plane or other linear subspace in many of the previously reported schemes but more generally complex varieties defined by polynomial constraints in an n-dimensional space) whose description, i.e., precise specification, is unknown to the holders of the private information. If there were no other constraints, a multilevel system would be trivial to realize for any set of k_i, since a simple shared secret scheme is possible for each k_i. To realize a multilevel system, the ith class could simply have its own separate and distinct k_i-out-of-ℓ_i shared secret scheme. This might be acceptable in some applications, but not in general. If, for example, a bank vault can be opened by either two VP's or three senior tellers, it would probably be unacceptable that one VP and two senior tellers not be able to open it. If the capabilities (private pieces of information) of members of the more privileged classes are to be usable when they cooperate with members of other less privileged classes, then the schemes are forced to be functionally related. We know how to do this in two ways, which leads into the discussion in the next paragraph of extrinsic and intrinsic shared secret schemes.

To illustrate an intrinsic shared secret scheme, assume that we have a 4-out-of-ℓ scheme in some n-dimensional space over GF(q). The private pieces of information are points in the space, i.e., n-tuples over GF(q), chosen so that any set of four of these points suffice to define the secret but any set of three or fewer will provide no information whatsoever about the secret. Clearly we could construct a 2-out-of-ℓ class by making the private pieces of information for the members of this more privileged class consist of pairs of the points out of the original set, i.e., two n-tuples. In fact, this is how Shamir proposed to realize what he called hierarchical control schemes. This type of construction of the private pieces of information is what we call an intrinsic scheme in which the value of a piece of private information (i.e., its contribution toward recovering the secret information) is internal to the private information itself. In an information theoretic sense, the more privileged pieces of information are more valuable simply because they contain more information about the shared secret. This means that the most privileged members would be responsible for the largest amounts of private information, and in the case of several levels with widely differing k_i perhaps responsible for infeasibly much information for them to handle (securely). Such hierarchical schemes have been discussed before, not only by Shamir [32], but by Ito, et al. [23], and other authors.

In an extrinsic scheme all the private pieces of information are alike in an information theoretic sense, say the coordinates of a single point in some n-dimensional space, and its value in recovering the secret is determined not by anything internal to that piece of information but rather by the functional relation between that particular piece of private information (point) and the private pieces of information (points) held by the other participants. In other words, the value is determined by something external to the private pieces of information. An extrinsic scheme does not penalize the more privileged classes by requiring them to handle more information than members of less privileged classes.

Prior to the results described here, there was no means known to realize either extrinsic multilevel control schemes or compartmented (multipart) schemes. Ito, Saito and Nishizeki [23] had devised an intrinsic general access control scheme which, however, can not be extended to an extrinsic scheme and all (k,ℓ) threshold schemes can be adapted in an obvious way to intrinsic (hierarchical) multilevel schemes similar to those Shamir proposed [32].

In a prepositioned shared secret scheme, say a simple k-out-of-ℓ scheme, the ℓ pieces of private information can all be placed in the hands of the participants in advance of when the scheme will be needed; with the added property that until the scheme is activated by providing some additional information, that even if all ℓ of the private pieces of information were to be exposed in violation of the protocol, the secret would not only not be exposed but it would be just as unlikely to be recovered, i.e., just as secure, as if none of the private pieces of information had been compromised. Only when the additional piece of information is made available does the system become activated, after which any set of k of the pieces of private information will allow the secret to be recovered. It is worth remarking that there is a trivial realization of a prepositioned shared secret scheme by simply making $\ell = k-1$, i.e., by designing a k-out-of-ℓ shared secret scheme, in which all of the private pieces of information when taken together are inadequate to recover the secret, but such that one more piece (the activating information) is required. We are not interested in such schemes since they fail to meet the most fundamental requirement of k-out-of-n systems, namely, avoiding the necessity to have to bring together a designated set of k private pieces of information in order to reconstruct the secret information. The main reason for being interested in prepositioned shared secret schemes is that the (relatively) large quantity of private information can be disseminated, authenticated, etc., in times of low stress and easily available communication and the small quantity of information needed to activate the scheme can be communicated under extreme duress -- such as a state of advanced alert for the military or even the outbreak of war.

A relatively new discovery is the possibility of setting up a prepositioned shared secret scheme, i.e., prepositioning the private pieces of information, with the additional property that there are several activating pieces of information available, each of which would lead to the recovery of a distinct secret piece of information. This could be a very valuable characteristic in some military applications where there are several different actions -- any one of which higher command might wish to enable -- but subject to a k-out-of-ℓ shared secret control in execution. The basic idea is that one needn't change the private pieces of information (which would require a great deal of communication, authentication, etc., and presents an enormous human factors problem) in order to change the secret protected by the shared secret scheme.

If the consequence of exercising a shared secret scheme is immediate -- for example, if after the VP's enter their private pieces of information, the bank vault door either opens or it doesn't -- then there is no need to provide a supplemental indication that the correct value for the secret has been recovered. If however the effect is distant, in either time or physical location, then it may be vital to the acceptability of the scheme that the participants have an immediate indication that the correct value of the secret has been reconstructed. If, for example, a shared secret scheme is to be used to control the enabling of a warhead in a missile, it is clearly desirable to have a confirmation that the correct value has been entered prior to launch as opposed to learning that the weapon had not been enabled after its arrival at the target. Providing an indication that the correct secret has been reconstructed is similar to the function of error detecting codes which, in probability, indicate when a received code word is in error, although we hasten to add that the functions are not identical. This last remark requires more discussion than is appropriate to an abbreviated description of the extended capabilities for shared secret schemes, but basically it is possible to cause a shared secret scheme to indicate when it has reconstructed the correct secret even though the secret itself was unknown prior to the reconstruction (and not available from any other source for direct comparison after reconstruction to determine its validity). This is similar to being able to verify a digital signature without being able to utter one. In general (but not in all cases which is the basis of the preceding remark), this costs one more piece of private information to achieve than is necessary for a simple shared secret scheme, i.e., k+1 instead of k inputs of private pieces of information.

If the capability discussed in the preceding paragraph was only similar in function to error detecting codes, the capability of recovering from erroneous inputs of private pieces of information is precisely the same as the function of error correcting codes. In other words, we can design shared secret schemes so that up to s of the inputs can be in error and not only will the correct value for the secret be found, but if we desire, a proof of correctness can be output to show that the right value has been reconstructed. Clearly this cannot be done for free, since if only k inputs are needed and s can be in error, k-s of the participants could collude and input their correct private pieces of information, after which any s random inputs would suffice to recover the secret. Roughly speaking (not so roughly as a matter of fact since the result is true within one required input) k+s+1 inputs of private pieces of information are needed to guarantee k-concurrence (i.e., k-man control), recovery from s erroneous entries, and a positive indication of the correctness of the secret value recovered.

Several authors have addressed the problem of detecting cheating (falsified inputs) in a secret sharing or threshold scheme [13,16,17,28,36]. McEliece and Sarrwate [28] actually construct a secret sharing scheme based on a Reed-Solomon error detecting and correcting code which can tolerate s incorrect entries. In their construction any set of k + 2s participants (holders of private pieces of information) will be able to correctly reconstruct the secret so long as at most e of the inputs are incorrect or falsified. Tompa and Woll [36] give a construction for an unconditionally perfect k-out-of-ℓ shared secret scheme. In both of these constructions the participants will (probably) be able to tell that cheating has occurred, but they cannot necessarily determine who the cheaters are. The combinatorial scheme of Brickell and Stinson [13] is also an unconditionally perfect k-out-of-ℓ scheme which also has the property that the cheater(s) will be identified in the process (with high probability).

Finally, in this list of capabilities, if k-1 inputs of correct private pieces of information are to provide no information whatsoever about the secret information, then every other piece of private information must appear completely random even though k-1 pieces are known. This says that an unknown n-tuple, if the setting is in an n-dimensional space over some GF(q), must itself appear random, not in all n coordinates, but effectively in α of them if the secret is α dimensional; by which we mean that the equivocation about the secret must be the same as the uncertainty of guessing a point in an α-dimensional space over GF(q). q must be large enough to provide the desired level of security against random picking of points. By present-day computational standards, 56 bits is regarded as barely large enough to be secure, witness the continuing debate over the long-term security of the DES, but 100 bits is unquestionably secure against a brute-force search of the key space. However even the modestly secure limit of 100 bits is a 20 alphanumeric character string that must appear totally random by the remarks above, which is beyond anyone but a stage memory expert's ability to recall. Since shared secret schemes are not communication channels, the standards for the security of a communications cryptographic key do not necessarily apply. But even at 56 bits or 12 alphanumeric characters as required for a DES key, it is still impossible for most people to recall a random string of this length as their private piece of information. Fortunately, there exists an approved mnemonic technique for generating a one-time key of sufficient length, using easily remembered private phrases or verses, to permit the secure recovery of something that can't be remembered (the random appearing private piece of information) from something that can (the private phrase).

There are a great many other technical aspects of shared secret schemes which need to be considered, however the main ones which we have been able to identify that affect the operational acceptability of these schemes have been described here.

The Basic Construction for Shared Secret Schemes

We illustrate the essential elements in the construction of shared secret schemes using the simplest possible example: a 2-out-of-ℓ scheme. Let the secret be a single numerical value, i.e., having a 1-dimensional uncertainty, which is equivalent to the identification of a point, p, on a line, L_d.

Figure 1.

If we now consider L_d to be embedded in the projective plane PG(2,q), and randomly choose any other line, L_i, in the plane, $L_i \neq L_d$, then the private pieces of information can be taken to be distinct points on L_i, none of which are the point p. L_i is kept secret, only the fact that such a line exists, etc., is public knowledge. For the purposes of this paper, L_d will be assumed to be known a priori. There are applications in which this is not the case, but we will not have time to discuss them here.

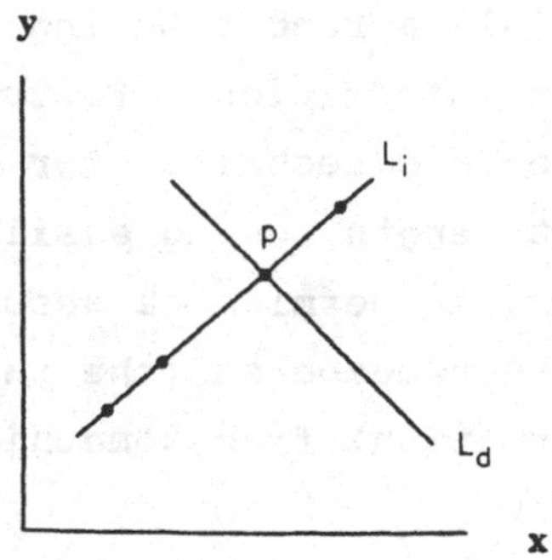

Figure 2.

Any pair of points on L_i determine the line and thence its intersection with L_d; the point p. On the other hand, knowing any one of the points, q, on L_i leaves p totally undetermined since for each choice of a point, r, on L_d there exists a unique line <q,r> lying on q and r which could (with equal probability) be the unknown line L_i -- in which case the (secret) point of intersection of L_i with L_d would be the arbitrary point r. Therefore every point on L_d is an equally likely candidate to be the secret point p given either no knowledge of the private pieces of information (points on L_i) or else of only one private piece. In this example, since p could (equally likely) be any point on the line, $P_d = 1/q+1$, while the number of participants, ℓ, can be as great as q, i.e., any point on L_i other than p could be used as a private piece of information.

It should be remarked that the point p, although it is unknown in advance of the 2-out-of-ℓ scheme being exercised, is not itself the secret. The secret is recovered by evaluating a predesignated function, f, at the point p: f could be as simple as one of the coordinate values of p or the distance of p from some reference point or it could be a much more complex function. Whatever the function is, it is assumed to be known a priori so that as soon as p is determined, so is the secret. There are restrictions on f that must be satisfied in order for it to be suitable for this sort of application. For example, if f were a simple parity check (on the coordinate values) mapping the points on L_d into the set (0,1), then the uncertainty about the secret would be at most one bit irrespective of how many different values p could take. For our purposes, we assume that f conserves entropy, i.e., that the uncertainty about f(p) is the same as the uncertainty about p.

Returning to the simple example shown in Figure 2; p was an (unknown) point in a larger set -- all of the points on the line L_d. The secret revealing function, f, is defined (at least) on all of the points in V_d and as mentioned above, conserves entropy. It is worth noting that it is immaterial (to the secret sharing scheme) whether f is also defined for points in the plane not on L_d. In our construction of shared secret schemes, the line L_d will be replaced by a more general type of geometrical object -- an algebraic variety, V_d, in some n-dimensional space: i.e., the set of points in L_d satisfying a set of specified polynomial constraints. This collection of points, any one of which could be the unknown point p, we will refer to as the domain (variety) for the function f hence the notation V_d. The line L_i can be thought of as "pointing" to the point p in L_d. In the most general formulation, the private pieces of information (points in the n-dimensional space) suffice to define a second algebraic variety, V_i, whose function it is to "point" to the point p in V_d. We will say that V_i is the indicator (variety) using the term indicator with its preferred meaning of pointing to or indicating a specific item, i.e., of pointing to the point p. p we will call the index. Without saying precisely how the private pieces of information determine the indicator, pictorially our shared secret schemes are of the form:

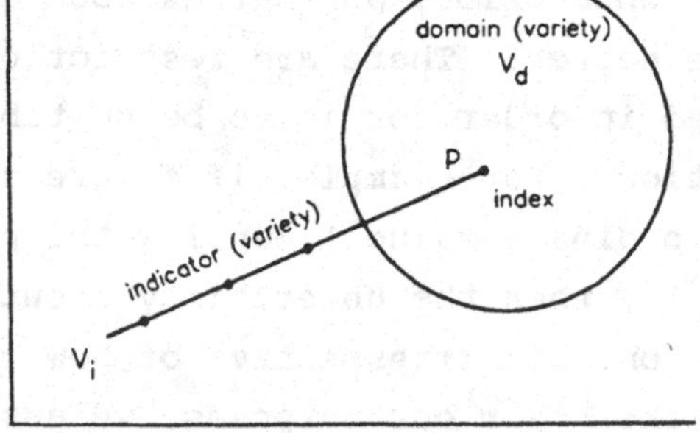

Figure 3.

where V_i and V_d are two algebraic varieties having only the single point p in common. The indicator V_i is shown as a line in Figure 3 to emphasize the fact that it is pointing to a unique point in V_d, but in general it can be any algebraic variety satisfying the conditions for a shared secret scheme. In order for the scheme to be acceptable, we will also require that any compromise (collusion) of less than the required number and types of private pieces of information will leave every point in V_d an equally likely candidate to be the unknown point p. As mentioned earlier Stinson has characterized shared secret schemes meeting this latter condition as perfect [33,34] and we will adopt that terminology also.

An example of a perfect 3-out-of-ℓ scheme is shown in Figure 4.

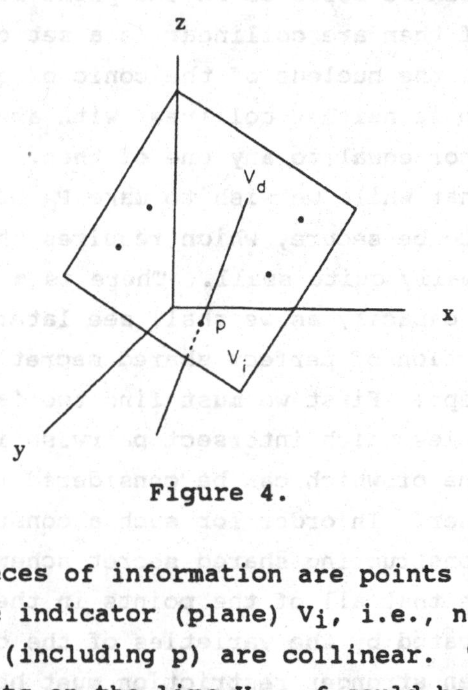

Figure 4.

The private pieces of information are points in general position in the indicator (plane) V_i, i.e., none of them are p and no three (including p) are collinear. The domain is the set of points on the line V_d. f could be any entropy conserving function defined on the points in V_d, say the value of the z-coordinate if V_d is chosen not to lie in a plane perpendicular to the z axis. To see that this scheme is perfect, consider the case in which two holders of private pieces of information collude in an attempt to cheat the system. The two points that they know defines a line, ℓ, in V_i which does not intersect V_d. Given any point r on V_d there exists a unique plane $\langle \ell, r \rangle$ lying on ℓ and r which is equally likely to be the unknown variety V_i as that determined by any other point on V_d. Consequently, for this collusion all points on V_d are equally likely candidates to be the unknown point p, and the scheme is perfect. It is worth remarking about the construction of Figure 4 that while the secret can be any one of q+1 points on the line, V_d, so that the security of the scheme is $P_d = 1/q+1)$, the number of participants, ℓ, could be as great as q or q+1 depending on whether q is odd or even, respectively. This follows from the well known result that the maximum number of points that can be selected in the plane PG(2,q) such that no three of them are collinear is a set of q+1 points on a conic (plus the nucleus of the conic of q is even) and that the point p is neither collinear with any pair of the private points nor equal to any one of them. The point of the remark is that while we wish to make P_d be small, i.e., for the scheme to be secure, which requires that q be very large, ℓ is normally quite small. There is a price exacted for this unused capacity as we shall see later.

The construction of perfect shared secret schemes proceeds in two steps. First we must find two families of algebraic varieties which intersect pairwise in single points, i.e., one of which can be considered to indicate a point in the other. In order for such a construction to be applicable to constructing shared secret schemes it must also be the case that all of the points in the domain variety can be indicated by the varieties of the other type, and in fact, the even stronger restriction must hold that each point in the domain is an equally likely index of the indicator (variety) as the indicator ranges over all possible values. The second step is: given two families of algebraic varieties satisfying these conditions, one must devise ways to define a unique member of one of these families that requires the specified level of concurrence on the part of the holders of the pieces of private information. In the two simple examples this took the form of 2-out-of-ℓ or

3-out-of-ℓ concurrence in order for the indicator (a line or a plane in the examples) to be reconstructed. In general, the required concurrence can be arbitrarily complex; for example, at least one member of each of n committees must be present for a vote to be binding or two, or three, etc. The point is that we want it to be possible to reconstruct the indicator variety only when the specified concurrence occurs and for it not only to be impossible in all other cases, but that the even stronger result will hold that every point in the domain, V_d, will be equally likely to be p in all other cases (collusions).

Our constructions will generally be based on a simple result from point geometry -- the rank formula:

$$r(S) + r(T) = r(S \cap T) + r(S \cup T) \tag{1}$$

which holds for all subspaces S and T of the n-dimensional projective space PG(n,q) or Q^n in short. For notational consistency the empty subspace is defined to have rank 0 and dimension -1. To illustrate how (1) applies, consider the following construction: π_1 and π_2 are planes in a 4-dimensional space, Q^4, which do not lie in a common 3-dimensional subspace. $\pi_1 \cup \pi_2 = Q^4$ in this case, and we have

$$\begin{aligned} r(\pi_1) + r(\pi_2) = 3 + 3 &= r(\pi_1 \cup \pi_2) + r(\pi_1 \cap \pi_2) \\ &= 5 + r(\pi_1 \cap \pi_2) \quad . \end{aligned}$$

Therefore,

$$r(\pi_1 \cap \pi_2) = 1$$

and

$$\pi_1 \cap \pi_2 = p \quad , \qquad p \text{ a point.}$$

Restated; in 4-dimensional space any pair of planes that do not lie in a common 3-dimensional subspace intersect in a point. Clearly this is a candidate construction for the pair of varieties we need to construct a shared secret scheme. We still have to show that the desired uniformity of intersection holds, i.e., that for a fixed π_1, as π_2 ranges over all of the planes in Q^4 that do not intersect π_1 in a line, each point of π_1 will occur equally often as the intersection $\pi_1 \cap \pi_2$. To see that this is true, fix π_1 and choose any line ℓ in Q^4 skew with respect to π_1. Let q be an arbitrary point in π_1, then $<q,\ell> = \pi$ is the unique plane lying on q and ℓ. If $\pi \cap \pi_1$ were a line ℓ^*, i.e., if $\pi \cup \pi_1$ is a 3-dimensional subspace of Q^4, then ℓ^* and ℓ are both in π and hence must intersect in a point. But this point would be in both ℓ and π_1 which contradicts the assumption that ℓ is skew to π_1. Therefore π and π_1 intersect in only the single point q. But q was an arbitrary point in π_1, hence for each skew (to π_1) line ℓ there is a unique plane on ℓ intersecting π_1 at point q. Now let ℓ range over all lines skew to π_1, etc.

We now show how the geometrical result of the preceding paragraph can be used to construct a 3-out-of-ℓ shared secret scheme to conceal a 2-dimensional secret. V_d is an arbitrary, but known a priori, plane in the 4-dimensional projective space Q^4. V_i is a randomly chosen plane which does not lie in any common 3-dimensional space with V_d. A possible selection procedure for V_i is to choose a point q, $q \not\in V_d$, and a point r, $r \not\in <V_d \cup q>$. Note that $q \not\in V_d$ implies by the rank formula that $<V_d \cup q>$ is 3-dimensional. $<q,r>$ is a line skew to V_d. Now choose (with a uniform probability distribution) a point $p \in V_d$ and define

$$V_i = <p, <q,r>> \quad .$$

The private pieces of information will be points in V_i none of which are p, and no three of which (including p) are collinear. Clearly this is a 3-out-of-ℓ shared secret scheme which can indicate any point p in V_d. A simple adaptation of the uniformity argument proves that the scheme is perfect even if two of the pieces of private information (points in V_i) are combined in an attempt to cheat the system. In this case the secret can be any one of the q^2+q+1 points in the plane V_d, so that $P_d = 1/(q^2+q+1)$, while ℓ is at most q or q+1 depending on whether q is odd or even as remarked earlier.

There are a couple of other important points to make about shared secret schemes in general. In the construction of a perfect 3-out-of-ℓ shared secret scheme to secure a 1-dimensional secret shown in Figure 4, the private pieces of information were points in a 3-dimensional space, i.e., 3-dimensional themselves. An alternative construction for a perfect 3-out-of-ℓ scheme which also secures a 1-dimensional secret is:

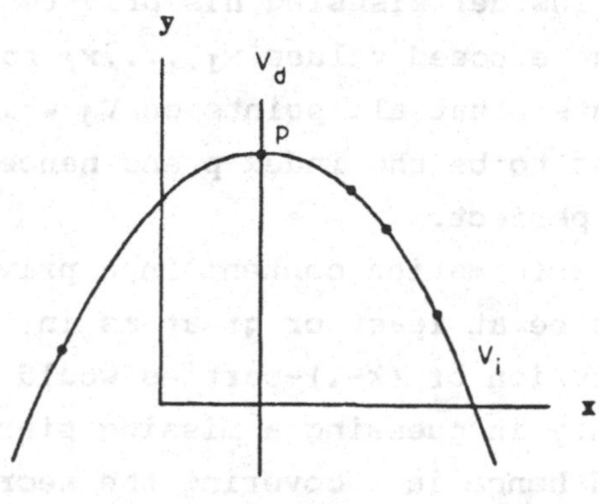

Figure 5.

where any three points on V_i suffice to define the quadratic curve and hence the point p at which it intersects V_d. The private pieces of information in this case are 2-dimensional, i.e, points in the plane. These two examples show that not only is a shared secret scheme not fixed by the specification of the level of concurrence (k-out-of-ℓ) and the dimension of the secret which is to be secured, but that even the dimension of the space in which the scheme is implemented -- and hence the dimension of the private pieces of information -- is not determined.

This leads to the second, and most important, observation: the information in the private pieces of information is not all of the same type in the sense of how it must be secured. To see this, consider the simple 2-out-of-ℓ scheme shown in Figure 2. The private pieces of information are points on the line L_i, i.e., 2-tuples of the form (x_j, y_j). It is not necessary to keep both of these coordinate values secret in order to protect the secret from improper recovery. One of the coordinate values can be kept secret, say y_j, which we indicate by $\textcircled{y_j}$, while the other need not be kept secret but only its integrity (against substitution, alteration, deletion, etc.) needs to be insured. It is easy to show that in the most damaging collusion possible for this scheme (an insider misusing his private information $(x_j, \textcircled{y_j})$ and the exposed values $x_1, \ldots, x_\ell$ for all of the other participants) that all points on V_d will be equally likely candidates to be the index p and hence that the scheme is still perfect.

Clearly the information content in a private piece of information must be at least or great as in the secret, otherwise a collusion of (k-1)-parties would be faced with a lesser uncertainty in guessing a missing piece of private information (and hence in recovering the secret) than the uncertainty they are assumed to have about the secret -- clearly a contradiction. In the example just given, $H(y_j) = H(p)$, i.e., the information content in the part of the private piece of information that has to be kept secret is exactly the same as the uncertainty about the secret itself. As we shall see for the constructions described here this is always possible. What does differ from one realization of a shared secret scheme to another (having the same specifications) is the amount of information in the private pieces of information which doesn't have to be kept secret.

Perfection: At What Price?

The reader has probably wondered why we introduced two varieties in our model for shared secret schemes, one of which was defined by the private pieces of information, and then defined the index to be their intersection instead of simply defining the index directly in terms of the private pieces of information; and whether both of the varieties are necessary. A discussion of the main reason for introducing the domain variety (in addition to the clearly essential indicator variety) will be deferred until a later paper however the simple answer to the question is that the domain can be dispensed with -- but only by sacrificing perfection for the shared secret schemes when $k < \ell$.

To illustrate the difficulty, consider the simplest possible example of a k-out-of-ℓ shared secret schemes, $k < \ell$, in which the private pieces of information directly determine the index shown in the construction in Figure 6. The index in this example is a point, p, in the plane and the private pieces of information are a pencil of lines on p.

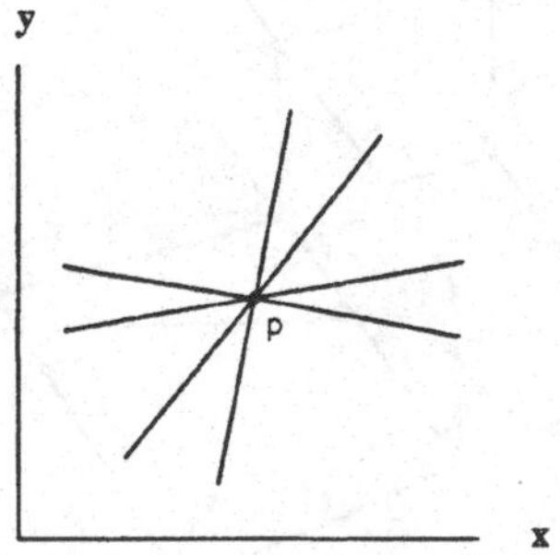

Figure 6.

Since any two of the lines determine p, while a knowledge of any one of them leaves p (linearly) indeterminate this is a 2-out-of-ℓ S^3 for a 1- (2?)-dimensional secret. The ambiguity as to the dimension of the secret is due to the fact that each insider knows that p must lie in the 1-dimensional variety which he knows and hence p is only 1-dimensional in uncertainty to him, while to an outsider p has 2-dimensional uncertainty since it could be any point in the plane.

Since k = 2 in this example, the only improper insider collusion possible is that of a lone individual trying to misuse his private piece of information. As a result, this example does not adequately illustrate what happens when k > 2 and the index is derived directly from the private pieces of information without the aid of an indicator. To show what happens in general, let the secret, p, be a point in a 3-dimensional space and the private pieces of information be a bundle of planes all containing p, but no three of which contain any common line. This is clearly a 3-out-of-ℓ shared secret scheme for a 3-dimensional (to outsiders) secret: any pair of the planes defines a line containing p which, since it isn't in any of the other planes, must intersect each of them at p.

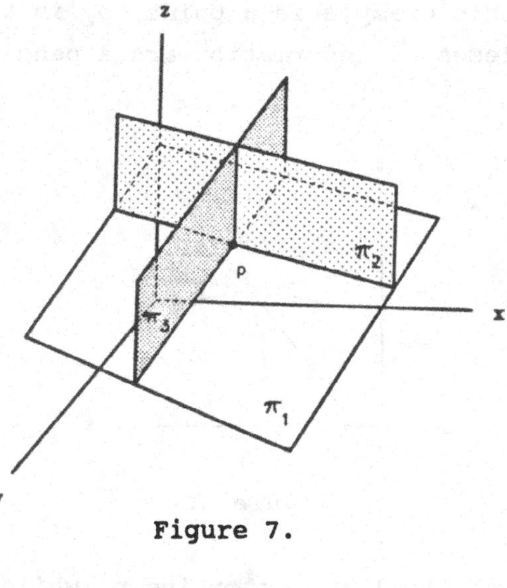

Figure 7.

However, the secret is only of 2-dimensional uncertainty to any single insider since he knows p must be in the plane which is his private piece of information and of only 1-dimensional uncertainty to any pair of insiders since they know p must be common to both their planes and must therefore be contained in the line of intersection of the two planes. The problem is that the index is contained in each of the private varieties in all of these examples (and in general in this type of shared secret schemes) and is identified by the intersection of sufficiently many of the private varieties to determine the index. As a result, the successive intersections define a sequence of, if not monotonically decreasing, at least nonincreasing (in dimension) varieties converging to the point p. It isn't possible to make this sequence of intersections be equal to the dimension (≥ 1) of the secret through the penultimate, (k-1)-st, step in the reconstruction of the secret and then on the final step at which the k-th private variety is introduced to suddenly become of dimension 0. This might be possible if the order in which the various pieces of private information had to be used could be specified in advance, but a shared secret scheme must be immune to compromise by all subsets of k-1 or fewer insiders and in whatever order they choose to collude. Hence this isn't possible. Consequently, erosion of the uncertainty about the index with increasing numbers of persons in a collusion is an inherent shortcoming of all shared secret schemes in which the index (set) is determined directly from the private pieces of information.

An interesting observation, though, is that this need not be true if $k = \ell$. For example, a perfect 2-out-of-2 shared secret scheme is easy to realize (for a secret of any dimension). One of the participants is given a random point, r, in V_d and the other the vector sum (Vernam encryption) of p with r, say p-r. Clearly this is a perfect 2-out-of-2 scheme irrespective of the dimension of V_d. Pictorially, if V_d is 1-dimensional, we have

Figure 8.

Both the secret and the private pieces of information are 1-dimensional. Their sum recovers the 1-dimensional secret, p. To extend this scheme to a perfect k-out-of-k 1-dimensional shared secret scheme, $k > 2$, it is only necessary to give k-1 of the insiders random numbers, r_i, as their private pieces of information and the Vernam cipher $p - \Sigma r_i$ to the k-th individual. In spite of the apparent asymmetry in this assignment procedure which appears to give more significant information to the holder of $p - \Sigma r_i$ than to the individuals whose private information is one of the r_i, this is not the case and any collusion of k-1 or fewer holders of private pieces of information will be totally uncertain of p in the sense that it could (equally likely) be any point in V_d. Obviously, by construction the sum of all k of the points is p. Consequently, not only are all of the pieces of private information equivalent (in uncertainty) but more importantly there is no erosion of the uncertainty about p until the k-th and final piece of information becomes available, at which point p is determined.

This construction for 1-dimensional k-out-of-k shared secret schemes in which there is no indicator but in which there is also no erosion of the uncertainty about the index, p, with the compromise of fewer than k of the private pieces of information can easily be extended to the concealment of secrets of any dimensionality. Let p be an m-dimensional secret (point in Q^n). Choose the k private pieces of information to be k-1 randomly chosen points, r_i, in V_d, and the point $p - \sum_{i=1}^{k-1} r_i$. The combining operation will be the vector sum -- component addition in the underlying finite field. Under these circumstances any subset of k-1 or fewer of the points will leave the index completely undetermined since it could be any point in V_d while the vector sum of all k will, by construction, be p. We remarked earlier that it wasn't possible to make the dimension of the secret and of the private pieces of information both be n in a perfect k-out-of-ℓ

shared secret schemes in the space Q^n if $k < \ell$. What we have seen in the constructions of this section is that there are perfect shared secret schemes in which an indicator doesn't appear and in which this common dimensionality is possible if $k = \ell$. We will utilize these perfect k-out-of-k shared secret schemes, later in a class of constructions for realizing compartmented shared secret schemes in which more than a single group of persons must concur in order for a controlled action to occur.

The emphasis on dimension in the preceding discussion is slightly misleading. While it is certainly true that for a fixed ground space Q, less information is needed to specify a point in Q^m than in Q^n, where $m < n$, as we have already pointed out, this information is not all equally costly to generate, distribute or to protect. In fact the expensive secret part of the private information can be made to be the same in all realizations for a particular set of specifications.

The application normally dictates the level of concurrence, k, required to provide the desired level of confidence in the proper execution of the controlled action and the number of participants, i.e., the number of private pieces of information that the scheme needs to accommodate. The application also dictates the maximum probability, P_d, that can be tolerated of someone (either outsiders or an improper collusion of insiders) guessing the shared secret on whose concealment the control scheme is predicated. If the values that the secret can assume are equiprobable, then the number of such values, i.e., the number of points in the domain, $|V_d|$, must be at least

$$|V_d| \geq \frac{1}{P_d} \quad .$$

There may also be other parameters involved. For example, as we have pointed out earlier, it may be natural to consider the secret information as having a dimension, d, etc.

In summary, both the indicator and domain varieties are essential to the realization of a perfect shared secret scheme. Given the basic construction (concept) of having one variety point to a point in the other at which the secret is defined, the geometrical nature of the resulting shared secret schemes is virtually forced. The problem is to devise ways to insure that the desired level(s) of concurrence will define the indicator and such that no lesser level of collusion will reveal anything about it. There are also important questions connected with making such schemes be practical such as minimizing the amount of secret information that needs to be protected by the holders of the private pieces of information, or of making such schemes robust against either deliberate or unintentional erroneous inputs. However, the basic principal for constructing shared secret schemes is the same in all cases.

An Application (and Two Realizations) of Compartmented Shared Secret Schemes

We consider first the simplest possible compartmented scheme: there are two parties (compartments) to the shared control, both of whom must concur for the controlled action to be initiated. Because of the sensitivity of the action, each party wishes to impose the requirement that at least two members of their control team must agree that the action should be initiated before their party's concurrence can be obtained. To be less abstract, assume that there is some treaty controlled action that requires U. S. and U.S.S.R. concurrence for its initiation. Each country has a team of its own representatives (controllers) at the site. Because the controllers are trusted -- but not unconditionally trusted -- to carry out their nation's commitment to the protocol, each country requires that at least two of their controllers must concur before their national input to the shared control scheme is to be possible. Clearly, this is quite a different control situation than occurs in a simple (k,ℓ) threshold scheme. In the present case, even if all ℓ of the Americans (ℓ could be a large number) and one of the Russians agree, the controlled action is to be inhibited! For simplicity, we will assume that the secret has 1-dimensional uncertainty, i.e., that it is equivalent to identifying a point on a line.

There are two approaches (using the construction for shared secret schemes described here) to constructing compartmented schemes. We will describe both of them and analyze their relative efficiencies in order to justify our choice of a preferred scheme. The first approach is to let the private information for each part(y) determine a subvariety V_j: ordinarily a (k_j-1)-dimensional subspace where the j-th part requires a k_j-out-of-ℓ_j control. These subvarieties are all chosen to be linearly independent subspaces of a common space, i.e., so that no pair of them have a point in common. The indicator variety is then the union of the required number of these subvarieties (both of them in the present example). V_d, as usual is a variety (subspace) any point of which could with equiprobability be p. We have in this case

$$V_i = V_1 \cup V_2$$

and

$$V_i \cap V_d = p \quad ,$$

where

$$\dim(V_i) = \dim(V_1) + \dim(V_2) + 1 \quad .$$

This is conceptually the simpler approach since the resulting compartmented scheme is essentially the same as we have already given for simple k-out-of-ℓ shared secret scheme. Because of the complexity of the general case, we will describe a construction of this type (for the simple two-part example) before describing the other type of construction for compartmented schemes.

We note that a 2-out-of-ℓ, $\ell > 2$, control scheme always determines a line in some space. If the line (shared variety) determined by the U. S. control team is to be independent of the line determined by the U.S.S.R. team, i.e., if the two lines are to be skew so that they do not intersect, then the subspace they span, the indicator V_i, will be 3-dimensional. The domain (variety), which is 1-dimensional from the problem statement, must be independent of the subspace spanned by the two shared varieties, hence the lowest dimensional space in which a scheme of the type we are considering could possibly be constructed would be 4-dimensional. This can be done as follows. Take as the two shared varieties a pair of skew lines, L_1 and L_2, in Q^4. The domain is a third line, V_d, skew to both L_1 and L_2. As usual in a 2-out-of-ℓ shared secret scheme, the private pieces of information will be points on the lines L_1 or L_2, subject to the side condition that none of them are on the unique line, ω, that intersects all three of the lines.[3] The points at which ω intersects the lines L_1, L_2 and V_d are q, r and p, respectively. The lines L_1 and L_2 span a 3-flat $V_i = \langle L_1, L_2 \rangle$ which does not contain V_d. Hence

$$V_i \cap V_d = p$$

which is the index for this particular shared secret scheme.

Since a clear understanding of how this scheme functions is essential to understanding the extensions to be described later, we rephrase in nonmathematical terms what has just been said geometrically. Any two members of the first group

3. *Note*: We prove rather more than is needed for the present construction. In Q^3 there is a unique line passing through a given point, p, and intersecting each of two skew lines L_1 and L_2, neither of which lies on p. To see this, note that p and L_1 determine a plane, π. L_2 intersects π in a point, q; $q \neq p$ by construction since L_2 does not lie on p. The line $\omega = \langle p,q \rangle$ is in π as is the line L_1, so they intersect in a point r. Hence ω is the unique line lying on p and intersecting L_1 and L_2 (in points q and r, respectively). Now consider any space Q^n, $n \geq 4$. Let L_1 and L_2 be a pair of skew lines in Q^n. L_1 and L_2 span a 3-dimensional subspace S of Q^n. Given an arbitrary (n-3)-dimensional subspace, T, of Q^n, independent of S, T intersects S in a single point, p, by the rank theorem. Let this point, p, be the point in the above construction, etc. We therefore have proven that in Q^n, $n \geq 4$, there is a unique line incident with each of a pair of skew lines and with an (n-3)-dimensional subspace independent of each of these lines.

can determine the line L_1 from their private pieces of information. Similarly any two members of the second group can determine the line L_2. Once L_1 and L_2 are known, it is easy to calculate the 3-flat they determine, in other words to determine the polynomial constraints that must be satisfied by all of the points in V_i. The domain V_d, which is assumed to be known a priori, is itself defined by a polynomial constraint. The index, p, is the unique point satisfying all of these constraints. The geometry of the construction guarantees that there is one and only one point satisfying both. Pictorially:

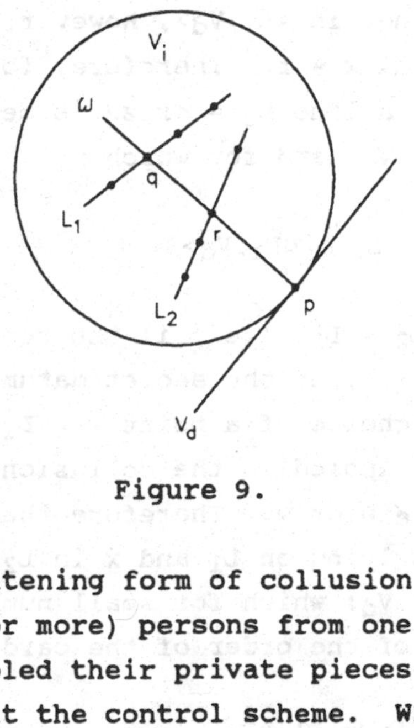

Figure 9.

The most threatening form of collusion for this scheme would be if two (or more) persons from one group and one from the other pooled their private pieces of information in an effort to defeat the control scheme. With no loss of generality, assume that L_1 has been compromised and one point, x, on L_2; $x \neq r$ by construction. To prove that the scheme is perfect we must show that every point on V_d is equally likely to be the secret datum under these circumstances. We extend Kerchoff's criteria from cryptography to shared secret schemes and assume that the geometrical nature of the scheme is known a priori to both insiders and outsiders, i.e., to all would-be cheaters. By this assumption a participant in a collusion knows that L_1, L_2 and V_d are skew lines in Q^4 and that the secret datum is the point of intersection of $V_i = \langle L_1, L_2 \rangle$ with the line V_d.

Choose any point, u, on V_d. An opponent knows that if u is to be the secret datum, it must be collinear with a point of L_1 (which has been exposed by the collusion) and a point on the line L_2. He doesn't know L_2 of course, only that it is a line lying on x and skew to both L_1 and V_d. Let w be an arbitrary point on L_1; not one of the exposed private points (pieces of information) since by construction none of these points are the (unknown) point of intersection, q, of ω with L_1. The line $\omega^* = \langle u, w \rangle$ lies in $\langle L_1, V_d \rangle$ since $u \in V_d$ and $w \in L_1$. x is not in $\langle L_1, V_d \rangle$, however, since $L_2 \cap \langle L_1, V_d \rangle = r$ and $x \neq r$. Therefore, for each point, z, on ω^*, $z \neq u$ or w, a line $L_2' = \langle x, z \rangle$ is determined which is independent of $\langle L_1, V_d \rangle$ and for which

$$L_2' \cap \langle L_1, V_d \rangle = z \quad .$$

Consequently, if $L_2 = L_2'$, i.e., if the constructed line, L_2', were the unknown L_2, then the secret datum would be u. This is true for every choice of a point $w \in L_1$, where w is not one of the points exposed in the collusion, and for all points z on ω^*, $z \neq u$ or w. Therefore the cardinality of the set of schemes lying on L_1 and x in L_2 is the same for all choices of $u \in V_d$; which for small numbers of colluders from group one is of the order of the cardinality of a 2-flat in Q^4.

Since the private points on L_1 were chosen to be different from q, a natural question to ask is whether the equivocation about p might be a function of the number of insiders from group one who join in the collusion. To see that this is not the case consider the most extreme case possible in which ℓ equals the number of points on the line less only the excluded point, q, and all ℓ of the private points are exposed in the collusion. By elimination in this case, q is unambiguously identified and exposed, and the only possible choice for w is w = q. For each choice of a point $u \in V_d$ the number of schemes on L_1, x and u is the number of points on a line less two, since $z \neq u$ or q. Therefore, even in this most extreme case of collusion, all points u on V_d are equally likely to be p insofar as the colluders can determine.

Any other collusion (the line L_1 (or L_2) or else a point on each line, $x \in L_1$ and $y \in L_2$ or else a point on only one of the lines $x \in L_1$ (or $y \in L_2$)) is less damaging than the case just analyzed, i.e., the probability of the collusion improperly determining the index p cannot be increased as a result of the opponent having less information about the scheme. Therefore this construction provides a perfect two-part scheme in which each part is a 2-out-of-ℓ scheme.

To summarize, a construction of the first type to realize a perfect two-party shared secret scheme to secure a 1-dimensional secret, in which each part is a 2-out-of-ℓ control scheme, is possible in four dimensions. Although we haven't described in detail how the private information is to be partitioned into the one part (dimension) which must be kept secret and another (three dimensions) which need not be, an obvious extension to the earlier discussion of the partitioning of the private information applies here as well.

The other approach to realizing a compartmented shared secret scheme is to let the subvarieties determined by the private pieces of information individually indicate points in a space containing V_d which can be treated as inputs to the overall concurrence scheme: in the present case 2-out-of-2 since both of the parties must concur. As we have already seen, k-out-of-k schemes are special so it should come as no surprise that the compartmented scheme is also special in this case (in the sense that it doesn't represent the general behavior of such schemes). Figure 10

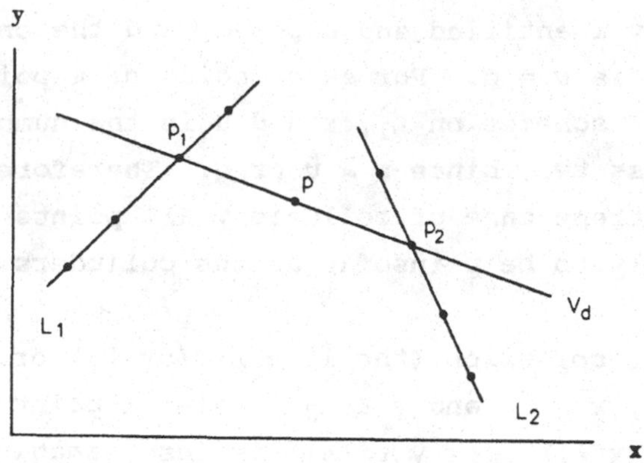

Figure 10.

shows a two-part scheme of the second type. L_1 is the subvariety (line) defined by the private pieces of information belonging to one party and L_2 is the other. The intersection of L_1 with V_d is a point p_1 which is treated as an input to the perfect 2-out-of-2 scheme defined on V_d. p_2 is determined similarly by L_2. Clearly this is a two-part scheme.

To prove that the scheme in Figure 10 is perfect, we introduce a method of proof which, while we have used it before, hasn't been explicitly stated. Given any shared secret scheme, simple, compartmented, multilevel, etc., it suffices to prove that the uncertainty about the index is the same for a more compromising collusion as it is for an outsider attack to simultaneously prove that it is the same for all lower levels of collusion dominated by the case under consideration. For simple k-out-of-ℓ schemes collusions are linearly ordered, so that it is only necessary to consider the most damaging collusion in order to prove perfection (a remark we made earlier). Compartmented and multilevel schemes however have a lattice (often partial) ordering on the collusions. For example the ordering on the five collusions C(0,0)-C(2,1)[4] is

4. The notation C(i,j) indicates a collusion in which i points from one private part and j from the other have been exposed. In a two-part scheme in which both parts require the same level of concurrence C(i,j) = C(j,i). C(0,0) is an outsider attack, etc. The notation generalizes to arbitrarily many parts in an obvious manner.

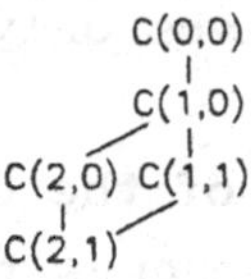

so that if the uncertainty about the index is the same for C(2,1) as it is for C(0,0), the scheme is perfect.

Now consider the scheme in Figure 10. p is only known a priori to be a point on V_d, i.e., of 1-dimensional uncertainty to collusion C(0,0). Similarly, if one of the input points, say p_1, is known and any other point on the indicator variety, say x, on L_2 is exposed -- $x \neq p_2$ by construction -- then, since for any point on V_d there is a unique line lying on it and x that could be the unknown (to the participants in the collusion) line L_2, every point on L_d is is an equally likely candidate to be p. p is therefore of 1-dimensional uncertainty to collusion C(2,1) and by the remark, to all of the other collusions as well. Hence, the shared secret scheme in Figure 10 is perfect.

The contrast between the two types of compartmented shared secret schemes is significant for the application we have been discussing and dramatic for other choices of parameters: in the present case the private information is 2-dimensional for the second type of scheme rather than 4-dimensional as was the case for the first type; and with no real difference in capability. The only difference is that in the first type, all of the points on the subvarieties which did not lie on the transversal ω were available for use as private pieces of information while in the second type, the points p_1 and p_2 had to be excluded. In both cases the part of the private information that has to be kept secret is only 1-dimensional. If there is no cost involved in insuring the integrity of the information that doesn't need to be kept secret the schemes are equally attractive, while if there is a cost the second type is the clear winner since it involves only half as much information in the private parts.

Unfortunately, because of the difference between k-out-of-k and k-out-of-ℓ schemes the construction for the second type of compartmented system for this example fails to illustrate a very important property of this class of schemes.

The smallest example which shows what happens in general is a scheme in which there are three parts, at least two of which must concur for the controlled action to be initiated. Each part, considered separately, is a 2-out-of-ℓ control scheme. The essential feature of this example over the one discussed earlier is that the highest level concurrence is a k-out-of-ℓ, $k < \ell$ scheme instead of a k-out-of-k scheme. It is trivial to extend the construction shown in Figure 9 to this case, or to any number of parts, $k \le q$ where the construction is in PG(4,q) for this example. To do this we simply choose (appropriately) another line, L_3, in the 3-dimensional subspace V_i to be the variety determined by the third party. By "appropriately" we mean that the three lines L_1, L_2 and L_3 must be skew by pairs so that any two of them span (determine) V_i and that they all intersect a common line ω in V_i lying on the point p. The points of intersection of ω with L_1, L_2 and L_3 -- q, r and s, respectively -- are not used as one of the private pieces of information, although any of the q other points on a line can be. This later requirement is imposed so that the proof of perfection given earlier will still hold for this case as well. Figure 11 shows the resulting construction.

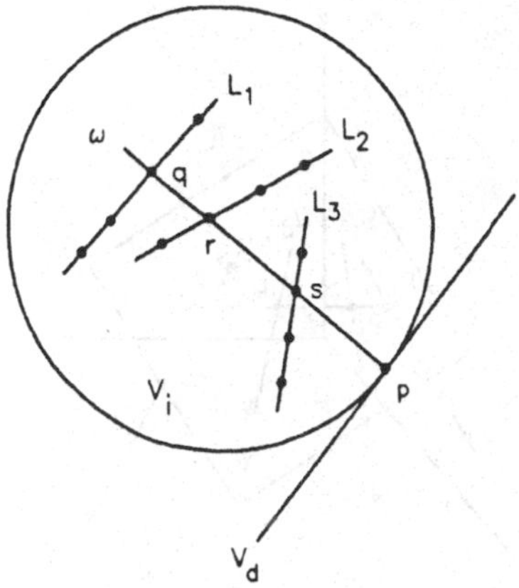

Figure 11.

We introduce the notation $\underset{c}{\cup} L_j$ or $\underset{c}{\cup} V_j$ to indicate the union of a designated concurrence of the individual parts: any two in this particular example.

$$\underset{c}{\cup} L_j = L_1 \cup L_2 = L_1 \cup L_3 = L_2 \cup L_3 = \cup L_j = V_i \quad .$$

The dimension of the space $\mathcal{S}$ in which the shared secret scheme is implemented is

$$\dim \mathcal{S} = \dim(V_d \cup V_i) = 4 \quad ,$$

and

$$V_i \cap V_d = p = (\underset{c}{\cup} L_j) \cap V_d$$

as was true in the construction given in Figure 9. Consequently, for the first type of construction, there is no significant effect in having gone from requiring a unanimous concurrence by the two parties to requiring only 2-out-of-ℓ, $\ell > 2$, concurrence. The second type of construction however is quite different from that shown in Figure 10 as is evident in Figure 12 where a 2-out-of-3 scheme is depicted.

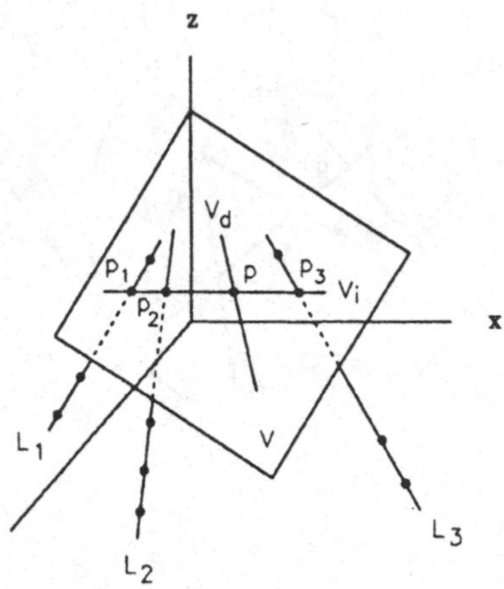

Figure 12.

A simple 2-out-of-3 scheme is implemented in the plane $V = V_i \cup V_d$. The index p is defined by the intersection of the lines V_i and V_d. What is different is that the points p_1, p_2 and p_3, any pair of which suffice to determine the indicator V_i, are themselves determined by the intersection of the lines L_1, L_2 and L_3, respectively, with the plane V, where the lines themselves are determined by any pair of the private points on them. The dimension of the containing space S in this construction has increased from 2 (for the 2-out-of-2 concurrence example) to 3. The fact that the dimension of the shared secret scheme of the first type remained fixed at 4 while the dimension of a scheme of the second type increased from 2 to 3 raises the question of whether there might be examples in which each type of scheme is the more efficient. We next show that this can never be the case.

In general, the first type of construction defines the indicator, V_i, by

$$V_i = \bigcup_{c} V_j$$

where the V_j are the varieties determined by the individual parts, irrespective of whether $k = \ell$ or $k < \ell$. In either case, the index is defined by

$$(1) \qquad V_i \cap V_d = p = (\underset{c}{\cup} V_j) \cap V_d \quad .$$

In general, to realize a scheme of the second type requiring a k-out-of-ℓ concurrence by the parts, we first define a space V,

$$V = V_i \cup V_d \quad ,$$

and embed a simple k-out-of-ℓ shared secret scheme in it. V_i is an indicator (variety or subspace) in V which intersects V_d in the index p, etc., and in which any k points in V_i suffice to determine it. V itself is then considered to be in a space $\mathcal{S}$ of a dimension adequate to allow each of the subvarieties, V_j, defined by the individual parts to intersect V in only a single point, p_j. Any k of these points of intersection will suffice to determine the indicator V_i and hence the point of intersection, p, of V_i and V_d to recover the secret.

The essential point to this construction is that

$$V_i = \underset{c}{\cup} p_i = \underset{c}{\cup} (V_j \cap V)$$

and

$$(2) \qquad V_i \cap V_d = p = (\underset{c}{\cup} (V_j \cap V)) \cap V_d \quad .$$

To simplify the comparison we first consider the case in which all of the parts require the same level of concurrence: k′-out-of-ℓ′. If the concurrence required of the individual parts is k-out-of-ℓ and the secret is d-dimensional, then the dimension of the containing space $\mathcal{S}$ is

$$(3) \qquad \dim(\mathcal{S}) = kk' + d - 1$$

for a scheme of the first type irrespective of whether $k = \ell$ or $k < \ell$. For a scheme of the second type,

(4) $$\dim(S) = k' + d - 1$$

if $k = \ell$, and

(5) $$\dim(S) = k' + k + d - 2$$

if $k < \ell$. For the example just analyzed, $k = k' = 2$ and $d = 1$ so that the dimension of the spaces were 4, 2 and 3, respectively. Since it must always be the case that $k \geq 2$ and $k' \geq 2$, it is easy to see that it is always possible to construct a shared secret scheme of the second type in a lower dimension space than is possible for a scheme of the first type. This is also true if the individual parts do not all require the same level of concurrence:

$$k_1' \geq k_2' \geq \ldots \geq k_\ell' \quad .$$

We then have, in analogy to the results above,

(3*) $$\dim(S) = \sum_{j=1}^{k} k_j' + d - 1$$

for a scheme of the first type irrespective of whether $k = \ell$ or $k < \ell$. For a scheme of the second type

(4*) $$\dim(S) = k_1' + d - 1$$

if $k = \ell$, and

(5*) $$\dim(S) = k_1' + k + d - 2$$

if $k < \ell$.

In summary, in spite of the simplicity of the first type of construction for compartmented shared secret schemes, it is never as efficient (in the usage of information) as schemes of the second type.

A Discussion of Exceptional Cases

It is almost as difficult to provide for unanimity in shared secret schemes as it is to secure it in real-life situations. In this section we will discuss several examples in which one or more of the parts requires unanimity of input and in which the overall control scheme may require either k-out-of-ℓ, $k < \ell$, or k-out-of-k concurrence.

The smallest -- not necessarily the simplest -- example is obtained by modifying the first problem we discussed: a two-part scheme in which each part required a 2-out-of-ℓ concurrence. If the concurrence required for one of the parts is changed from a 2-out-of-ℓ scheme to a 2-out-of-2 scheme, it isn't obvious how to construct a compartmented scheme of the first type. Recall that in this type of construction the indicator, V_i, is a subspace spanned by the varieties determined by the individual parts. In this case, since there are two parts -- both of whom must concur in order for the secret to be recovered -- V_i would be the union of the line, say L_2, determined by the 2-out-of-ℓ scheme and presumably the point, p_1, determined by the 2-out-of-2 scheme. V_i must then be a plane

$$V_i = <L_2, p_1>$$

The private pieces of information for the second part are points on the line L_2, etc. The problem is: where are the two points (private pieces of information) q_1 and r_1 that define p_1 for the 2-out-of-2 scheme. They can't be confined to the plane V_i, otherwise V_i would be determined by L_2 and only one of the points q_1 or r_1. Hence if the system is to be perfect, it must be the case that the two points, q_1, and r_1, lie in a 3-space which contains V_i. Pictorially:

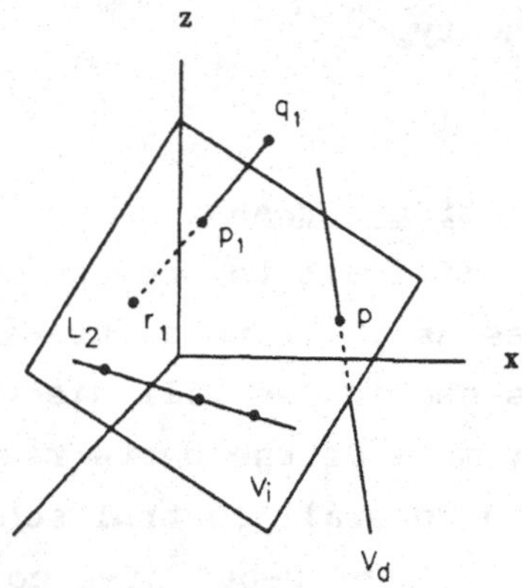

Figure 13.

p_1 cannot be a point on L_2 nor collinear with p and any point on L_2 used as one of the private pieces of information for the second part. The first condition is to insure that when the concurrence conditions are satisfied that V_i and hence p will be determined. The second is to insure that a collusion consisting of p_1 and one point on L_2 will not reveal the secret.

The construction in Figure 13 illustrates one (of the many) problems associated with k-out-of-k schemes. In this case the dimensionality of the containing space $\mathcal{S}$ suddenly ceases to obey the counting formula given earlier. If part one were a 3-out-of-3 or a 4-out-of-4 or in general a k_1-out-of-k_1, $k_1 \le q$, concurrence scheme and part two remained a 2-out-of-ℓ scheme, $\mathcal{S}$ would still only need to be 3-dimensional; exactly as shown in Figure 13. In other words, we seem to have lost the functional dependence between the minimum dimension for the containing space $\mathcal{S}$ and the concurrence level k_1 which we had identified earlier.

Now consider a compartmented scheme of the second type for the same example. Recall that in this type of scheme, the individual parts determine indicators that point to points in an intermediate subspace V in which the overall shared secret scheme is embedded. Since this highest level scheme is a 2-out-of-2 concurrence for this example, V need only be a line as shown in Figure 8. L_2 must be a line which intersects $V = V_d$ in a single point p_2. p_1 of course is also a point on V_d; for which $p = p_1 + p_2$. The question is: where must the points q_1 and r_1 be located? There is no reason for them to be outside of the plane determined by L_2 and V_d, $\pi = \langle L_2, V_d \rangle$, but is there any restriction on where they can be located in π? For example, the following construction in which q_1 and r_1 are constrained to lie on V_d satisfies the conditions to be a perfect two-part shared secret scheme, etc.

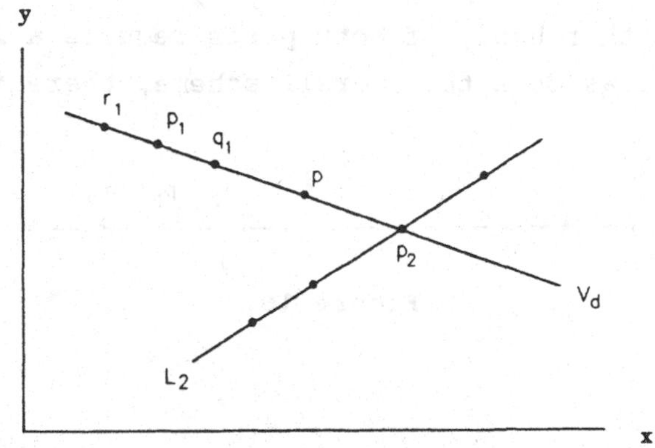

Figure 14.

However, so does the construction

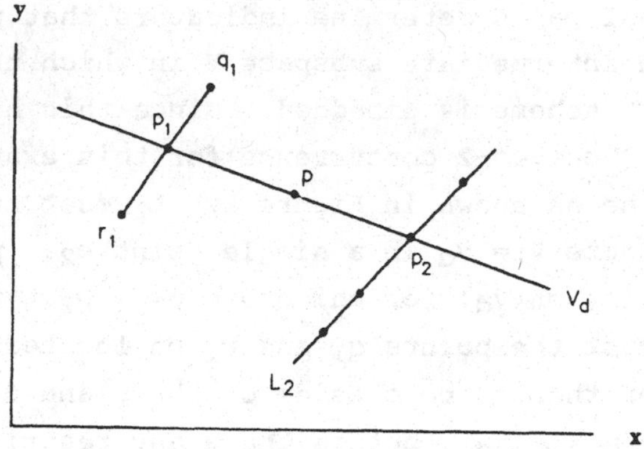

Figure 15.

In both of these constructions, the dimension of the containing space S is two so that it doesn't appear to make any difference where the points q_1 and r_1 are located.

On the other hand, if both parts require a 2-out-of-2 concurrence, as does the overall scheme, there is a difference:

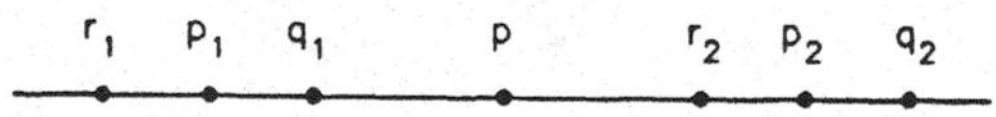

Figure 16.

versus

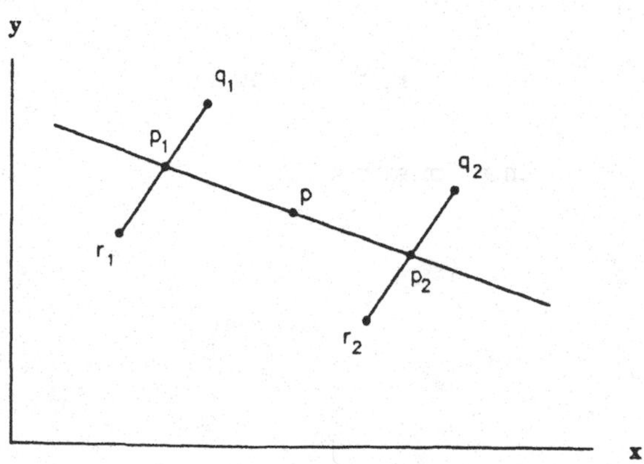

Figure 17.

The question is, which of these is the proper, i.e., logically consistent, generalization for the type of constructions we've used earlier. Increasing the number of parts from 2 to k doesn't differentiate between the two constructions either so long as the overall scheme requires unanimous agreement by the separate parts.

We consider next a three-part scheme in which the overall scheme requires the concurrence of only 2-out-of-3 parts and in which two of the parts are 2-out-of-2 schemes. The other part is a 2-out-of-ℓ scheme. In this case, a construction of the first type is given by:

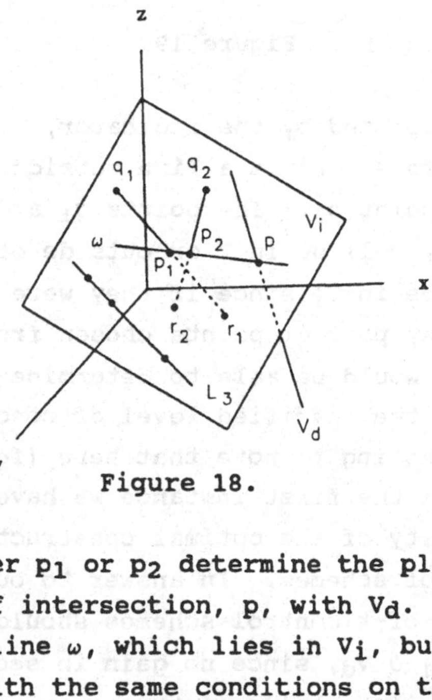

Figure 18.

where L_3 and either p_1 or p_2 determine the plane V_i and hence its point of intersection, p, with V_d. Points p_1 and p_2 determine the line ω, which lies in V_i, but which intersects V_d at p. With the same conditions on the choices for p_1 and p_2 that had to be imposed on the choice of p, in the construction of Figure 13 (and for the same reasons) this is a perfect shared secret scheme of the first type satisfying the problem specifications.

A construction of the second type is shown in Figure 19.

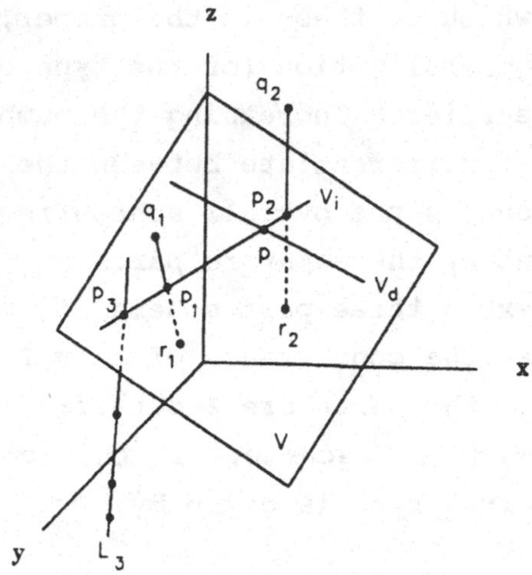

Figure 19.

V is the plane spanned by the indicator, V_i, and the domain, V_d, etc., as before. L_3 is a line outside of V which intersects V at the point p_3. The points q_1 and v_1, and q_2 and v_2 could equally well be in V or outside of V. They cannot be confined to be in V_i since if they were then a collusion consisting of any pair of points chosen from the set (p_3, q_1, r_1, q_2, r_2) would be able to determine V_i and hence p, in violation of the specified level of concurrence.

It is interesting to note that here (for this particular example) we have the first instance we have seen in which the dimensionality of the optimal constructions are the same for both types of schemes. In answer to our earlier question, the k-out-of-k control schemes should be confined to the space $V = V_i \cup V_d$, since no gain in security is achieved by letting them lie outside of this space, and one dimension (to S) may -- for some choices of specifications -- be saved by this restriction. We have already seen this in the degenerate case shown in Figures 16 and 17 -- degenerate because there is no V_i, so that $V = V_d$. To see this in the present case, assume that all three parts require 2-out-of-2 concurrence but that the overall scheme is 2-out-of-3. An obvious modification to L_3 in Figure 19 yields a 3-dimensional solution. However a 2-dimensional solution is possible in exact analogy to the construction in Figure 16.

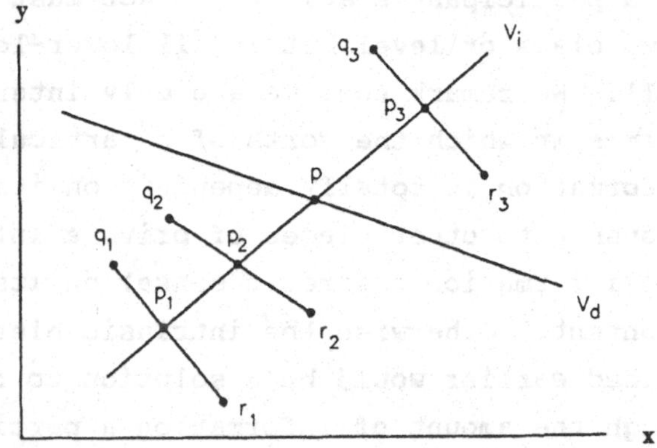

Figure 20.

It is interesting to note that in this final example the constructions are identical for both type one and type two schemes.

After all of this discussion of exceptional cases, our conclusion is the same as it was before: one cannot do better (in terms of the efficient use of information) in constructing compartmented shared secret schemes than to base them on constructions of the second type; in other words, to let the individual parts determine subindicators that point to points in a space V that define an indicator for the overall k-out-of-ℓ concurrence scheme.

An Application (and Realization) of Multilevel Shared Secret Schemes

In the brief discussion given earlier of the various extended capabilities to shared secret schemes, we described one scenario in which any two vice presidents of a bank were authorized to approve an electronic funds transfer (up to some maximum amount) or in which any three senior tellers could do so. As we remarked then, for this application it would almost certainly be unacceptable that one vice president and two senior tellers not be able to approve a transfer. In other words, in this and many other real-world applications, a participant's ability to act must hold not only in his own class or level but in all lower-level classes as well. We remark that we are only interested in extrinsic schemes in which the worth of a particular piece of private information is totally dependent on its functional relationship to other pieces of private information, and not (in an information theoretic sense) on its own information content. Otherwise the intrinsic hierarchical schemes described earlier would be a solution to the problem, even though the amount of information a participant has to protect (keep secret) might be so great as to make the solution totally infeasible for practical application. In other words, all the pieces of private information should consist of n bits of information, even though some may be several times more effective in recovering the secret than others.

Figure 21 shows a perfect shared secret scheme for the electronic funds transfer problem.

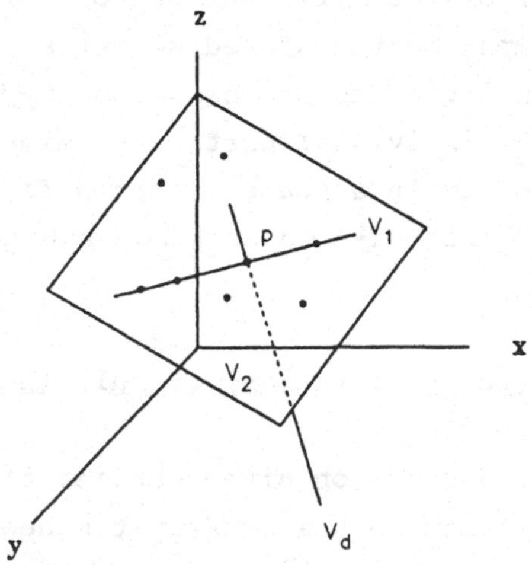

Figure 21.

Vice presidents know points on the line V_1 which intersects V_d at the point p so that any two of them can determine V_1

and hence p, etc. Senior tellers know points in general position in the plane V_2 -- not on V_1 -- and no two of which are collinear with any of the private points chosen on V_1 nor with p. Any three of them can determine V_2 and hence p, etc. Clearly any point on V_1 taken with a pair of the points in V_2 define V_2 as desired, since no such triple of points is collinear by construction.

By now the reader should be very familiar (and comfortable) with the way in which shared secret systems are constructed. For example, if we wished to conceal a 2-dimensional secret instead of a 1-dimensional secret in a 2-level scheme in which level one is a 2-out-of-ℓ scheme, we could use the same geometrical construction that was used earlier to construct a simple 3-out-of-ℓ scheme to conceal a 2-dimensional secret. Two planes, V_2 and V_d, are chosen in a 4-dimensional space such that they do not lie in a common 3-dimensional subspace. This forces them to have a single point, p, in common. In fact, we can use the same procedure used earlier to construct V_2, given V_d, so that a desired index p is the point of intersection. An arbitrary point, q, in V_2, $q \neq p$, is chosen and the line $V_1 = \langle p,q \rangle$ used to determine the points for the first class participants. The second class participants receive points in general position in V_2 none of which are on V_1 and no pair of which are collinear with either p or any point from L_1 assigned to one of the first class participants. Pictorially:

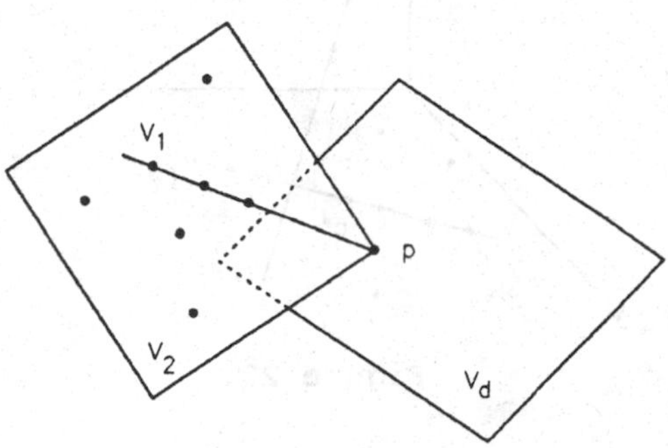

Figure 22.

An obvious extension to these constructions will accommodate an arbitrary sequence of concurrence levels, k_i, and/or a d-dimensional secret. It would appear, therefore, that this completely solves the problem of multilevel schemes.

It is only necessary to examine the construction in Figure 21 a little more critically to realize that there is more to the problem (and solution) than we have suggested. We remarked earlier that the amount of information that had to be kept secret in the private pieces of information was the same as the information contained in the secret itself. In the scheme shown in Figure 21 **S** is 3-dimensional while the secret is only 1-dimensional. One might think that in analogy to what was done with the private pieces of information in the 2-dimensional scheme shown in Figure 2 where one coordinate value was kept secret and one was exposed, that one coordinate value could be kept secret in this case as well, say z, and two exposed: $(x_j, y_j, \textcircled{z_j})$. If this is done however, the secret is revealed to even outsiders -- not just to a collusion of insiders.

Anyone knowing the nonsecret parts of the private pieces of information, i.e., their projection (along the z axis) onto the xy plane,

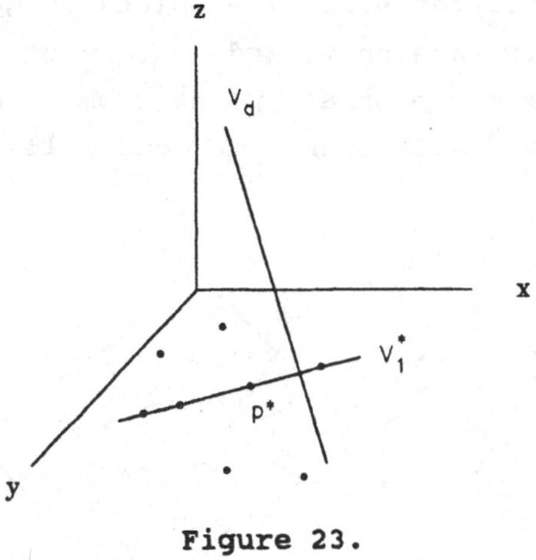

Figure 23.

also knows the projection of the line V_1 into the line V_1^*. This is easy to determine by finding any set of three or more collinear points in the projection. The line V_1 is therefore known to be in the plane π which is parallel to the z axis and includes the line V_1^*.

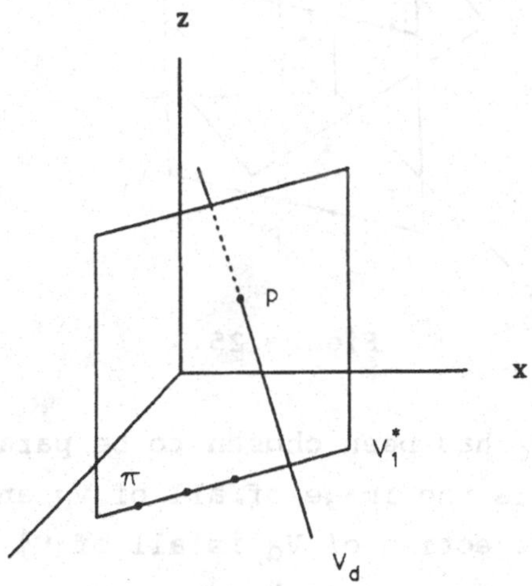

Figure 24.

In fact the unknown subvariety V_1 must be one of the pencil of lines with common point p at which the line V_d intersects π. The important point is that it isn't necessary to identify V_1, only its intersection, p, with V_d. Consequently p (and the secret) is revealed from only a knowledge of the nonsecret parts of the private pieces of information unless V_d satisfies some additional constraints. The problem goes away if the projection of V_d onto the xy plane is in the line V_1^*, in other words, if V_d is a line in π. In the extreme case V_d could be a line parallel to the z axis so that the entire line projects into a point p^* in V_1^*. p^* is the image of p under the projection along the z axis: $proj_z(p) = p^*$. V_1 and V_d are therefore distinct lines in π, at least one of which must project into the entire line V_1^*. The plane V_2 is not the same as π and in fact cannot be parallel to the z axis, hence its projection is the whole of the xy plane.

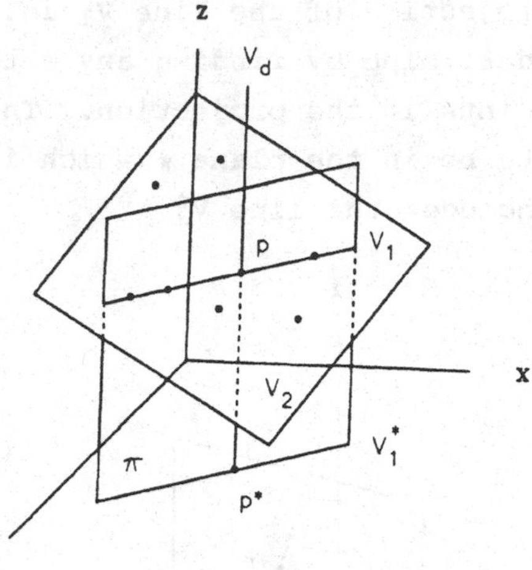

Figure 25.

In this figure V_d has been chosen to be parallel to the z axis so that p^* is the image of all of V_d and hence known. Otherwise the projection of V_d is all of V_1^* and the projection of p, p^*, would be unknown.

The problem that we encountered in partitioning the private information into a secret part and a nonsecret part for the multilevel scheme shown in Figure 21 without compromising the security of the secret is common to almost all multilevel schemes. The solution for that particular case, while suggestive of the general method of solution, is not definitive. To better illustrate the general case, we next consider the two-level scheme shown in Figure 22. $\mathcal{S}$ was 4-dimensional in that case and the secret was 2-dimensional so that the private information would be (if the previous examples are any guide) of the form $(x_j, y_j, \textcircled{z_j}, \textcircled{w_j})$. The line V_1 projects into a line V_1^* in the xy plane. In this case, corresponding to the plane π that was defined on V_1^* in the construction in Figure 24, there is a 3-space, S, parallel to the z and w axes which includes the line V_1^*. Since $\mathcal{S}$ is only 4-dimensional, the plane V_d either intersects S in a line or else contains S. By the rank formula, $\mathcal{S}$ would have to be 6-dimensional for the two subspaces to be skew and 5-dimensional for them to intersect in only a point. If $S \cap V_d = \ell$, ℓ a line, then the scheme cannot be perfect since the equivocation about the secret would be only of $O(q)$ instead of $O(q^2)$ using the exposed (nonsecret) parts of the private pieces of information. It must therefore be the case that $V_d \subset S$. This does not say that

$$\mathrm{proj}_{z,w}(V_d) = \mathrm{proj}_{z,w}(V_1)$$

but merely that

$$\mathrm{proj}_{z,w}(V_d) \subset \mathrm{proj}_{z,w}(V_1) \quad .$$

This is analogous to the previous case in which $\mathrm{proj}_z(V_d)$ was either the point p^* (in the line $L_1^\dagger$) or else all of $L_1^\dagger$.

Although it is possible to formulate general conditions on the subspaces which will insure that these problems are avoided -- even if the subspaces are chosen almost at random -- there is no gain in security nor a compensating increase in capability to justify this additional freedom of choice. Instead, in the first example we may as well take V_d to be a line parallel to the z axis so that $\mathrm{proj}_z(V_d) = p^*$, $p^* \in V_1^\dagger$, and in the second to take V_d to be parallel to the z and w axes so that $\mathrm{proj}_{z,w}(V_d) = p^*$, $p^* \in \mathrm{proj}_{z,w}(V_1) = V_1^\dagger$ in this case also. If we construct the domain V_d in this manner, the secret part of the private information will be totally lost in the projection, i.e., in the disclosure of the nonsecret part, and the scheme will be secure.

Finally, given a d-dimensional secret which is to be secured in a t-level scheme, where the concurrence required at level j is k_j,

$$k_t > k_{t-1} > \ldots > k_1 \quad ,$$

and in which a participant at the j-th level is to be able to function at all lower levels (having however only the capability associated with that level) we can construct a perfect multilevel control scheme with these characteristics. We start with an n-dimensional space, $\mathcal{S}$ = PG(n,q), where $n = d+k_t-1$. V_d is a d-dimensional subspace of $\mathcal{S}$ parallel to the coordinates $x_d, x_{d-1}, \ldots, x_1$. Given the secret point p in V_d, we construct a (k_t-1)-dimensional subspace, V_t, of $\mathcal{S}$ that intersects V_d only in the point p. We next choose a $(k_{t-1}-1)$-dimensional subspace V_{t-1} of V_t lying on the point p. This procedure is repeated to finally yield a chain of nested subspaces

$$V_t \supset V_{t-1} \supset \ldots \supset V_1$$

of dimensions $k_t-1, k_{t-1}-1, \ldots, k_1-1$, respectively, all of which lie on the point p.

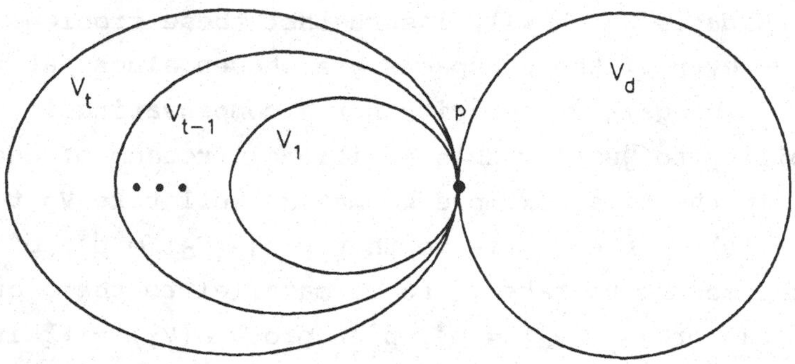

Figure 26.

The private pieces of information are to be chosen so as to have rank k_j in V_j and to not lie in any of the higher order subspaces. In other words, in the construction shown in Figure 21, the points in V_2 were chosen not to lie on V_1 and such that no two were collinear with any of the private points chosen on V_1 nor with the index, p. In general, this says that the points in the j-th class are to be chosen in general position in $V_j \setminus \bigcup_{i=1}^{j-1} V_i$, such that the rank of any set of k_j points drawn from among all of the private points in $\bigcup_{i=1}^{j} V_i$ and the index, p, will be k_j. Under these conditions clearly any participant can act as a member of any lower class. The private information will be of the form

$$(x_n, \ldots, x_{d+1}, \textcircled{x_d}, \ldots, \textcircled{x_2}, \textcircled{x_1})$$

and the scheme will be perfect since the secret information is totally lost in the projection along the first d coordinates.

Conclusion

In view of the length of this paper we merely remark in conclusion that the two types of partitioning of secret information which have been described here can be combined to form hybrid control schemes involving simple, multipart and multilevel controls. For example, it would be easy to devise a two-part control scheme in which both the U.S. military command and the U.S.S.R. military command had to concur in order for the controlled event to be initiated. The U. S. could choose to use a multilevel scheme, say one in which two or more generals, three or more colonels (or generals) five or more lieutenent colonels (or colonels or generals) had to concur in order for the U. S. input to be made. The U.S.S.R. on the other hand might have entirely different requirements; for example they might require the unanimous concurrence of three of their general staff in order for the U.S.S.R. input to be made. The constructions described here are sufficiently general to accommodate both arbitrary concurrence of the parties and arbitrary multi-level concurrence within the individual parts. There are concurrence schemes, however, that can't be satisfied by schemes of the type described here, but it appears unlikely that any such scheme will be of practical interest: one such example would be if participants A and B together could cause an event to be initiated but A, B and C together could not.

Bibliography

Note: This bibliography includes all of the papers on shared secret or threshold schemes which the author is aware of. Although only a few of the references given are cited in this paper, it has been included for its own value to other researchers.

[1.] C. A. Asmuth and G. R. Blakley, "Pooling, Splitting and Reconstituting Information to Overcome Total Failure of Some Channels of Communication," Proc. IEEE Computer Soc. 1982 Symp. on Security and Privacy, Oakland, CA, April 26-28, 1982, pp. 156-169.

*[2.] C. Asmuth and J. Bloom, "A Modular Approach to Key Safeguarding," IEEE Trans. Info. Theory, Vol. IT-29, No. 2, March 1983, pp. 208-210.

[3.] J. C. Benaloh, "Secret Sharing Homomorphisms: Keeping Shares of a Secret Secret," Crypto'86, Santa Barbara, CA, Aug. 11-15, 1986, Advances in Cryptology, Vol. 263, Ed. by A. M. Odlyzko, Springer-Verlag, Berlin, 1986, pp. 251-260.

*[4.] A. Beutelspacher and K. Vedder, "Geometric Structures as Threshold Schemes," Proceedings of the 1987 IMA Conference on Cryptography and Coding Theory, Cirencester, England, Oxford University Press, to appear.

[5.] A. Beutelspacher, "Enciphered Geometry: Some Applications of Geometry to Cryptography," Proceedings of Combinatorics'86, Annals of Discrete Mathematics, 37, North-Holland, 1988, pp. 59-68.

[6.] G. R. Blakley and R. D. Dixon, "Smallest Possible Message Expansion in Threshold Schemes," Crypto'86, Santa Barbara, CA, Aug. 11-15, 1988, Advances in Cryptology, Vol. 263, Ed. by A. M. Odlyzko, Springer-Verlag, Berlin, 1986, pp. 266-274.

[7.] G. R. Blakley and C. Meadows, "Security of Ramp Schemes," Crypto'84, Santa Barbara, CA, Aug. 19-22, 1984, Advances in Cryptology, Vol. 196, Ed. by G. R. Blakley and D. Chaum, Springer-Verlag, Berlin, 1985, pp. 411-431.

[8.] G. R. Blakley and L. Swanson, "Security Proofs for Information Protection Systems," Proc. IEEE Computer Soc. 1981 Symp. on Security and Privacy, Oakland, CA, April 27-29, 1981, pp. 75-88.

[9.] G. R. Blakley, "One-time Pads are Key Safeguarding Schemes, Not Cryptosystems: Fast Key Safeguarding Schemes (Threshold Schemes) Exist," Proc. IEEE Computer Soc. 1980 Symp. on Security and Privacy, Oakland, CA, April 14-16, 1980, pp. 108-113.

*[10.] G. R. Blakley, "Safeguarding Cryptographic Keys," Proc. AFIPS 1979 Nat. Computer Conf., Vol. 48, New York, NY, June 1979, pp. 313-317.

[11.] J. R. Bloom, "A Note on Superfast Threshold Schemes," preprint, Texas A&M Univ., Dept. of Mathematics, 1981.

*[12.] J. R. Bloom, "Threshold Schemes and Error Correcting Codes," Am. Math. Soc., Vol. 2, 1981, pp. 230.

[13.] E. F. Brickell and D. R. Stinson, "The Detection of Cheaters in Threshold Schemes," preprint (available from authors).

[14.] D. Chaum, Claude Crepeau and I. Damgard, "Multiparty Unconditionally Secure Protocols," 4th SIAM Conference on Discrete Mathematics, San Francisco, CA, June 13-16, 1988, abstract appearing in SIAM Final Program Abstracts: Minisymposia, #M-28/3:20pm, pp. A8.

[15.] D. Chaum, "How to Keep a Secret Alive: Extensible Partial Key, Key Safeguarding, and Threshold Systems," Crypto'84, Santa Barbara, CA, Aug. 19-22, 1984, Advances in Cryptology, Vol. 196, Ed. by G. R. Blakley and D. Chaum, Springer-Verlag, Berlin, 1984.

[16.] D. Chaum, "Computer Systems Established, Maintained, and Trusted by Mutually Suspicious Groups," Memo. No. UCB/ERL/M79/10, Univ. of Calif, Berkeley, ERL 1979; also, Ph.D. dissertation in Computer Science, University of California, Berkeley, 1982.

[17.] B. Chor, S. Goldwasser, S. Micali and B. Awerbuch, "Verifiable Secret Sharing and Achieving Simultaneity in the Presence of Faults," Proc. 26th IEEE Symp. Found. Comp. Sci., Portland, OR, October 1985, pp. 383-395.

[18.] G. I. Davida, R. A. DeMillo and R. J. Lipton, "Protecting Shared Cryptographic Keys," Proc. IEEE Computer Soc. 1980 Symp. on Security and Privacy, Oakland, CA, April 14-16, 1980, pp. 100-102.

*[19.] M. De Soete and K. Vedder, "Some New Classes of Geometric Threshold Schemes," Proc. Eurocrypt'88, May 25-27, 1988, Davos, Switzerland, to appear.

*[20.] A. Ecker, "Tactical Configurations and Threshold Schemes," preprint (available from author).

[21.] Paul Feldman, "A Practical Scheme for Non-interactive Verifiable Secret Sharing," Proc. 28th Annual Symp. on Foundations of Comp. Sci., Los Angeles, CA, Oct. 12-14, 1987, IEEE Computing Soc. Press, Washington, D.C., 1987, pp. 427-437.

[22.] S. Harari, "Secret Sharing Systems," Secure Digital Communications, Ed. by G. Longo, Springer-Verlag, Wien, 1983, pp. 105-110.

*[23.] M. Ito, A. Saito and T. Nishizeki, "Secret Sharing Scheme Realizing General Access Structure," (in English) Proc. IEEE Global Telecommunications Conf., Globecom'87, Tokyo, Japan, 1987, IEEE Communications Soc. Press, Washington, D.C., 1987, pp. 99-102. Also to appear in Trans. IEICE Japan, Vol. J71-A, No. 8, 1988 (in Japanese).

*[24.] M. Ito, A. Saito and T. Nishizeki, "Multiple Assignment Scheme for Sharing Secret," preprint (available from T. Nishizeki).

*[25.] E. D. Karnin, J. W. Greene and M. E. Hellman, "On Secret Sharing Systems," IEEE International Symposium on Information Theory, Session B3 (Cryptography), Santa Monica, CA, February 9-12, 1981, IEEE Trans. Info. Theory, Vol. IT-29, No. 1, January 1983, pp. 35-41.

*[26.] S. C. Kothari, "Generalized Linear Threshold Scheme," Crypto'84, Santa Barbara, CA, Aug. 19-22, 1984, Advances in Cryptology, Vol. 196, Ed. by G. R. Blakley and D. Chaum, Springer-Verlag, Berlin, 1985, pp. 231-241.

[27.] K. Koyama, "Cryptographic Key Sharing Methods for Multi-groups and Security Analysis," Trans. IECE Japan, Vol. E66, No. 1, 1983, pp. 13-20.

*[28.] R. J. McEliece and D. V. Sarwate, "On Sharing Secrets and Reed-Solomon Codes," Com. ACM, Vol. 24, No. 9, September 1981, pp. 583-584.

[29.] M. Merritt, "Key Reconstruction," Crypto'82, Santa Barbara, CA, Aug. 23-25, 1982, Advances in Cryptology, Ed. by D. Chaum, R. L. Rivest and A. T. Sherman, Plenum Press, New York, 1983, pp. 321-322.

[30.] M. Mignotte, "How to Share a Secret," Workshop on Cryptography, Burg Feuerstein, Germany, March 29-April 2, 1982, Cryptography, Vol. 149, Ed. by T. Beth, Springer-Verlag, Berlin, 1983, pp. 371-375.

[31.] R. von Randow, "The Bank Safe Problem," Discrete Applied Mathematics, 4, 1982, pp. 335-337.

*[32.] A. Shamir, "How to Share a Secret," Massachusetts Inst. of Tech. Tech. Rpt. MIT/LCS/TM-134, May 1979. (See also Comm. ACM, Vol. 22, No. 11, November 1979, pp. 612-613.

*[33.] D. R. Stinson and S. A. Vanstone, "A Combinatorial Approach to Threshold Schemes," Crypto'87, Santa Barbara, CA, Aug. 16-20, 1987, Advances in Cryptology, Ed. by Carl Pomerance, Springer-Verlag, Berlin, 1988, pp. 330-339.

*[34.] D. R. Stinson and S. A. Vanstone, "A Combinatorial Approach to Threshold Schemes," SIAM J. Disc. Math, Vol. 1, No. 2, May 1988, pp. 230-236. (This is an expanded version of the paper appearing in Advances in Cryptology: Proceedings of Crypto'87, Vol. 293, Ed. By Carl Pomerance, Springer-Verlag, Berlin, 1988.)

*[35.] D. R. Stinson, "Threshold Schemes from Combinatorial Designs," submitted to the Journal of Combinatorial Mathematics and Combinatorial Computing.

[36.] M. Tompa and H. Woll, "How to Share a Secret with Cheaters," Crypto'86, Santa Barbara, CA, Aug. 19-21, 1986, Advances in Cryptology, Vol. 263, Ed. by A. M. Odlyzko, Springer-Verlag, Berlin, 1986, pp. 261-265.

*[37.] H. Unterwalcher, "A Department Threshold Scheme Based on Algebraic Equations," Contributions to General Algebra, 6, Dedicated to the memory of Wilfried Nöbauer, Verlag B. G. Teubner, Stuttgart (GFR), to appear December 1988.

*[38.] H. Unterwalcher, "Threshold Schemes Based on Systems of Equations," Österr. Akad. d. Wiss, Math.-Natur. Kl, Sitzungsber. II, Vol. 197, 1988, to appear.

*[39.] H. Yamamoto," On Secret Sharing Schemes Using (k,L,n) Threshold Scheme," Trans. IECE Japan, Vol. J68-A, No. 9, 1985, pp. 945-952, (in Japanese) English translation available from G. J. Simmons.

[40.] H. Yamamoto, "Secret Sharing System Using (k,L,n) Threshold Scheme," Electronics and Communications in Japan, Part 1, Vol. 69, No. 9, 1986, pp. 46-54; translated from Tsushin Denshi Gakkai Ronbunshi Vol. 68-A, No. 9, Sept. 1985, pp. 945-952.

*[41.] T. Uehara, T. Nishizeki, E. Okamoto and K. Nakamura, "Secret Sharing Systems with Matroidal Schemes," Trans. IECE Japan, Vol. J69-A, No. 9, 1986, pp. 1124-1132, (in Japanese; English translation available from G. J. Simmons) presented at the 1st China-USA International Conference on Graph Theory and Its Applications, Jinan, China, June 1986. English summary by Takao Nishizeki available as Tech. Rept. TRECIS8601, Dept. of Elect. Communs., Tohoku University, 1986.

Session 11

Linear Complexity

Chair: T. Berson, Anagram Laboratories

The Strict Avalanche Criterion: Spectral Properties of Boolean Functions and an Extended Definition

Réjane Forré
Inst. for Communication Technology *

Abstract

A necessary and sufficient condition on the Walsh-spectrum of a boolean function is given, which implies that this function fulfills the Strict Avalanche Criterion. This condition is shown to be fulfilled for a class of functions exhibiting simple spectral symmetries. Finally, an extended definition of the Strict Avalanche Criterion is proposed and the corresponding spectral characterization is derived.

1 Introduction

The "Strict Avalanche Criterion" (SAC) was introduced by A.F. Webster and S.E. Tavares. They write [1]: "If a function is to satisfy the strict avalanche criterion, then each of its output bits should change with a probability of one half whenever a single input bit x is complemented to $\overline{x}$." The cryptographic significance of the SAC is highlighted by considering the situation where a cryptographer needs some "complex" mapping f of n bits onto one bit. Although the expression "complex" has no precise mathematical definition here, an information-theoretical approach can help assigning it an intuitively pleasant meaning. Maximizing the entropy $H(f([x_1, x_2, \ldots, x_n]))$ yields zero-one balanced functions, but this alone certainly does not ensure the "complexity" of a function. Maximizing the conditional entropy $H([f(x_1, \ldots, \overline{x_i}, \ldots, x_n]) \mid f([x_1, \ldots, x_i, \ldots, x_n]))$ for all i, $1 \leq i \leq n$, leads to SAC-fulfilling boolean functions, according to the definition in [1]. It is proposed here to go even further, by keeping one or more input bits of f constant, and making the obtained "subfunctions" complex as well. It is worthwhile pointing out the fact that any function f' of $n-1$ bits will be a relatively bad approximation of f if f fulfills the SAC. Indeed, the output of the best possible f' will differ from

*Sternwartstr. 7, ETH-Zentrum, 8092 Zürich, Switzerland

the output of f with a probability of $\frac{1}{4}$. This lack of accuracy of lower-dimensional approximations is a wishable property of cryptosystems: the existence of some (relatively accurate) lower-dimensional approximation of an enciphering transformation could reduce the amount of work for an exhaustive search according to the dimension of the domain of the approximation. Functions for which flipping one input bit always flips the output of course are still more difficult to approximate (the best lower-dimensional approximation is inaccurate in 50% of the cases), but their conditional entropy $H([f(x_1,\ldots,\overline{x_i},\ldots,x_n]) \mid f([x_1,\ldots,x_i,\ldots,x_n]))$ is zero.

In the first part of this paper, Boolean functions $f(\underline{x})$ with n bits input and one bit output are considered. The Walsh-transform has shown to be very useful for the analysis of (statistical) properties of boolean functions. It is shown that a boolean function $f(\underline{x})$ fulfills the SAC if and only if, for all $i \in \{1,2,\ldots,n\}$, its Walsh transform $\hat{F}(\underline{w})$, $\underline{w} = [w_1, w_2, \ldots, w_n]$, fulfills

$$\sum_{\underline{w} \in Z_2^n} (-1)^{w_i} \cdot \hat{F}^2(\underline{w}) = 0,$$

where Z_2^n denotes the n-dimensional vector space over the finite field GF(2). This set of conditions is shown to be fulfilled for a class of functions $\hat{F}(\underline{w})$ that exhibits certain "visible symmetries" arising from equalities of the form $\hat{F}(\underline{w}) = \hat{F}(\underline{w} \oplus \underline{c})$.

In the second part of the paper, the requirements on a boolean function are made stronger, introducing the concept of "SAC of higher order". The corresponding spectral conditions are then established.

2 Walsh-Spectrum of SAC-fulfilling Functions

2.1 Spectral Characterization of Functions Fulfilling the SAC

First, a few basic definitions, lemmas and theorems are needed.

Definition 1 *[2,3,5] If $f(\underline{x})$ is any real-valued function whose domain is the vector space Z_2^n, the* **Walsh transform** *of $f(\underline{x})$ is defined as:*

$$F(\underline{w}) = \sum_{\underline{x} \in Z_2^n} f(\underline{x}) \cdot (-1)^{\underline{x} \cdot \underline{w}}, \tag{1}$$

where $\underline{w} \in Z_2^n$ and $\underline{x} \cdot \underline{w}$ denotes the dot-product of $\underline{x}$ and $\underline{w}$, defined as

$$\underline{x} \cdot \underline{w} = x_1 w_1 \oplus x_2 w_2 \oplus \ldots \oplus x_n w_n. \tag{2}$$

The function $f(\underline{x})$ can be recovered from $F(\underline{w})$ by the **inverse Walsh transform***:*

$$f(\underline{x}) = 2^{-n} \sum_{\underline{w} \in Z_2^n} F(\underline{w}) \cdot (-1)^{\underline{x} \cdot \underline{w}}. \tag{3}$$

The Walsh transform and its inverse (both defined for real-valued functions) may be applied to boolean functions if their values are viewed as the real values 0 and 1.

Very often, it is easier to work with boolean functions that take values in the range $\{1,-1\}$. The function $\hat{f}(\underline{x})$ is defined as

$$\hat{f}(\underline{x}) = (-1)^{f(\underline{x})} \quad \text{or} \quad \hat{f}(\underline{x}) = 1 - 2f(\underline{x}). \tag{4}$$

The relationship between the Walsh transforms of $f(\underline{x})$ and $\hat{f}(\underline{x})$ is stated in the following lemma [2,3].

Lemma 1 *If $\hat{f}(\underline{x}) = (-1)^{f(\underline{x})}$, then*

$$\hat{F}(\underline{w}) = -2F(\underline{w}) + 2^n\delta(\underline{w}), \tag{5}$$

which is equivalent to

$$F(\underline{w}) = 2^{n-1}\delta(\underline{w}) - \frac{1}{2}\hat{F}(\underline{w}), \tag{6}$$

where

$$\delta(\underline{w}) = \begin{cases} 1, & \text{for } \underline{w} = \underline{0} \\ 0, & \text{else.} \end{cases} \tag{7}$$

Let $\underline{x}$ and $\underline{x}_i$ denote two n-bit vectors, such that $\underline{x}$ and $\underline{x}_i$ differ only in bit i, $1 \le i \le n$. Z_2^n denotes the n-dimensional vector space over $\{0,1\}$. The function $f(\underline{x}) = z$, $z \in \{0,1\}$ fulfills the SAC if and only if

$$\sum_{\underline{x} \in Z_2^n} f(\underline{x}) \oplus f(\underline{x}_i) = 2^{n-1}, \quad \text{for all } i \text{ with } 1 \le i \le n. \tag{8}$$

If we denote by $\underline{c}_i$ the n-dimensional unit-vector with a one at the i-th place and zeroes elsewhere, condition (8) may be alternatively written as

$$\sum_{\underline{x} \in Z_2^n} f(\underline{x}) \oplus f(\underline{x} \oplus \underline{c}_i) = 2^{n-1}, \quad \text{for all } i \text{ with } 1 \le i \le n. \tag{9}$$

We now wish to express the SAC for the case of an $\hat{f}$-function (with range $\{1,-1\}$). The following Lemma yields an alternative definition of the SAC.

Lemma 2 *$f(\underline{x})$ fulfills the SAC if and only if the function $\hat{f}(\underline{x}) = (-1)^{f(\underline{x})}$ fulfills*

$$\sum_{\underline{x} \in Z_2^n} \hat{f}(\underline{x}) \cdot \hat{f}(\underline{x} \oplus \underline{c}_i) = 0, \tag{10}$$

for all $\underline{c}_i$ with Hamming-weight one.

This lemma is easily derived, considering that if a function $f(\underline{x})$ fulfills the SAC, exactly half the $\underline{x} \in Z_2^n$ satisfy $f(\underline{x}) \neq f(\underline{x} \oplus \underline{c}_i)$, for all $i \in 1, 2, \ldots, n$. This means that the function $\hat{f}(\underline{x}) = (-1)^{f(\underline{x})}$ satisfies

$$\hat{f}(\underline{x}) \cdot \hat{f}(\underline{x} \oplus \underline{c}_i) = -1 \quad \text{for half the } \underline{x} \in Z_2^n \text{, and} \tag{11}$$

$$\hat{f}(\underline{x}) \cdot \hat{f}(\underline{x} \oplus \underline{c}_i) = 1 \quad \text{for the other half.} \tag{12}$$

Summing up over all the $\underline{x} \in Z_2^n$ thus yields (10). The term on the left-hand side of equation (10) can also be represented by the convolution of $\hat{f}(\underline{x})$ with itself:

$$\sum_{\underline{x} \in Z_2^n} \hat{f}(\underline{x}) \cdot \hat{f}(\underline{x} \oplus \underline{c}) = [\hat{f} * \hat{f}](\underline{c}). \tag{13}$$

From the well-known convolution theorem, which states that

$$h(\underline{x}) = \sum_{\underline{y} \in Z_2^n} f(\underline{y}) \cdot g(\underline{y} \oplus \underline{x}) \Longleftrightarrow H(\underline{w}) = F(\underline{w}) \cdot G(\underline{w}), \tag{14}$$

we see that the left-hand side of (10) is also the inverse Walsh-transform of $\hat{F}(\underline{w}) \cdot \hat{F}(\underline{w}) = \hat{F}^2(\underline{w})$, and with (3) we get:

$$[\hat{f} * \hat{f}](\underline{c}_i) = 2^{-n} \sum_{\underline{w} \in Z_2^n} \hat{F}^2(\underline{w}) \cdot (-1)^{\underline{c}_i \cdot \underline{w}} \tag{15}$$

$$= 2^{-n} \sum_{\underline{w} \in Z_2^n} \hat{F}^2(\underline{w}) \cdot (-1)^{w_i}, \tag{16}$$

where we made use of the fact that $\underline{c}_i$ is of the form $[0, 0, \ldots, 0, c_i = 1, 0, \ldots, 0]$. This, together with (5), proves the following theorem.

Theorem 1 *A function $\hat{f}(\underline{x}) : Z_2^n \longrightarrow \{1, -1\}$ fulfills the SAC if and only if its Walsh-transform $\hat{F}(\underline{w})$ satisfies*

$$\sum_{\underline{w} \in Z_2^n} (-1)^{w_i} \cdot \hat{F}^2(\underline{w}) = 0 \tag{17}$$

for all $i \in \{1, 2, \ldots, n\}$. Equivalently, the Walsh-transform $F(\underline{w})$ of $f(\underline{x}) = \frac{1}{2}(1 - \hat{f}(\underline{x}))$ has to fulfill

$$\sum_{\underline{w} \in Z_2^n} (-1)^{w_i} \cdot F^2(\underline{w}) = 2^n F([0, \ldots, 0]) - 2^{2n-2} \tag{18}$$

for all $i \in \{1, 2, \ldots, n\}$.

Note that $F([0, \ldots, 0])$ equals the number of ones in the truth table of $f(\underline{x})$.

Example 1:
Consider the function $f(\underline{x}) : Z_2^3 \longrightarrow \{0, 1\}$, the corresponding $\hat{f}(\underline{x}) = (-1)^{f(\underline{x})}$ and their respective Walsh-transforms $F(\underline{w})$ and $\hat{F}(\underline{x})$ given by the following table:

x_1 / w_1	x_2 / w_2	x_3 / w_3	$f(\underline{x})$	$\hat{f}(\underline{x})$	$F(\underline{w})$	$\hat{F}(\underline{w})$
0	0	0	0	1	4	0
0	0	1	1	-1	0	0
0	1	0	1	-1	-2	-4
0	1	1	0	1	-2	-4
1	0	0	0	1	0	0
1	0	1	0	1	0	0
1	1	0	1	-1	2	4
1	1	1	1	-1	-2	-4

It is easily checked that flipping the bit x_1 flips the output $f(\underline{x})$ in 50% of the cases. That is true for x_3 too, but not for x_2: flipping x_2 always changes $f(\underline{x})$. Therefore,

$$H(f([x_1, \overline{x_2}, x_3]) \mid f([x_1, x_2, x_3])) = 0$$

and this function does not fulfill the SAC. Indeed, when we compute $\sum_{\underline{w} \in Z_2^n} (-1)^{w_i} \cdot \hat{F}^2(\underline{w})$ for $i = 1, 2$ and 3, we get zero for $i = 1$ and $i = 3$ and -64 for $i = 2$, which does not satisfy the requirements of theorem 1.

Example 2:
Next, we examine another function of three bits, $g(\underline{x})$.

x_1 / w_1	x_2 / w_2	x_3 / w_3	$g(\underline{x})$	$\hat{g}(\underline{x})$	$G(\underline{w})$	$\hat{G}(\underline{w})$
0	0	0	0	1	4	0
0	0	1	0	1	-2	-4
0	1	0	0	1	-2	-4
0	1	1	1	-1	0	0
1	0	0	0	1	-2	-4
1	0	1	1	-1	0	0
1	1	0	1	-1	0	0
1	1	1	1	-1	2	4

The reader can check that flipping any of the three input bits involves an output change in 50% of the cases. Therefore, this function fulfills the SAC and the requirements of theorem 1 can be checked to hold for $i = 1, 2$ and 3.

It should be pointed out that if a function fulfills the SAC, it does not imply that it is zero/one balanced, as can be seen from the following example.

Example 3:

x_1 / w_1	x_2 / w_2	x_3 / w_3	$h(\underline{x})$	$\hat{h}(\underline{x})$	$H(\underline{w})$	$\hat{H}(\underline{w})$
0	0	0	0	1	2	-4
0	0	1	0	1	0	0
0	1	0	0	1	0	0
0	1	1	1	-1	2	4
1	0	0	1	-1	0	0
1	0	1	0	1	-2	-4
1	1	0	0	1	-2	-4
1	1	1	0	1	0	0

$h(\underline{x})$ takes on six times the value zero and only twice the value one, which doesn't prevent it from fulfilling the SAC.

2.2 Construction of SAC-Fulfilling Functions

A geometrical interpretation of theorem 1 can be introduced if we look at the n-tuples $[w_1, w_2, \ldots, w_n]$ as the corners of an n-dimensional cube with edges of length one. Let's attach to each corner $\underline{w} = [w_1, w_2, \ldots, w_n]$ a weight $m_{\underline{w}}$ equal to $\hat{F}^2(\underline{w})$. The center of gravity of this n-dimensional body has the coordinates $[\overline{w_1}, \overline{w_2}, \ldots, \overline{w_n}]$ with

$$\overline{w_i} = \frac{\sum_{\underline{w} \in Z_2^n} m_{\underline{w}} \cdot w_i}{\sum_{\underline{w} \in Z_2^n} m_{\underline{w}}} = \frac{\sum_{\underline{w}\,:\,w_i=1} \hat{F}^2(\underline{w})}{\sum_{\underline{w} \in Z_2^n} \hat{F}^2(\underline{w})}, \tag{19}$$

for $1 \le i \le n$. If a function $\hat{f}(\underline{x}) : Z_2^n \longrightarrow \{1, -1\}$ fulfills the SAC, we know by theorem 1 that

$$\sum_{\underline{w}\,:\,w_i=0} \hat{F}^2(\underline{w}) - \sum_{\underline{w}\,:\,w_i=1} \hat{F}^2(\underline{w}) = 0 \tag{20}$$

$$\Longrightarrow \sum_{\underline{w}\,:\,w_i=0} \hat{F}^2(\underline{w}) = \sum_{\underline{w}\,:\,w_i=1} \hat{F}^2(\underline{w}). \tag{21}$$

And in that case we have

$$\overline{w_i} = \frac{\sum_{\underline{w}\,:\,w_i=1} \hat{F}^2(\underline{w})}{\sum_{\underline{w} \in Z_2^n} \hat{F}^2(\underline{w})} = \frac{\sum_{\underline{w}\,:\,w_i=0} \hat{F}^2(\underline{w})}{\sum_{\underline{w} \in Z_2^n} \hat{F}^2(\underline{w})}, \tag{22}$$

which shows that the coordinate $\overline{w_i}$ of the center of gravity of the considered cubic body remains unchanged if all the weights on one "face" of the cube (face with $w_i = 0$) are moved to the opposite "face" (face with $w_i = 1$) and conversely. Therefore, we can state that a function $\hat{f}(\underline{x})$ fulfills the SAC if and only if the n-cube with weights equal to $\hat{F}^2(\underline{w})$ attached to its corners has a center of gravity which is *equidistant from any two opposite "faces" of the cube, and thus from all the corners of the cube.* The center of gravity of the body associated to the Walsh-spectrum of an SAC-fulfilling function therefore has the coordinates $[\frac{1}{2}, \frac{1}{2}, \ldots, \frac{1}{2}]$.

Example 4:
The 3-dimensional cube associated to the function $g(\underline{x})$ of example 2 is represented on the right-hand side of Fig. 1. The dark circles designate weights of magnitude $\hat{F}^2(\underline{w}) = 16$. The exchange of "faces" may be performed in three ways:

$$\begin{aligned} \hat{G}_1^2(\underline{w}) &= \hat{F}^2(\underline{w} \oplus [1,0,0]), \\ \hat{G}_2^2(\underline{w}) &= \hat{F}^2(\underline{w} \oplus [0,1,0]), \\ \hat{G}_3^2(\underline{w}) &= \hat{F}^2(\underline{w} \oplus [0,0,1]), \end{aligned}$$

all of them yielding the same body, namely the one represented on the left-hand side of Fig. 1.

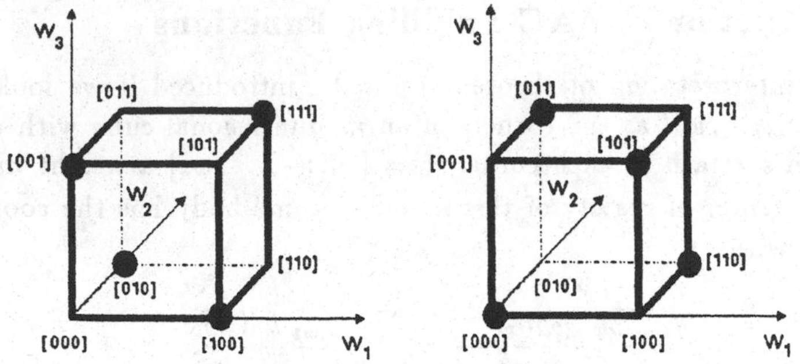

Figure 1: The 3-dimensional cubic body associated to the function $\hat{g}(\underline{x})$ of example 2 and its associated body obtained by exchanging "faces".

The idea that now naturally arises is to use this as a construction for new SAC-fulfilling functions from known ones. The pitfall is that $\hat{F}(\underline{w})$ might be taken as $\pm\sqrt{\hat{F}^2(\underline{w})}$ for each one of the 2^n $\underline{w}$'s. For the worst case where all 2^n $\underline{w}$'s are associated to nonzero values of $\hat{F}^2(\underline{w})$, this will yield 2^{2^n} possible choices for the mapping $\hat{F}(\underline{w})$, not all of them having valid boolean functions (i.e. 1/-1 valued) as inverse Walsh-transforms. In fact, a function $\hat{f}(\underline{x})$ is a boolean (1/-1 valued) function if and only if

$$\hat{f}^2(\underline{x}) = 1, \text{ for all } \underline{x} \in Z_2^n. \tag{23}$$

By the convolution theorem, we see that this is equivalent to

Theorem 2 *[2, p.167] $\hat{F}(\underline{w})$ is the Walsh-transform of a boolean function $\hat{f}(\underline{x})$: $Z_2^n \longrightarrow \{1,-1\}$ if and only if*

$$\sum_{\underline{w}\in Z_2^n} \hat{F}(\underline{w}) \cdot \hat{F}(\underline{w} \oplus \underline{s}) = 2^n \delta(\underline{s}) = \begin{cases} 2^n & \text{for } \underline{s} = [0,\ldots,0], \\ 0 & \text{otherwise.} \end{cases} \tag{24}$$

Let π be an operator on Z_2^n which, when applied to $\underline{x}$, permutes its indices [2, p.165]:

$$\underline{x} = [x_1, x_2, \ldots, x_n] \implies \pi\underline{x} = [x_{\pi_1}, x_{\pi_2}, \ldots, x_{\pi_n}]. \tag{25}$$

π^{-1} is the inverse operator such that

$$\pi^{-1}(\pi\underline{x}) = \underline{x}. \tag{26}$$

We write

$$\underline{y} = [y_1, y_2, \ldots, y_n] \implies \pi^{-1}\underline{y} = [y_{\pi'_1}, y_{\pi'_2}, \ldots, y_{\pi'_n}]. \tag{27}$$

Example 5:
If the permutation $\pi = [\pi_1, \pi_2, \pi_3] = [2,3,1]$ is applied to $\underline{x} = [x_1, x_2, x_3]$, one gets $\pi\underline{x} = [x_2, x_3, x_1]$. The inverse operator $\pi^{-1} = [\pi'_1, \pi'_2, \pi'_3]$ in this case equals $[3,1,2]$, since $\pi^{-1}(\pi\underline{x})$ must equal $\underline{x}$.

If a function $\hat{f}(\underline{x})$ fulfills the SAC, it is easy to see that this property is preserved under any permutation π of the input bits. Thus, $\hat{g}(\underline{x}) = \hat{f}(\pi\underline{x})$ fulfills the SAC too. Furthermore, $\hat{g}(\underline{x}\oplus\underline{c}) = \hat{h}(\underline{x})$ has $(-1)^{\underline{c}\cdot\underline{w}}\cdot\hat{G}(\underline{w}) = \hat{H}(\underline{w})$ as Walsh-transform (by the translate theorem), and this implies $\hat{H}^2(\underline{w}) = \hat{G}^2(\underline{w})$ for all $\underline{w} \in Z_2^n$. Consequently, $\hat{H}(\underline{w})$ satisfies equation (17) and the following theorem holds.

Theorem 3 *If $\hat{f}(\underline{x}) : Z_2^n \longrightarrow \{1,-1\}$ fulfills the SAC, then $\hat{g}(\underline{x}) = \hat{g}(\pi\underline{x}\oplus\underline{c})$ fulfills it too, for any permutation operator π and any constant $\underline{c} \in Z_2^n$.*

For symmetry reasons, the following lemma is easily seen to be true.

Lemma 3 *The function $\hat{g}(\underline{x}) = -\hat{f}(\underline{x})$ (resp. $g(\underline{x}) = \overline{f(\underline{x})}$) fulfills the SAC if and only if $\hat{f}(\underline{x})$ (resp. $f(\underline{x})$) fulfills the SAC.*

At this point, we already dispose of some tools to construct SAC-fulfilling boolean functions, and the question arises whether it is possible to construct all SAC-fulfilling functions with those tools. Computer experiments were carried out, in order to find such functions

(i) by exhaustive testing of all the 2^{2^n} existing boolean functions of n bits ($n = 3$ and $n = 4$),

(ii) by making use of Theorem 3 and Lemma 3 (but without trying out all possible assignations $\hat{G}(\underline{w}) = \pm\sqrt{\hat{F}^2(\underline{w})}$).

This established the fact that the above construction does not generate all the SAC-fulfilling functions, but only subclasses of them. We call the attention of the reader to the redundancy of the described synthesis rules: nothing ensures us that a newly obtained function will be different from the starting one or from a formerly constructed one.

Example 6:
Let $\hat{g}(\underline{x}) = \hat{f}(\underline{x} \oplus [1,0,1])$, where $\hat{f}(\underline{x})$ is defined through the following table.

x_1	x_2	x_3	$\hat{f}(\underline{x})$	$\hat{g}(\underline{x})$
0	0	0	1	1
0	0	1	1	1
0	1	0	1	1
0	1	1	-1	-1
1	0	0	.1	1
1	0	1	1	1
1	1	0	-1	-1
1	1	1	1	1

We notice that $\hat{g}(\underline{x}) = \hat{f}(\underline{x})$ for all $\underline{x} \in Z_2^3$. The reason is that $\hat{f}(\underline{x})$ is *partially symmetric* in x_1 and x_3 [4, p.123], that is $\hat{f}([x_1, x_2, x_3]) = \hat{f}([x_3, x_2, x_1])$ for all $[x_1, x_2, x_3] \in Z_2^3$.

2.3 Spectral Symmetries of SAC-Fulfilling Functions

We now introduce the definition of the 50%-dependence of boolean functions with respect to one of their input bits. The concept is not new: it was implicitely used in the definition of the SAC.

Definition 2 *A function $\hat{f} : Z_2^n \longrightarrow \{1,-1\}$ (resp. $f : Z_2^n \longrightarrow \{0,1\}$) is said to be* **50%-dependent of its i-th input bit** x_i *if and only if any two n-tuples $\underline{x}$ and $\underline{x}_i$ that differ only in bit i are mapped onto two different values with probability 1/2 and onto the same value with the same probability of 1/2. Or formally*

$$\sum_{\underline{x}\in Z_2^n} \hat{f}(\underline{x}) \cdot \hat{f}(\underline{x} \oplus \underline{c}_i) = 0, \tag{28}$$

for $\{1,-1\}$-valued functions, and

$$\sum_{\underline{x}\in Z_2^n} f(\underline{x}) \oplus f(\underline{x} \oplus \underline{c}_i) = 2^{n-1} \tag{29}$$

for $\{0,1\}$-valued functions.

We thus see that a boolean function fulfills the SAC if and only if it is 50%-dependent of each of its input bits.

The following theorem gives a *sufficient* condition for a function to be 50%-dependent of *one or more* of its input bits.

Theorem 4 *If for some nonzero $\underline{c} \in Z_2^n$ and for all $\underline{w} \in Z_2^n$*

$$\hat{F}^2(\underline{w}) = \hat{F}^2(\underline{w} \oplus \underline{c}) \tag{30}$$

holds, and if $\underline{c}$ has Hamming-weight m ($c_{i_1} = c_{i_2} = \ldots = c_{i_m} = 1$, $1 \le m \le n$), then $\hat{f}(\underline{x})$ is 50%-dependent of the input bits $x_{i_1}, x_{i_2}, \ldots, x_{i_m}$.

Proof:

According to the value of the subvector $\underline{w}' = [w_{i_1}, w_{i_2}, \ldots, w_{i_m}]$, the vector space Z_2^n can be divided into 2^m disjoint subsets $S_{\underline{w}'}$. To each of these subsets $S_{\underline{w}'}$ one can uniquely associate the subset $S_{\underline{v}'}$ where $\underline{v}' = [\overline{w_{i_1}}, \overline{w_{i_2}}, \ldots, \overline{w_{i_n}}]$, and because of (30) one can write

$$\sum_{\underline{w}\in S_{\underline{w}'}} \hat{F}^2(\underline{w}) = \sum_{\underline{w}\in S_{\underline{v}'}} \hat{F}^2(\underline{w}) \tag{31}$$

for each choice of $\underline{w}' \in Z_2^m$. Consequently, we have the following set of 2^{m-1} equations:

$$\begin{aligned}
\sum_{\underline{w}\in S_{[0,0,\ldots,0]}} \hat{F}^2(\underline{w}) &= \sum_{\underline{w}\in S_{[1,1,\ldots,1]}} \hat{F}^2(\underline{w}) \\
\sum_{\underline{w}\in S_{[0,\ldots,0,1]}} \hat{F}^2(\underline{w}) &= \sum_{\underline{w}\in S_{[1,\ldots,1,0]}} \hat{F}^2(\underline{w}) \\
&\vdots \\
\sum_{\underline{w}\in S_{[0,1,\ldots,1]}} \hat{F}^2(\underline{w}) &= \sum_{\underline{w}\in S_{[1,0,\ldots,0]}} \hat{F}^2(\underline{w}).
\end{aligned}$$

Summing up the left-hand side terms and the right-hand side terms respectively, we get

$$\sum_{\underline{w}\,:\,w_{i_1}=0} \hat{F}^2(\underline{w}) = \sum_{\underline{w}\,:\,w_{i_1}=1} \hat{F}^2(\underline{w}), \tag{32}$$

or equivalently

$$\sum_{\underline{w}\in Z_2^n} (-1)^{w_{i_1}} \cdot \hat{F}^2(\underline{w}) = 0, \tag{33}$$

which means that $\hat{f}(\underline{x})$ is 50%-dependent of x_{i_1}. For symmetry reasons, we get the same result for $x_{i_2}, \ldots, x_{i_m}$.

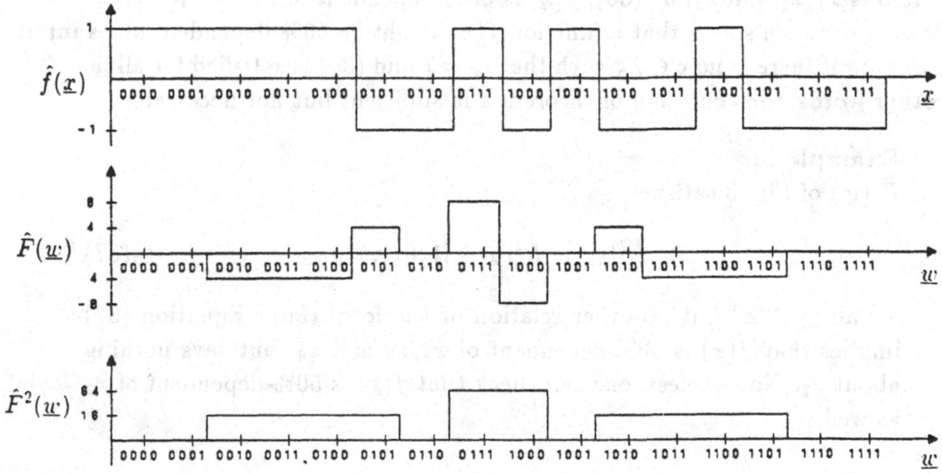

Figure 2: An SAC-fulfilling function $\hat{f}(\underline{x})$ of 4 bits whose squared Walsh-spectrum satisfies (34)

For the special case

$$\hat{F}^2(\underline{w}) = \hat{F}^2(\underline{w} \oplus [1, \ldots, 1]) = \hat{F}^2(\overline{\underline{w}}), \tag{34}$$

theorem 4 asserts that $\hat{f}(\underline{x})$ is 50%-dependent of all its input bits, or, in other words, that $\hat{f}(\underline{x})$ fulfills the SAC. This is interesting from a practical point of view, because the equality (34) is easily noticeable when looking at the squared Walsh-spectrum $\hat{F}^2(\underline{w})$.

Example 7:
The function $\hat{f}(\underline{x}) : Z_2^4 \longrightarrow \{1,-1\}$ takes on the following values (from the top to the bottom of the truth table): 1,1,1,1,1,-1,-1,1,-1,1,-1,-1,1,-1,-1,-1. Fig. 2 shows this function, its Walsh-spectrum and its squared Walsh-spectrum. The discrete points where the functions are defined are connected by lines to make the diagrams more easily readable. We observe a symmetrical form of $\hat{F}^2(\underline{w})$ according to (34) and $\hat{f}(\underline{x})$ therefore fulfills the SAC.

But (34) is not a necessary condition for a function to fulfill the SAC. If, for example, $\hat{f}(\underline{x})$ is such that its squared Walsh-transform satisfies

$$\hat{F}^2(\underline{w}) = \hat{F}^2(\underline{w} \oplus [1,1,1,0,\ldots,0]) \quad (35)$$
$$\text{and } \hat{F}^2(\underline{w}) = \hat{F}^2(\underline{w} \oplus [1,1,0,1,\ldots,1]) \quad (36)$$

we know, by theorem 4 that $\hat{f}(\underline{x})$ fulfills the SAC (by (35), $\hat{f}(\underline{x})$ is 50%-dependent of the bits x_1, x_2 and x_3, by (36), $\hat{f}(\underline{x})$ is 50%-dependent of $x_1, x_2, x_4, \ldots, x_n$). The following example shows that a function $f(\underline{x})$ might be 50%-dependent of its input bit x_i even if there is no $\underline{c} \in Z_2^n$ such that $c_i = 1$ and (30) is satisfied for all $\underline{w} \in Z_2^n$. In other words, the condition of theorem 4 is sufficient but not necessary.

Example 8:
$\hat{F}^2(\underline{w})$ of Fig. 3 satisfies

$$\hat{F}^2(\underline{w}) = \hat{F}^2(\underline{w} \oplus [1,0,1,1]) \quad (37)$$

for all $\underline{w} \in Z_2^4$ but no other relation of the form (30). Equation (37) implies that $\hat{f}(\underline{x})$ is 50%-dependent of x_1, x_3 and x_4, but says nothing about x_2. Nonetheless, one can check that $\hat{f}(\underline{x})$ is 50%-dependent of x_2 as well.

3 Strict Avalanche Criterion of Higher Order

3.1 Definitions

As mentioned in the introduction, the SAC is cryptographically relevant because it maximizes the conditional entropy $H([f(x_1,\ldots,\overline{x_i},\ldots,x_n]) \mid f([x_1,\ldots,x_i,\ldots,x_n]))$ and it assures that the best possible lower-dimensional space approximation of a mapping yields an erroneous result in 25% of the cases. We consider now a mapping of n bits onto one bit that fulfills the SAC. If one or more of its input bits are kept constant, the question arises whether it is possible to find some accurate approximation of this reduced mapping (reduced in the sense that it is defined only on a subspace of Z_2^n). If this is possible, the exhaustive search over the considered subspace can be reduced (compared with the exhaustive search over the full space

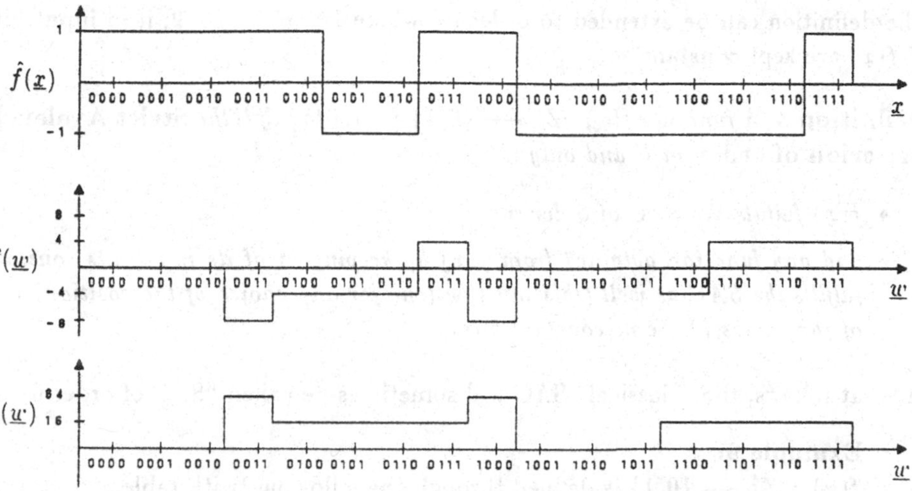

Figure 3: An SAC-fulfilling function that does not satisfy any equation of the form $\hat{F}^2(\underline{w}) = \hat{F}^2(\underline{w} \oplus [c_1, c_2 = 1, c_3, c_4])$ but nevertheless is 50%-dependent of the second input bit.

without approximation). In a chosen-plaintext attack, the opponent has the opportunity to perform such tests where one or more input bits are kept constant. For this reason, we now extend the definition of the SAC in order to cover situations like the one just described.

Let $f(\underline{x})$ be a function which maps Z_2^n onto $\{0,1\}$ and which fulfills the SAC. It is well-known that $f(\underline{x})$ can be written as

$$f(\underline{x}) = x_i \cdot f_{i,1}(x_1, \ldots, x_{i-1}, x_{i+1}, \ldots, x_n) \oplus \overline{x_i} \cdot f_{i,0}(x_1, \ldots, x_{i-1}, x_{i+1}, \ldots, x_n) \quad (38)$$

for every $i \in \{1, 2, \ldots, n\}$. The function $f_{i,1}$ (resp. $f_{i,0}$) is obtained from $f(x)$ by keeping the i-th bit of $\underline{x}$ constant and equal to 1 (resp. to 0). We now consider the 50%-dependence of the output of $f_{i,1}$ and $f_{i,0}$ with respect to each of their $n-1$ input bits.

Definition 3 *A function $f(\underline{x}) : Z_2^n \longrightarrow \{0,1\}$ is said to fulfill the* **Strict Avalanche Criterion of order 1** *if and only if*

- *$f(\underline{x})$ fulfills the SAC,*
- *and every function obtained from $f(\underline{x})$ by keeping the i-th input bit constant and equal to c fulfills the SAC as well (for every $i \in \{1, 2, \ldots, n\}$, and for $c = 0$ and $c = 1$).*

The definition can be extended to order m, where $1 \leq m \leq n-2$, if m input bits of $f(\underline{x})$ are kept constant.

Definition 4 *A function* $f(\underline{x}) : Z_2^n \longrightarrow \{0,1\}$ *is said to fulfill the* **Strict Avalanche Criterion of order** m *if and only if*

- $f(\underline{x})$ *fulfills the SAC of order* $m-1$,
- *and any function obtained from* $f(\underline{x})$ *by keeping* m *of its input bits constant fulfills the SAC as well (this must be true for any choice of the positions and of the values of the* m *constant bits).*

In what follows, the "classical" SAC will sometimes be called "SAC of order 0".

Example 9:
$f(\underline{x}) : Z_2^4 \longrightarrow \{0,1\}$ is defined through the following truth table.

x_1	0	0	0	0	0	0	0	0	1	1	1	1	1	1	1	1
x_2	0	0	0	0	1	1	1	1	0	0	0	0	1	1	1	1
x_3	0	0	1	1	0	0	1	1	0	0	1	1	0	0	1	1
x_4	0	1	0	1	0	1	0	1	0	1	0	1	0	1	0	1
$f(\underline{x})$	0	0	0	1	0	1	1	1	1	0	0	0	0	0	0	1

Keeping the bit x_1 equal to 0, we get a function $f_{1,0} : Z_2^3 \longrightarrow \{0,1\}$ (left-hand half of truth table of $f(\underline{x})$) which can be checked to fulfill the SAC. To check whether $\hat{f}(\underline{x})$ fulfills the SAC of order one, we must go further and control all eight functions of three bits obtained by keeping each input bit of $f(\underline{x})$ fix (equal to zero resp. to one); they are listed in the following table. All of them fulfill the SAC.

y_1	y_2	y_3	$f_{1,0}$	$f_{1,1}$	$f_{2,0}$	$f_{2,1}$	$f_{3,0}$	$f_{3,1}$	$f_{4,0}$	$f_{4,1}$
0	0	0	0	1	0	0	0	0	0	0
0	0	1	0	0	0	1	0	1	0	1
0	1	0	0	0	0	1	0	1	0	1
0	1	1	1	0	1	1	1	1	1	1
1	0	0	0	0	1	0	1	0	1	0
1	0	1	1	0	0	0	0	0	0	0
1	1	0	1	0	0	0	0	0	0	0
1	1	1	1	1	0	1	0	1	1	1

Therefore, $f(\underline{x})$ fulfills the SAC of order one. Keeping each pair (x_i, x_j) constant and equal to $(0,0)$, $(0,1)$, $(1,0)$ and $(1,1)$ respectively, one gets

$\binom{4}{2} \cdot 4 = 6 \cdot 4 = 24$ functions of 2 bits and each of them fulfills the SAC. $f(\underline{x})$ thus even satisfies the SAC of order 2. It makes of course no sense to consider the SAC of order 3 for this function, since keeping three input bits constant yields functions of one variable for which the SAC is not defined.

3.2 Spectral Characterization for SAC of Higher Order

From example 9, it is clear that a boolean function of n bits can fulfill the SAC of order *at most* $n-2$.

We are interested in a spectral characterization of boolean functions that fulfill some SAC of higher order. We again consider $\hat{f}(\underline{x}) = (-1)^{f(\underline{x})}$ rather than $f(\underline{x})$. The following equation is quite similar to (38).

$$\hat{f}(\underline{x}) = x_i \cdot \hat{f}_{i,1}([x_1, \ldots, x_{i-1}, x_{i+1}, \ldots, x_n]) + \overline{x_i} \cdot \hat{f}_{i,0}([x_1, \ldots, x_{i-1}, x_{i+1}, \ldots, x_n]) \quad (39)$$

and can be written for each $i \in \{1, 2, \ldots, n\}$. The "subfunctions" $\hat{f}_{i,1}$ and $\hat{f}_{i,0}$ map Z_2^{n-1} onto $\{1, -1\}$, and all $2n$ subfunctions $\hat{f}_{i,j}$ must fulfill the SAC of order zero if $\hat{f}(\underline{x})$ is to fulfill the SAC of order 1. We introduce

$$\hat{f}_{i,I}(\underline{x}) = x_i \cdot \hat{f}_{i,1}([x_1, \ldots, x_{i-1}, x_{i+1}, \ldots, x_n]) \quad \text{and} \quad (40)$$

$$\hat{f}_{i,II}(\underline{x}) = \overline{x_i} \cdot \hat{f}_{i,0}([x_1, \ldots, x_{i-1}, x_{i+1}, \ldots, x_n]) \quad (41)$$

and we compute their Walsh-transforms.

$$\hat{F}_{i,I}(\underline{w}) = \sum_{\underline{x} \in Z_2^n} x_i \cdot \hat{f}_{i,1}([x_1, \ldots, x_{i-1}, x_{i+1}, \ldots, x_n]) \cdot (-1)^{\underline{x} \cdot \underline{w}} \quad (42)$$

$$= \sum_{\underline{x} : x_i = 1} \hat{f}_{i,1}([x_1, \ldots, x_{i-1}, x_{i+1}, \ldots, x_n]) \cdot (-1)^{w_i \oplus [x_i, \ldots, x_{i-1}, x_{i+1}, \ldots, x_n] \cdot [w_1, \ldots, w_{i-1}, w_{i+1}, \ldots, w_n]} \quad (43)$$

With the substitutions

$$\underline{x}' = [x_1, \ldots, x_{i-1}, x_{i+1}, \ldots, x_n], \ \underline{x}' \in Z_2^{n-1} \quad \text{and} \quad (44)$$

$$\underline{w}' = [w_1, \ldots, w_{i-1}, w_{i+1}, \ldots, w_n], \ \underline{w}' \in Z_2^{n-1} \quad (45)$$

we obtain

$$\hat{F}_{i,I}(\underline{w}) = \sum_{\underline{x}' \in Z_2^{n-1}} \hat{f}_{i,1}(\underline{x}') \cdot (-1)^{w_i} \cdot (-1)^{\underline{x}' \cdot \underline{w}'} \quad (46)$$

$$= (-1)^{w_i} \cdot \hat{F}_{i,1}(\underline{w}'), \quad (47)$$

where $\hat{F}_{i,1}(\underline{w}')$ designates the Walsh-transform of $\hat{f}_{i,1}(\underline{x}')$. Similarly, we get

$$\hat{F}_{i,II}(\underline{w}) = \sum_{\underline{x} : x_i = 0} \hat{f}_{i,0}(\underline{x}') \cdot (-1)^{\underline{x}' \cdot \underline{w}'} = \hat{F}_{i,0}(\underline{w}'). \quad (48)$$

Because of the linearity of the Walsh-transform and the fact that "+" in expression (39) can be considered as integer addition (because always one of both terms on the right-hand side of (39) equals zero) we get:

$$\hat{F}(\underline{w}) = (-1)^{w_i} \cdot \hat{F}_{i,1}(\underline{w}') + \hat{F}_{i,0}(\underline{w}'), \quad \text{for all } i \in \{1,2,\ldots,n\} \tag{49}$$

or equivalently

$$\hat{F}(\underline{w}) = \begin{cases} \hat{F}_{i,1}(\underline{w}') + \hat{F}_{i,0}(\underline{w}') & \text{for } w_i = 0, \\ -\hat{F}_{i,1}(\underline{w}') + \hat{F}_{i,0}(\underline{w}') & \text{for } w_i = 1. \end{cases} \tag{50}$$

Adding, respectively subtracting both equations gives

$$\hat{F}_{i,0}(\underline{w}') = \frac{1}{2}[\hat{F}(\underline{w}) + \hat{F}(\underline{w} \oplus \underline{c}_i)] \tag{51}$$

$$\hat{F}_{i,1}(\underline{w}') = \frac{1}{2} \cdot (-1)^{w_i} \cdot [\hat{F}(\underline{w}) - \hat{F}(\underline{w} \oplus \underline{c}_i)] \tag{52}$$

where $\underline{c}_i = [0,0,\ldots,0,c_i = 1,0,\ldots,0]$. By theorem 1, $\hat{f}(\underline{x})$ will fulfill the SAC of order 1 if and only if

$$\sum_{\underline{w}' \in Z_2^{n-1}} (-1)^{w'_j} \cdot \hat{F}^2_{i,0}(\underline{w}') = 0 \quad \text{and} \tag{53}$$

$$\sum_{\underline{w}' \in Z_2^{n-1}} (-1)^{w'_j} \cdot \hat{F}^2_{i,1}(\underline{w}') = 0 \tag{54}$$

for all $i, j \in \{1,2,\ldots,n\}$ with $i \neq j$. Replacing $\hat{F}_{i,0}$ in (53) by its equivalent form from (51) gives

$$\sum_{\underline{w}\,:\,w_i=0} \frac{1}{4}[\hat{F}(\underline{w}) + \hat{F}(\underline{w} \oplus \underline{c}_i)]^2 \cdot (-1)^{w'_j} = 0, \quad j' \neq i \tag{55}$$

or

$$\frac{1}{4} \sum_{\underline{w}\,:\,w_i=0} [\hat{F}^2(\underline{w}) + \hat{F}^2(\underline{w} \oplus \underline{c}_i)] \cdot (-1)^{w'_j} + \frac{1}{2} \sum_{\underline{w}\,:\,w_i=0} \hat{F}(\underline{w})\hat{F}(\underline{w} \oplus \underline{c}_i) \cdot (-1)^{w'_j} = 0. \tag{56}$$

The first sum in (56) can be written as $\sum_{\underline{w} \in Z_2^n} \hat{F}^2(\underline{w}) \cdot (-1)^{w'_j}$ and therefore equals zero since $\hat{f}(\underline{x})$ fulfills the SAC of order 0 (necessary condition for fulfilling the SAC of order one). Thus

$$\sum_{\underline{w}\,:\,w_i=0} \hat{F}(\underline{w}) \cdot \hat{F}(\underline{w} \oplus \underline{c}_i) \cdot (-1)^{w_{j'}} = 0, \quad (j' \neq i) \tag{57}$$

which implies

$$\sum_{\underline{w} \in Z_2^n} \hat{F}(\underline{w}) \cdot \hat{F}(\underline{w} \oplus \underline{c}_i) \cdot (-1)^{w_{j'}} = 0, \quad (j' \neq i). \tag{58}$$

Inserting (52) in (54) also leads to (58). Theorem 5 follows.

Theorem 5 *A function $\hat{f}(\underline{x}) : Z_2^n \longrightarrow \{1,-1\}$ fulfills the SAC of order 1 if and only if it fulfills the SAC of order zero and*

$$\sum_{\underline{w}\in Z_2^n} \hat{F}(\underline{w})\hat{F}(\underline{w}\oplus \underline{c}_i)\cdot(-1)^{w_j} = 0 \tag{59}$$

for all $i,j \in \{1,2,\ldots,n\}$ with $i \neq j$.

To verify whether a function of n bits fulfills the SAC of order 1 or not, at most $\overbrace{n}^{\text{SAC order 0}} + \overbrace{n\cdot(n-1)}^{\text{SAC order 1}}$ checks are therefore required. The spectral characterizations of the SAC of order 2 and of higher orders can be derived in a similar way and are given without proof in the following two theorems.

Theorem 6 *A function $\hat{f}(\underline{x}) : Z_2^n \longrightarrow \{1,-1\}$ fulfills the SAC of order 2 if and only if it fulfills the SAC of orders 0 and 1, and*

$$\sum_{\underline{w}\in Z_2^n} \hat{F}(\underline{w})\hat{F}(\underline{w}\oplus \underline{c}_{i,j})\cdot(-1)^{w_k} = 0 \tag{60}$$

for all distinct $i,j,k \in \{1,2,\ldots,\}$, and with $\underline{c}_{i,j}$ denoting the $n-$tuple with a one at the $i-$th and $j-$th place and zeroes elsewhere.

Verifying whether the SAC of order 2 is fulfilled or not thus requires at most $n + n(n-1) + \binom{n}{2}(n-2)$ checks.

Theorem 7 *A function $\hat{f}(\underline{x}) : Z_2^n \longrightarrow \{1,-1\}$ fulfills the SAC of order m, $0 \leq m \leq n-2$, if and only if*

$$\sum_{\underline{w}\in Z_2^n} \hat{F}(\underline{w})\hat{F}(\underline{w}\oplus \underline{c}_s)\cdot(-1)^{w_k} = 0 \tag{61}$$

for all $\underline{c}_s \in Z_2^n$ with Hamming-weights $s = 0,1,2,\ldots,m$ and for all $k \in \{1,2,\ldots,n\}$ such that the $k-$th bit of $\underline{c}_s$ is zero.

Verifying whether the SAC of order m is fulfilled or not requires at most $n + n(n-1) + \binom{n}{2}(n-2) + \binom{n}{3}(n-3) + \ldots + \binom{n}{m}(n-m)$ checks.

Example 10:
If $\hat{f}(\underline{x})$ is a boolean function of five bits, the following sums have to be checked:

SAC order 0 $\left\{\sum_{\underline{w}\in Z_2^n} \hat{F}^2(\underline{w})\cdot(-1)^{w_j},\ j\in\{1,2,\ldots,n\},\right.$

$$\text{SAC order 1}\ \left\{\begin{array}{ll} \sum_{\underline{w}\in Z_2^n} \hat{F}(\underline{w})\hat{F}(\underline{w}\oplus[00001])\cdot(-1)^{w_j}, & j\in\{1,2,3,4\},\\ \sum_{\underline{w}\in Z_2^n} \hat{F}(\underline{w})\hat{F}(\underline{w}\oplus[00010])\cdot(-1)^{w_j}, & j\in\{1,2,3,5\},\\ \vdots & \vdots\\ \sum_{\underline{w}\in Z_2^n} \hat{F}(\underline{w})\hat{F}(\underline{w}\oplus[10000])\cdot(-1)^{w_j}, & j\in\{2,3,4,5\}, \end{array}\right.$$

$n \longrightarrow$	2	3	4	5	6
no SAC	8	192	61408	?	?
SAC order 0	8	48	3808	?	?
SAC order 1	-	16	288	?	?
SAC order 2	-	-	32	?	?
SAC order 3	-	-	-	64	?
SAC order 4	-	-	-	-	128

Table 1: Number of functions that fulfill the SAC of some given order

$$\text{SAC order 2} \begin{cases} \sum_{\underline{w}\in Z_2^n} \hat{F}(\underline{w})\hat{F}(\underline{w}\oplus[00011])\cdot(-1)^{w_j}, & j\in\{1,2,3\},\\ \sum_{\underline{w}\in Z_2^n} \hat{F}(\underline{w})\hat{F}(\underline{w}\oplus[00101])\cdot(-1)^{w_j}, & j\in\{1,2,4\},\\ \vdots & \vdots\\ \sum_{\underline{w}\in Z_2^n} \hat{F}(\underline{w})\hat{F}(\underline{w}\oplus[11000])\cdot(-1)^{w_j}, & j\in\{3,4,5\}, \end{cases}$$

$$\text{SAC order 3} \begin{cases} \sum_{\underline{w}\in Z_2^n} \hat{F}(\underline{w})\hat{F}(\underline{w}\oplus[00111])\cdot(-1)^{w_j}, & j\in\{1,2\},\\ \sum_{\underline{w}\in Z_2^n} \hat{F}(\underline{w})\hat{F}(\underline{w}\oplus[01011])\cdot(-1)^{w_j}, & j\in\{1,3\},\\ \vdots & \vdots\\ \sum_{\underline{w}\in Z_2^n} \hat{F}(\underline{w})\hat{F}(\underline{w}\oplus[11100])\cdot(-1)^{w_j}, & j\in\{4,5\}. \end{cases}$$

Exhaustive computer search through functions of 2, 3 and 4 bits allowed to count how many boolean functions fulfill the SAC of a given order. The results are listed in table 1. One can check that the columns for $n = 2, 3$ and 4 sum up to 2^{2^n}. Notice that no function is counted twice, although in fact each function that fulfills the SAC of some order m by definition also fulfills the SAC of orders $m-1, m-2, \ldots, 1, 0$.

3.3 Construction of Functions Fulfilling the SAC of Maximum Order

The method used to count the SAC-fulfilling functions of maximum order $n-2$ for $n = 5$ and $n = 6$ is a constructive one. The definition of the SAC of order m implies the following lemma.

Lemma 4 *A boolean function $f(\underline{x})$ of n bits fulfills the SAC of order m if and only if*

- *$f(\underline{x})$ fulfills the SAC of order 0, and*
- *any function obtained from $f(\underline{x})$ by keeping one input bit constant (equal to 0 or to 1) fulfills the SAC of order $m-1$.*

This gives rise to the idea of using functions of $n-1$ bits that fulfill the SAC of order $n-3$ as basic elements for the synthesis of functions of n bits that fulfill the SAC of order $n-2$.

Example 11:
The eight functions of two bits that fulfill the SAC of order zero are listed below.

x_1	x_2	$f_1(\underline{x})$	$f_2(\underline{x})$	$f_3(\underline{x})$	$f_4(\underline{x})$	$f_5(\underline{x})$	$f_6(\underline{x})$	$f_7(\underline{x})$	$f_8(\underline{x})$
0	0	0	0	0	0	1	1	1	1
0	1	0	0	1	1	0	0	1	1
1	0	0	1	0	1	0	1	0	1
1	1	1	0	0	1	0	1	1	0

We can define $f(\underline{x}) : Z_2^3 \longrightarrow \{0,1\}$ as

$$f(\underline{x}) = f([x_1, x_2, x_3]) = x_1 \cdot f_i([x_2, x_3]) + \overline{x_1} \cdot f_j([x_2, x_3]) \qquad (62)$$

with $i, j \in \{1, 2, \ldots, 8\}$, $i \neq j$ and we get $\binom{8}{2} = 28$ functions $f(\underline{x})$; sixteen of them can be checked to fulfill the SAC of order 1. We can be sure that no other function of three bits satisfies the SAC of order 1, since any such function necessarily is decomposable according to (62) (by Lemma 4).

The procedure used in example 11 can be applied to the sixteen functions of three bits that fulfill the SAC of order 1, and it yields the 32 functions of four bits that fulfill the SAC of order 2, and so on.

4 Conclusion

The Strict Avalanche Criterion of order m has been introduced which corresponds to a generalized definition of the known SAC. It has been shown that the SAC of any order can be easily characterized in the Walsh-domain. This representation was used for the construction of further SAC-fulfilling boolean functions. The application of SAC-fulfilling functions for cryptosystem-design has still to be studied. An application would be, for instance, to use such functions for the synthesis of S-boxes in substitution/permutation (SP) block-ciphers. Since an S-box has many inputs and n outputs, n SAC-fulfilling functions should be chosen and combined in some adequate manner. For example, statistical dependencies between output bits should be avoided. Statistical independencies between input $m-$tuples and the output of boolean functions is known as $m-$th order correlation-immunity. It might be interesting to examine whether there are restrictions in the compatibility of correlation-immunity and SAC of order m. Any boolean function that is

$m-$th order correlation-immune [6] has vanishing values of $F(\underline{w})$ for all $\underline{w}$'s with Hamming-weigths between one and m [5]. Exhaustive search for functions of three and four bits showed that eight functions of three bits as well as ninety-six functions of four bits are first-order correlation-immune and fulfill the SAC of order 1 at the same time.

Acknowledgements

The author is grateful to Thomas Siegenthaler for many constructive discussions and for his suggestions to improve this paper. She also wishes to thank Othmar Staffelbach for his helpful comments.

References

[1] A.F. Webster and S.E. Tavares, *"On the Design of S-Boxes"*, Advances in Cryptology: Crypto'85 proceedings, Springer, 1986.

[2] R.C. Titsworth, *"Correlation Properties of Cyclic Sequences"*, Thesis, California Institute of Technology, Pasadena, California, 1962.

[3] Th. Siegenthaler, *"Methoden für den Entwurf von Stream Cipher-Systemen"*, Diss. ETH No. 8185, Dec. 1986.

[4] S.C. Lee, *"Modern Switching Theory and Digital Design"*, Prentice-Hall, 1978.

[5] G.Z. Xiao, J.L. Massey, *"A Spectral Characterization of Correlation-Immune Combining Functions"*, to be published in IEEE Tr. on Information Theory.

[6] Th. Siegenthaler, *"Correlation-immunity of Nonlinear Combining Functions for Cryptographic Applications"*, IEEE Tr. on Information Theory, vol. IT-30, pp. 776-780, Oct. 1984.

On the Linear Syndrome Method in Cryptanalysis

Kencheng Zeng   Minqiang Huang

Data and Communications Security Research Center

Graduate School of USTC, BX 100039-08, Beijing, China

The linear syndrome (LS) method is elaborated for the purpose of solving problems encountered in cryptanalysis, which can be reduced to the following mathematical setting. Suppose the cryptanalyst has at his hand a sufficiently long segment of the binary sequence

$$B = A + X ,$$

where A is a linear sequence with known feedback polynomial f(x) and X is a sequence with unknown or very complicated algebraic structure, but is sparse in the sense that, if we denote its signals by x(i), $i > 0$, then we shall have

$$s = \text{prob}(\ x(i) = 1\ ) = 1/2 - \varepsilon \ , \qquad 0 < \varepsilon < 1/2 \ .$$

We call s the error rate of the sequence A in the sequence B, and the job of the cryptanalyst is to recover the former from the captured segment of the latter.

One way for tackling this problem is to make use of the ideas of error correction, especially when s is comparatively small. In doing this we consider, for some fixed integer $r \geqslant 3$, a finite collection of r-nomials of the form

$$g(x) = 1 + x^{i_1} + x^{i_2} + \ldots + x^{i_{r-1}},$$

and compute, for every $i \geqslant \max\{\ \deg g(x)\ \}$ and all g(x), the syndromes

$$\sigma_{i,k}(g) = \sum_{p=0}^{r-1} b(i - i_k + i_p) \ , \qquad 0 \leqslant k \leqslant r-1 \ ,$$

$b(i)$, $i \geq 0$, being the signals of the sequence B. The LS method is based on the following

Lemma 1. If $f(x)$ divides $g(x)$, then

$$\text{prob}(\,\sigma_{i,k}(g) = x(i)) = 1/2 + (1 - 2s)^{r-1}/2. \qquad (1)$$

Proof. Denote the signals of A by $a(i)$, $i \geq 0$. Since $f(x) \mid g(x)$, we have

$$\begin{aligned}\sigma_{i,k}(g) &= \sum_{p=0}^{r-1} b(i - i_k + i_p) \\ &= \sum_{p=0}^{r-1} a(i - i_k + i_p) + \sum_{p=0}^{r-1} x(i - i_k + i_p) \\ &= \sum_{p=0}^{r-1} x(i - i_k + i_p) .\end{aligned}$$

Thus we see $\sigma_{i,k}(g) = x(i)$ if and only if an even number of the signals

$$x(i - i_k) \;,\ldots,\; x(i - i_k + i_{k-1}) \;,\; x(i - i_k + i_{k+1}) \;,\ldots,\; x(i - i_k + i_{r-1})$$

are " 1 ", and hence we have

$$\begin{aligned}\text{prob}(\,\sigma_{i,k}(g) = x(i)) &= \sum_{2\mid p} C_{r-1}^{p}\, s^{p}(1 - s)^{r-p-1} \\ &= 1/2 + (1-2s)^{r-1}/2 \\ &= 1/2 + (2\epsilon)^{r-1}/2.\end{aligned}$$

This simple lemma suggests that it will be wise for the cryptanalyst to behave as follows. Choose the r-nomials $g(x)$ to be multiples of the given polynomial $f(x)$, take into consideration $2m + 1$ of the syndromes provided by these r-nomials,and revise the signals of the sequence B in accordance with the following rule of majority logic decision,

$$b(i) \longrightarrow b'(i) = \begin{cases} b(i) + 1, & \text{if at least } m + 1 \text{ syndromes are "1"}, \\ b(i), & \text{if otherwise}, \end{cases}$$

in the hope that the error rate s' of the sequence A in the resulting sequence B' will be less than the initial error rate s.

In order to see, under which conditions this will be the case, we write

$$p = p(s) = (1 - (1 - 2s)^{r-1})/2, \quad q = 1 - p,$$

and prove the following

Theorem 1. If the number of syndromes used in making the majority logic decision is $n = 2m + 1$, then the error rate of the sequence A in the sequence B' which results from one round of revision will be

$$s' = T_m = p - (1 - 2p) \sum_{k=0}^{m-1} C^k_{2k+1} (pq)^{k+1}. \qquad (2)$$

Proof. It is easy to see from the revision algorithm, that $b'(i) = a(i)$ if and only if at least $m + 1$ syndrome values are different from $x(i)$. But, by lemma 1, the probability for a given syndrome value to be different from $x(i)$ is p, so we have

$$s' = T_m = p^n + C^1_n p^{n-1} q + \dots + C^m_n p^{m+1} q^m .$$

Further, we have

$$\begin{aligned} T_m &= T_m(p + q) \\ &= p^{n+1} + C^1_n p^n q + \dots + C^m_n p^{m+2} q^m \\ &\qquad + p^n q + \dots + C^{m-1}_n p^{m+2} q^m + C^m_n p^{m+1} q \\ &= (p^{n+1} + C^1_{n+1} p^n q + \dots + C^m_{n+1} p^{m+2} q^m)(p + q) + C^m_n p^{m+1} q^{m+1} \\ &= (p^{n+2} + C^1_{n+2} p^{n+1} q + \dots + C^m_{n+2} p^{m+3} q^m) + C^m_{n+1} p^{m+2} q^{m+1} + C^m_n p^{m+1} q^{m+1} \\ &= T_{m+1} - (C^{m+1}_{n+2} - C^m_{n+1}) p^{m+2} q^{m+1} + C^m_n p^{m+1} q^{m+1}. \end{aligned}$$

But

$$C^{m+1}_{n+2} = C^{m+1}_{n+1} + C^m_{n+1} = C^m_{n+1} + 2C^m_n ,$$

so we have the following recursive relation

$$T_{m+1} = T_m - (1 - 2p) C^m_{2m+1} (pq)^{m+1} ,$$

which, together with $T_0 = p$, gives rise to (2).

Now, it is easy to see from (2) that, s being fixed, s' decreases as m increases. Furthermore, since

$$T_0 = 1/2 - (2\epsilon)^{r-1} > 1/2 - \epsilon = s$$

and

$$\lim_{m\to\infty} T_m = p - (1 - 2p) \sum_{k=0}^{\infty} C_{2k+1}^{k} (pq)^{k+1}$$

$$= p - (1 - 2p)((1 - 4pq)^{-\frac{1}{2}} - 1)/2$$

$$= p - (1 - 2p)((1 - 2p)^{-1} - 1)/2 = 0 ,$$

we see that for each possible initial error rate s there is a critical number mc = mc(s), such that s' will be less than s if and only if m > mc. The following is a table of critical numbers computed for practically tractable values of s, for the case r = 3, where the LS method works the best.

s	mc
0.22	3
0.28	4
0.32	5
0.35	6
0.37	7
0.38	8
0.40	9

(II) Iterated revision and its convergence

The above analysis shows also, that the error rate of the sequence A can be made arbitrarily small, when we make use of a large enough number of syndromes. But such an approach is quite impractical in view of the difficulty in finding the necessary collection of r-nomials, divisible by f(x) and of degrees not too large. A better alternative is to fix the number of syndromes but apply the revision algorithm iteratedly to the segment under consideration, and the problem is that the convergence of such an iterative revision procedure has to be considered.

In order to settle the problem just raised, we consider the polynomial

$$T_m(x) = x - (1 - 2x) \sum_{k=0}^{\infty} C_{2k+1}^{k} (x(1 - x))^{k+1}$$

and prove a couple of simple lemmas about the function $p(s)$ mentioned before as well as about the function

$$s' = f(x) = T_m(p(s)).$$

Lemma 2.The function $p(s)$ is increasing on $(0 , 1/2)$ and maps this interval onto itself.

Proof. In fact, we have

$$p'(s) = (r - 1)(1 - 2s)^{r-2} > 0$$

and

$$p(0) = 0 \quad , \quad p(1/2) = 1/2 ,$$

as expected.

Lemma 3. The derivative of the polynomial $T_m(x)$ is

$$T'_m(x) = (m + 1)C_{2m+1}^{m}(x(1 - x))^{m} .$$

Proof. First, as we have noticed before

$$x = (1 - 2x) \sum_{k=0}^{\infty} C_{2k+1}^{k} (x(1 - x))^{k+1} ,$$

whenever $|x| < 1$. So we have

$$\begin{aligned} T_m(x) &= x - (1 - 2x) \sum_{k=0}^{m-1} C_{2k+1}^{k}(x(1 - x)) \\ &= (1 - 2x) \sum_{k=m}^{\infty} C_{2k+1}^{k}(x(1 - x))^{k+1} \\ &= C_{2m+1}^{m} x^{m+1} \quad (\bmod x^{m+2}) \quad . \end{aligned}$$

and hence

$$T'_m(x) = (m + 1)C_{2m+1}^{m} x^{m} \quad (\bmod x^{m+1}) \ .$$

Further, we have the functional relation

$$T_m(1 - x) = 1 - x + (1 - 2x) \sum_{k=0}^{m-1} C_{2k+1}^{k}(x(1 - x))^{k+1}$$

$$= 1 - T_m(x) .$$

By differentiation on both sides we have

$$T_m'(x) = (m + 1)C_{2m+1}^{m}(1 - x)^{m} \quad (\bmod (1 - x)^{m+1}).$$

But $T_m'(x)$ is a polynomial of degree 2m, so we conclude that

$$T_m'(x) = (m + 1)C_{2m+1}^{m}(x(1 - x))^{m} .$$

Lemma 4. There is a number $\alpha \in (0 , 1/2)$ such that

$$f(s) < s \quad , \quad \text{if} \quad 0 < s < \alpha$$

and

$$f(s) > s \quad , \quad \text{if} \quad \alpha < s < 1/2.$$

Proof. Consider the auxiliary function

$$w(s) = f(s) - s .$$

We see from

$$p(0) = o \quad , \quad p(1/2) = 1/2$$

and the expression for $T_m(x)$ that

$$w(0) = w(1/2) = 0 . \tag{3}$$

Further, we see from

$$w'(s) = T_m'(p(s))p'(s) - 1$$

and

$$T'(0) = 0 \quad , \quad p'(1/2) = 0$$

that

$$w'(0) = w'(1/2) = -1 \qquad (4)$$

(3) and (4) taken together imply that w(s) has at least one zero in the interval (0 , 1/2). On the other hand, if w(s) has in this interval two or more zeroes, then as can be easily seen from (3) and the mean value theorem of differential calculus, w''(s), too, will have at least two zeroes in it. But by direct manipulation we have

$$\begin{aligned} w''(s) &= (r-1)(1-2s)^{r-3}[(r-1)T_m''(p(s))(1-2s)^{r-1} - 2(r-2)T_m'(p(s))] \\ &= (r-1)(1-2s)^{r-3}[(r-1)T_m''(p(s))(1-2p(s)) - 2(r-2)T_m'(p(s))] \\ &= (r-1)(1-2s)^{r-3}K(p(s)), \end{aligned}$$

where

$$K(x) = (m+1)C_{2m+1}^{m}(x(1-x))^{m-1}(ax^2 - ax + b)$$

and

$$a = 4m(r-1) + 2(r-2), \quad b = m(r-1),$$

so we see, by noticing the statement of lemma 2, that w''(s) has only one zero β in the interval (0 , 1/2), satisfying

$$p(\beta) = 1/2 - (1 - 4b/a)^{\frac{1}{2}}/2 .$$

This conclusion means that the function w(s) has only one zero in (0 , 1/2). If we denote this unique zero of w(s) by α , then, by returning to (3) and (4) again, we see w(s) is negative on (0 , α) and positive on (α, 1/2). But this is just what we wanted to prove.

Now we are in a position to prove the convergence theorem for the procedure of iterated revision.

Theorem 2. If we denote by s the error rate of the sequence A in the sequence, which results from the i-th round of revision, then the number sequence { s_i} will decrease to 0 if m > mc, and increase to 1/2 if m < mc.

Proof. Suppose m > mc. Then we have by the definition of mc

$$s_1 = f(s_0) < s_0.$$

So we see from lemma 4 that

$$s_1 < s_0 < \alpha .$$

By applying the same lemma to s we have

$$s_2 = f(s_1) < s_1 < \alpha .$$

By going along with the same argument we have

$$\alpha > s = s_0 > s_1 > \dots > s_i > \dots > 0,$$

so we must have

$$\alpha > \lim_{i \to \infty} s_i = s^* \geq 0 ,$$

and s, being a zero of the function w(s) met in the proof of lemma 4, can be nothing else than 0.

The case m < mc,where iterated revision will lead to disastrous garble, can be discussed in exactly the same manner.

(III) An example of applying the LS method

The above analysis of the LS method is by no means rigorous in view of the assumptions made tacitly in computing the probabilities. For a really convincing justification of this method we, in the last run,have to resort to its usefulness in solving concrete problems. Practical problems encountered in cryptanalysis may not yield to the LS method immediately, but can in some cases be reduced to a suitable form,so as to make the method applicable.The following example,though artificial in nature,will be sufficient as an illustration for what we say here.

In the laboratory of the DCS-center people produced a stream X of digital speech by the method of code excited linear prediction followed by vector quantization and turned it, as an experiment, into a stream

$$Y = A + X$$

of incomprehensible enciphered speech, by the help of a linear sequence A with generating polynomial

$$f(x) = 1 + x^4 + x^{39} .$$

Now, suppose the cryptanalyst knows the polynomial $f(x)$, but has no concept about how to make use of the specific properties of the stream X itself, then for the purpose of recovering it he has to test one after another the $2^{39} - 1$ possible initial states of A,a task far beyond the reach of today's technique.

We show,it is the specific structure of the plaintext stream X, that makes the stream Y easily breakable. In fact, as a result of the slow variation nature of the speech data flow and the imprudent way of encoding it, there exists a sort of betraying correlation between the frames

$$F_0 , F_1 , \ldots , F_i , \ldots$$

of X . A closer examination shows that if we denote the number of " 1 "s in the frame F by w(F) and denote the frame length by l, then for most of the adjacent frame pairs F_i , F_{i+1} we have

$$w(F_i + F_{i+1}) \leqslant l/4 .$$

And here is the clue we need.In fact,if the cryptanalyst proves to be clever enough to think of going from the stream Y over to the transformed stream

$$Y' = Y + LY = (A + LA) + (X + LX) = A' + X' ,$$

L being the l-step shift to the left,then he will find himself in the typical situation discussed in the present paper, where A' is linear with the same generating polynomial f(x) as A, while X' is sparse with s = 1/4 Experimentation shows, that by making use of the 9 syndromes provided by the trinomials

$$1 + x^4 + x^{39} , \quad 1 + x^8 + x^{78} , \quad 1 + x^{16} + x^{156},$$

four rounds of iterated revision applied to Y' suffice to recover A' from a captured segment of length $n < 1500$, and after that the plaintext X can be determined easily by

$$X = Y + (I + L)^{-1} A'.$$

A tape record has been prepared by the same lab for this simple, but instructive instance of successful codebreaking. This example reminds us, in particular, that in order to guarantee safety in communication, not only the algorithm for generating the enciphering signals, but also the data flow to be enciphered, as well as the problem about the suitable way of encoding and enciphering, should be considered carefully.

References

[1] Zeng Kencheng," Phenomena of Key-entropy Leak in Cryptosystems ", unpublished report presented to " Symposium on Problems of Cryptanalysis " Beijing,1986.

Aperiodic Linear Complexities of de Bruijn Sequences

Richard T.C.Kwok, M.Sc., and Maurice Beale, B.Sc., Ph.D.,
Electrical Engineering Department,
University of Manchester,
Manchester, M13 9PL,
England.

Extended Abstract.

Binary de Bruijn sequences of period 2^n bits have the property that all 2^n distinct n-tuples occur once per period. To generate such a sequence with an n-stage shift-register requires the use of nonlinear feedback. These properties suggest that de Bruijn sequences may be useful in stream ciphers. However, any binary sequence can be generated using a linear-feedback shift register (LFSR) of sufficient length. Thus, the *linear complexity* of a sequence, defined as the length of the shortest LFSR which generates it, is often used as a measure of the unpredictability of the sequence. This is a useful measure, since a well-known algorithm[1] can be used to successfully predict all bits of any sequence with linear complexity C from a knowledge of 2C bits. As an example, an m-sequence of period 2^n -1 has linear complexity C=n, which clearly indicates that m-sequences are highly predictable.

Now, the widely used definition of linear complexity stated above is open to different interpretations. We distinguish here between the *periodic linear complexity* (PLC) - the length of the shortest LFSR which generates the given sequence and then repeats it cyclically - and the *aperiodic linear complexity* (ALC) - the length of the shortest LFSR which generates the given sequence followed by any arbitrary sequence of bits. This distinction is not made in the literature on de Bruijn sequences, but it has important practical consequences. In a stream cipher, it is clearly undesirable for keystream sequences to be allowed to repeat. Consequently, no more than P bits (where P is the sequence period) will ever be used, which implies that it is the ALC, not the PLC, which is of real concern. Unfortunately, all of the published results on the linear complexity of de Bruijn sequences (e.g [2] and [3]) relate only to the PLC not the ALC. The research described here goes some way towards addressing this imbalance.

Having decided that the ALC is the most useful measure of unpredictability, we note however that a large value of the ALC of an entire sequence (one period) is not , by itself, a sufficient condition for high randomness. We also require the ALC of all sub-sequences of the given sequence to be as large as possible. For a given sequence of P bits, we are therefore interested in the ALC of the first k bits of the sequence, as a function of k ($1 \leq k \leq P$). The importance of this function was identified by Rueppel [4], who referred to it as the linear complexity profile (LCP) of the sequence. Note that, in general, the LCP of a sequence depends on the starting point within the sequence. Thus, if we consider all P cyclic shifts of any given sequence of P bits, some cyclic shifts may have notably better LCPs than others.

Although statistical results on the LCP of random binary sequences have been obtained [4], these authors are unaware of any published results on the LCPs of any class of finite deterministic sequences. The LCPs of de Bruijn sequences were therefore investigated and the results of this study are summarized below. For comparison, it is interesting to note that the expected value of the LCP of a random binary sequence (with equiprobable ones and zeros), as a function of the sub-sequence length k, is given by [4] :-

$$E[C] = k/2 + [(4+R_2(k)]/18 - 2^{-k}(k/3 + 2/9), \ldots\ldots\ldots\ldots(1)$$

which rapidly approaches k/2 as k increases. (Here, $R_2(k)$ denotes the remainder when k is divided by 2). On the basis of this result and other observations, Rueppel[4] proposed that a "good random sequence" for cryptographic use should have a LCP which closely, but irregularly, follows the k/2 line. Also note that a linear complexity of k/2 for a sub-sequence of length k is sufficient to foil an attack based on the Berlekamp-Massey algorithm[1].

Now consider the aperiodic LCP of a de Bruijn sequence of period 2^n. As noted earlier, the LCP depends on the cyclic shift of the sequence under consideration. However, if we take the average value of the LCP of a de Bruijn sequence over all 2^n cyclic shifts, it is readily seen that for sub-sequence lengths $k \le n$, equal numbers of all 2^k possible sub-sequences have been included in the averaging process. For such k, the average LCP is therefore identical to the expected value for random sequences, since the latter is also an ensemble average over all choices of sub-sequence of length k. Hence, we have the following :

Theorem : For $k \le n$, the average LCP of any de Bruijn sequence of length 2^n, over all cyclic shifts, is identical to the average LCP for random sequences given by eqn (1).

For $k > n$, when all cyclic shifts of a fixed de Bruijn sequence are considered, only a subset of 2^n of the possible 2^k k-bit sub-sequences occur. Which of the 2^k possible sub-sequences occur depends on the de Bruijn sequence being considered. For this case, it has proved difficult to derive analytical results concerning the LCP. However, extensive numerical investigations have been carried out on the sets of de Bruijn sequences generated by Fredricksen's 'cross-join' algorithm[5]. Although this algorithm (in common with all other practical algorithms) generates only subsets of de Bruijn sequences, these subsets are large, containing 2^{2n-5} or 2^{2n-6} sequences of length 2^n, for odd and even n, respectively. Furthermore, an implementation in the form of a programmable nonlinear-feedback shift-register can be derived from this algorithm[6], making it an attractive choice for applications. The LCP investigations were carried out over all de Bruijn sequences generated by this algorithm for all $n \le 12$.

Consider again the average LCP over all cyclic shifts of a given de Bruijn sequence. As a measure of the non-randomness of a sequence, we can take the difference between this average LCP and the ensemble average for random sequences in eqn(1). For each of the de Bruijn sequences investigated, the fluctuations themselves appear to be random, and show no tendency to increase or decrease as a function of the sub-sequence length k. Typical results for de Bruijn sequences of length 512 and 4096 are shown in Figs.1 and 2. Of course the fluctuations are identically zero for $k \leq n$, as predicted by the previous Theorem, although this is clearly visible only in Fig.1 due to the scale. It is also apparent that the magnitude of the fluctuations decreases, albeit slowly, as the de Bruijn sequence length is increased.

Now, although the average LCP over all cyclic shifts of a sequence is of some interest for comparison with the ensemble average in eqn(1), an issue of greater practical concern is the LCP behaviour of fixed cyclic shifts of a sequence. In particular, one would like to know if there are any 'bad' cyclic shifts of a de Bruijn sequence which ought to be avoided. The results of our investigations suggest that this question can be answered in the negative, at least for all the de Bruijn sequences generated by Fridricksen's 'cross-join' algorithm for all $n \leq 12$. The LCPs of these de Bruijn sequences are relatively insensitive to the choice of cyclic shift and all appear to satisfy Rueppel's criterion for closely, but irregularly, following the k/2 line. A typical example of the LCP of a de Bruijn sequence of length 256 is shown in Fig.3. In this case, the cyclic shift chosen was that beginning with the all-zeros n-tuple. The steps in the LCP are all of the order of n or less in magnitude; indeed, the inevitable step associated with the all-zeros n-tuple is the largest present. This result appears to hold in general.

To illustrate the insensitivity of the LCP to the choice of cyclic shift of a de Bruijn sequence, Fig.4 shows typical results for the average, maximum and minimum values of the LCP, over all cyclic shifts of a 512-bit de Bruijn sequence. As expected, the average LCP is indistinguishable from the k/2 line. An interesting feature, which appears to hold in general, is that the maxima and minima of the LCP show a remarkable symmetry about its average value. More importantly, the peak deviation from the average (and, in effect, from the k/2 line) is small relative to the sequence length, and shows no tendency to increase or decrease as a function of the sub-sequence length k.

References

[1] Massey, J.L : " Shift-Register synthesis and BCH decoding". *IEEE Trans on IT*, Vol. IT-15,Jan,1969, pp.122-127.

[2] Etzion, T. & Lempel,A : "On the distribution of de Bruijn sequences of given complexity". *IEEE Trans on IT*, Vol.IT-30, no.4, July 1984, pp.611-614.

[3] Chan, A.H , Games, R.A & Key, E.L : " On the complexities of de Bruijn sequences". *J.Comb. Theory*, (Ser A), Vol.3, No.3, Nov.1982, pp.223-246.

[4] Rueppel, R.A: " New Approaches to Stream Ciphers". D.Sc. dissertation, No.ETH-7714, Swiss Federal Institute of Technology, Zurich, 1984.

[5] Fredricksen, H.M : " A survey of full length non-linear shift register cycle algorithms". *SIAM Review*, Vol.24, Apr.1982, pp.195-221.

[6] Beale, M., Cochrane.S.D & Lau ,S.M.S : "A programmable de Bruijn sequence generator for stream ciphers". *Proc.IEE Int. Conf.on Secure Communication Systems*, Oct. 1986,pp.69-73.

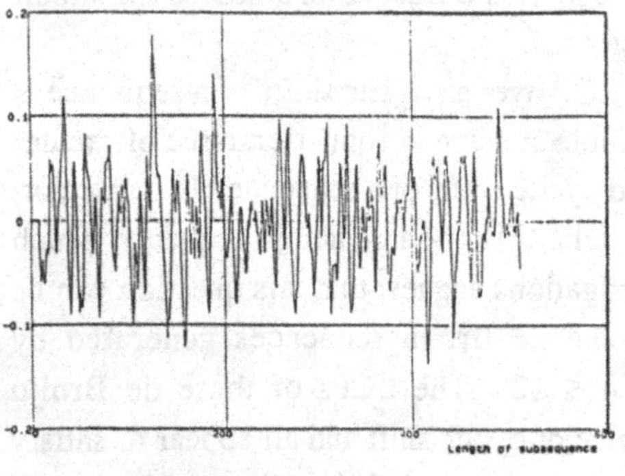

Fig.1 : Fluctuations of the average LCP (over all cyclic shifts) of a 512-bit de Bruijn sequence from the expected value for random sequences.

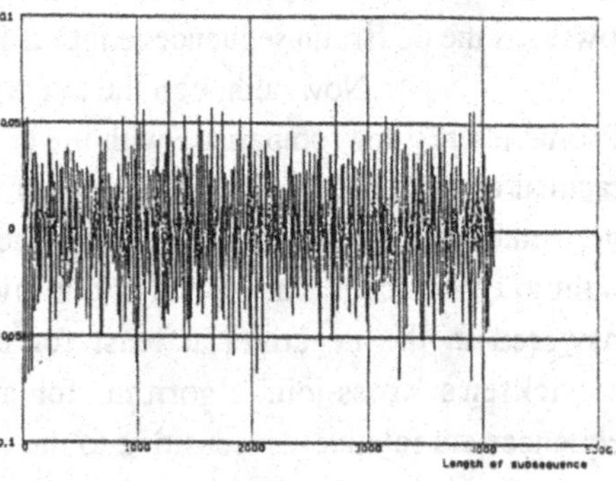

Fig.2 : As Fig.1, but for a de Bruijn sequence of length 4096.

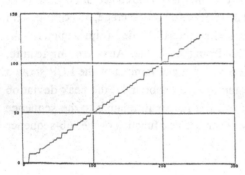

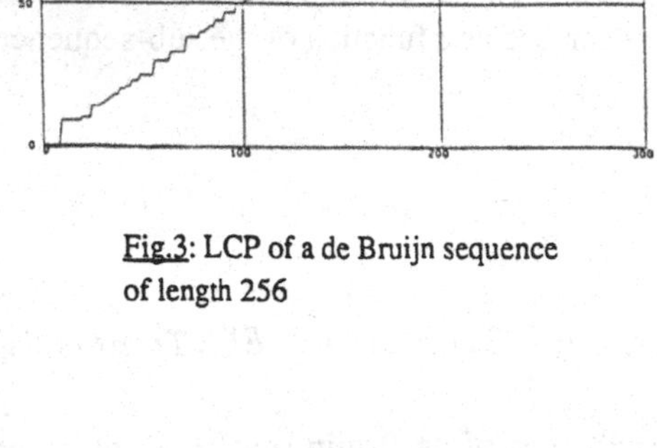

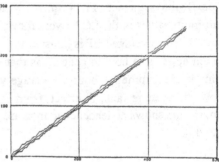

Fig.3: LCP of a de Bruijn sequence of length 256

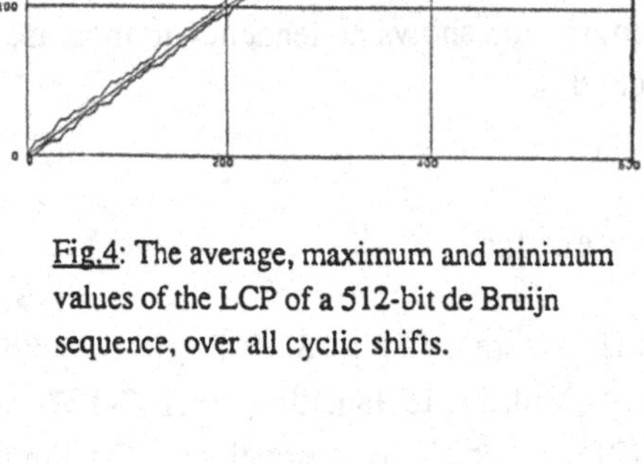

Fig.4: The average, maximum and minimum values of the LCP of a 512-bit de Bruijn sequence, over all cyclic shifts.

Session 12

Systems

Chair: T. Berson, Anagram Laboratories

THE APPLICATION OF SMART CARDS FOR RSA DIGITAL SIGNATURES IN A NETWORK COMPRISING BOTH INTERACTIVE AND STORE-AND-FORWARD FACILITIES

J.R. Sherwood and V.A. Gallo
Computer Security Limited
Brighton, UK

Abstract. *Smart card technology is relatively new but offers an economic and convenient solution to the problems of user-authentication. This paper discusses the requirements for user authentication and digital signature in complex networks and examines the problems of integrating a smart-card sub-system. It proposes some design approaches for providing a useful lifetime for a smart card and for handling the computations required for 512-bit RSA digital signatures.*

Environment

Many data communications networks are known to be based on interactive working between a user on a workstation and a remote central application on a mainframe host. Many other networks are based on store-and-forward facilities for message or file transfer using the mail-box principle. Increasingly, large corporate networks are offering both of these facilities in an integrated data communications network. At the same time, network services providers and users are becoming more conscious of the need to implement security in these environments.

Introduction

This paper addresses the problems of security sub-system design in a network environment of some complexity. It outlines a design approach which is suited to both interactive and store-and-forward working, bringing together a wide range of cryptographic techniques and system components. The paper explores some of the design considerations that are relevant to the development of an integrated solution for system security and provides a framework within which various cryptographic techniques can be interwoven. In particular, it discusses the applicability of each technique and examines the contribution that each can make to the overall design.

Smart cards are relatively new and as yet there are few systems which exploit their potential as a user token providing an automated logon protocol. This is one area that the paper explores in some detail; we develop some existing ideas on the use of a one-way function to encrypt a random challenge for constructing a logon protocol and in particular some techniques are described for ensuring attack-resistance over the expected lifetime of the smart-card token.

Security Requirements

The type of network under consideration here comprises both interactive and store-and-forward mailbox facilities, based on personal computer workstations connected to mainframe computers.

Broadly speaking, the security requirements can be summarised as follows:-

1. *Access control over local workstations and their applications.*

2. *Access control over remote hosts and their applications.*

3. *Privacy of communications over data networks.*

4. *Integrity checks on the contents of communications (message authentication).*

5. *Proof of message origin.*

To achieve these requirements a security sub-system is required which integrates into all possible configurations of the network, and which is applicable to both store-and-forward and interactive environments. It must also be capable of integrating successfully with existing security sub-systems, such as that provided in IBM SNA networks [1] [2] involving the use of the SNA encrypt/decrypt facility, and such as the access control facilities of RACF [3] and ACF2 [4]. Where SNA encrypt/decrypt is employed, the host mainframes are either equipped with an IBM 3848 Cryptographic Unit [5] or with the IBM Programmed Cryptographic Facility Program Product [6] with ACF/VTAM [7].

System Requirements

The system component requirements for constructing a suitable security sub-system are as follows:

1. *User tokens*

2. *A suitable collection of cryptographic algorithms for implementing personal authentication, data encryption, message authentication and digital signature.*

3. *A cryptographic key management architecture which provides both security and manageability.*

4. *Cryptographic units (possibly with tamper resistance* [8] [9]*) or cryptographic programs.*

Additionally, the security sub-system should not unduly affect response times for network users nor should it present an unfriendly and complex user interface.

Security Sub-System Overview

The security sub-system architecture proposed here uses a smart card as the user token. This provides the basic mechanism for access control, and also stores user-specific cryptographic keys. Local access control is effected by PIN-protection of the smart card. Hence the user needs to possess both the card and knowledge of the PIN. Remote access control is achieved by means of a challenge-response protocol designed to be thoroughly resistant to cryptanalytic attack for the entire lifetime of the card. For the encryption of data and the generation of message authentication codes, symmetrical encryption algorithms such as DES [10] [11] are used. Digital signatures are generated using the RSA asymmetric public key algorithm [12], and each authorised signatory carries a personal RSA key pair as part of the data on the smart card user token [13].

Top-level key management protocols are also implemented using RSA, which has the advantage of enabling a fully automated and therefore very manageable key distribution scheme.[14]. To maintain acceptable response times the cryptographic units and smart card readers are all equipped with a digital signal processor providing "fast RSA" processing facilities.[15]

To overcome the problems of mutual trust between a smart card and an intelligent cryptographic smart card reader, one of two approaches is possible. The card and the reader can mutually authenticate one another using a zero-knowledge proof protocol such as Fiat-Shamir,[16] [17], or the card can delegate some of the heavier computations to the card reader without disclosing its secret information.[18]

System Detail

Local Access Control

Every workstation is provided with an integral smart card reader, into which an authorised user inserts a smart card. The smart card itself is only activated for further functions if the correct user PIN is supplied. The user is prompted by the local application to supply this PIN, which is then submitted to the smart card for validation. If the PIN validation is successful, the card may then enter into a mutual authentication process with the smart card reader using the Fiat-Shamir protocol. The card reader is equipped with a digital signal processor which performs all cryptographic processing in that unit. At this stage, the system has achieved the following authentication:

1. *User to smart card*

2. *Smart card to reader*

3. *Reader to smart card*

By implication, the user has also been authenticated to the reader, and hence to the application which is driving it. The smart card now provides the data required for remote access control and for digital signature using RSA.

Also stored on the smart card is a user privileges profile which is sent to the application and which controls the range of application facilities to which the user is to be granted access. If the privileges profile data is too great to store on the card it can be stored encrypted on the workstation database under a secret DES storage key which is generated and held on the card.

Remote Access Control

One-time passwords are used for logons which involve plaintext transmission across the network. This prevents an eavesdropper from capturing a useful password. For interactive working a challenge-response protocol [19] is used to authenticate the user to a remote host and its applications. If RACF or ACF2 are in use, these packages provide "exits" via which one-time password sub-systems can be interfaced. The one-time password is checked at the host and if valid, the user is granted access. In challenge-response mode this is achieved by the host security module generating a 128-bit random number, which is sent across the network to the workstation and from there to the smart card. A one-way function is now applied to the challenge to obtain the response. The smart card encrypts the random number under a secret user (DES) key which we shall call KU. The 128-bit ciphertext output from DES ECB mode [11] is then subjected to another algorithm which selects individual bits and combines them to form a 96-bit output. The mask for bit selection constitutes the user key for this selection algorithm. We shall call this key KS. There are 96 bits to be selected from 128. Hence the keyspace for KS is ${}^{128}C_{96}$ or 2^{218}.

The 96-bit output from the selection algorithm is transmitted back across the network to the host, where it is processed by the host security module to verify it against the issued challenge.

Considering the possible attacks on this challenge-response system there are two possible threats - firstly that an opponent will collect transmitted challenge-response pairs to build a dictionary and secondly that the opponent will construct a DES engine to perform a brute-force attack on KU by using known plaintext/ciphertext pairs.

The dictionary attack must be judged against the expected number of logons over the required lifetime of the smart card. Assuming an average of one logon per day for a period of three years this will give an opponent 365 x 3 = 1095 matching challenge-response pairs. This is reasonable since it includes all weekends and holidays and will hence allow for multiple logons on some days. For convenience we approximate this value to $1024 = 2^{10}$. We now examine the probability of the opponent having the necessary dictionary entry for a given challenge at the end of this three year period. Meyer and Matyas [20] have shown that the probability (p) of finding a correct look-up table entry is :-

$p = 1 - e^{-mn/N}$; $n/2N < 1$

where N = total number of possible response values

n = number of challenge-response pairs available to an opponent

m = number of exhaustive trials to obtain equivalent values of the keys that will generate the same output.

If we examine the use of 64-bit DES alone and for the present time neglect the effect of the selection algorithm, this gives the following values:-

$N = 2^{64}$
$n = 2^{10}$
$m = 2^{56}$
$n/2N = 2^{-55}$

Hence $mn/N = 2^2 = 4$

$p = 1 - e^{-4} = 0.98$

This is a totally unacceptable probability, but if we now include the selection component the picture changes dramatically. Any one of the possible 2^{56} DES keys could have been used to generate the final 96-bit output. This means that on any trial it will always be possible to find a pair of equivalent values of KU and KS which generate the observed output, hence:

$m = 1$
$N = 2^{96}$
$n/2N = 2^{-87}$

Hence $mn/N = 2^{-86}$

and p is negligible

If we were to extend the required lifetime of the card to (say) 2^{20} logons it makes no substantial impact on the value of p, and hence we have a scheme that is for practical purposes completely resistant to a dictionary-style attack.

Now let us examine the brute-force attack on KU and KS. Since a given output is possible with all values of KU we have completely decoupled the DES process from any direct attack on sets of matched plaintext/ciphertext. The opponent must search every value of KS for each and every value of KU, making a complete search of 2^{274}.

We estimate that it is feasible to build a DES engine that would work at 10^{12} tests per second and would perform an average DES key search in one day. Assuming that we employ the same engine to search for KS and that the selection algorithm operations are 100 times faster than DES operations, a complete search of KS and KU would take 2^{212} days, which of course is infeasible.

The incorporation of the selection algorithm extends the resistance to attack well beyond the resources of an opponent and adds virtually nothing to the cost of implementation. It could be used equally effectively with a non-DES preprocessor, thus extending the scope of its applications to organisations which are prohibited the use of DES.

Host-End Management for Remote Access Control

Every issued smart card has a matching user record on the host database. When an authentication request is received at the host, the appropriate record is retrieved and sent into the host security module. The secret part of the record is stored on the host database encrypted under a storage DES key, and storage keys are changed regularly to reduce their exposure to attack. The record is processed and then written back to the database encrypted under the latest storage key. A dummy request is also provided to enable the host application to refresh infrequently used records which would otherwise fall out-of-date with the storage key window.

A new "session" starts with a request for a random number challenge. The host security module provides a session sequence number and sets up a temporary store of session variables, including the values of KS, KU and the random challenge. On receipt of the encrypted challenge the host security module uses the session sequence number to index the appropriate block of session variables and hence verify the response.

If the remote logon is not interactive but forms part of a batch submission, the same user key is used to generate the next one-time password in a psuedo random sequence. [21] This sequence is tracked at the host-end and the password is validated when the batch job is processed. Password windows are used to improve system resilience, and database management is as before.

Additional resilience is incorporated by providing dual host security modules. The units each have their own unique RSA key pair and the public keys are used to organise encrypted, certified replication of storage DES keys between the two units via the host application.

Issue Authority

All smart cards for use in the network are issued at one central point. They are loaded with keys and PINs and then mailed to users. PINs are secretly printed inside special envelopes and mailed to users under separate cover. The card is loaded with a newly generated RSA key pair (512-bit keylength is used), the public key of which is certified by means of a digital signature made with an issue authority secret RSA key. [22] The certified public key is entered into the system directory and is available for reference by all other users. The card-issue function and directory function are performed on a PC acting as a key distribution centre (KDC). The KDC is available to all network users as a central reference library for certified public keys. Each PC or host application can obtain these via the network, and can store local directories of frequently used keys. Every system node also has a copy of the issue authority public key with which the certified public keys can be authenticated at any time by validating the issue authority RSA digital signature. Hence only authorised users' public keys can be used.

Privacy of Communications

In addition to the user RSA keys held on the smart cards, each cryptographic unit in the system has a unique RSA key pair, issued to the unit in certified form just as for the smart cards, and also stored on the public system directory. It is therefore possible to have the following relationships:-

1. *User to user*

2. *User to application (and its crypto unit)*

3. *Application to application (and their crypto units)*

When messages are to be encrypted for transmission to protect against eavesdropping, a data key (DES) is generated at the originating user or application inside the cryptographic unit. For duplex communications different data keys are used for each direction. A DES key is encrypted under the RSA public key of the destination unit and signed with the RSA secret key of the source. The encrypted, signed data key is sent with the message and at the destination it is recovered by using the public key of the source to validate the signature, and the secret key of the destination to decrypt the data key. All RSA processing is performed on fast RSA processors (such as a digital signal processor) to maintain acceptable response times.

The Texas Instruments TMS 32010 DSP with suitable software can provide a 512-bit RSA secret key operation in approximately 2.8 seconds. This can be substantially improved with the TMS 32020 to approximately 2 seconds and with the TMS 320C25 to approximately 800 milliseconds. We estimate that using two Motorola 56000 DSPs an execution time of less than 50 milliseconds can be achieved.

Key Management Protocol

The transmission of keys requires a suitable protocol at the application level. This protocol must exchange messages, the contents of which are encrypted keys, unit identities, key counters and other control information. One such protocol suitable for this purpose is the group of cryptographic service messages (CSMs) described in ANSI X9.17. [23] The CSMs defined in the standard accommodate only single and double length DES keys, with no provision being made for RSA keys or RSA-encrypted DES keys. However, it is not difficult to extend the CSM set to include new field-tags and new field definitions that can handle these RSA blocks; proprietary implementations of ANSI X9.17 CSM protocol do make these facilities available, but of course their precise definition does not conform to a standard since non exists.

The key management technique described above is applicable to both interactive and store-and-forward networks. However, in the case of store-and-forward mailbox systems, it may be a function of the mail server to broadcast messages to all system users or to closed user groups. In this case the mail server is equipped with its own cryptographic unit which performs only key translation services. Broadcast messages have their data key encrypted by the source under the RSA public key of the mail server, and the cryptographic unit on the server then translates these key blocks under the RSA public key of all authorised recipients, placing the translated key block into the mailbox of each. The data key remains unchanged and so the key translation unit does not need to process the message itself.

Message Authentication

Message contents are authenticated by generating at source and validating at destination a message authentication code (MAC). This can be of the type defined in ANSI X9.9 [24] using DES. The data authentication key is carried in exactly the same way as described above for data encryption keys, using RSA for authenticating both source and destination, and using a key translation server for broadcast messages.

Digital Signatures

When a message is sent encrypted and/or authenticated as described above the data keys are already signed using RSA. However, for additional security the MAC itself is signed by the source RSA secret key, thus providing a digital signature that can be validated using the source RSA public key from the system directory.

The RSA secret key which is used to generate the digital signature belongs either to an application or to a user. In the case of it being the property of an application it is stored securely in a tamper-resistant crypto-unit which is attached to the host machine. However, where the RSA key belongs to a user it is stored in the personal smart card and is carried around by that user.

Smart card technology does not at present support 512-bit RSA processing to meet the response time requirements, and so this must be performed on the digital signal processor in the smart card reader. To achieve this the smart card must give up its secret RSA key to the reader, which is why the mutual authentication process between these two components is so important. Additionally, the reader can be made to be tamper resistant [8] [9] to protect secret keys during their residence in the unit.

Alternatively, if the speeding-up techniques discussed by Matsumoto, Kato and Imai [18] can be successfully implemented to achieve acceptable response times there is no need for the Fiat-Shamir mutual authentication protocol. In this case the smart card will not surrender its secret information to the reader and this latter device need not be either trusted or tamper resistant

Integration with IBM SNA Environments

The IBM SNA encrypt/decrypt facilities do not include RSA key management. However, on each IBM host an additional cryptographic unit with fast RSA processing capability is provided so that master DES keys can be moved around the network between hosts. PCs are equipped with plug-in cryptographic boards that emulate IBM 3848 capability [5] and also provide the additional RSA key management layer. The applications on the IBM hosts are responsible for organising the automated management of master keys under the RSA layer, using a protocol similar to that described above using ANSI X9.17 CSMs. [23]

Integration with non-IBM Interactive Environments

IBM SNA is somewhat unique in its provision of an encrypt/decrypt feature within the network architecture. Other proprietary network architectures leave cryptography largely to the implementers of the applications. The approach described in this paper is ideally suited to this latter environment, since the application drives both the key management protocols and the service requests to the presentation layer for encryption/decryption facilities. Integration into these environments therefore poses no substantial problems, since a standard interface can be defined which requires the application to incorporate only the necessary message handler for requesting and receiving cryptographic services.

Summary

The system solution described here provides a multi-layer security architecture using unified key management and multi-purpose system components to support a wide range of environments and can be integrated with both IBM-style and other proprietary security sub-systems. User participation is limited to carrying a secure token and supplying a PIN, after which the layered authentication processes are automated. System response time is maintained at acceptable levels for the user, and system management is eased by the use of automated techniques. Above all an elegant and highly secure end-to-end solution is created.

Bibliography

[1] Anura GURUGE, "*SNA - Theory and Practice*", Pergamon Infotech, 1984

[2] "*IBM Cryptographic Subsystem Concepts and Facilities*", (GC22-9063), 1985

[3] "*IBM RACF*", Datapro Reports on Information Security, Report No. IS52-504, McGraw Hill, 1987

[4] "*CA-ACF2*", Datapro Reports on Information Security, Report No. IS52-187, McGraw Hill, 1988

[5] "*IBM 3848 Cryptographic Unit Product Description and Operating Procedures*", (GA22-7073), 1982

[6] *"IBM OS/VSI and OS/VS2 MVS Programmed Cryptographic Facility, General Information"*, (GC28-0942), 1980

[7] *"ACF/VTAM General Information"*, (GC38-0254), 1980

[8] Andrew CLARK, *"Physical Protection of Cryptographic Devices"*, Eurocrypt '87, Amsterdam, 1987

[9] Andrew CLARK, *"Physical Protection of Cryptographic Devices (Revised)"*, Proc. of Corporate Computer Security '88 Conference, Brighton, UK 1988

[10] *"Information Processing - Data Encryption Algorithm"*, ANSI X3.92, 1981

[11] *"Data Encryption Algorithm - Modes of Operation"*, ANSI X3.106, 1983

[12] Ronald RIVEST, Adi SHAMIR and Leonard ADLEMAN, *"A Method of Obtaining Digital Signatures and Public Key Cryptosystems"*, Comm. of ACM, Vol.21, No.2 Feb 1978

[13] John SHERWOOD, *"Digital Signature Schemes Using Smart Cards"*, Proc. of Smart Card '88 Conference, London, 1988

[14] John SHERWOOD, *"Automatic Key Management for Transparent Security on Corporate Data Networks"*, Proc. of International Systems Security Conference, London, 1986

[15] Paul BARRETT, *"Implementing the RSA Public Key Encryption Scheme on a Digital Signal Processor"*, Proc. of Crypto '86, Springer-Verlag, 1986

[16] Amos FIAT and Adi SHAMIR, *"How to prove yourself: Practical Solutions to Identification and Signature Problems"*, Proc. of Crypto '86, Springer-Verlag, 1986

[17] Amos FIAT and Adi SHAMIR, *"Unforgeable Proofs of Identity"*, 5th SECURICOM, Paris, 1987

[18] Tsutomu MATSUMOTO, Koki KATO and Hideki IMAI, *"Speeding-Up Secret Computations with Insecure Auxiliary Devices"*, CRYPTO '88 - to appear

[19] Raymond WONG, Thomas BERSON and Richard FEIERTAG, *"Polonius: An Identity Authentication System"*, Proc. of IEEE Symposium on Security and Privacy, 1985

[20] Carl MEYER and Stephen MATYAS, *"Cryptography: A New Dimension in Computer Data Security"*, Wiley, 1982

[21] Raymond EISELE, *"Host Access Security"*, Presented at Interact '86, Orlando, Florida, 1986

[22] Vince GALLO and Andrew CLARK, *"Issue Authority"*, 2nd Nordic Conference on Information Security, Stockholm, 1988

[23] *"Financial Institution Key Management (Wholesale)"*, ANSI X9.17, 1985

[24] *"Financial Institution Message Authentication (Wholesale)"*, ANSI X9.9, 1984

Speeding Up Secret Computations with Insecure Auxiliary Devices

Tsutomu MATSUMOTO
Koki KATO
Hideki IMAI
Division of Electrical and Computer Engineering
YOKOHAMA NATIONAL UNIVERSITY
156 Tokiwadai, Hodogaya, Yokohama, 240 Japan

Abstract This paper deals with and gives some solutions to the problem of how a small device such as a smart card can efficiently execute secret computations using computing power of auxiliary devices like (banking-, telephone-, ...) terminals which are not necessarily trusted. One of the solutions shows that the RSA signatures can be practically generated by a smart card.

1. Introduction

Small devices such as smart cards or IC cards are easy to be carried and have the ability to compute, memorize, and protect data. Such convenient *ultimate personal computers*[1] have been useful tools for constructing various information systems and now they are expected to be utilized in much wider applications. Unfortunately, smart cards now available are not so powerful, still the jobs we want them to execute are liable to hard for them. For example, many want to realize public key cryptographic algorithms in smart cards. But these are not easy tasks for them. Even if future smart cards would be more powerful, the gap would not be filled, because we would require them more intelligence.

An easy and usual way to overcome this situation is the use of auxiliary computers such as (network-, POS-, banking-, telephone-, facsimile-, ...) terminals for supplying the short-computing power of smart cards. In the following we use the terms '*client*' and '*server*'. A *client* denotes the main device such as a smart card which has a secret computation and can execute the computation by itself but takes long time because of the lack of computing resources. A *server* denotes the auxiliary device which has enough computing resources.

If a *server* is trustworthy and will not leak the secrets, the *client* can pass the *server* the description of the secret computation and can ask the *server* to perform it and to tell the result. As an example, in a public key signature scheme, a *client* sends the message to be signed and the secret key to a *server* for generating a signature on behalf of the *client*.

But *servers* are not always trustworthy. A terminal in a public telephone booth or a POS terminal in a supermarket, etc., may be a *server* which may be equipped with a wiretapping device or might be infected by some computer viruses. When the *server* is insecure, the *client* has to protect its secret from the *server* during the interaction.

How a *client* can securely accelerate secret computations by using untrustworthy *servers* ? This is the problem to be solved in this paper. We believe this problem will be very important for our future's daily life. In our prior paper [2], we have pointed out the importance of the problem and presented some primal considerations. In the following sections we demonstrate several protocols solving the problem for (1) matrix computation, (2) modular equations, and (3) the RSA cryptosystem.

2. Related Works and Assumptions

There are some other researches [3][4][5][6][7][8] looking like ours.

Privacy Homomorphisms, proposed by Rivest-Adleman-Dertouzos [3] and recently examined by Brickell-Yacobi [4] and by Ahituv-Lapid-Neumann [5], are cryptographic functions preserving some operations. For example, when two data a and b are stored in a database in the form of ciphertexts $f(a)$, $f(b)$, if the enciphering function f is homomorphic with respect to an operation $\circ$ in the domain of f and an operation $\bullet$ in the codomain of f, then the ciphertext of the data $a \circ b$ can be obtained as $f(a) \bullet f(b)$ without deciphering and re-enciphering.

Similar notion called the *Directly Transformed Link Encryption* has been proposed by Matsumoto-Okada-Imai [6] in the field of network security. In each node of a communication network with the link encryption, each ciphertext c comming from an input link i is deciphered into a plaintext $m = D_i(c)$ and then, after a routing, enciphered into another ciphertext $c' = E_j(m)$ to be emitted into an output link j. Here D_i and E_j are deciphering and enciphering algorithms associated with the input link i and the output link j, respectively. The core idea of the directly transformed link encryption is to use instead of D_i and E_j an algorithm H_{ij} which directly transforms c into c' so that the security of the plaintext in the node is enhanced. Cryptosystems based on power functions are examples for those applicable to the directly transformed link encryption.

Though the notions of privacy homomorphism and the directly transformed link encryption are attractive, they don't suffice for our porposes.

On the other hand, Feigenbaum [7] and Abadi-Feigenbaum-Kilian [8] studied the problem of so-called *Computing with Encrypted Data*. Their problem is very similar to ours. However, their stance seems to be a little bit different from ours. Since their interest was focused on the theory, if we use our terminology, they assumed that the *client* has probabilistic polynomial time computing power and that the *server* has unlimited computing power and they derived interesting conclusion that hard functions are also difficult to be securely encrypted. Their work is very interesting but not sufficient to our practical problem.

To make clear the differences, we summarize here our assumptions specifically:

Assumptions A computation originally owned by a *client* is a feasible one with respect to an ordinary computer. In typical situations, it is probabilistic polynomial time computable. However, when the size of the input is small and then the computation is tractable by some ordinary computer, the computation might be outside the probabilistic polynomial time. *Servers* adopted for the speed up are such ordinary computers. That is, there are limitations to the computational complexity with which the *servers* can cope. *Servers* might leak the data treated in interactions with a *client*, but do not refuse the jobs given by the *client* nor send false answers to the *client*. Each *client* is sufficient to protect its secret from resource bounded enemies. But the computational complexity for that purpose used by the *client* should not beyond the computational complexity for executing the secret computation by itself.

3. General Idea of Speed Up

Let a client want to obtain the value $y = g(x)$ of a computable function g on input x. Assume that there is an algorithm (circuit) C_g to compute g which is practically tractable by a *server* but not by the *client*.

Our general idea of speed up is as follows.

Protocol P [9]

(0) The *client* randomly decompose the algorithm C_g into three algorithms (circuits) I, M, F such that (i) the consecurive applications of I and M and F compute the function g and (ii) the *client* can execute I and F in enough speed. For many practical applications, it is worth while considering a simplar version such that the *client* is restricted to select M from a predetermined set of algorithms.

(1) The *client* applies I to have $u = I(x)$, and sends $[M, u]$ to the *server*.

(2) The *server* applies M to u and sends $v = M(u)$ back to the *client*.

(3) The *client* obtains y by applying F to v as $y = F(v)$.

Efficiency and Security: Let $Comm(\alpha)$, $Comp_C(\beta)$, and $Comp_S(\gamma)$ denote the time to transfer α between the *client* and the *server*, the time to execute algorithm β in the *client*, and the time to execute algorithm γ in the *server*, respectively. The total time to execute steps (1),(2),(3) in the protocol P is

$$T(\mathrm{P}) = Comp_C(I) + Comm([M, u]) + Comp_S(M) + Comm(v) + Comp_C(F).$$

Thus the speed up effect of P is $Comp_C(C_g)/T(\mathrm{P})$. Given upper bound B_C of the computing time of the *client*, the security of the protocol P is roughly measured by the ambiguity

$$A(\mathrm{P}) = \#\{[I, F] | Comp_C(I) + Comp_C(F) \leq B_C\}.$$

Variations: The protocol P can be generalized into two directions by decomposing M further. One is to have a series of algorithms and adopt more interactions. The other is to have a set of algorithms executable in parallel by independent *servers*.

4. Demonstrating the Speed Up Protocols

We show now some of the basic protocols to demonstrate our idea of enhancing smart cards.

4.1 Speed Up via Coordinate Permutation

In usual computing environments, multiplying matrices is recognized as rather light computation. But it is not so easy for today's smart cards. On the other hand, permuting rows and columns of matrices can be performed by only changing indices of the components of matrices. We have verified through an experiment [2] that this technique actually works well for programs in smart cards. The following three examples are based on this fact. The coordinate permutation is a very useful tool for constructing speed up protocols.

[a] Matrix Multiplications

Target: A *client* has two secret matrices A and B and wants to obtain the product $C = AB$.
Assumption: The *client* can efficiently permute rows and columns of matrices. There is a *server* which can multiply matrices overwhelmingly faster than the *client* can.

Protocol MM

(0) The *client* randomly generates permutation matrices P, Q and R.
(1) The *client* permutes A and B along with $[P, Q]$ and $[Q^{-1}, R]$ to have $A' = PAQ$, $B' = Q^{-1}BR$, and sends $[A', B']$ to the *server*.
(2) The *server* computes and sends back to the *client* $C' = A'B'$.
(3) The *client* obtains C by permuting C' along with $[P^{-1}, R^{-1}]$ as $C = P^{-1}C'R^{-1}$.

Remark: This protocol can be applied to the speed up of evaluating a tuple of multivariate polynomials.

[b] Linear Equations

Target: A *client* has a secret non-singular matrix A and a secret matrix B and wants to obtain the solution X of the equation $AX = B$.
Assumption: The *client* can efficiently permute rows and columns of matrices. There is a *server* which can solve linear equations overwhelmingly faster than the *client* can.

Protocol LE

(0) The *client* randomly generates permutation matrices P, Q and R.
(1) The *client* permutes A and B along with $[P, Q]$ and $[P, R]$ to have $A' = PAQ$, $B' = PBR$, and sends $[A', B']$ to the *server*.
(2) The *server* solves the equation $A'X' = B'$ for X' and sends X' back to the *client*.
(3) The *client* obtains X by permuting X' along with $[Q, R^{-1}]$ as $X = QX'R^{-1}$.

Remark: A slightly modified version of this protocol can be applied to the ***linear programming*** problem.

Evaluation: We have organized software experiments for Protocol MM and LE. The conclusion is that these protocols are effective since, as described above, the permutations can be done faster than the matrix multiplications and the amount of communication between the *client* and the ***server*** is only three matrices. However, by these protocols, the ***server*** might acquire some statistical information. Indeed, they cannot protect values of the functions not affected by permutations. An example is the determinant of C in Protocol MM. But we think there are many practical applications to which these protocols are useful.

[c] Graph Isomorphisms

Target: A *client* has two secret graphs a and b of the same number of verteces and edges with their adjacency matrices A and B, respectively, and wants to know whether these graphs are isomorphic or not. And if they are isomorphic, the *client* also wants to obtain the isomorphism, which is the permutation x of coordinates determined by the permutation matrix X satisfying the equation $AX = XB$.
Assumption: The *client* can efficiently permute rows and columns of matrices. There is a ***server*** which can quickly solve the graph isomorphism problem. But the ***server*** may not quickly solve the graph nonisomorphism problem.

Protocol GI

(0) The *client* randomly generates permutations p and q, to which correspond permutation matrices P, Q, respectively.
(1) The *client* permutes rows and columns of A and B along with p and q to have $A' = PAP^{-1}$, $B' = QBQ^{-1}$, and sends $[A', B']$ to the ***server***.
(2) The ***server*** tries to decide , in a period of time, whether the graphs with adjacency matrices A' and B' are isomorphic or not. If they are decided to be isomorphic, the ***server*** computes the permutation x' corresponding to the isomorphism and sends x' to the *client*. Otherwise, the ***server*** sends '#' to the *client*.
(3) If x' is sent, the *client* obtains x by transforming x' with p^{-1} and q as $x = p^{-1}x'q$. If '#' is sent, the *client* decides that that a and b are not isomorphic.

Remark: This protocol can be applied to the speed up of ***pattern recognition*** based on the graph isomorphisms.

4.2 Modular Equations

Univariate polynomial equations over finite fields can be solved by polynomial time algorithms. And as Rabin [10] shows, there are efficient probabilistic algorithms for them. Main jobs of these algorithms are to take the greatest common divisor of two polynomials by applying the well known extended Euclidean algorithm. However, these are not so easy tasks for ordinary smart cards.

Target: A *client* has secret integers k and $a_0, a_1, a_2, \ldots, a_{m-1}$ such that the modular equation

$$a_0 + a_1x + a_2x^2 + \cdots + a_{m-1}x^{m-1} + x^m \equiv 0 \pmod{k}$$

is solvable in x, and wants to obtain a solution x of this equation.
Assumption: The *client* can efficiently execute multiplication mod$k$ and division mod$k$. There is a *server* which can solve modular equations overwhelmingly faster than the *client* can.

Protocol ME

(0) The *client* randomly selects an integer r such that $\gcd(r, k) = 1$.
(1) The *client*and computes $[b_0, b_1, b_2, \ldots, b_{m-1}]$ by

$$c_0 = 1, \text{ and } c_i = rc_{i-1} \bmod k, \quad b_{m-i} = c_i a_{m-i} \bmod k \text{ for } i = 1, \ldots, m$$

and sends $[b_0, b_1, b_2, \ldots, b_{m-1}, k]$ to the *server*.
(2) The *server* obtains a solution y of the equation

$$b_0 + b_1y + b_2y^2 + \cdots + b_{m-1}y^{m-1} + y^m \equiv 0 \pmod{k}$$

and sends back y to the *client*.
(3) The *client* obtains x as $x = yr^{-1} \bmod k$.

Remark: This protocol can be generalized to fit any system of multivariate polynomial equations over any commutative ring.

Evaluation: Protocol ME is very effective, because in the protocol the computation the *client* has to do is only $2m$ multiplications mod$k$ and one division mod$k$ and the amount of communication between the *client* and the *server* is only $m + 2$ integers while the *server* could investigate nothing on the secret of the *client*.

5. Speeding Up the RSA Transformations

Is it possible to securely implement the RSA cryptosystem[11] with smart cards and terminals ? We think the answer is *yes*. For the RSA public transformation, a speed up protocol is described in [2]. For the RSA secret transformation, we show below two of the developed protocols.

Target: A *client* has integers x, d, n and wants to obtain the integer $y = x^d \bmod n$. The integer d is the secret of the *client*, while the integers n and e such that $ed \equiv 1 \pmod{\lambda(n)}$ are made public. Here n is the product of two large secret primes p, q $(p \neq q)$, and $\lambda(n)$ is the secret integer $\mathrm{lcm}(p-1, q-1)$.

For simplicity, the integer x may be known to the *server*. (It is an easy task to modify the following protocols with slightly adding the complexity so that x is also hidden from the *server*.)

5.1 Secret Powering 1

Assumption: The *client* can execute several multiplications mod$n$. There is a *server* which is equipped with a device which can implement the RSA secret transformation overwhelmingly faster than the naked *client* can.

Protocol RSA-S1

(0) The *client* randomly generates an integer vector $D = [d_1, d_2, \ldots, d_M]$ and a binary vector $F = [f_1, f_2, \ldots, f_M]$ such that

$$d \equiv f_1 d_1 + f_2 d_2 + \cdots + f_M d_M \pmod{\lambda(n)}$$

and $1 \leq d_i < n$ and $Weight(F) = \sum_{i=1}^{M} f_i \leq L$, where M and L are some integers.

(1) The *client* sends n, D, and x to the *server*.

(2) The *server* computes and sends back to the *client* $Z = [z_1, z_2, \ldots, z_M]$ such that

$$z_i = x^{d_i} \bmod n.$$

(3) The *client* obtains y by computing $y = y_M$ as follows:

$$y_0 = 1, \quad y_i = y_{i-1} z_i \bmod n \ \text{ if } \ f_i = 1; \quad y_i = y_{i-1} \ \text{ if } \ f_i = 0, \quad (i = 1, 2, \ldots, M).$$

Variation: We have a more general protocol if we exclude '*binary*' from the condition to F.

Complexity: Since the step (0) can be precomputed, for each x it is sufficient for the *client* to do at most $L-1$ multiplications mod n. The amount of communication is $2(M+1)$ integers of size at most $\log n$ bits.

Security: If the RSA cryptosystem is secure, the protocol could be broken only by searching true d via the exhaustion of

$$\sum_{i=1}^{L} \binom{M}{i} > \binom{M}{L}$$

possibilities.

Remark: If e is hidden from the *server*, Protocol RSA-S1 is applicable also to the case where $\lambda(n)$ can be readily computed from n. Secret powering over a finite field is an example.

Though such property are not preserved, we can have a more efficient protocol by utilizing the Chinese Remainder Theorem:

5.2 Secret Powering 2

Assumption: The *client* can execute several multiplications mod$p$ and mod$q$. The *client* has computed integers w_p and w_q such that

$$w_p = q(q^{-1} \bmod p), \quad w_q = p(p^{-1} \bmod q).$$

There is a *server* which is equipped with a device which can implement the RSA secret transformation overwhelmingly faster than the naked *client* can.

Protocol RSA-S2

(0) The *client* randomly generates an integer vector $D = [d_1, d_2, \ldots, d_M]$ and two binary vectors $F = [f_1, f_2, \ldots, f_M]$ and $G = [g_1, g_2, \ldots, g_M]$ such that

$$d \equiv f_1 d_1 + f_2 d_2 + \cdots + f_M d_M \pmod{p-1}$$

$$d \equiv g_1 d_1 + g_2 d_2 + \cdots + g_M d_M \pmod{q-1}$$

and $1 \leq d_i < n$ and $Weight(F) + Weight(G) = \sum_{i=1}^{M} f_i + \sum_{j=1}^{M} g_j \leq L$, where M and L are some integers.

(1) The *client* sends n, D, and x to the *server*.

(2) The *server* computes and sends back to the *client* $Z = [z_1, z_2, \ldots, z_M]$ such that

$$z_i = x^{d_i} \bmod n.$$

(3) The *client* obtains y by computing y as follows:

$$y = (y_{pM} w_p + y_{qM} w_q) \bmod n,$$

$$y_{p0} = 1, \quad y_{pi} = y_{p,i-1} z_i \bmod p \ \text{ if } \ f_i = 1; \quad y_{pi} = y_{p,i-1} \ \text{ if } \ f_i = 0,$$

$$y_{q0} = 1, \quad y_{qi} = y_{q,i-1} z_i \bmod q \ \text{ if } \ g_i = 1; \quad y_{qi} = y_{q,i-1} \ \text{ if } \ g_i = 0,$$

for $i = 1, 2, \ldots, M$.

Variation: We have a more general protocol if we exclude '*binary*' from the condition to F and G.

Complexity: Since the step (0) can be precomputed, for each x the amount of commputation the *client* has to do is equivalent to at most $3L/2$ multiplications mod p or mod q. The amount of communication is $2(M+1)$ integers of size at most $\log n$ bits.

Security: If the RSA cryptosystem is secure, the protocol could be broken only by searching true d via the exhaustion of

$$\sum_{j=1}^{L} \sum_{i=0}^{j} \binom{M}{i} \binom{M}{j-i} > \binom{M}{L/2}^2$$

possibilities.

Examples: Using a *i*8086 (5*MHz*) (30 *msec* / 256-*bit* modular multiplication) or Z-80 (6*MHz*) (300 *msec* / 256-*bit* modular multiplication) as a smart card with a single 64-*Kbps* (non-contact type) or 9600-*bps* serial link and the RSA hardwares (chips) [13] with speed 32-*Kbps* or 4800-*bps*, Protocol RSA-S2 can be accomplished about 4 to 30 times faster than the case where the microprocessor does the whole computation with the method due to Quisquater-Couvreur [12] (see Table A).

6. Conclusion

With several demonstrating examples, we have presented an important research problem of how to supply short-computing power of smart cards. The described protocols are all very simple but can be actually utilized in a system consists of smart cards and auxiliary computers. Other protocols and problems to be developed are described in [9].

Acknowledgment

The authors wish to thank Susumu Inomata for fruitful discussions and Kenji Koyama for providing Ref[13]. They also thank many who take interests in this work and anonymous refrees for their sugessions to improve this paper. Part of this work was supported by the Ministry of Education, Science and Culture under Grant-in-Aid for Encouragement of Young Scientists #62750283 and #63750316.

Table A. Examples of Protocol RSA−S2
[Processing time for a 512-bit message block]

	i8086 (5MHz)	Z80 (6MHz)
Serial Link 64Kbps + RSA hardware 32Kbps	L=20, M=50 1.7 sec <13.5 times faster>	L=12, M=142 7.7 sec <30 times faster>
Serial Link 9600bps + RSA hardware 4800bps	L=32, M=37 5.5 sec <4.2 times faster>	L=18, L=58 14.4 sec <16 times faster>
conventional method [12]	23 sec	230 sec

Modulus n : 512 bit   Security $> 10^{20}$

References

[1] Svigals,J., *Smart Cards: The Ultimate Personal Computer*, Macmillan, 1985.

[2] Matsumoto,T., Kato,K. and Imai,H., "Smart cards can compute secret heavy functions with powerful terminals," (*written in Japanese with an English abstract*) *Proc. of 10th Symposium on Information Theory and Its Applications*, Enoshima-Island, Japan, pp.17-22, Nov.19-21, 1987.

[3] Rivest,R., Adleman,L. and Dertouzos,M., "On databanks and privacy homomorphisms," *Foundations of Secure Computation*, Demillo,R.A. *et al.*, editors, Academic Press, pp.168-177, 1978.

[4] Brickell,E.F. and Yacobi,Y., "On privacy homomorphisms," *Advances in Cryptology - EUROCRYPT'87*, Chaum,D. and Price,W.L. editors, Springer-Verlag, pp.117-125, 1988.

[5] Ahituv,N., Lapid,Y. and Neumann,S., "Processing encrypted data," Communications of the ACM, Vol.30, No.9, pp.777-780, Sep. 1987.

[6] Matsumoto,T., Okada,T. and Imai,H., "Directly transformed link encryption," (*in Japanese*) Trans. of IECE Japan, Vol.J65-D, No.11, pp.1443-1450, Nov. 1982.

[7] Feigenbaum,J., "Encrypting problem instances, or, ..., Can you take advantage of someone without having to trust him ? " *Advances in Cryptology - CRYPTO'85*, Williams,H.C. editor, Springer-Verlag, pp.477-488, 1986.

[8] Abadi,M., Feigenbaum,J. and Kilian,J., "On hiding information from an oracle," to appear in *Journal of Computer and System Sciences.* An extended abstract appeared in *Proc. of 19th Symposium on Theory of Computation*, pp.195-203, May, 1987.

[9] Matsumoto,T. and Imai,H., "How to use servers without releasing privacy – Making IC cards more powerful – ," (*in Japanese*) *IEICE Technical Report (ISEC)*, Vol.88, No.33, pp.53-59, May 1988.

[10] Rabin,M.O., "Probabilistic algorithms in finite fields," *SIAM J. Comput.*, Vol.9, No.2, pp.273-280, May 1980.

[11] Rivest,R., Shamir,A. and Adleman,L., "A method of obtaining digital signatures and public key cryptosystems," *Comm. of ACM*, Vol.21, No.2, pp.120-126, Feb. 1978.

[12] Quisquater,J.J. and Couvreuer,C., "Fast decipherment algorithm for RSA public-key cryptosystem," *Electron. Lett.* Vol.18, No.21, pp.905-907, Oct. 1982.

[13] Koyama,K., Table 1.(Developments of hardwares for the RSA cryptosystem), in "Information Security for Communications," (*in Japanese*) to appear in *Journal of the Institute of Television Engineers of Japan*, Dec. 1988.

Developing Ethernet Enhanced–Security System

B.J. Herbison
Secure Systems
Digital Equipment Corporation

Abstract

The Ethernet Enhanced–Security System (EESS) provides encryption of Ethernet frames using the DES algorithm with pairwise keys, and a centralized key distribution center (KDC) using a variation of the Needham and Schroeder key distribution protocol. This paper is a discussion of some practical problems that arose during the development of this system. Section 1 contains an overview of the system and section 2 provides more detail on the system architecture. The remaining sections discuss various problem that were considered during the development and how they were resolved.

1 Overview of the System

The Ethernet Enhanced–Security System (EESS) consists of Digital Ethernet Secure Network Controllers and VAX Key Distribution Center software. DESNC controllers are encryption devices that provide node authentication and data confidentiality and integrity on an Ethernet[1] (or IEEE 802.3) local area network (LAN). The VAX KDC software manages the DESNC controllers on a LAN and enforces a LAN access control policy.

DESNC controllers are store-and-forward communication devices that sit between nodes and the Ethernet. Each controller has four ports for nodes and one port that is connected to the LAN. Communication among these five ports is restricted by the controller according to the LAN access control policy.

When Ethernet frames are exchanged between two nodes that are connected to two different DESNC controllers, the frames are encrypted by one controller and decrypted by the other controller. This encryption occurs at the Data Link layer of the network and is transparent to higher network protocol layers. Nodes can use any network protocols that normally work over Ethernet (e.g., DECnet or

The following are trademarks of Digital Equipment Corporation:
DESNC, VAX KDC, DECnet, VAX, and VMS.

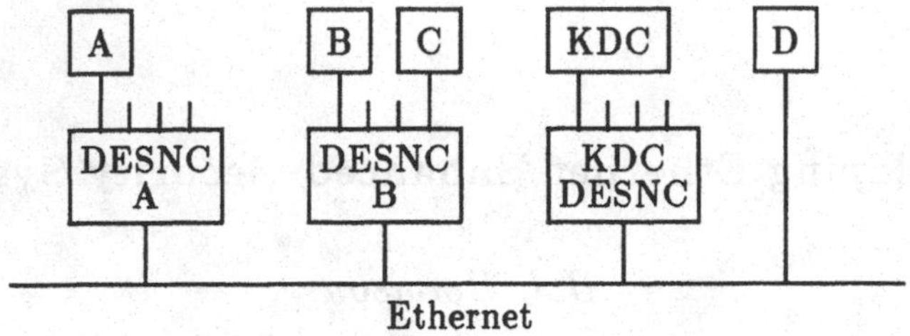

Figure 1: Sample Secure Ethernet

TCP/IP) without modification, and any device that conforms to the Ethernet or IEEE 802.3 standard can be attached to a DESNC controller.

DESNC controllers are managed by VAX KDC software running under VMS on specially designated KDC nodes on the Ethernet. Each KDC node must be attached to a DESNC controller that assists the KDC node; this controller is called a KDC controller. A KDC node and the attached KDC controller are collectively referred to as a KDC.

KDC nodes provide a user interface for the network security manager to control the security of the network. Through the interface the network security manager informs the KDC and the DESNC controllers of the configuration of the LAN, enters the LAN access control policy, determines the status of controllers on the LAN, and controls the network security auditing on the LAN.

It is possible to have multiple KDC nodes on one LAN. A large extended LAN can be supported with only a few KDC nodes, but having more than one KDC node improves the availability of the LAN by eliminating a single point of failure.

A sample secure Ethernet is shown in figure 1. The format used for encrypted node frames transmitted between DESNC controllers is shown in figure 2. Everything from the sequence number to the manipulation detection code (MDC) is encrypted. Most of the fields in the message are present as a result of the issues discussed in this paper. The section references indicate the locations in this paper where the fields are discussed.

2 System Architecture

2.1 Encryption Keys

Messages exchanged among DESNC controllers and KDCs are encrypted using the Data Encryption Standard (DES) encryption algorithm[2,3]. The messages are encrypted using the Cipher Block Chaining (CBC) mode of DES. When a message is encrypted, the encrypted portion of the message is padded to a multiple of the

Fields		Size	Section
Destination Address		6 bytes	
Source Address		6 bytes	
IEEE 802 Header		10 bytes	5.2
Message Type		2 bytes	3
Encryption Identifier		2 bytes	3
Original Header		10 bytes	5.3
Sequence Number	*	4 bytes	4
Message Type Copy	*	2 bytes	3
Original Header	*		
Original Data Field	*		
Padding	*	0-7 bytes	2.1
MDC	*	2 bytes	2.2
Ethernet FCS		4 bytes	

* marks the encrypted fields

Figure 2: Encrypted Frame Format

DES block length (8 bytes).

Several different types of DES encryption keys are used by controllers and the KDC software.

VAX KDC Master Key: This key is used to encrypt controller master keys that are stored on KDC nodes. This encryption prevents an intruder from compromising the security of a LAN by merely obtaining a copy of the information stored on the KDC node (for example, reading a BACKUP tape from the KDC node).

This key is only known by the network security manager and the KDC controllers.

Key Generation Key: This key is used as part of the process that generates encryption keys. This key is only known to the network security manager and a KDC controller.

Initialization Key: These keys, one per controller, are used to distribute the master and service keys for a controller, and are then discarded. Each initialization key is known only by the network security manager, the controller initialized with that key, and the KDC that initializes the controller.

Master Key and Service Key: These encryption keys are used to communicate between controllers and KDCs. A different pair of keys is used for each controller. The keys for a controller are only known by the controller and the KDCs, and they are only stored in encrypted form on KDC nodes. These keys are never handled in unencrypted form by any person.

Association Key: These keys are used to encrypt communication between nodes protected by controllers. A different association key is used for each pair of nodes that communicate. Association keys are distributed by KDCs when controllers request associations. Association keys are only known by the controllers involved and by a KDC controller. These keys are never stored on a KDC node or handled by any person.

KDC controllers will generate encryption keys as needed by the KDC nodes, or the network security manager can have the KDC nodes acquire the keys that they need from a user-supplied key source. A small amount of user programming is required to use a user-supplied key source, as well as a large supply of keys.

2.2 Modification Detection

When messages are encrypted, a manipulation detection code (MDC) is appended to the end of the message before encryption. The MDC field, produced by using a 16-bit CRC, is part of the encrypted portion of the message. When messages are decrypted, the MDC function of the message is calculated again and compared with the value sent with the message to determine if the message was modified as it was sent over the LAN.

2.3 Initializing Controllers

Before a DESNC controller can operate, it must be initialized. To initialize a controller, the following steps are required:

- The network security manager enters information about the controller into a KDC node. This information includes the Ethernet address of the controller, the Ethernet addresses of the nodes protected by the controller, and the access control policy for those nodes.

  The access control policy is specified by assigning an access class range to each node on the LAN. The access class ranges are from a Bell and LaPadula[4]/Biba[5] secrecy and integrity lattice, with 256 secrecy and integrity levels and 64 secrecy and integrity categories.

- On the request of the network security manager, the KDC node prints out an initialization key for the controller. The key is either generated by the KDC controller or taken from a supplied key source.

- The network security manager enters the initialization key in the controller through a keypad on the controller's front panel.

- The controller communicates with the KDC and receives its master key. The master key exchange is encrypted with the initialization key that was entered into the controller. After this step the initialization key is erased and not used again.

- The controller communicates with the KDC and receives the information that it needs to operate. This information includes:
  - The lifetime for association encryption keys.
  - The name of the firmware that the controller should be using, and a cryptographic checksum for the firmware image.
  - The addresses of the key distribution centers on the LAN.
  - The addresses of the nodes supported by the controller.
  - Information about the supported nodes.
  - A list of the events that the controller should audit.

  All of this information is encrypted under the controller's master key when it is distributed over the LAN.

After these steps, the controller is operational. The controller can now communicate with any KDC on the LAN. Once a controller is initialized it is not necessary to enter any additional information manually. If the distributed information needs to be changed, the changes can be made remotely from any KDC. DESNC controllers retain the distributed information during power-off and over power interruptions, but the information will be erased if a DESNC controller is opened.

Operational controllers request association keys from KDC nodes as necessary, and encrypt and decrypt Ethernet frames sent by nodes using those keys.

2.4 Downline Loading Controllers

The operational firmware image used by DESNC controllers is downline loaded over the Ethernet using the same mechanism employed by other Digital products. This allows the controllers to be downline loaded by the same downline load servers that load other products on the Ethernet. These images are not encrypted and the servers are not necessarily KDCs. The integrity of the images (and the security of the LAN) is protected in the following manner:

1. When a new firmware image is installed on a KDC and downline load servers, the KDC generates an encryption key and a cryptographic checksum for the image. The KDC generates a different key and checksum for each controller on the Ethernet.

2. During controller initialization, the KDC distributes the name of the firmware image and the appropriate checksum information to each controller. If a new image is installed after a controller is initialized, any KDC may distribute the new image name and checksum information to the controller.

3. When a controller needs to be downline loaded, it requests the appropriate image. After it receives the image from a downline load server, the controller calculates the checksum for the image and compares the value against the stored value. If the received image does not have the correct checksum then that image is ignored and a new image is requested.

2.5 Associations

When two nodes try to communicate by exchanging Ethernet frames over the LAN, controllers will not allow the communication unless the *association* is allowed by a KDC. If allowed, these associations are granted upon demand by KDCs.

There are three different types of associations:

- Associations between two nodes protected by different controllers. Frames sent under these associations are secured through encryption while they are on the Ethernet.

  An example of this type of association would be an association between node A and node B in the LAN shown in figure 1.

- Associations between two nodes protected by the same controller. Frames sent under these associations are never sent on the Ethernet so there is no need for them to be encrypted. The controller only sends the frame to the node port where the destination node is attached, so this type of association is secure.

  An example of this type of association would be an association between node B and node C in the LAN shown in figure 1.

- Associations between a node protected by a controller and a node not protected by a controller. Frames sent under these associations are not encrypted (because there is no second controller to decrypt the frame), but communication is not allowed unless approved by a KDC.

  An example of this type of association would be an association between node C and node D in the LAN shown in figure 1.

When communication occurs between two nodes not protected by DESNC controllers, controllers and KDCs are not involved in the communication.

2.5.1 Encrypted Association Set-Up

The protocol exchange used between DESNC controllers and KDCs to encrypted set-up associations is similar to the protocols described in Needham and Schroeder[6] and Voydock and Kent[7]. Here is an example of how an association would be established between two nodes that are both connected to DESNC controllers.

Consider the LAN shown in figure 1. Setting up an association between node A and node B involves the following steps:

1. Node A sends an Ethernet frame to node B.

2. Controller A receives the Ethernet frame and verifies the source address of the frame.

3. Controller A requests an association from the KDC.

4. The KDC checks the access control policy, determines that nodes A and B are allowed to communicate, and sends an Association Open message to controller A. The Association Open message is encrypted with the master key of controller A. The message contains an association key, either generated by the KDC controller or taken from a supplied key source.

5. Controller A sends an Association Forward message to controller B. The Association Forward message is encrypted with the master key of controller B. This message was generated by the KDC and included in the Association Open message sent to controller A.

6. Controllers A and B communicate and determine that they share a common association key.

7. Controller A encrypts the frame sent in step 1 with the association key and sends the encrypted frame to controller B.

8. Controller B receives the encrypted frame, decrypts the frame, checks the manipulation detection code, and transmits the frame to node B.

Once the association is established, no further interaction with the KDC is required and all communication between nodes A and B is encrypted with the association key until the association expires. If an association is active and approaching expiration, the controller that originally requested the association (controller A in this example) will request another association before the first association expires.

The duration of associations is determined by the network security manager, and the information is distributed to controllers when they are initialized.

2.5.2 Unencrypted Association Set-Up

Associations that involve only one DESNC controller (either because both nodes are attached to the same controller or because only one node is attached to a controller) do not require an encryption key and do not involve synchronization between two controllers. Setting up these associations is simpler than setting up encrypted associations, several steps can be omitted.

For example, if node B in figure 1 wants to communicate with node C or node D, the following steps are required:

1. Node B sends an Ethernet frame.

2. Controller B receives the Ethernet frame and verifies the source address of the frame.

3. Controller B requests an association from the KDC.

4. The KDC checks the access control policy, determines that the nodes are allowed to communicate, and sends an Association Open message to controller B. The Association Open message is encrypted with the master key of controller B. No encryption keys are included in the message.

5. Controller B sends the frame received in step 1 to the appropriate destination (either to the correct port or to the LAN).

As in the previous case, communication between the node pair continues without KDC intervention for the duration of the association.

2.6 Trust

With any security system, it is important to know which components must be trusted, and the degree of trust required. The EESS architecture was designed to limit the degree to which an individual DESNC controller needs to be trusted.

The compromise of a DESNC controller may compromise the nodes protected by the controller, but will not compromise any other controllers or nodes on the LAN. This means that a controller must be protected as well as any of the nodes protected by the controller.

If multiple nodes are connected to the same node port of a controller, the nodes can masquerade as each other. This means that those nodes must be mutually trusting. If this is level of trust is not appropriate, a site can use DESNC controllers with only one node attached to each of the four node ports.

If a KDC node or the controller that supports a KDC node is compromised, the security of the LAN can be compromised. This means that KDC nodes and KDC controllers must be protected as well as any node on the LAN. While KDC nodes can be used for multiple purposes, the security of the network is improved if the KDC nodes are limited to network management functions and access to the nodes is limited to trusted individuals.

3 Determining the Encryption Key

Unless a cryptographic system uses a single encryption key to encrypt all messages exchanged, it is necessary to determine whether a message is a control message or an encrypted node frame and which key should be used to decrypt a particular message. While, in many cases, it is possible to determine the correct key from the context and from the source and destination addresses, the choice is occasionally ambiguous. Placing the information in the message explicitly avoids any ambiguity and is also more efficient to handle. To prevent modification attacks on the protocol messages, it is necessary to guarantee that any modifications of this information are detected.

The messages exchanged between the components of an Ethernet Enhanced-Security System contain two fields that are used to identify the encryption key. Each message contains:

Message Type: This field identifies the type of the message and, in particular, whether the frame is a control frame or an encrypted node frame.

This field is protected against modification by including a duplicate copy of the field in the encrypted portion of the frame (protected by the manipulation detection code). These copies are compared after the frame is decrypted.

Encryption Identifier: Once the type of message is known, this field uniquely identifies the encryption key.

Rather than use an explicit check, this field is implicitly verified. If the field is modified, then the wrong encryption key will be used to decrypt the frame and the manipulation detection check will fail.

For encrypted node frames, DESNC controllers are designed to allow rapid determination of the association key from the encryption identifier.

4 Sequence Numbers

When Ethernet frames are encrypted, sequence numbers are used for two purposes:

- To prevent attacks that involve the replay or reflection (exchange of source and destination addresses) of encrypted Ethernet frames.
- To whiten messages to prevent intruders from inspecting two encrypted Ethernet frames and determining if the original frames (or an initial portion) were identical.

Sequence numbers are used to protect both encrypted node frames transmitted between DESNC controllers and the control frames exchanged among the controllers and between controllers and KDCs.

Sequence numbers are 4 bytes long and contain a 31 bit count value and a 1 bit direction.

4.1 Sequence Numbers Versus Timestamps

While timestamps are commonly used to detect replay attacks, the EESS architecture uses sequence numbers. Using sequence numbers avoids then problem of synchronizing clocks, and sequence numbers were found to be easier to generate and compare than timestamps. In particular, detecting replay attacks while allowing for out-of-order frames it is easier and requires less storage with sequential sequence numbers than with timestamps.

However, sequence numbers do have their own synchronization problems. A pair of components can loose synchronization if one component sends a large number of messages while the other component is not working or is otherwise out of communication. For example, this may occur if a KDC is unable to communicate with some controllers for several days.

The architecture provides a way to securely resynchronize sequence numbers when these problems occur. KDCs and controllers synchronize their sequence numbers when they exchange status information. This synchronization allows any sequence number mismatch to be corrected. To avoid any possible replay attacks, sequence numbers are only raised during synchronization, never lowered.

4.2 Sequence Number Use

The EESS architecture uses a separate sequence number stream for each encryption key used. The keys used to encrypt node frames are distributed upon demand by KDCs and are used for at most a few days. The encryption keys used for control messages are used for longer periods.

Each time a message is transmitted using a particular encryption key, the DESNC controller (or KDC node) transmitting the message increments the sequence number associated with that encryption key. When a message is received, the recipient controller checks the sequence number and rejects the message if the sequence number is significantly lower than the highest sequence number received, or if another message has been received with the same sequence number.

Messages are accepted out of order, and no attempt is made to reorder the messages or to guarantee that all messages are delivered. (These functions are not normally provided by the Data Link layer, and should be provided by higher protocol layers.)

5 Interoperability

Because encrypted Ethernet traffic will probably be using the same Ethernet cable as unencrypted traffic, an Ethernet security system must be a 'good neighbor' on the LAN. This implies that the system must:

- Follow the Ethernet physical standards, including the restriction on maximum frame length,
- Follow standard Ethernet packet formats by including valid frames headers on all transmitted frames, and
- Allow the LAN to be maintained, or at least not prevent standard LAN maintenance operations.

The implications of each of these restrictions are discussed below.

5.1 Frame Length

For several reasons, including the addition of sequence numbers to frames, it is necessary for the size of frames to be increased when the frames are encrypted by DESNC controllers. However, to satisfy the Ethernet standard, the Ethernet frames transmitted by DESNC controllers must not be more than 1518 bytes long. There are two possible resolutions to these two requirements: Either restrict controllers to only encrypting frames that are short enough to be encrypted without exceeding the length restriction, or fragment long Ethernet frames when they are encrypted.

We chose to fragment long Ethernet frames when they are encrypted by a controller, and to reassemble them transparently when they are decrypted by the recipient controller (before they are transmitted to the destination). When one long Ethernet frame is transmitted by a node, two separately encrypted Ethernet frames are sent from one DESNC controller to the other. The recipient controller checks each frame and uses the two frames to rebuild the original Ethernet frame. The frame received by the destination node is identical to the frame transmitted by the source node.

Fragmentation affects the performance for long frames because it is necessary to send twice as many frames. But it is possible for network users to voluntarily reduce the length of Ethernet frames they transmit. This reduction avoids the need for fragmentation for the applications that can handle a reduced maximum frame length, but the fragmentation allows any existing application to continue to work correctly even if it sends maximum length Ethernet frames.

5.2 Frame Header

When frames are encrypted, the source and destination addresses are left unencrypted and the rest of the original header is replaced by an IEEE 802 header with a protocol identifier that identifies the frame as an encrypted frame. The reasons for this are:

- If the addresses were encrypted, then Ethernet bridges would no longer be useful in filtering network traffic and there would be a significant performance penalty because it would be necessary for every node to decrypt each frame to determine if it is the intended recipient.

  The addresses are authenticated by the encryption key used, so there is no loss of node authentication due to plaintext addresses.

- If the original header (other than the addresses) was encrypted, the header would no longer have have a valid format (i.e., it would probably have an unassigned Ethernet protocol type, or an incorrectly formatted IEEE 802 header), thereby confusing LAN monitoring tools.

- If the original header is left unchanged, the rest of the message would look malformed (for a message with that header) because it was encrypted. This would also confuse LAN monitoring tools. (Also, an integrity check would be necessary for the header.)

- A distinct header provides an easy way to determine if a message needs to be decrypted when it is received.

Therefore, even though the extra header increases the overhead of encrypting the Ethernet frames, the header is added because it simplifies the processing of the frames and prevents confusion over the contents of the frame.

5.3 Network Maintenance

While it is necessary to replace the headers of encrypted Ethernet frame with headers containing protocol identifiers that identify the frame as being encrypted, the original header is also useful to network management tools. Tools that can examine this header can determine how a LAN is being used.

For this reason, DESNC controllers include the original header of the frame (except for the addresses) in unencrypted form after the header that identifies the frame as being encrypted. The header is also included in the encrypted portion of the message so that attempted modifications to the message can be detected.

When the frame is encrypted the DESNC controller examines the start of the frame and determines the frame format and the size of the header fields (excluding the addresses). DESNC controllers distinguish between:

- Ethernet format frames (with only a 2 byte protocol type),

- IEEE 802 format frames (with 5 or 6 bytes of header fields), and
- IEEE 802 format frames with protocol identifier (with 10 bytes of header fields).

The original header fields are copied into a 10 byte field in the encrypted message. This field is zero-padded if the header fields are shorter than 10 bytes.

References

[1] *The Ethernet, A Local Area Network, Data Link Layer and Physical Layer*, Version 2.0, (Digital, Intel, and Xerox), November 1982.

[2] *Data Encryption Standard*, Federal Information Processing Standards Publication 46 (FIPS PUB 46), National Bureau of Standards, 15 January 1977.

[3] *DES Modes of Operations*, Federal Information Processing Standards Publication 81 (FIPS PUB 81), National Bureau of Standards, 2 December 1980.

[4] D.E. Bell and L.J. LaPadula, *Secure Computer Systems: Unified Exposition and Multics Interpretation*, ESD-TR-75-306, MITRE Corporation, March 1976.

[5] K.J. Biba, *Integrity Considerations for Secure Computer Systems*, ESD-TR-76-372, MITRE TR-3153, MITRE Corporation, April 1977.

[6] R.M. Needham and M.D. Schroeder, "Using Encryption for Authentication in Large Networks of Computers", *Communications of the ACM*, December 1978.

[7] V.L. Voydock and S.T. Kent, *Security in Higher Level Protocols: Approaches, Alternatives and Recommendations*, Report No. ICST/HLNP-81-19, National Bureau of Standards, September 1981.

A Secure Audio Teleconference System

D. G. Steer, L. Strawczynski,
W. Diffie, M. Wiener
BNR,
P.O. Box 3511, Station C, Ottawa, Canada
K1Y 4H7.

ABSTRACT

Users of large communication networks often require a multi-party teleconferencing facility. The most common technique for providing secure audio teleconferencing requires the speech of each participant to be returned to clear form in a bridge circuit where it is combined with the speech of the other participants. The combined signal is then re-encrypted for distribution to the conferees. This introduces a security weakness as the bridge works with the clear speech and the cipher keys for all of the participants. In this paper we describe secure conferencing systems in which some of the bridge functions are distributed among the users so that no additional security weakness is introduced. The network conference bridge knows the addresses of the participants and receives and distributes the encrypted speech without modification. The conference system can be used with a number of encryption algorithms and the system is suitable for deployment on digital networks such as ISDN. High quality and robust secure voice communications can be provided with this technique.

Introduction

As the use of secure communication techniques becomes more widespread, the need will grow for convenient secure conferencing facilities. Conferences allow a number of separated callers to participate in a group discussion. Some facilities also include video, data, and graphics facilities for these conferences. To provide a secure conference call, the conference facility is augmented with security features. The intent is to provide privacy (through encryption) and authentication of the the participants.

In this paper we will outline some practical considerations for providing secure audio conference facilities within the public switched telephone network (PSTN). The basic problem to be resolved is to find a practical means to apply the mechanisms of encryption and authentication protocols to a multi-party conference without compromise of the security but within the constraints of the network. The goal is to achieve a conference facility that can be operated by the network supplier without concern for security features and thus leave the customers free to select the security systems to meet their individual needs.

Firstly, we will outline the requirements for a conference facility and point out the limitations of certain implementation techniques. The principal concerns are the non-linear coding of the speech in the PSTN and the requirements for limited bandwidth and minimum delay. We will introduce the concept of a distributed conference system and will review a distributed authentication scheme. A further conferencing technique, suitable for networks with long transmission delays, will also be outlined. We will conclude that secure conference systems are practical in the PSTN, and that they can be safely provided to meet the user's requirements. The Integrated Services Digital Network (ISDN) is especially well adapted to providing these secure services.

Basics of Audio Conferencing

Figure 1 shows the typical configuration of a conference system and in this section we will review some of the basic operations. This section is included to provide background on the operation of conference circuits. The users (conferees) are connected by the communications network (usually the PSTN) to a "conference bridge" circuit. This circuit functions to sum together the speech from all the participants and to distribute this sum (which is the conference signal) to all the conferees. The network connections may be established with the assistance of a conference operator (who may be a participant in the conference) or it may be a "dial-in" conference where participants dial a special directory number to reach the bridge and be included in the conference. Usually an audible signal or a verbal protocol is used to announce when people enter or leave the conference.

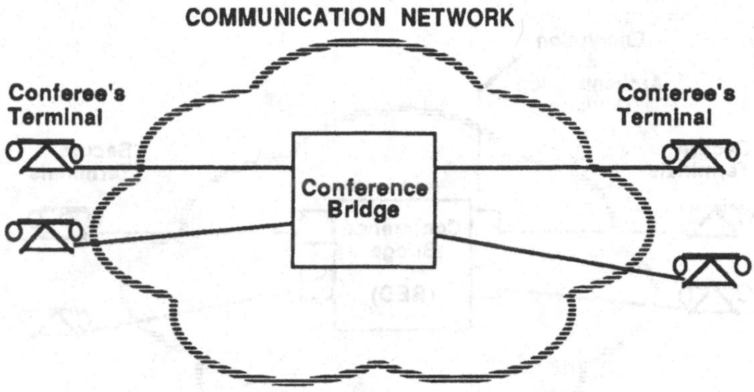

FIGURE 1 - AUDIO CONFERENCE BRIDGE

In concept, the bridge circuit simply sums the speech signals from each participant and distributes this as the conference signal for all to hear. In practice the process is a bit more complicated. One complication is that the signal sent to a participant should not include the participant's own speech. If it does, then transmission delays can cause objectionable echos. Thus the conference sum sent to speakers differs from that sent to passive listeners.

The accumulation of noise is another important consideration in an analogue network. In order to avoid the buildup of noise, the bridge will usually monitor the incoming signals and only distribute the loudest, or perhaps a weighted sum of the 2 loudest, speakers to all the others (see for example [1]). As an alternative, the bridge may include a threshold, and only include a conferee's signal in the conference if the level exceeds the threshold. Noise from the idle channels is kept to a minimum with this arrangement.

Many bridges also include an automatic gain control (AGC) function to equalize the speech volumes of distant and near conferees and of loud and soft speakers.

It is important to keep in mind that in a digital system voice signals are encoded with non-linear codes designed to increase the dynamic range and minimize the effects of quantization noise. The standard PTSN codes for speech are called μ-law (in North America) and A-law (in Europe) [2]. Speech signals encoded with these codes must be converted to a linear representation before the bridge can process them.

There are thus a number of functions required of the bridge to support conferencing:

- summation of speech signals
- code conversions (μ-law or A-law to linear)
- automatic gain control (AGC)
- identification

The concept of a bridge circuit can be extended to a secure conference as shown in Figure 2. Since the speech signals are now encrypted the bridge circuit can not simply sum the signals as the encryption is typically a non-linear process. (Here we define "non-linear" to mean that two encrypted signals cannot be summed to yield a third signal that can be decrypted to meaningful speech.)

Figure 2 shows one possible arrangement for a secure conference. In this case the telephone sets have been augmented with encryption and authentication devices. In normal two-party use these sets provide end-to-end encryption protection. The bridge is also fitted with like devices at its ports. Thus the bridge can receive a secure call from each participant, decrypt the speech, perform the conference operations (linearize, threshold, summation), and then encrypt the result for distribution to the conferees. This is a simple extension to the concept of a conference bridge.

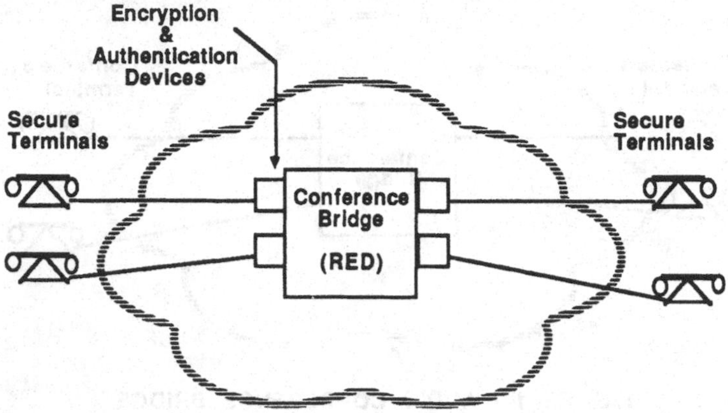

Figure 2 - Conventional Conference Bridge with Security

This method requires the bridge circuit to know the encryption keys for all parties and clear speech is contained within the unit. Thus, the conference bridge must be included when determining the security of the communications system. This is often referred to as a "red" bridge because its internal signals are unprotected by encryption, and it introduces a point of weakness in the system. Some users may not wish to trust the security of a bridge operating outside their direct control. It is thus desirable to make a conference unit that can operate without requiring the speech to be in clear form. This would be referred to as a "black" bridge because all of its signals are protected by encryption.

One possible approach to providing a black bridge has been provided by Brickell, Lee, and Yacobi [3]. In this method, an encryption process is used with certain linear properties which allow the bridge circuit to sum and distribute the signals without the need for decryption. Unfortunately this limits the number of applicable encryption techniques and not all users would be willing to trust these schemes. This method also restricts the speech coding techniques allowed, produces bandwidth expansion, and requires synchronization (in time) of the signals from all of the conferees. This scheme, for example, is unable to use the standard μ-law coded speech common in the PSTN. It is thus desirable to design a method which can function independently of the encryption process and the speech coding, and also does not require any special synchronization.

The Distributed Bridge

Returning for a moment to Figure 2, note that one place where speech must be in clear form is at the users' telephone terminals. Decrypted speech must be provided at these points for the conferees to hear. The security weakness inherent in the red conference bridge will be eliminated if the conference operation is migrated out to the telephone sets where clear speech must always be available for the users.

One way to accomplish this is shown in Figure 3. In this case each telephone set is equipped with a simple three-party conference circuit in addition to two encryption/decryption units. The sets are connected to the digital switching network by two logical circuits. These access connections would typically be made using digital loop technology. A chain-like connection pattern is established between the parties to form the conference. Each party receives the signals from its neighbours, sums these with its local speech input, and forwards the result along the chain. Each party hears the speech of all the others as the signal is passed down the chain. There is no centralized bridge. The function is distributed among the participants. No additional security risk is introduced by this distributed bridge as clear speech must be available at each phone for the conferees to hear anyway. Modern signal processing chips can be used to perform the bridge functions in each user's terminal.

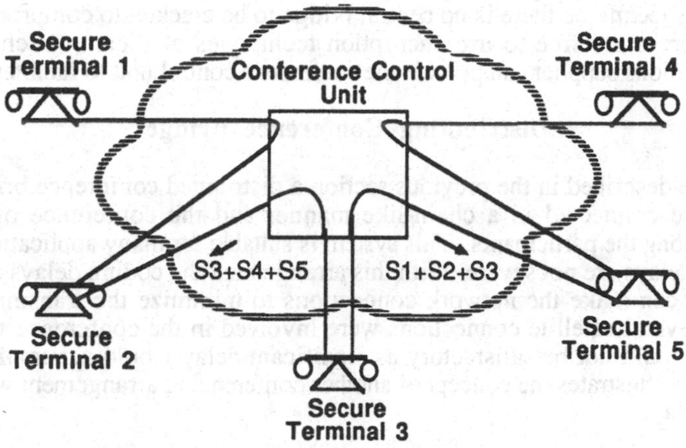

Figure 3 - Distributed Secure Conference Bridge

One of the difficulties with a chain-like conference connection of this type is that anyone dropping out of the conference breaks the chain. The conference would thus be split into two disconnected parts. This may be overcome by incorporating a conference control unit (CCU) in the network. This CCU keeps the list of network addresses of the (active) participants and is responsible for maintaining the connections through the communications network to all the parties, and automatically reconfiguring the connections when the list of participants changes. This is illustrated in Figure 3 by the rectangle labeled conference control unit. In normal operation, the CCU is under the direction of a conference operator. This may be a network resource person, or it may be simply a participant acting as chairperson and controlling the CCU via a separate communications channel. If parties wish to be added or deleted from the conference, the CCU will automatically reconnect the network to the new configuration. Thus, while we technically have a chain-like connection, operationally it behaves like the common star topology for conference systems. Note that this conference control unit could be implemented as a software program operating in the switching nodes of the communications network.

Figure 3 shows two connections from each phone to the network (except for the ends of the chain). While this could be implemented as two physical connections, modern digital techniques allow the multiplexing of two speech channels on a single pair of wires. The Integrated Services Digital Network (ISDN), for example, directly supports two 64 kbps speech channels (the "B" channels) plus a signaling channel (the "D" channel) on a single subscriber pair. Thus the distributed conference configuration could easily be implemented in an ISDN environment, and in fact we have tested such an arrangement on the BNR ISDN facilities. Other multiplexing techniques can also be used to maintain a single subscriber access connection and to reduce the number of network connections.

In analogue communication networks, the chain-like connection of Figure 3 would be impractical due to the accumulation of transmission and idle channel noise. (Each party receives the accumulated sum of all the noise sources along the chain.) In digital transmission, noise does not accumulate on tandem connections, and bit errors can be corrected by the use of suitable error correcting codes. Inherent in any digital speech coding system is quantization and coding noise. However, with suitable speech coding techniques, such as the standard companded PCM, ADPCM [5], and the CCITT wideband audio standard [6], this noise will not accumulate in the tandem connections used in this conference arrangement. The conferencing algorithm at each terminal can be designed with a local speech detector and AGC function. The speech detector would only include local speech input when speech is present to eliminate the accumulation of idle channel noise. The ACG function would equalize the conference levels.

With this technique there is no central bridge to be cracked to compromise security, and the conferees are free to use encryption techniques of their own choosing. The telecommunications supplier can provide the conference control unit to facilitate the service.

Distributing Conference Bridge

We have described in the previous section a distributed conference bridge in which the parties are connected in a chain-like manner and the conference operations are distributed among the participants. This system is suitable for many applications where the transmission delays are not severe. With this arrangement the coding delays are negligible and the CCU can make the network connections to minimize the transmission delays. However if several satellite connections were involved in the conference the chain-like arrangement would not be satisfactory as significant delay would accumulate along the chain. Figure 4 illustrates the concept of another conferencing arrangement which is not as sensitive to delay.

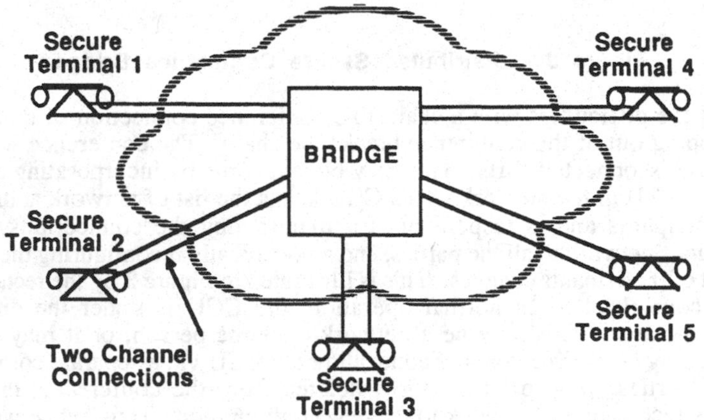

Figure 4 - Distributing Bridge

In this arrangement, each secure telephone set is connected to the PSTN with 2 channels. The use of multiplexing and digital access technology makes this arrangement practical with a single wire pair. Participants in the conference are connected by the network to the distributing bridge. During operation, each terminal sends to the bridge two signals. One of these signals is the encrypted speech from the participant. The second signal is information giving the average volume level of the speech for the preceding time period of say, 4 milliseconds. The bridge unit would examine the volume level information from all the participants, and distribute to each conferee the two loudest encrypted speech signals (other than the conferee's own signal) using the two channel connections. The terminals would receive these two encrypted speech signals, decrypt them and present the sum to the local listener. As the speaker activity in the conference changed, the volume level information would reflect the new speakers, and the bridge would distribute the new speakers. Initially, cryptographic session keys would be developed among all participants by means of a protocol such as the one described in the next section section.

In Figure 5 we give a simple illustration of how this might be implemented. Three terminals are involved in this example conference. The diagram illustrates the signals sent by terminals 1 and 2 towards the bridge. These are the encrypted speech signals and the volume level information. In this example, speaker 1 is followed by speaker 2. The encryption process uses a cipher-feedback mode of operation. The listener at terminal 3 will receive a short pause to signify the change of speakers and allow a change of crypto-key if required. The beginning of the new speaker's encrypted speech would be preceded by a short period of extra data to allow the encryption process to resynchronize. This synchronization period would be muted by the receiver.

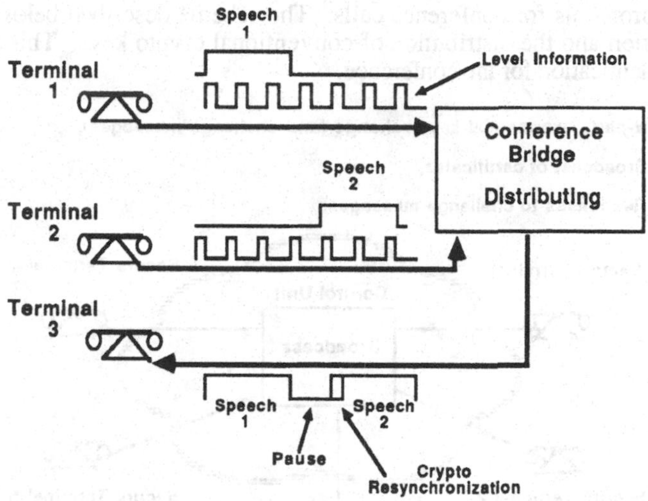

Figure 5 - Operation of Distributing Bridge

This form of conference bridge handles only encrypted signals and acts only as a distributing center. The users must give out extra information in the form of the average speech level. In the simplest implementation this information would be in the clear. However, it could be sent to the bridge using an encrypted channel . (If the bridge is to remain truly black there is no advantage in doing this.) This average level does not reveal the speech signal but it does represent a (small) leak of information that some users may find unacceptable. This system does not suffer from significant problems with long delay paths as it uses a star topology for the network connections.

Key Distribution and Authentication

The operation of security features in a large network is dependent upon the proper distribution of encryption keys and the authentication of terminals and users. Practical techniques to provide for two-party end-to-end encryption and authentication systems in the public network are available [4]. These use a "hybrid cryptosystem" with public key techniques being used for key distribution and authentication, and conventional ciphers being used for encryption of the speech. There are two basic requirements. One is for the secure distribution of the conventional encryption keys (also known as session keys), and the second is for the authentication of the participants.

These protocols are designed to provide a new session key for each connection to ensure privacy. These session keys are generated by a random process at each terminal and exchanged between terminals using the exponential key exchange technique of Diffie and Hellman [7]. The terminals are identified by means of certificates. These certificates are prepared by the central key management facility (KMF) using public key signature techniques and are unique for each terminal. By exchanging these certificates and verifying the signature of the KMF, the terminals are assured of the identity of their correspondents. Finally to guard against active intruders in the circuit, challenge messages are exchanged between the terminals. Typically the terminal is asked to sign with its private key a message sent by the other terminal. An intruder would be unable to respond to such a challenge with the needed signature. These techniques can be extended to multi-party conference systems and we will briefly outline below a simple key distribution and authentication scheme that is suitable for use with the distributed bridge. Note that while this description is in the context of an audio conference, the protocols are general and could be applied to data or video conferences. Brickell, Lee and Yacobi [3] have also proposed authentication protocols for conference calls. The scheme described below provides for both authentication and the distribution of conventional crypto keys. This provides both privacy and authentication for the conference.

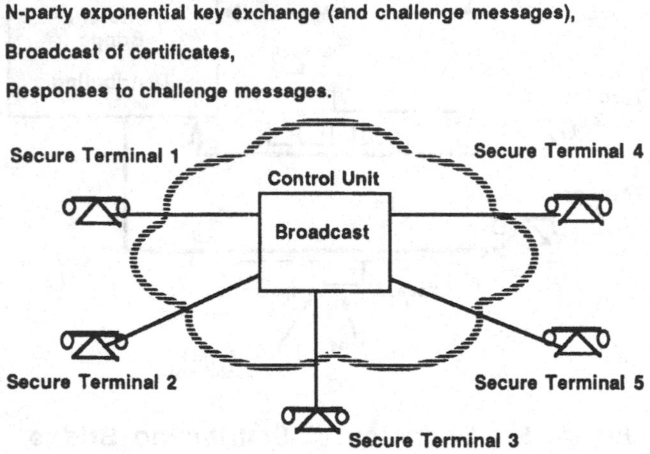

Figure 6 - Distributed Authentication

The key distribution and authentication phase of the conference set up would take place at the beginning of the conference before the onset of speech or data communications. The process would be repeated each time a party is added to the conference. During this set up phase, the conference control unit operates in a broadcast mode so that messages sent to it are distributed to all participants. This is illustrated in Figure 6.

The process for distributed authentication in a conference with **n** participants is divided into six steps as follows:

1) Each terminal, i, generates a random number x_i and calculates the exponential $y_i \leftarrow \alpha^{x_i}$ and sends this to the CCU for broadcast to the other terminals. (α is the pre-established common base and calculations are performed modulo a large prime **p**.)

2) Each terminal now has the set of y_i. This set can be used to provide an order for the terminals. A convenient order would assign the terminals an index based on the magnitude of the y_i. (Terminal 1 generated the smallest of the y_i etc. In the unlikely event that two of the y_i were equal then the terminals would be expected

to submit new values.) The order could also be established by the CCU based on the chain topology of the connections.

3) Each of the terminals will now calculate an exponential of the following form :

$$X \leftarrow \alpha^{(x_n \alpha^{(x_{n-1} \alpha^{(... \alpha^{(x_3 \alpha^{(x_1 x_2)})})})})}$$

Terminals 1 and 2 can calculate X based on their knowledge of the y_i and their own x_i. However, terminals 3 to **n** require further information to calculate X. This information comes in the form of the numbers $w_1 ... w_{n-1}$ calculated by the terminals and broadcast to the other terminals. The terminals proceed as follows :

Terminal 1 performs the following operations :
$X \leftarrow x_1$,
for j = 2 to **n** do the following :
$X \leftarrow y_j^X$.

Terminal i (for i = 2 to **n-1**) performs the following :
receive w_{i-1} via the CCU
(Terminal 2 uses $w_1 \leftarrow y_1$),
$X \leftarrow w_{i-1}^{x_i}$,
$w_i \leftarrow \alpha^X$,
broadcast w_i to all other terminals via the CCU,
for j = i+1 to **n** do the following :
$X \leftarrow y_j^X$.

Terminal **n** performs the following operation :
$X \leftarrow w_{n-1}^{x_n}$.

4) All terminals now share the number X which can be used to derive a common session key for the conference. This key would be used to encrypt all further transmissions. It is not necessary to use a common session key. The number X could also be used by pairs of terminals to form a unique session key for their connections.

5) Each terminal broadcasts (via the CCU) its certificate, and verifies the signature on all received certificates.

6) Each terminal signs with its secret key a suitable hash function of the set of numbers y_i and broadcasts this (via the CCU) to all other terminals. Each terminal verifies the signature of these messages.

Step 1 requires a total of **n** broadcast messages. Step 3 requires **n-2** messages. Steps 5 and 6 each require **n** broadcast messages. This gives a total of of **4n-2** broadcast messages. Note, however, that the certificates and response messages (5 and 6) could be combined into a single message for transmission. In this case the total number of broadcast messages is **3n-2**. The number of exponentiation operations to be performed by each terminal for key generation depends on the terminal's location in the sequence. Terminals 1 and 2 must perform **n** exponentiations. Verifying the certificates, and responding to the challenge for authentication would require an additional 2 operations and thus a total of 3**n** exponentiations could be required. Note that this grows linearly with **n**. Once the authentication phase is completed, the conference control unit can switch from the broadcast to the chain mode. With this procedure, the authentication process is distributed and thus no party need rely on a centralized bridge for the security of the call. However, a trusted key management facility is required for the creation of the user's certificates.

The procedure for adding or deleting parties to the conference will depend on the security policy of the participants. Some policies may require a new session key and a complete re-authentication of all parties each time the network configuration changes. In the simple scheme described above, if a common session key is used, the chain may be reconnected (with an appropriate mechanism for crypto resynchronization) when a party leaves. An additional party may be accomodated by simply labeling it as terminal n+1 and appending its input (y_{n+1}) to the accumulated chain calculation. This would result in a new common number X' and a new session key. Changes in the participation in the conference can thus be accomodated with a minimum of interruption.

Thus it is practical to provide a true black bridge for secure teleconferencing in the public or other large networks. A wide variety of digital encryption processes can be used. The authentication mechanisms can function with many public key systems. Users are free to choose whatever system they feel comfortable with. They are also free to change the operation of the security system without consultation with the network provider. These are significant benefits for the widespread use of and confidence in the system.

Conclusions

We have described secure teleconferencing systems which can be operated in the public communications network. In these systems, advantage is taken of the fact that clear speech must be available at each telephone set for the local participant, and this can be used with a distributed conference circuit to provide a multi-party conferencing system. The network conference control unit knows the addresses of all the participants but its role is limited to providing channel connections and demultiplexing and distributing unaltered signals from the participants. The conference unit does not need to know any security information. The system can operate with a number of security methods and network configurations, and the users are free to select the techniques that they are most comfortable with. The system is suitable for implementation in digital networks such as ISDN, and can provide extremely high quality and robust operation.

References

[1] John Ellis, Bruce Townsend; *State of the art in Teleconferencing*. Telesis 1987 one pp23-31.

[2] David R. Smith; *Digital Transmission Systems.* Van Nostrand Reinhold Co., New York. 1985 pp77-88.

[3] E.F. Brickell, P,J, Lee, Y. Yacobi; *Secure Audio Teleconference*. Advances in Cryptology - Crypto'87 (Proceedings), pp418-426. Carl Pomerance (Ed.), Springer-Verlag, Berlin Heidelberg, 1988.

[4] W. Diffie, B. O'Higgins, L. Strawczynski, D.G. Steer; *Secure Communications with the Integrated Services Digital Network (ISDN).* Proc 3rd Annual Symposium on Physical/Electronic Security, Philadelphia Chapter AFCEA pp 34.1-34.5, August 1987.

[5] CCITT Recommendation **G.721** (ADPCM).

[6] P. Mermelstein; **G.722,** *A new CCITT Coding Standard for Digital Transmission of Wideband Audio Signals.* IEEE Communications, Volume 26, No 1, pp 8-15 Jan 1988.

[7] W.Diffie, M.Hellman, *New Directions in Cryptography.* IEEE Trans Information Theory IT-22 : pp 644-654, 1976.

Rump Session

Short Presentations

Chair: W. Diffie, Bell Northern

Diffie-Hellman is as Strong as Discrete Log for Certain Primes

Bert den Boer *

Centre for Mathematics and Computer Science
Kruislaan 413 1098SJ Amsterdam The Netherlands

Abstract

Diffie and Hellman proposed a key exchange scheme in 1976, which got their name in the literature afterwards. In the same epoch-making paper, they conjectured that breaking their scheme would be as hard as taking discrete logarithms. This problem has remained open for the multiplicative group modulo a prime P that they originally proposed. Here it is proven that both problems are (probabilisticly) polynomial-time equivalent if the totient of P-1 has only small prime factors with respect to a (fixed) polynomial in ${}^2logP$.

There is no algorithm known that solves the discrete log problem in probabilistic polynomial time for the this case, i.e., where the totient of P-1 is smooth. Consequently, either there exists a (probabilistic) polynomial algorithm to solve the discrete log problem when the totient of P-1 is smooth or there exist primes (satisfying this condition) for which Diffie-Hellman key exchange is secure.

Introduction

Let P be a prime and g be a generator of the multiplicative group $GF(P)^*$. Let f be an element of $GF(P)^*$. Solving for x

(1) $$f = g^x \bmod P,$$

*Supported in part by the Netherlands Organization for Scientific Research (NWO).

is called the *Discrete Logarithm Problem* (D.L. problem). Until now, no general purpose algorithm has appeared that computes $x \bmod P\text{-}1$ in probabilistic polynomial time (see [Odl]). In [Poh], an algorithm is given which solves the D.L. problem in polynomial time for the case where all prime factors of $P\text{-}1$ are smaller than some constant B. The time remains polynomial if we take $B = q({}^2\log P)$, for some fixed polynomial $q(.)$. Numbers with this property will be called *smooth*(with respect to P and q).

Based on the difficulty of solving the general D.L. problem a key exchange protocol is proposed in [DH]: The first step is to establish a prime P and a generator g of $GF(P)^*$ common for all participants. Each party j chooses randomly a secret number z_j and computes $f_j = g^{z_j}$. The f_j are made public. The public numbers are sufficient to establish a secret and common key for each pair of participants. For example let us assume that the first party has secret x and computed

(2) $$f_1 = g^x \bmod P.$$

and similarly the second party has secret y and computed

(3) $$f_2 = g^y \bmod P.$$

The first party computes $f_2{}^x$ and the second party computes $f_1{}^y$. Thus both parties obtain

(4) $$f_3 = g^{xy} \bmod P,$$

which they can use as a common secret key. For a passive eavesdropper to determine this key, he must find f_3 satisfying (4) with suitable x and y satisfying (2) and (3) where the gathered data is just

$$(P, g, f_1, f_2).$$

This problem will be called the *D.H. problem.* An obvious way to solve it is to solve a D.L. problem with input (P, g, f_1) and then to use the output x

to compute f_3 by $f_3 = f_2^x$. The critical question is can f_3 be found without obtaining (much) information about x or y in the process. In [DH] the authors conjecture that solving x or y is basicly the only way to solve the D.H. problem. More precisely they conjecture that the D.H. problem is hard if the D.L. problem is hard. A sufficient way to prove this is showing there exists an algorithm to solve the D.L. problem comprising of a (probabilistic) polynomial number of calls to an algorithm which solves the D.H. problem (here after called a *D.H. oracle*) and a (probabilistic) polynomial number of "elementary" operations.

In this paper we present such an algorithm for the case where the totient of P-1 is smooth with respect to P and $q(.)$. In the literature no probabilistic time algorithm $\mathfrak{R}$ exists to solve the D.L. problem for primes P for which the totient of P-1 is smooth (notice that P-1 itself has to be smooth for Pohlig-Hellman algorithm to be polynomial-time, see[Poh]). From our design follows that either such an algorithm $\mathfrak{R}$ exists or primes P exist for which the D.H. problem is hard. In the latter case primes P for which the biggest prime factor of P-1 is bigger than P^e for some fixed positive real number e serve as candidates (while the double totient of P is smooth).

Main idea

Let B be a number which depends polynomially on the logarithm of P. It always holds that P-1 has a unique factorization MN, where the prime factors of M are smaller than B and the prime factors of N are bigger (or equal) than B. Assume also that the totient of P-1 is smooth. Given a generator g of the multiplicative group modulo P and a member f of that group we are interested in solving x from (1). It suffices to compute x modulo M and x modulo N and use them in a Chinese Remainder Algorithm. A solution z of the equation

$$f^N = g^{zN} \bmod P$$

is equal to x mod M. This number can be computed in polynomial time with techniques similar to Pohlig-Hellman[Poh].

Replacing N by M in the last equation and solving the new equation would give us x modulo N. At this moment it is not (publicly) known how to solve this in polynomial time. What we do is exploiting the D.H.oracle to solve that new equation. Because the totient of P-1 is smooth, each prime factor Q of N appears with multiplicity 1 (otherwise Q would be a prime factor of the totient of P-1) and moreover it holds that Q-1 is smooth. We proceed computing x modulo N by computing x modulo Q for each prime divisor of N.

This algorithm requires a generator h of the multiplicative group $GF(Q)^*$, which can be found in probabilistic polynomial time, [Rie] (in a practical setting h is already constructed while constructing the big prime P). Either $x = 0$ mod Q or $x = h^y$ mod Q for some y. The first possibility appears iff $f^L = 1$ mod P, where $LQ = P\text{-}1 = MN$. In this (unlikely) case we are lucky, having found the answer (x mod Q) already. In the other case we proceed by computing the unknown y mod Q-1.

To compute y it is sufficient to compute y modulo any prime power divisor of Q-1 because we can combine these answers with the Chinese Remainder Algorithm to get y modulo Q-1. We will only show how to compute y modulo a prime divisor p of Q-1 .

Let l be $(Q\text{-}1)/p$ and compute $U = g^{x^l}$. This can be done by less than $2\ {}^2log(l)$ calls to the D.H.oracle. This indeed can be done because on input (g, g^{x^a}, g^{x^b}) the D.H.oracle should output $g^{x^{a+b}}$. Next one computes U^L and compares this with $g^{Lh^{la}}$ for $0 \leq a < p$. There is bound to be equality for one of the choices of a and for this a it holds that $a = y$ modulo p .

The above mentioned equality occurs because it is equivalent to the equality $Lh^{yl} = Lh^{al}$ mod P-1 and this is equivalent to the equality $h^{yl} = h^{al}$ mod Q. This last equality is equivalent to $yl = al$ mod Q-1 and finally this is equivalent to $y = a$ mod p.

Combining these answers for all prime divisors of Q-1 with the Chinese Remainder Algorithm we get a solution for y mod Q-1 and we compute $x = h^y \bmod Q$. Repeating this for each prime divisor Q of N and using the the Chinese Remainder Algorithm again we establish x mod N. Combining this with the already established value for x mod M we have arrived at the solution of equation (1). In the actual algorithm in the next section we do not mind about multiplicities of prime factors and we will also exploit the birthday attack. Furthermore we solve y modulo $Q-1$ without the Chinese Remainder Algorithm.

The main algorithm

In what follows we describe our algorithm to solve the D.L. problem ,an algorithm which is allowed to make a polynomial number of calls to a D.H. oracle and a polynomial number of elementary steps. At the end of this section we will be specific about these numbers.

Let $q(.)$ be a fixed polynomial. Let P be prime and $P\text{-}1 = MN$, where M, resp. N has small, resp. big prime factors with respect to $B = q(log(P))$ (like in the previous section). We want to solve equation (1). We already remarked that solving x modulo M can be done in polynomial time. In practical settings we can assume that the factorization of N is given (otherwise finding a generator modulo P would be difficult). If the factorization of N is not given we can find the factors of N in probabilistic polynomial time using Pollard-p-1 method [Pol] in an adapted form (details left to the reader). As we remarked in the previous section N is squarefree.

Let Q be prime divisor of N. We will establish w defined by x modulo Q. Let us define $L =(P\text{-}1)/Q$. The algorithm to compute w requires a generator h of the multiplicative group $GF(Q)^*$, which can be found in probabilistic polynomial time, [Rie] (in a practical setting h is already constructed while constructing the big prime P). Either $w = 0$ mod Q or $w = h^y \bmod Q$. The first possibility appears iff $f^L = 1$ mod P. In this (unlikely) case we are done. In the other case we proceed by computing the unknown y mod $Q\text{-}1$.

Now we will describe the residues which has to be computed in our algorithm. Given is the factorization $Q\text{-}1 = \prod_{j=1}^{k} p_j$, where $p_i \le p_{i+1}$ and $p_k \le B$. Define for $j = (-1),(0),1,\ldots,k\text{-}1,(k)$

$$l_0 = 1,$$

$$l_j = \prod_{i=1}^{j} p_i \quad ,$$

$$m_j = (Q\text{-}1)/p_{j+1}\,,$$

$$D_j = \prod_{i=j+2}^{k} p_i\,,$$

$$D_{k\text{-}1} = 1\,,$$

$$r_j = \lceil \sqrt{p_j}\,\rceil,$$

$$U_j = g^{x^{l_j}} \text{ modulo } P,$$

$$t_{bj} = (p_{j+1} - br_{j+1})m_j \text{ and}$$
$$T_{bj} = g^{Lh^{t_{bj}}} \text{modulo } P$$
$$\text{for } 0 \le b \le \lfloor (p_j/r_j) \rfloor,$$

$$S_{aj} = h^{am_j} \text{ modulo } Q \text{ for } 0 \le a < r_j,$$

$$y_{k\text{-}1} = 0\,,$$

$$C_j = h^{y_j l_j} \text{modulo } Q.$$

The algorithm starts with computing U_{j+1} from U_j using the D.H.oracle. All intermediate U_j have to be stored. The residues T_{bj} need to be stored for each different prime divisor p_j of Q-1. The residues S_{aj} may need to be stored (especially if p_j^2 is a divisor of Q-1). We compute z_j starting with $j = k$-1 down to 0 and y_j starting from $j = k$-2 down to -1 as follows. For $j = k$-1 down to 0 we search for a and b for which the equation

$$(5) \qquad (U_j^{C_j^L})^{S_{aj}} = T_{bj}$$

holds. Define z_j by $a+br_{j+1}$ and y_{j-1} by $y_j+z_jD_j$. Equation (5) always has a solution and finally we can compute w because it is equal to $h^{Q-1-y_{-1}}$ modulo Q.

Now we will briefly sketch why this algorithm computes x mod Q. After we checked that Q does not divide x we may assume that $x = h^y + Qr$ mod MN for some y and r. By induction we will show that $y + y_i = 0$ mod D_i for $i = k$-1,...,-1.

This is trivially true for $i = k$-1. Consider the left side of (5). This is equal

$$g^{(h^y + Qr)^{l_j} h^{y_j} L h^{am_j}} .$$

The exponent can be written as

$$(h^{yl_j} + Qt)\, h^{y_j l_j} L\, h^{am_j}$$

for some t. The right side of (5) is equal to

$$g^{L\, h^{(p_{j+1} - b\, r_{j+1})}} .$$

Equality (5) holds iff the exponents are equal mod MN. This holds iff

$$h^{y\, l_j + y_j\, l_j + a\, m_j} = h^{(p_{j+1} - b\, r_{j+1}) m_j} \bmod Q.$$

This last equality holds iff

(6) $$l_j(y + y_j) + a\, m_j = (p_{j+1} - b\, r_{j+1})\, m_j \bmod Q\text{-}1.$$

Under the induction hypothesis $y + y_j = c\, D_j$ for some c. Because $l_j\, D_j$ equals m_j equation (6) holds iff $c + a = p_{j+1} - b\, r_{j+1} \bmod p_{j+1}$. Because of the chosen range for (a,b) such an equality will occur. For this pair (a,b) we define $z_j = a + b_j$ and $y_{j-1} = y_j + z_j\, D_j$. Now it holds that

$$\begin{aligned} y + y_{j-1} &= y + y_j + (a + b\, r_{j+1})\, D_j = \\ &= (c + a + b\, r_{j+1})\, D_j = \\ &= 0 \bmod p_{j+1}\, D_j. \end{aligned}$$

Because $p_{j+1}\, D_j = D_{j-1}$, this proves our induction step and ends the sketch of the proof.

We repeat this algorithm for each (big) divisor Q of N and use the Chinese Remainder Algorithm to find x modulo N. After that another combination with x modulo M establishes the solution x (modulo P-1) of our original problem (equation (1).)

In the following theorem we assume that the factorization of P-1 is given and also that generators of the multiplicative groups $GF(Q)^*$ are given for each big prime factor Q of P-1. Furthermore we require that the D.H. oracle gives the right answer for each question.

Theorem : An algorithm to solve the D.L. problem for a prime P for which the totient of P-1 is smooth with respect to P and q requires order $log^2(P)$ calls to the D.H.oracle and order $log^2(P)\sqrt{q(log(P))}$ multiplications. This algorithm requires order $log^2(P)\sqrt{q(log(P))}$ of bits of memory space.

The first two assumptions of the three assumptions in the paragraph just preceding our theorem can be dropped and we still have (probabilistic) polynomial-time equivalence. This is also the case if we weaken the last of the three assumptions (because of the algebraic structure we can construct "majority answers").

Conclusion

At this moment no efficient algorithm to compute the Discrete Logarithm is known for the case where the totient of P contains a prime factor bigger than P^e, where e is some fixed positive real number. So we could safely use a Diffie-Hellman key exchange for the subcase where the biggest prime factor of the double-totient is smaller than some fixed polynomial in the logarithm of P

An interesting property of our algorithm to solve the D.L. problem using a D.H. oracle is that it mimics the algorithm to solve the D.L. problem in polynomial time without a D.H. oracle for the case where $P-1$ itself is smooth. It is conceivable that a polynomial-time algorithm to solve the D.L. problem for the case where the totient of $P-1$ (the *double totient case*) is smooth may enable us to design an algorithm to solve the D.L. problem in the triple totient case in polynomial time and a polynomial number of calls to a D.H. problem. If this goes on for higher and higher totients we at last have proved that either the general D.L. problem has a polynomial-time solution or there exists primes for which the D.H. problem is hard.

References

[DH] Diffie, W. and M.E. Hellman, New directions in cryptography, IEEE Trans. Inf. Theory, IT-22, pp. 644-654, Nov. 1976.

[Odl] Odlyzko, A.M., Discrete logarithms in finite fields and their cryptographic significance, Advances in Cryptology: Proc. Eurocrypt '84, Lecture Notes in Computer Science 209, Springer, Berlin etc., pp. 224-314, 1985

[Poh] Pohlig, S.C. and M.E. Hellman, An improved algorithm for computing logarithms over GF(p) and its cryptographic significance, IEEE Trans. Inf. Theory, IT-24, pp.106-110, Jan 1978.
[Pol] Pollard, J.M., Theorems on Factorization and Primality testing, Proc. Cambr. Philos. Soc., 76, pp 521-528, 1974
[Rie] Riesel, H., Primality Testing and Factorisation, Birkhauser, Boston, 1985

SECRET ERROR-CORRECTING CODES (SECC)

Tzonelih Hwang

National Cheng Kung University
Institute of Information Engineering
Tainan, Taiwan, R.O.C.

T.R.N. Rao

University of Southwestern Louisiana
The Center for Advanced Computer Studies
Lafayette, Louisiana 70504

Abstract. A *secret error-correcting coding* (SECC) scheme is one that provides both data secrecy and data reliability in one process to combat with problems in an insecure and unreliable channel. In an SECC scheme, only the authorized user havingsecretly held information can correct channel errors systematically. Two SECC schemes are proposed in this paper. The first is a block encryption using Preparata based nonlinear codes; the second one is based on block chaining technique. Along with each schemes can be secure.

Key words. Algebraic-Code Cryptosystem, Block Chaining, Ciphertext-Only Attach, Chosen-Plaintext Attach, Cryptanalysis, Cryptographic Parameter, Cryptography, Cryptology, Cryptosystems, Data Authenticity, Data Integrity, Data Reliability, Data Secrecy, Data Security, JEEC, Known-Plaintext Attach, SECC, Work Factor.

1. Introduction

The demand for **reliable, secure** and **efficient** digital data transmission and storage system has been accelerated by the emergence of large-scale and high speed communication networks. In 1948, Shannon demonstrated that errors induced by a noisy channel or storage medium can be reduced to any desirable level by proper encoding of the information [Shannon 48]. Since Shannon's work, a great deal of developments have contributed toward achieving *data reliability* and the use of coding for error control has become an integral part in the design of modern communication systems and digital computers.

Information transmitted through communication channel or stored in storage system is particularly vulnerable to eavesdropping and tampering. Although information can be protected by several ways (e.g., physical control -- data are stored in physically secure place; or computer system control -- the operating system provides access control mechanisms to check user's authentication), data encryption is the most cost-effective way to provide *data secrecy* [Diffie 76, Wood 81, Denning 82].

As computer communications are expanding to many applications, assurance of both data reliability and data secrecy becomes an important issue. To achieve this purpose, conventionally the first step is to encipher a plaintext (M) into a ciphertext and the second step is to encode the ciphertext into a codeword (C). To recover the plaintext (M), the receiver decodes the received word (C' = C + noise) first and then deciphers the ciphertext (see Figure 1.) Combining these two steps into one may obtain faster and more efficient implementations.

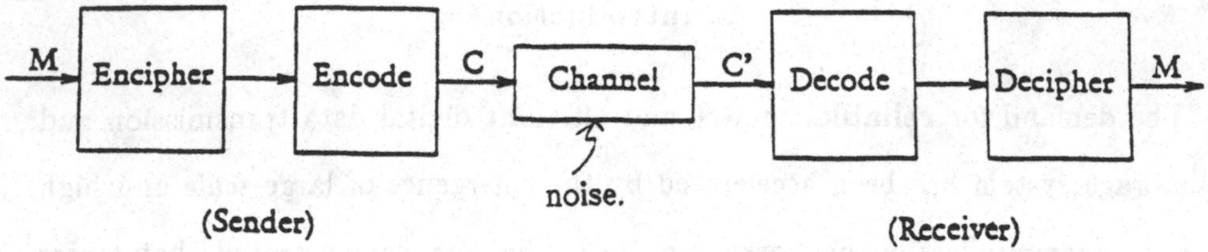

Fig. 1 Conventional approach for data reliability and data secrecy.

1.1 *Joint Encryption and Error Correction (JEEC) Scheme*

In his public-key cryptosystem, McEliece applied error-correcting capability of Goppa codes to provide data secrecy [McEliece 78]. His idea is to introduce a random error vector to each encoded plaintext before transmission. The Hamming weight (t') of the error vector is equal to the number (t) of errors the code can correct. Therefore, the receiver can remove the error vector and recover the plaintext by applying the decoding of the code.

If $t' < t$, then up to $t - t'$ errors may occur in the channel and these errors can be corrected by the receiver. Thus, the system provides both data secrecy and data reliability simultaneously. Since the system becomes less secure if t' is small but provides less error correcting capability if t' is large, there is a trade-off between data secrecy and data reliability. This approach, to obtain both data secrecy and data reliability while providing a trade-off between them, is called the *Joint Encryption and Error Correction* (JEEC) scheme [Rao 85].

Definition 1. The JEEC Scheme

A scheme that combines data encryption with data encoding into one process while providing a trade-off between data secrecy and data reliability is called a JEEC scheme.

1.2 Secret Error-Correcting Codes (SECC)

Conventional approach to obtain both data reliability and data secrecy has the disadvantage of inefficiency in the implementation because data encoding and data enciphering are implemented as two different steps. JEEC scheme combines both transformations into one process while providing only a trade-off between data reliability and data secrecy. Large distance and also large block length codes are required in JEEC to combat with problems in an insecure and unreliable channel. However, such codes have low information rates and a relatively high amount of decryption overhead. Therefore, they may not be cost-effective. This leads us to introduce the SECC scheme which may use simple algebraic codes (e.g., $d_{\min} \leq 6$) and also provides both data reliability and data security in one process. The SECC scheme can be defined as follows (see Figure 2).

Definition 2. The SECC Scheme

A scheme that combines data encryption with data encoding into one process to obtain both data secrecy and data reliability, while retaining the **full** error correction capability of the introduced code for possible channel errors, is called an SECC scheme. Also in an SECC scheme, the cryptanalyst is unable to correct channel errors *systematically.* By that we mean it is computationally infeasible for the cryptanalyst to correct channel errors without decoding keys.

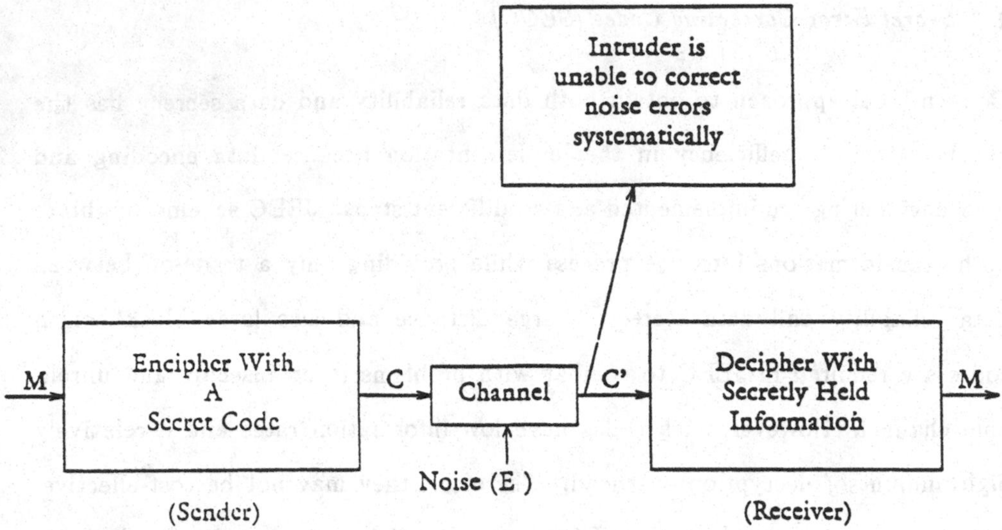

Fig. 2 The SECC scheme

Note that JEEC scheme preserves only partial error-correcting ability, whereas SECC scheme preserves full error-correcting ability of the code. Therefore SECC scheme can provide better error correcting capabilities than JEEC does under the use of the same algebraic codes. In a noisy channel, before the plaintext can be recovered the cryptanalyst has to correct channel errors (if any) first. If he cannot correct these errors, then he cannot also recover the plaintext. This is because any uncorrected error in the received ciphertext C' will only generate an M' totally different from the plaintext M (due to the so-called "Avalanche effect" in any good cryptosystem.) Therefore, the presence of noise errors would only increase the security of the system. However, the strength of an SECC system should not depend on the presence of channel errors because they are random in nature. On the other hand, in a conventional system since the coding scheme is public, the cryptanalyst is able to correct channel errors. Therefore, the presence of channel errors doesn't increase the security of the system. We will study an SECC scheme using nonlinear codes in Sec. 2 and an SECC scheme using block chaining technique in Sec. 3. Along with each scheme discussed, we investigate various cryptanalytic attacks to show how the scheme can be secure.

2. SECC Scheme Using Nonlinear Codes

Nonlinear codes with high degree of nonlinearity and whose decoding highly depends on the structure of the codeword, are particularly promising to construct SECC systems. In this section, we investigate Preparata-based nonlinear codes to construct SECC systems. Nonlinear codes, such as Vasil'yev nonlinear codes [Vasil'yev 62] which have only one nonlinear bit in each codeword, are not very useful in this application. We begin with a brief introduction to Preparata codes. Then, we review a code construction technique to construct nonlinear codes with large minimum distances from old codes. Finally, we propose an SECC scheme using nonlinear codes and investigate its security level.

2.1 Preparata Nonlinear Codes [Preparata 68]

Preparata has constructed a class of nonlinear double error-correcting $(2^m-1, 2^m-2m)$ codes, for each even $m \geq 4$, with some interesting features. They contain twice as many codewords as the double error-correcting BCH codes of the same length and they are optimal. Moreover, their decoding can be based on the calculation of syndrome-like quantities and thus the complexity is comparable to the corresponding BCH codes. The encoding and decoding are given here without proof. However, they can be found in [Preparata 68].

Assume that all polynomials discussed here belong to the algebra of polynomials modulo $(x^{2^{m-1}-1}+1)$ over GF(2). Let $\mathbf{B} = \{m(x)\}$ be a single-error-correcting BCH code generated by a primitive polynomial $g_1(x)$ of degree $m-1$ that has a primitive element α as its root. Let $\mathbf{C} = \{s(x)\}$ be the BCH code whose generator polynomial has roots α, α^3, and 1. The polynomial $u(x)$ will denote $(x^{2^{m-1}-1}+1)/(x+1)$. Consider a linear code $\mathbf{C}_n$ given by the vectors of the form

$$\mathbf{v} = [m(x),\ i,\ m(x)+(m(1)+i)u(x)+s(x)], \text{ where } i \in GF(2).$$

$\mathbf{C}_n$ can be shown to be a $(2^m-1, 2^m-3m+1)$ linear code of minimum distance 6.

Let $\phi(x) = (x^{2^{m-1}-1}+1)/g_1(x)$. Then, there exists an s $(0 \le s \le 2^{m-1}-2)$ such that $x^s\phi(x) = (x^s\phi(x))^2$. Let $f(x) = x^s\phi(x)$ and $q(x)$ be a monomial of degree less than or equal to $2^{m-1}-2$. $m(x)$, $s(x)$, i, and $q(x)$ are independently chosen. Then, the code $\mathbf{K}_n$ of the form

$$\mathbf{w} = [m(x)+q(x),\ i,\ m(x)+q(x)f(x)+(m(1)+i)u(x)+s(x)]$$

is an $(n, k) = (2^m-1, 2^m-2m)$ Preparata nonlinear code of distance 5. To encode a (2^m-2m)-bit information, the first $(2^{m-1}-m)$ bits are encoded into $m(x)$; the next $(2^{m-1}-2m)$ bits are encoded into $s(x)$; the following one bit is interpreted as i and the last $(m-1)$ bit are encoded into $q(x)$.

For decoding, assume that the vector $\mathbf{w}$ was sent and that the vector

$$\mathbf{r} = [r_0(x),\ r,\ r_1(x)] = \mathbf{w} + [e_0(x),\ e,\ e_1(x)]$$

is received. Given the following definitions

$$H_1 = [\alpha^{2^{m-1}-2}, \alpha^{2^{m-1}-3}, \cdots, \alpha, 1]$$

$$H_3 = [(\alpha^3)^{2^m-1-2}, (\alpha^3)^{2^m-1-3}, \ldots,(\alpha^3), 1]$$
$$U = [1, 1, \ldots,1, 1]$$

the syndrome $S = (\sigma_0, \sigma_1, \sigma, d)$ of **r** can be computed in the following manner.

$$\sigma_0 = r_0(x)H_1^T = a\alpha^p + e_0(\alpha)$$
$$\sigma_1 = r_1(x)H_1^T = a\alpha^p + e_1(\alpha)$$
$$\sigma = (r_0(x) + r_1(x))H_3^T = a\alpha^3+e_0(\alpha^3)+e_1(\alpha^3)$$
$$d = r + r_1(x)U^T = e+e_1(1)$$

where $q(x) = ax^p$ is the monomial in the codeword.

Let $\rho = \sigma + (\sigma_0 + \sigma_1)^3$. If $\rho = \sigma_j{}^3$ $(j = 0,1)$ and $d=0$, then **r** is a member of the nonlinear code. If the above condition is not true, then let $\mathbf{c} = [e_0(x), e, e_1(x)]$ be the "correction" vector that can be added to **r** to get the codeword **w**. The vector **c** can be found by the following rules, if j is taken modulo 2.

Rule 1: If $\rho=\sigma_j{}^3$ and $\rho\neq\sigma_{j+1}$ then $e_{j+1}(x)=x^h$ where $\alpha^h = \sigma_0 + \sigma_1$ and $e = d + e_1(1)$.

If $\rho\neq\sigma_j^3$, then we have the following rules.

Rule 2: If $d = 1$ then $e = 0$ and $e_j(x) = x^{k_j}$ where $\alpha^{k_j} = \sigma_{j+1} + \sqrt[3]{\sigma + \sigma_0\sigma_1(\sigma_0 + \sigma_1)}$.

Rule 3: If $d = 0$ and $\sigma_0+\sigma_1\neq0$ then set $e = 0$, $e_j(x) = 0$, and $e_{j+1}(x) = x^{k_1} + x^{k_2}$ where α^{k_1} and α^{k_2} are the solutions of

$$z^2+(\sigma_0 + \sigma_1)z + (\rho + \sigma_j^3)/(\sigma_0+\sigma_1) = 0$$

Rule 4: If $d = 0$ and $\sigma_0 + \sigma_1 = 0$ then **r** is at a distance ≥ 3 from any codeword.

If Preparata codes are to be used in the cryptosystems, then the only practical values of m are 6, 8 and 10. Therefore, the codes that will be considered here are (63, 52), (255, 240) and (1023, 1004) Preparata codes.

2.2 Construction of New Code From Old Codes [MacWilliams 77]

We have given a very brief introduction to Preparata nonlinear codes which can only correct double errors. In this section, we review a code construction technique to construct nonlinear codes that can correct more than two errors.

Let $\mathbf{K}_i$ be an (n_i, N_i, d_i) code where n_i is the code length, N_i is the number of codewords and d_i is the minimum distance between any two codewords of the code $(1 \leq i \leq 2)$. A new code $\mathbf{K}^*$ can be constructed from both $\mathbf{K}_1$ and $\mathbf{K}_2$, called the base codes of $\mathbf{K}^*$ here, as follows.

$$\mathbf{K}^* = \left\{(\mathbf{E}_1(M_1),\ \mathbf{E}_1(M_1)+\mathbf{E}_2(M_2))\right\},$$

where $\mathbf{E}_i$ is the encoding of $\mathbf{K}_i$ and $M=(M_1, M_2)$ is a plaintext block which is divided into two subblocks M_1 and M_2 over GF(2). Then, $\mathbf{K}^*$ is a $(2(max\{n_1, n_2\}), N_1 \cdot N_2, d=min\{2d_1, d_2\})$ code [MacWilliams 77]. If $n_1 \neq n_2$, then enough zeros can be added to the end of the shorter code. Note that $\mathbf{K}^*$ is a linear code if and only if both $\mathbf{K}_1$ and $\mathbf{K}_2$ are linear codes, otherwise it is a nonlinear code. This procedure can be iterated to construct nonlinear codes with large minimum distances. Here, we suggest the use of Preparata codes as base codes to construct new nonlinear codes, or we assume that either $\mathbf{K}_1$ or $\mathbf{K}_2$ or both are nonlinear codes constructed from Preparata codes. The decoding of the newly developed code is rather straightforward and is omitted here.

2.3 Encryption and Decryption of SECC Scheme Using Nonlinear Codes

The SECC scheme using nonlinear codes is a block encryption and error correction combined into one that also preserves the full error correction capability of the code for possible channel errors. Each block is enciphered and deciphered independently under this scheme.

Encryption

Let $\mathbf{E}_{K^*}$ denote the encoding of a nonlinear code that encodes a k-bit information into an n-bit codeword. Let Ψ be an invertible function that transforms a k-bit block into a k-bit block in either a linear or nonlinear manner. The matrix $\mathbf{P}$ is a random permutation matrix of size n. A k-bit plaintext block (M) is enciphered into an n-bit ciphertext (C) by the following equation

$$C = \mathbf{E}_{K^*}(\Psi(M)) \cdot \mathbf{P}. \tag{1}$$

The cryptographic parameters that are secretly held in the system are Ψ, $\mathbf{P}$ and $\mathbf{E}_{K^*}$. Since ciphertext-only attacks are much weaker attacks than known-plaintext or chosen-plaintext attacks, constructing a cryptosystem which can withstand ciphertext-only attacks is considered to be much easier than constructing a cryptosystem which can withstand either known-plaintext or chosen-plaintext attacks. In the proposed scheme, we assume that the function Ψ can withstand ciphertext-only attacks and may be broken by a known-plaintext attack. Hence the security of the scheme should depend on the strength of the combination of functions Ψ, $\mathbf{E}_k$ and $\mathbf{P}$ and not on the strength of either Ψ or $\mathbf{E}_k$ or $\mathbf{P}$ alone. This also illustrates the difference between SECC and the conventional approach to provide both data secrecy and data reliability.

Decryption

Let $\mathbf{D}_{K^*}$ be the decoding of the nonlinear code and E_i be a correctable error vector which occurs due to channel noise when the i-th ciphertext block is transmitted. The deciphering procedure is given below.

(1) Remove the permutation matrix $\mathbf{P}$ ($\mathbf{P}^T$ is the transpose matrix of $\mathbf{P}$).
$(C+E_i)\cdot\mathbf{P}^T = \mathbf{E}_{K^*}(\Psi(M))+E_i\cdot\mathbf{P}^T$.

(2) Decoding.
$\mathbf{D}_{K^*}(\mathbf{E}_{K^*}(\Psi(M))+E_i\mathbf{P}^T) = \Psi(M)$.

(3) Recover the plaintext M.
$M = \Psi^{-1}(\Psi(M))$.

Notice that the error-correcting capability of the code is fully preserved to correct channel errors (E_i's) as a property required in an SECC scheme. Since the decoding of Preparata codes highly depends on the structure of the codeword, cryptanalyst cannot correct channel errors without knowing the matrix $\mathbf{P}$. This is another property required in an SECC scheme.

2.4 Security of SECC Scheme Using Preparata-Based Nonlinear Codes

We have discussed both the enciphering and deciphering of the SECC scheme using Preparata-based nonlinear codes. What remains to be studied is the security of the scheme. For simplicity we investigate the security of the SECC scheme using Preparata codes mainly. The security of the SECC scheme using extended nonlinear codes follows directly. Let $\mathbf{E}_p$ and $\mathbf{D}_p$ represent the encoding and decoding of a Preparata code respectively.

As we mentioned earlier that the function Ψ can either be a linear or a nonlinear transformation. If the system using a linear function Ψ could provide an acceptable level of security ($\approx 2^{60}$ operations), then the system could provide even a better security if Ψ is a nonlinear function.

First, we consider the case that both Ψ and **P** are removed from the original scheme. In the following lemma, we shall show that the simplified scheme can be broken by a known-plaintext attack. For this discussion, we assume that no error occurs in the channel.

Lemma 1.

The encryption scheme

$C = \mathbf{E}_p(M)$

can be broken by a known-plaintext attack in $O(n^2)$ *bit operations.*

<Proof>

The generator polynomial $g_1(x)$ can be derived from a pair of plaintext and ciphertext as follows. The cryptanalyst obtains $q(x)$ from the last m-1 bits of the plaintext. Hence, $m(x)$ can be computed from the first part (2^{m-1}-1 bits) of the ciphertext. Subsequently, he can derive $g_1(x)$ from $m(x)$ and the first (2^{m-1}–m) bits of the plaintext under a known-plaintext attack. Obviously, this requires only $O(n^2)$ bit operations.

Q.E.D

Let $N(g_1)$ denote the number of primitive polynomials ($g_1(x)$'s) in a class of Preparata codes of a given code length (n). Then, $N(g_1)$ can be computed by the formula $N(g_1)=[\frac{\theta(2^{m-1}-1)}{m-1}]$, where θ is the Euler totient function and m-1 is the degree of the primitive polynomial $g_1(x)$. Therefore, we have

$N(g_1)=2$ if $n=15$,

$N(g_1)=6$ if $n=63$,

$N(g_1)=18$ if $n=255$,

$N(g_1)=48$ if $n=1023$.

The number of choices of the primitive polynomials $g_1(x)$'s in Preparata codes of practical lengths is too small for the simplified scheme to be secure.

We may introduce a secret, linear function Ψ to scramble the plaintext before encoding. However, the modified system is still insecure under a chosen-plaintext attack as can be shown in the following lemma.

Lemma 2.

The encryption scheme

$$C = \mathbf{E}_p(\Psi(M))$$

can be broken by a chosen-plaintext attack in $O(n^2)$ *bit operations.*

<Proof>

Let M_1, M_2 and M_3 be the three plaintext blocks to be enciphered in the system where $M_1=M_2+M_3$. Let C_1, C_2 and C_3 be their ciphertexts respectively. Then,

$$C_1+C_2+C_3 = (q_1(x)+q_2(x)+q_3(x),\ 0,\ q_1(x)f(x)+q_2(x)f(x)+q_3(x)f(x)),$$

where $q_i(x)$ is a monomial whose power j is taken from the decimal equivalent of the last $m-1$ bits of the scrambled plaintext $\Psi(M_i)$. Let $q_i(x)=0$ if $j=2^{m-1}-1$. Consequently, $f(x)$ can be derived from the first $2^{m-1}-1$ bits and the last $2^{m-1}-1$ bits of the ciphertext in $O(n^2)$ bit operations. Once $f(x)$ is obtained, $g_1(x)$ can be derived easily (see Sec. 2.1). Therefore, the security of the system totally depends on the strength of the function Ψ which, unfortunately, can be broken by a known-plaintext attacks as mentioned previously.

Q.E.D.

The simplified scheme in Lemma 2 is insecure because the structure of the code is revealed. Therefore, the cryptanalyst can remove the linear component of the codewords and then break the system. In order to avoid this weakness, a permutation matrix may be introduced to scramble the structure of the code. However, the following lemma shows that the modified scheme can be broken by a chosen-plaintext attack if the function Ψ is not introduced to the scheme.

Lemma 3.

The encryption scheme

$$C = \mathbf{E}_p(M) \cdot \mathbf{P}$$

can be broken by a chosen-plaintext attack in $O(n^3)$ ***operations.***

<Proof>

Let $\mathbf{M} = \begin{bmatrix} 0 & \cdots & 0 & 0\,0 & \cdots & 0\,1 \\ 0 & \cdots & 0 & 0\,0 & \cdots & 1\,0 \\ 0 & \cdots & 0 & 0\,0 & \cdots & 1\,1 \\ & & & \cdots & & \\ 0 & \cdots & 0 & 1\,1 & \cdots & 1\,1 \end{bmatrix}_{(2^{m-1}-1)\times(2^m-2m)}$ be a matrix of plaintexts.

Let M_i $(1 \le i \le 2^{m-1}-1)$, the i-th row in $\mathbf{M}$, be the i-th plaintext to be enciphered in the system. Let $\mathbf{C}_M$ denote the matrix of ciphertexts of $\mathbf{M}$. Then, the following relation holds where $\mathbf{T}_M$ is the matrix of codewords of $\mathbf{M}$ encoded by the Preparata code.

$$\mathbf{C}_M = \mathbf{T}_M \cdot \mathbf{P} = \begin{bmatrix} 0\,0 & \cdots & 0\,1\,0 & f(x) \\ 0\,0 & \cdots & 1\,0\,0 & x^1 f(x) \\ & \cdots & & \cdots \\ 1\,0 & \cdots & 0\,0\,0 & x^{2^{m-1}-2} f(x) \end{bmatrix}_{(2^{m-1}-1)\times(2^m-1)} \cdot \mathbf{P}.$$

Notice that the columns in the matrix $\mathbf{T}_M$ are all distinct. Therefore, by trying all possible $f(x)$'s (i.e., $N(g_1)$ of them), the cryptanalyst can obtain both the function $f(x)$ and the permutation matrix $\mathbf{P}$ used in the system. The work factor of this attack is dominated by the overhead of enciphering $2^{m-1}-1$ chosen plaintexts, i.e. $O(n^3)$.

Q.E.D.

From these lemmas, we see if both the function Ψ and the permutation matrix $\mathbf{P}$ are introduced to the system as a portion of the key, then these attacks cannot break the resulting scheme.

Since there are only a small number of primitive polynominals for a given code length n, the cryptanalyst may try to guess the generator polynomial $g_1(x)$ used in the system. However, the work factor to check the correctness of each guess involves a very large amount of overhead to figure out both functions Ψ and $\mathbf{P}$. That is a hopeless task.

The SECC scheme using Preparata codes is a block encryption and error correction combined into one that also preserves the full error correction capability of the code for possible channel errors. This is a major distinction from McEliece's scheme, which has no error correcting capability or has only a partial error correcting capability when used as JEEC. While somewhat simpler SECC schemes given by Lemmas 1-3 are shown to be breakable under known-plaintext or chosen-plaintext attacks, the proposed scheme with both functions of Ψ and $\mathbf{P}$ appears to be secure. It would be a challenge indeed to find cryptanalytic attacks to break this scheme.

These attacks are performed under the assumption that there is no error occurs in the channel. If there exist channel errors, then it will be much more difficult to perform these attacks against the SECC system. Therefore, the presence of channel errors introduces additional level of data security to the system as required in an SECC scheme.

There are several types of cryptanalytic attack against algebraic-code cryptosystems discussed in [Rao 87, Struik 87]. These attacks are performed based on the linearity of the system. They will not be applicable for this nonlinear coding scheme.

3. SECC Scheme Using Block Chaining Technique

In this section, we proposed an SECC scheme based on block chaining technique. In this scheme because each ciphertext is a function of all previous plaintexts, decoding error of one ciphertext will propagate all the way through the last block. This "error propagation" property can be applied to detected any illegal modification to the ciphertext thus provides data integrity [Meyer 82]. Therefore, this scheme can provide not only data reliability and data secrecy but also data integrity in one enciphering. But any decoding error requires the retransmission of all blocks chained together.

3.1 Encryption and Decryption of the Proposed Scheme

Rao and Nam have suggested a private-key algebraic code cryptosystem (Rao-Nam scheme) using *simple* linear codes [Rao 87]. By *simple* codes we mean small distance codes, i.e. $d_{min} \leq 6$. In this scheme, a k-bit plaintext block M_i is enciphered to an n-bit ciphertext block C_i by the following equation.

$$C_i = (M_i \mathbf{SG} + Z_i)\mathbf{P},$$

where

$\mathbf{S}$: an arbitrary $(k \times k)$ nonsingular matrix,

$\mathbf{G}$: an (n, k) code generator matrix,

$\mathbf{P}$: a random $(n \times n)$ permutation matrix,

Z_i : an error vector of length n randomly selected from a predetermined syndrome-error table.

$\mathbf{S}$, $\mathbf{G}$ and $\mathbf{P}$ are private keys.

Struik and van Tilburg proposed chosen-plaintext attacks (ST-type attacks) on Rao-Nam scheme. Their attacks are based on estimating the rows of the encipher matrix $\mathbf{G}' = \mathbf{SGP}$ by constructing unit vectors from the chosen plaintext or by solving a set of linear equations [Struik 87]. They also proposed a modified scheme to withstand these attacks. In their modified scheme, the matrix **S** in Rao-Nam scheme is replaced by an invertible, nonlinear function f such that $M_i = f^{-1}(f(M_i, Z_i), Z_i)$. In the modified scheme, M_i is enciphered into C_i by the following equation.

$$C_i = f(M_i, Z_i)\mathbf{GP} + Z_i .$$

These schemes are proposed mainly for providing data secrecy. They are not designed to realize JEEC or SECC and therefore do not provide data reliability. However, by modifying the way the error vectors (Z_i's) is introduced, an SECC system can be constructed. Block chaining technique will be applied to facilitate this construction. The proposed system is described below and is shown in Figure 3.

Encryption.

The cryptographic parameters (that are secretly held) for this scheme are

f : an *invertible*, nonlinear function which transforms a k-bit block to a k-bit block,

G : an (n, k) code generator matrix,

g : a k-bit to n-bit block expanding function.

The following symbols are used for this scheme.

X_i : the i-th output of f, $(i=1,2,\ldots)$.

Z_i : the i-th error vector, $Z_{i+1} = g(X_i)$. Z_1 is a correctable error randomly generated by the system.

E_i : error vector due to channel noise occurs when the i-th block is transmitted.

$C_i' = C_i + E_i$ is the i-th block received at the receiver end.

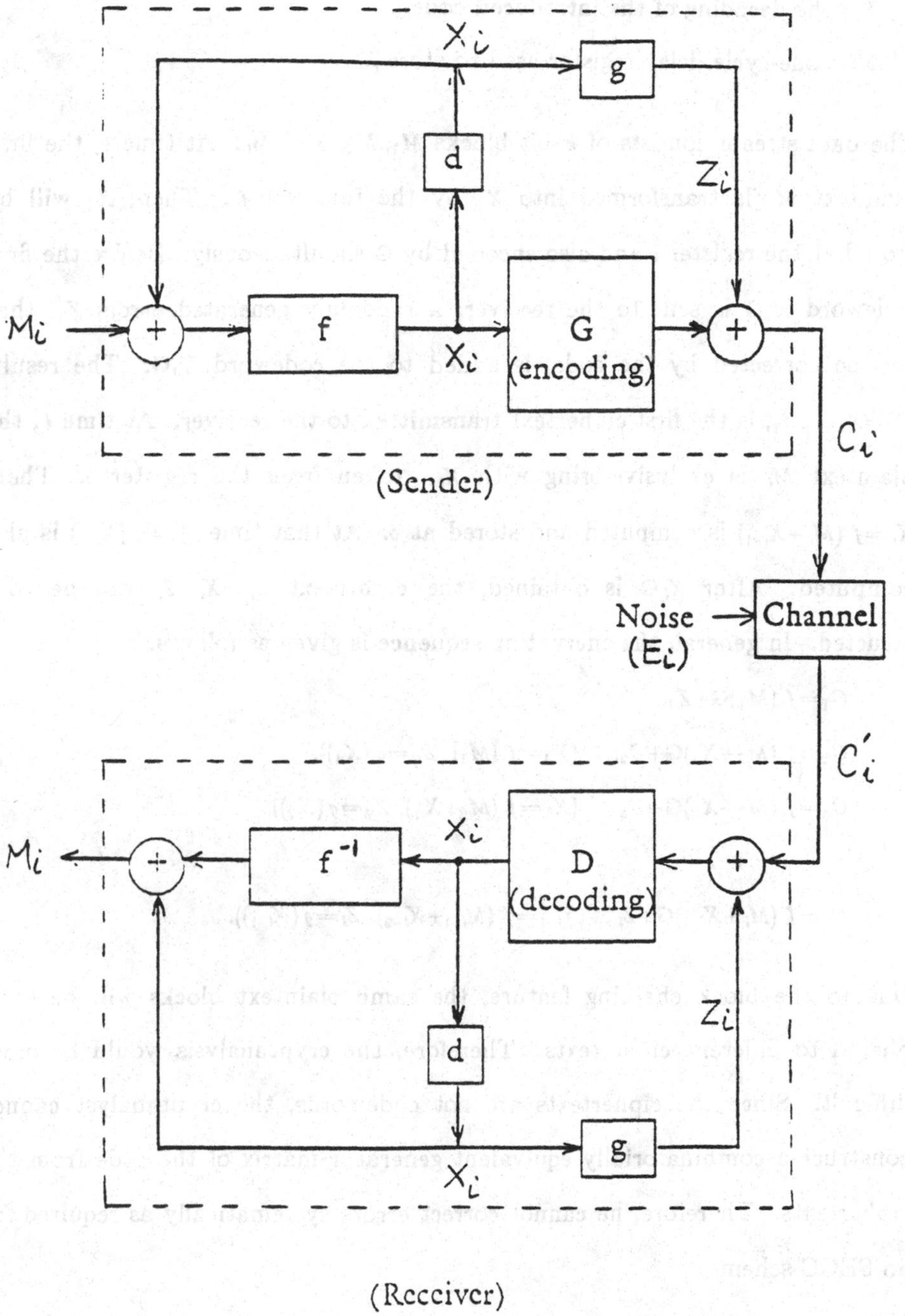

Fig.3 SECC Scheme Using Block Chaining Technique

D : the decoding of the introduced code.

δ : a one-cycle delay register used to store X_i.

The data stream consists of k-bit blocks $M_1, M_2, \cdots M_l$. At time 1, the first plaintext M_1 is transformed into X_1 by the function f. Then, X_1 will be stored at the register δ and also encoded by **G** simultaneously. Before the first codeword $X_1\mathbf{G}$ is sent to the receiver, a randomly generated errors Z_1, that can be corrected by the code, is added to the codeword $X_1\mathbf{G}$. The result, $C_1=X_1\mathbf{G}+Z_1$, is the first ciphertext transmitted to the receiver. At time i, the plaintext M_i is exclusive-oring with X_{i-1} taken from the register δ. Then, $X_i=f(M_i+X_{i-1})$ is computed and stored at δ. At that time, $Z_i=g(X_{i-1})$ is also computed. After $X_i\mathbf{G}$ is obtained, the ciphertext $C_i=X_i+Z_i$ can be constructed. In general, the encryption sequence is given as follows.

$$C_1=f(M_1)\mathbf{G}+Z_1,$$

$$C_2=f(M_2+X_1)\mathbf{G}+Z_2, \quad (X_1=f(M_1),\ Z_2=g(X_1)),$$

$$C_3=f(M_3+X_2)\mathbf{G}+Z_3, \quad (X_2=f(M_2+X_1),\ Z_3=g(X_2)),$$

$$\cdots$$

$$C_l=f(M_l+X_{l-1})\mathbf{G}+Z_l, \quad (X_{l-1}=f(M_{l-1}+X_{l-2}),\ Z_l=g(X_{l-1})).$$

Due to the block chaining feature, the same plaintext blocks will be enciphered to different ciphertexts. Therefore, the cryptanalysis would be more difficult. Since the ciphertexts are not codewords, the cryptanalyst cannot construct a combinatorially equivalent generator matrix of the code from the ciphertexts. Therefore, he cannot correct errors systematically as required for an SECC scheme.

Decryption.

Here, we assume that the receiver could synchronize with the sender on the sequence of vectors X_i and Z_i added to both plaintext and the corresponding codeword respectively. Furthermore, we assume that the decoding is correctly carried out. The decryption sequence is given below.

$$D(C_1')=X_1=f(M_1), \quad f^{-1}(X_1)=M_1,$$

$$D(C_2'+g(X_1))=X_2=f(M_2+X_1), \quad f^{-1}(X_2)+X_1=M_2,$$

$$\cdots$$

$$D(C_i'+g(X_{i-1}))=X_i=f(M_i+X_i), \quad f^{-1}(X_i)+X_{i-1}=M_i.$$

Because the errors introduced deliberately at the sender end can be removed at the receiver end by this synchronization, the error-correcting capability of the code is fully preserved for possible channel errors. By this chaining feature, errors due to intruder's tampering which cannot be corrected by the code will propagate all the way through the last block. However, this may serve as a checksum to detect illegal modification to the ciphertext by the intruder [Denning 82]. Hence the proposed scheme provides not only data reliability and data secrecy but also data integrity (data authenticity) [Meyer 82]. That is, the SECC scheme using block chaining technique can provide two levels of error control. The first level is the correction of channel errors; the second level is the detection of uncorrectable modification to the ciphertext by the intruder. But, the presence of such errors requires the retransmission (or reenciphering) of all blocks chained together.

3.2 Security of the Proposed SECC Scheme

If we define errors in one block of binary information as the bits different from the original block sent by the sender, then both channel errors and intruder's tampering are regarded as errors. However, the manner of the errors introduced by channel noise is different from that of intruder's tampering. The errors introduced by intruder's tampering are primarily multiple errors. In binary symmetric channels, the probability of multiple random errors is very small [Lin 83]. Algebraic codes are designed for correcting random errors due to channel noise. They are not designed to correct multiple errors due to intruder's tampering. In the presence of multiple errors, erroneous decoding might occur. Consequently, to combat with problems in an *insecure and unreliable* channel, a scheme which is capable of hiding information, correcting channel errors and also detecting any illegal modification to the ciphertext is desirable. The SECC scheme based on block chaining technique could provides these characteristics and hence is very useful in an insecure and unreliable environment.

The SECC scheme withstands ST-type chosen-plaintext attacks because of the plaintext is transformed by a nonlinear function f before encoding and also because of the chaining feature. This prevents the cryptanalyst from constructing unit vectors from the chosen plaintexts to derive **G**.

Simplified versions of the SECC Scheme

To show how this scheme provides a high level of security, we may consider two simplified versions of the original one. First, if X_i, the output of f, is fed forward to the function g only (i.e., X_i is not fed back to f), then the encryption sequence is given as follows.

$$C_1 = f(M_1)\mathbf{G} + Z_1,$$

$$C_2 = f(M_2)\mathbf{G} + g(f(M_1)),$$

$$\dots$$

$$C_l = f(M_l)\mathbf{G} + g(f(M_{l-1})).$$

A chosen-plaintext attack can break **G** if g is a public linear function that has a left inverse. For example, let $M_i = M_{i+1}$ and $M_{i+2} = M_{i+3}$. Then

$$C_{i+1} + C_{i+2} = (f(M_{i+1}) + f(M_{i+2}))\mathbf{G}, \text{ and}$$
$$C_{i+2} + C_{i+3} = g(f(M_{i+1})) + g(f(M_{i+2})).$$

Thus $(f(M_{i+1}) + f(M_{i+2})) = g^{-1}(C_{i+2} + C_{i+3})$. If the cryptanalyst could obtain k such distinct pairs, then $\mathbf{G}$ can be derived. However, if g is a secret nonlinear function or g has no left inverse then this attack does not work.

On the other hand, if X_i is fed back to f only (i.e., X_i is not fed forward to g), then the encryption sequence is given as follows.

$$C_1 = f(M_1)\mathbf{G},$$
$$C_2 = f(M_2 + X_1)\mathbf{G}, \quad (X_1 = f(M_1),$$
$$\cdots$$
$$C_l = f(M_l + X_{l-1})\mathbf{G}, \quad (X_{l-1} = f(M_{l-1} + X_{l-2})).$$

To attack the scheme, the cryptanalyst may find the equivalent ciphertexts. For example, if $C_i = C_j$, then $f(M_i + X_{i-1}) = f(M_j + X_{j-1})$ i.e., $X_i = X_j$. If f is a linear transformation, then $C_{i+1} + C_{j+1} = f(M_{i+1})\mathbf{G} + f(M_{j+1})\mathbf{G}$. Thus, $f \cdot \mathbf{G}$ can be figured out by a known plaintext attack.

If f is a nonlinear transformation, then this line of attack may not work. However, the cryptanalyst could collect k linearly independent codewords to construct a generator matrix ($\hat{\mathbf{G}}$) which is combinatorially equivalent to $\mathbf{G}$. Let $\hat{\mathbf{G}} = \mathbf{S}^{-1}\mathbf{G}$ for any nonsingular matrix $\mathbf{S}$ of rank k. Since the number of nonsingular matrices of rank k is about $0.3 \times 2^{k^2}$, it is computationally infeasible to estimate the matrix $\mathbf{G}$ used if k is large enough. Thus, the scheme appears secure. But, the cryptanalyst may be able to correct channel errors if t is small (e.g. $t \leq 3$). Thus, it is important to feed X_i forward to g in order to construct an SECC system. As a result, the SECC scheme can be very secure if f is an invertible, nonlinear function and g is a nonlinear, one-way function. It will be a challenge to design other lines of attack to break this scheme.

4. Conclusion

For the very first time, we introduce the concept of secret error-correcting codes in this paper. An SECC scheme combines data encoding with data encryption into one process and enables the system to correct channel errors as well as conceal information from unauthorized user simultaneously. The main purpose of this research is to construct SECC schemes to facilitate a *reliable, secure* and *efficient* digital transmission.

We have proposed two SECC schemes to realize this new concept. The first one is a block encryption using Preparata-based nonlinear codes. In this scheme, each block can be enciphered and deciphered independently.

The other SECC scheme is based on block chaining technique. This scheme provides not only data secrecy and data reliability but also data integrity due to the chaining feature. However, the decryption of each ciphertext cannot be carried out independently. The decoding error in one block requires retransmission of all blocks chained together.

Although we have investigate various cryptanalytical attacks against these schemes, they are still not fully proven systems. Several problems relating to the proposed schemes, such as the key generation and key management problems, still remain unsolved. Furthermore, there may exist other good techniques to realize the SECC concept. These indeed require further research.

References

[Denning 82] Dorothy E. Denning, *Cryptography and Data Security* Addison Wesley, 1982.

[Diffie 76] Whitfield Diffie and Martin E. Hellman, "New Directions in Cryptography," *IEEE Trans. on Information Theory,* Vol. IT-22, No. 6, pp. 644-654, Nov. 1976

[Lin 83] Shu Lin and Daniel J. Costello, Jr., *Error Control Coding: Fundamentals and Applications,* Prentice-Hall, 1983.

[MacWilliams 77] F.J. MacWilliams and J.J.A. Sloane, *The Theory of Error-Correcting Codes,* North-Holland, Amsterdam, 1977.

[McEliece 78] R.J. McEliece, "A Public-Key Cryptosystem Based on Algebraic Coding Theory", *DSN Progress Report,* Jet Propulsion Laboratory, CA., Jan. & Feb. 1978, pp. 42-44.

[Meyer 82] Meyer C.H. and Matyas S.M., *Cryptography: A New Dimension in Computer Data Security,* John Wiley & Sons, Inc., 1982.

[Peterson 72] W. Wesley Peterson and E.J. Weldon, Jr., *Error-Correcting Codes,* Second Edition, The MIT Press, 1972.

[Preparata 68] F.P. Preparata, "A Class of Optimum Nonlinear Double-Error-Correcting Codes," Inform. and Control, 13, pp. 378-400, 1968.

[Rao 85] T.R.N. Rao, "Cryptosystems Using Algebraic Codes," *IEEE International Symp. on Info. Theory,,* Brighton, England, June, 1985.

[Rao 87] T.R.N. Rao and K.H. Nam "A Private-Key Algebraic-Coded Cryptosystem", Advances in Cryptology CRYPTO '86, editor A.M. Odlyzko, New York, Springler Verlag, pp. 35-48, 1987.

[Shannon 48] C.E. Shannon, "A Mathematical Theory of Communication," *Bell Syst. Tech. J.,* 27, pp. 379-423 (Part I), 623-656 (Part II), July 1948.

[Struik 87] R. Struik and van Tilburg J., "The Rao-Nam Scheme is Insecure Against a Chosen-plaintext Attack," Advances in Cryptology CRYPTO '87, pp. 445-457, 1987.

[Wood 81] Charles C. Wood, "Future Application of Cryptography," *Proc. of the 1981 Symposium on Security and Privacy,* pp. 70-74, Apr. 1981.

[Vasil'yev 62] Vasil'yev, Jr. L. "Nongroup Close-Packed Codes", *Probl. Cybernet.* (USSR) 8, pp. 337-339, 1962.

The Detection of Cheaters in Threshold Schemes

E. F. Brickell
Sandia National Laboratories
Albuquerque, NM 87185

D. R. Stinson
Department of Computer Science
University of Manitoba
Winnipeg, Manitoba R3T 2N2 Canada

Abstract

Informally, a (t, w)-*threshold scheme* is a way of distributing partial information (*shadows*) to w participants, so that any t of them can easily calculate a *key* (or *secret*), but no subset of fewer than t participants can determine the key. In this paper, we present an unconditionally secure threshold scheme in which any cheating participant can be detected and identified with high probability by any honest participant, even if the cheater is in coalition with other participants. We also give a construction that will detect with high probability a dealer who distributes inconsistent shadows (shares) to the honest participants. Our scheme is not perfect; a set of t – 1 participants can rule out at most $1 + \binom{w-t+1}{t-1}$ possible keys, given the information they have. In our scheme, the key will be an element of GF(q) for some prime power q. Hence, q can be chosen large enough so that the amount of information obtained by any t – 1 participants is negligible.

1. Introduction

Informally, a (t, w)-*threshold scheme* is a way of distributing partial information (*shadows*) to w participants, so that any t of them can easily calculate a *key* (or *secret*), but no subset of fewer than t participants can determine the key. Threshold schemes are also known as *secret sharing schemes*. A *perfect* threshold scheme is one in which no subset of fewer than t participants can determine any partial information regarding the key.

Threshold schemes were first described independently by Blakley [2] and Shamir [7] in 1979. Since then, many constructions have been given for threshold schemes. More recently, various researchers have considered the problem of guarding against the presence of cheaters in threshold schemes. It is conceivable that any subset of the participants may attempt to *cheat*, that is, to deceive any of the other participants by lying about the shadows they possess. There is also the possibility that the person distributing the shadows (the *dealer*) may attempt to cheat. The dealer might distribute an inconsistent set of shadows, so that the key cannot be determined correctly, or so that different subsets of t participants would calculate different keys from the shadows they possess. If this is done without the knowledge or co-operation of any of the participants, we refer to this form of cheating as *disruption*. However, if this cheating is done in co-operation with one or more of the participants, we call it *collusion*.

A threshold scheme is said to be *unconditionally secure* (against cheating) if the probability of cheating successfully is independent of the computational resources available to the cheaters. Under the assumption that the dealer is honest, several constructions have been given for threshold schemes which are unconditionally secure against cheating [3, 6, 8, 9]. We now briefly summarize the properties of these threshold schemes.

As far as the authors are aware, the first researchers to address the problem of cheaters in threshold schemes were McEliece and Sarwate in [6]. They use an error-correcting code to construct a threshold scheme in which any group of $t + 2e$ participants which includes at most e cheaters can correctly calculate the key.

Tompa and Woll [9] proceed as follows. The dealer specifies a subset K_0 of the set of possible keys K. A key will be accepted as authentic only if it is an element of K_0. If a set of t participants calculate the key to be an element of $K \setminus K_0$, then they realize that one of them is cheating. The probability of successful cheating is at most $1 - |K_0| / |K|$, even if $t - 1$ participants conspire to to cheat another participant. However, even though participants can detect when cheating has occurred, they cannot determine who is cheating.

The construction of Simmons [8] is more general, in that it can be applied to most existing threshold schemes. This method detects cheating only if at least $t + 1$ participants exchange their shadows. Define a set S of at least t shadows to be *consistent* if all t-subsets of S determine the same key. Then, a key is accepted as authentic only if there is a consistent subset of at least $t + 1$ shadows which determine it. If $t + e$ participants exchange shadows and there are at most $e - 1$

cheaters among them, then they possess a consistent subset of at least $t + 1$ shadows. Unfortunately, the only known method to determine the existence of a consistent set of $t + 1$ shadows is an exhaustive search.

Finally, Chaum [3] has suggested the following approach. For *each* bit b to be communicated to the ith participant, the dealer chooses $2w - 2$ large random numbers r_{j0} and r_{j1} $(1 \le j \le w, j \ne i)$. For each j, r_{j0} and r_{j1} are given to participant j. The dealer gives to the ith participant the bit b and all r_{jb} $(1 \le j \le w, j \ne i)$. Then, r_{jb} is used to authenticate the bit b (as 0 or 1, respectively) to participant j. This procedure is used for every bit communicated to each participant.

In the schemes discussed above, it is assumed that the dealer is honest. Also, the Tompa and Woll scheme and the Simmons construction require that the participants be able to simultaneously release their shadows, in order to ensure that no participant is able to obtain partial information about the shadows of the other participants before releasing his own shadow. Simultaneous release of shadows is *not* required in the Chaum scheme.

Threshold schemes which provide protection against dealer disruption have been presented by Chor, Goldwasser, Micali and Awerbuch in [5] and by Benaloh in [1]. These schemes provide *computational security* only, since they rely on computational assumptions regarding certain encryption schemes. Chaum, Crepeau and Damgard [4] use threshold schemes as a building block in unconditionally secure multiparty protocols. They tolerate both dealer disruption and collusion, but require that less than one third of the participants cheat. Under these assumptions, they describe a scheme that is unconditionally secure and which allows the key to be determined correctly by the honest participants.

The threshold scheme we present provides unconditional security and gives the honest participants the ability to *identify* cheaters, assuming the dealer is honest. Also, we do *not* require that the participants simultaneously release their shadows. The properties of our construction can be summarized as follows.

1) The key is an element of $GF(q)$, and each shadow is a t-dimensional vector over $GF(q)$ (q will be some large prime power).

2) Any participant who attempts to cheat will be identified by any honest participant with probability $1 - 1 / (q - 1)$.

3) Even if there is only one honest participant and the remaining $w-1$ participants form a coalition in order to deceive him, their probability of cheating successfully is only $(w-t+1)/(q-1)$.

4) The scheme is nearly perfect. A group of $t-1$ participants can eliminate at most $1+\binom{w-t+1}{t-1}$ possible keys, and can obtain no other partial information about the key. If q is large, this will cause no difficulty in practice.

5) The scheme can also protect against dealer disruption, by using a "cut-and-choose" technique similar to that of [4].

2. The construction

Our construction is a modification of Blakley's threshold scheme [2], which we now review briefly. Suppose the participants are denoted A_i, $1 \le i \le w$, and the dealer is denoted by D. Let V be a t-dimensional vector space over GF(q), where q is some large prime power. First, D fixes a line ℓ in V. This line is made known to all the participants. There are q possible keys, namely the q points on ℓ. If D wants to distribute shadows corresponding to a key p, he first constructs a random $(t-1)$-dimensional subspace H that meets ℓ in a point. Then, he constructs the hyperplane $H_p = H + p$. (Note that $H_p \cap \ell = p$.) Finally, he picks w random points on H_p, denoted s_i $(1 \le i \le w)$, such that the points in the set $\{p\} \cup \{s_i\colon 1 \le i \le w\}$ are in general position (that is, no j of them lie on a flat of dimension $j-2$, if $j \le t$). The point s_i is the shadow that D gives to A_i.

Any t participants can uniquely determine the hyperplane H_p, and then obtain p by calculating $H_p \cap \ell = p$. However, a subset of t' (< t) participants know only that H_p contains the flat F of dimension $t'-1$ generated by the shadows they possess. For any p' on ℓ, there is a hyperplane H_p containing F and p'. Hence, they have no information as to the point p. Thus, the scheme is indeed a (t, w)-threshold scheme.

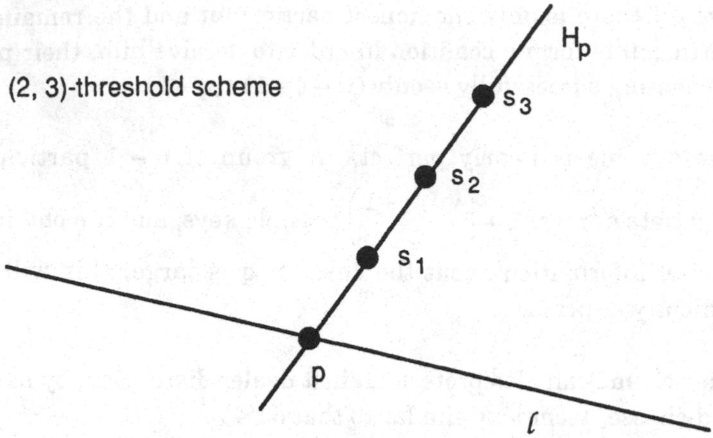

In order to guard against cheating, we modify the threshold scheme. D will distribute extra information to the participants, along with the shadows. For ease of exposition, we first discuss the case $t = 2$. In this case, H is a 1-dimensional subspace and the hyperplane H_p is a line. D constructs w random 1-dimensional subspaces, denoted H_i ($1 \le i \le w$), each of which is distinct from H. We do *not* require that the subspaces H_i ($1 \le i \le w$) be distinct. D gives to each A_j the $w - 1$ parallel lines $H_{ji} = H_j + s_i$, $1 \le i \le w$, $i \ne j$. These lines H_{ji} are called *supershadows*. Note that H_{ji} is given only to A_j.

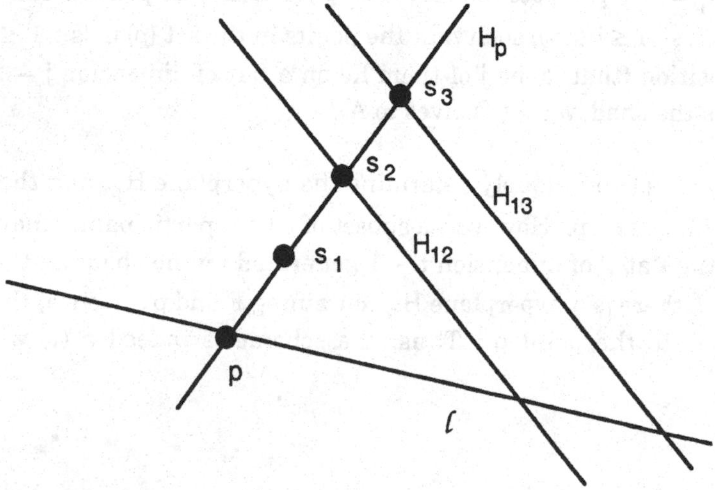

We must first show that knowledge of the supershadows does not enable any one participant to determine the key. Let's consider A_1. He knows that $s_2 \in H_{12}$. This *does* give him some partial information, namely that the key $p \neq H_{12} \cap \ell$. For, if $p = H_{12} \cap \ell$, then $p = s_2$, which is not allowed. Similarly, A_1 knows that $p \neq H_{1i} \cap \ell$, for any i, $2 \leq i \leq w$. As well, $p \neq H_{11} \cap \ell$, where H_{11} denotes the line through s_1 parallel to the H_{1i}'s. For, this would require that $H_p = H_{11}$, but $s_2 \notin H_{11}$. Thus, A_1 has ruled out w possibilities for p. However, the key, p, could be any point p_0 on ℓ other than these w points, since the line p_0s_1 will intersect each H_{1i} in a point. Each of these $q - w$ possibilities for p is equally likely to occur.

Hence, each participant can rule out w possibilities for the key, and knows that the key is equally likely to be one of the $q - w$ remaining possibilities. Thus, the scheme is no longer perfect. However, if q is large relative to w, this will cause no difficulty in practice. (A variation of this scheme, described in Section 4, allows only one possible value to be ruled out for the key in the case $t = 2$.)

Next, we consider the possibility that certain participants will cheat, by lying as to what shadows they possess. In the worst case, $w - 1$ participants, say A_i $(2 \leq i \leq w)$ will form a coalition in order to try to convince A_1 that the key is some value $p' \neq p$. We will assume that $w \geq 3$, so that the coalition can determine the line H_p and the key p before attempting to deceive A_1. Note that they can also calculate s_1, since $s_1 = H_p \cap H_{21}$, for example.

Suppose A_2 tells A_1 that his shadow is some point s_2' rather than s_2. A_2 will not choose s_2' to be any point on ℓ, or any point on the line through s_1 parallel to ℓ, since A_1 would then realize that A_2 is lying. Also, A_2 will not choose s_2' to be a point on H_p, since this would not deceive A_1 as to the value of p. Hence, he will choose s_2' to be one of the remaining $q^2 - 3q + 2$ points. For any such choice of s_2', there is a unique line H_{12}' joining s_2' and s_2. A_1 will be deceived if and only if $H_{12}' = H_{12}$. Since $H_{12} \neq H_p$, there are $q - 1$ possibilities for H_{12}, all equally likely. Each of these $q - 1$ lines through s_2 contains $q - 2$ of the $q^2 - 3q + 2$ points mentioned above. Thus, the chance that A_2 deceives A_1 is $1 / (q - 1)$.

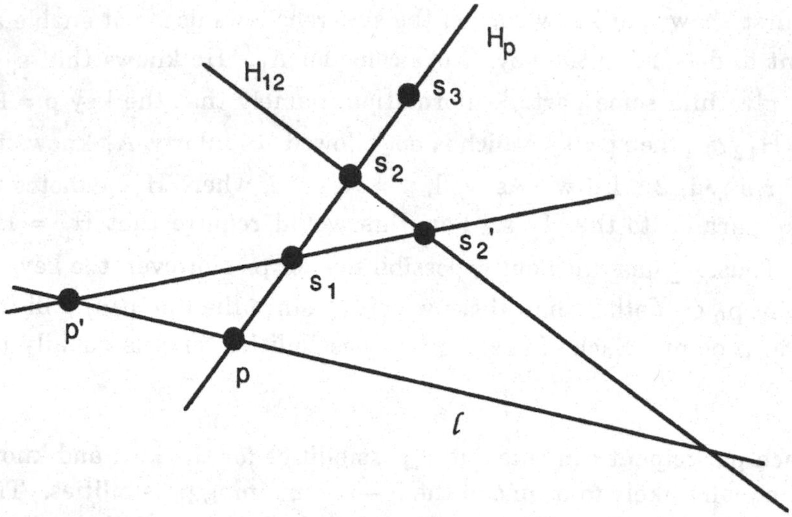

If all the other A_i $(2 \le i \le w)$ independently try to deceive A_1 in a similar fashion, the probability that at least one of them succeeds is

$$1-\left(1-\left(\frac{1}{q-1}\right)\right)^{w-1} \le \frac{w-1}{q-1}.$$

Their best strategy is to conspire; if they ensure that no two of the lines $s_i's_i$ are parallel, then A_1 will be deceived by one of them with probability equal to $(w-1)/(q-1)$. This will be a negligible quantity if q is large compared to w.

If w = 2, then the analysis is slightly different. Suppose A_2 attempts to deceive A_1. If A_2 can obtain the value of s_1, then the arguments proceed as before, and A_2 can deceive A_1 with probability $1/(q-1)$. (This could happen if A_1 reveals s_1 to A_2 before A_2 reveals s_2 to A_1, for example.) If A_2 cannot obtain the value of s_1, then his probability of deceiving A_1 is decreased to $1/q$, since he might choose s_2' to be a point on H_p.

Let's now consider the general case $t \ge 3$. Recall that H is a $(t-1)$-dimensional subspace and H_p is a hyperplane. D constructs w random $(t-1)$-dimensional subspaces, denoted H_i $(1 \le i \le w)$. We require that the intersection of H with $j-1$ of these H_i's is a subspace of dimension $t-j$, if $j \le t$. (In the case t = 2, this condition reduces to the previous requirement that the H_i's $(1 \le i \le w)$ be distinct from H.) The $w-1$ supershadows D gives to each A_j are the parallel hyperplanes $H_{ji} = H_j + s_i$, $1 \le i \le w$, $i \ne j$.

One way to select the H_i's is as follows. First, choose w subspaces of H, denoted K_i $(1 \leq i \leq w)$, each of dimension $t - 2$, in general position. Then select w points not in H, denoted q_i $(1 \leq i \leq w)$. These points need not be distinct. Finally, define H_i to be the subspace spanned by K_i and q_i $(1 \leq i \leq w)$.

First, we show that knowledge of the supershadows does not enable any $t - 1$ participants to determine the key. Suppose that participants A_i, $1 \leq i \leq t-1$, attempt to determine the key. They know that H_p contains F, the $(t - 2)$-dimensional flat generated by $s_1, \ldots, s_{t-1}$. They know also that a shadow s_j $(t \leq j \leq w)$ occurs on the line ℓ_j which is the intersection of the H_{ij}, $1 \leq i \leq t-1$. (Since ℓ_j meets H_p in a point, it has dimension one and is indeed a line.) Notice that any two of these lines ℓ_j are parallel, since the hyperplanes H_{ij} are parallel (for any fixed i).

We claim that for any j, $t \leq j \leq w$, ℓ_j and F generate the whole n-dimensional space (consequently, $\ell_j \cap F = \emptyset$). This is seen as follows. Suppose ℓ_j and F are contained in some hyperplane H', for some j, $t \leq j \leq w$. Since $s_j \in \ell_j$ and $s_1, \ldots, s_{t-1} \in F$, $H' = H_p$. Then $\ell_j \subseteq H_p \cap H_{1j} \cap H_{2j} \cap \ldots \cap H_{(t-1)j}$. It follows that $H \cap H_1 \cap H_2 \cap \ldots \cap H_{(t-1)}$ has dimension at least one, which is ruled out by the way in which the hyperplanes H_i were chosen.

Next, we observe that $F \cap \ell = \emptyset$. It is impossible that $\ell \subseteq H_p$ since $H_p \cap \ell = \{p\}$ and $F \subseteq H_p$. Also, F and ℓ cannot intersect in a point, for this point would have to be p, which would contradict the requirement that the shadows are in general position with respect to p.

It is now easy to verify that there is a unique point p' on ℓ such that the hyperplane determined by F and p' is parallel to each ℓ_j, $t \leq j \leq w$. Then, the key p $\neq$ p'. For, if p = p', then $H_p \cap \ell_j = \emptyset$; but $s_j \in H_p \cap \ell_j$, a contradiction. This enables the participants A_i $(1 \leq i \leq t-1)$ to rule out one possible value for the key.

There are in fact other points that can be ruled out as possible values for the key. We saw earlier that when t = 2, the w − 1 points $\ell \cap \ell_j$ $(t \le j \le w)$ can also be eliminated as possible values for p. In general, the number of possible keys that can be ruled out (other than the point p') is $\binom{w-t+1}{t-1}$.

We can see this as follows. Let $j_1, \ldots, j_{t-1}$ be distinct integers such that $t \le j_i \le w$ $(1 \le i \le t-1)$, and let U be the flat spanned by the ℓ_{j_i} $(1 \le i \le t-1)$. Since the lines ℓ_j are all parallel, U has dimension at most t − 1. The flat T spanned by the points s_{j_i} $(1 \le i \le t-1)$ has dimension t − 2, and is contained in $U \cap H_p$. As well, $\ell_j \cap H_p = \{s_j\}$, for any j, $t \le j \le w$. It follows that the dimension of U is exactly t − 1 and $T = U \cap H_p$.

Next, we observe that it is impossible that $\ell \subseteq U$. Since $\ell \cap H_p = \{p\}$, this would force $p \in T$. But then the t − 1 shadows $s_{j_1}, \ldots, s_{j_{t-1}}$ and p would then be contained in the flat T having dimension at most t − 2. Hence, either $\ell \cap U$ is empty, or $\ell \cap U$ is a point, say r. In the latter case, r cannot be the key, since (as before) the t − 1 shadows $s_{j_1}, \ldots, s_{j_{t-1}}$ and r would then be contained in the flat T.

Hence, it is possible that t − 1 participants can rule out as many as $1 + \binom{w-t+1}{t-1}$ possible values for the key.

Example: Suppose we have a (3, 5)-threshold scheme over GF(q), for some large prime q. Suppose ℓ is the line $(b, 0, 0)$ $(b \in GF(q))$, $s_1 = (1, 1, 2)$ and $s_2 = (1, 1, 6)$. Thus, F is the line $(1, 1, b)$ $(b \in GF(q))$. Suppose also that ℓ_3 is the line $(1 + a, 3 - a, 2)$ $(a \in GF(q))$, ℓ_4 is the line $(1 + a, -a, 1)$, and ℓ_5 is the line $(8 + a, -a, 3)$ (these three lines are parallel, having direction vector $(1, -1, 0)$). A_1 and A_2 would analyze the situation as follows. Suppose the key is $p = (x_0, 0, 0)$. Then, H_p is the plane $x + y(x_0 - 1) = x_0$. This plane intersects ℓ_3, ℓ_4, and ℓ_5 if and only if $x_0 \ne 2$. Thus, (2, 0, 0) is ruled out as the key. Three other points can also be ruled out. For example, ℓ_3 and ℓ_4 generate the plane U having equation $x + y - 3z = -2$. U meets ℓ in the point (−2, 0, 0). If −2 were the key, then H_p would have equation $x - 3y = -2$. Hence, it would follow that $s_3 = (5/2, 3/2, 2)$ and $s_4 = (1/4, 3/4, 1)$ (all arithmetic being done in GF(q)). Then s_3, s_4, and p are all collinear, a contradiction. In a similar manner, −4 is ruled out by consideration of ℓ_3 and ℓ_5, and −5 / 2 is eliminated by consideration of ℓ_4 and ℓ_5.

The last topic we examine in this section is the probability of successful cheating. Suppose $w-1$ participants, say A_i ($2 \leq i \leq w$) form a coalition in order to try to convince A_1 that the key is some value $p' \neq p$. Their best strategy is to leave $t-2$ of their shadows unchanged, and lie about the remaining $w-t+1$ shadows. The probability that A_1 will detect that any particular shadow is a forgery is $1/(q-1)$, as in the $t = 2$ case. The chance that A_1 is fooled by at least one of the $w-t+1$ altered shadows is at most $(w-t+1)/(q-1)$.

3. A cut-and-choose procedure to eliminate dealer disruption

We can eliminate the possibility of the dealer disruption by using a *cut-and-choose* procedure, as in [4] and [1]. Let K be some security parameter (say $K = 50$). Suppose H_p is the hyperplane $ax^T = c$, where the superscript "T" denotes transpose. The following protocol will be repeated K times.

1. D generates a random non-singular matrix M and a random t-tuple b. D then computes $s_i' = s_i M^T + b$ and gives s_i' to A_i, $1 \leq i \leq w$. (So, the s_i' are obtained from the s_i by a random affine transformation.)

2. Depending on a coin flip f, D performs a) or b).

   a) if f = "heads", then D reveals M and b, and each A_i verifies that $s_i' = s_i M^T + b$.

   b) if f = "tails", then D computes $a' = aM^{-1}$ and $c' = c + a'b^T$, and reveals a' and c'. Then, each A_i verifies that $a'(s_i')^T = c'$.

If the dealer can answer *both* challenges a) and b), then it must be the case that $c = as_i^T$, $1 \leq i \leq w$. That is, the shadows all lie on a hyperplane. If the dealer attempts to cheat, he can answer only one of the two challenges in any given round of the protocol. Hence, the probability of the dealer fooling any given set of t honest participants after K rounds is 2^{-K}.

It is also easy to see that no information is revealed to the participants by this protocol. If operation 2a) is performed in any round of the protocol, then the participants learn only the affine transformation used in that round. This is of no use in determining the key. If 2b) is performed, then the participants obtain the hyperplane $a'x^T = c'$. This tells them nothing about H_p, since any hyperplane can be mapped to any other hyperplane by means of an affine transformation.

Notice that we require the existence of a *broadcast channel* in step 2) of this protocol. This is a channel in which it is guaranteed that every participant receives the *same* information from the dealer (i.e. the values of M and b in 2a); or a' and c' in 2b)). If a broadcast channel is not used, then the dealer could attempt to cheat during this protocol by giving different information to different participants.

We can also do a cut-and-choose procedure on the supershadows. Here, the object is to convince each participant A_i that $s_j \in H_{ij}$, $i \neq j$, without revealing s_j. Suppose the hyperplane H_{ij} is given by the equation $a_i \cdot x = b_{ij}$, $1 \leq i, j \leq w$, $i \neq j$. The following protocol will be repeated K times.

1. For $1 \leq j \leq w$, D generates a random t-tuple s_j', and gives s_j' to A_j. D then computes $b_{ij}' = a_i \cdot s_j'$ and gives b_{ij}' to A_i, $1 \leq i, j \leq w$, $i \neq j$.

2. Depending on a coin flip f, D performs a) or b).

   a) if f = "heads", then D reveals all s_j', $1 \leq j \leq w$, and each A_i verifies that $b_{ij}' = a_i \cdot s_j'$.
   b) if f = "tails", then D reveals all $s_j + s_j'$, $1 \leq j \leq w$, and each A_i verifies that $a_i \cdot (s_j + s_j') = b_{ij} + b_{ij}'$, $1 \leq j \leq w$.

The analysis of dealer disruption is similar to the previous situation. If the dealer can answer *both* challenges a) and b) in any given round of the protocol, then it must be the case that $a_i \cdot s_j = b_{ij}$, $1 \leq i, j \leq w$, $i \neq j$. That is, the shadow s_j lies on the hyperplane $a_i \cdot x = b_{ij}$. As before, the probability of the dealer fooling any t honest participants in all K rounds is 2^{-K}.

Next, we consider whether any information about the shadows is released by this protocol. As before, if operation 2a) is performed in any round of the protocol, then clearly no information about the shadow is released. If operation 2b) is done, then A_i learns all values $s_j + s_j'$, but this tells him nothing about any s_j.

Finally, observe that we require a broadcast channel in step 2), as in the previous protocol.

Although the protocol protects against dealer disruption, we cannot guard against collusion of the dealer and any participant. For suppose D colludes with participant A_1. D can tell A_1 all the supershadows H_{i1}, and all the shadows s_i, $2 \leq i \leq w$. No collusion can be detected in the cut-and-choose procedure, since A_1 never reveals any information. Then, suppose a group of t participants including A_1, say $\{A_i: 1 \leq i \leq t\}$, attempt to determine the key. A_1 can compute the intersection ℓ_1 of the $t-1$ hyperplanes H_{i1}, $2 \leq i \leq t$. Note that ℓ_1 is a line. If A_1 claims that his shadow is any point on ℓ_1 other than s_1, then the other $t-1$ participants will not detect that he is cheating, and they will calculate an incorrect key. In this way, A_1 can make the other $t-1$ participants believe the key is any value he desires.

4. Remarks

There are many variations of this threshold scheme. For example, the threshold scheme could be implemented in a projective space rather than in an affine space. In the case $t = 2$, less partial information is revealed in a projective setting. D would fix a line ℓ in a projective plane P. As before the key p would be a point on ℓ. D also picks a random line H intersecting ℓ in p, and distribute points on $H \setminus \{p\}$ as the shadows. Supershadows are obtained as follows. For each participant A_i, D picks a point $q_i \in \ell \setminus \{p\}$ (these points need not be distinct). The supershadow H_{ij} is the line $s_j q_i$. With supershadows defined in this way, each participant A_i can only rule out the point q_i as the key (note that A_i can compute q_i as the intersection of any two of the supershadows he possesses).

It is an interesting open question to determine if there is a *perfect* threshold scheme satisfying all the other properties of our scheme (i.e. one in which *no* possible keys can be ruled out).

Another question is the amount of computation required. The dealer must verify certain conditions, including that the shadows are in general position. This is not difficult for small t and w, but could require a lot of time if t and w are large. Is there a scheme which is still computationally efficient for large t and w? (Note that the Shamir scheme [7] *is* computationally efficient; but it is not clear how to modify it to detect cheating.)

Yet another issue is the amount of (secret) information that needs to be communicated, in the form of shadows and supershadows. We ask if a scheme can be constructed which requires less information to be distributed.

Finally, we ask if it is possible to construct a threshold scheme that provides unconditional security against collusion of the dealer and one or more participants.

Acknowledgements

This research was discussed with David Chaum and his colleagues at C. W. I. We would like to thank them for their helpful observations and comments. Thanks also to Marijke De Soete for useful comments.

Added in proof

After writing this paper, we discovered that Tal Rabin was working independently on a related problem. Her results were presented at CRYPTO '88, in a paper entitled "Robust sharing of secrets when the dealer is honest or cheating". The techniques she employs can also be used to solve the problem we consider in our paper. Our approach requires that less secret information be communicated, but is slightly less efficient computationally.

References

1. Josh Cohen Benaloh, *Secret sharing homomorphisms: keeping shares of a secret secret*, Advances in Cryptology – CRYPTO 86 Proceedings, pp. 251-260, Lecture Notes in Computer Science, vol. 263, Springer-Verlag, Berlin, 1987.

2. G. R. Blakley, *Safeguarding cryptographic keys*, Proc. N. C. C., vol. 48, AFIPS Conference Proceedings 48 (1979), 313-317.

3. David Chaum, personal communication.

4. David Chaum, Claude Crepeau and Ivan Damgard, *Multiparty unconditionally secure protocols*, to appear in Proceedings of the 20th ACM Symposium on the Theory of Computing, 1988.

5. B. Chor, S. Goldwasser, S. Micali and B. Awerbuch, *Verifiable secret sharing and achieving simultaneity in the presence of faults*, Proc. 26th IEEE Symp. on Foundations of Computer Science, 1985, 383-395.

6. R. J. McEliece and D. V. Sarwate, *On sharing secrets and Reed-Soloman codes*, Comm. of the ACM 24 (1981), 583-584.

7. A. Shamir, *How to share a secret*, Comm. of the ACM 22, (1979), 612-613.

8. G. Simmons, *An introduction to shared secret schemes and their applications*, Sandia Report SAND88-2298, 1988.

9. M. Tompa and H. Woll, *How to share a secret with cheaters*, J. of Cryptology 1 (1988), 133-138

On the Power of 1-way Functions (Abstract)

Stuart A. Kurtz* University of Chicago

Stephen R. Mahaney A. T. & T. Bell Laboratories

James S. Royer† University of Chicago

The earliest definition of *1-way function* is due to Berman [Ber77], who considered polynomial-time computable, length-increasing, 1-1 functions that do not have a polynomial-time computable inverses. Recently, more powerful notions are considered, e.g., polynomial-time computable, length-increasing, 1-1 functions f such that the probability that a BPP algorithm can compute x from $f(x)$ for a randomly selected x is superpolynomially small [CYa82]. Whatever definition is used, these functions are necessarily easy invert on *some* inputs:

Proposition 1 *If f is a polynomial-time computable, length-increasing, 1-1 function, and if p is a polynomial, then there is a polynomial time algorithm that for sufficiently large n inverts f on at least $p(n)$ strings of length less than n. Therefore, the range of every such function must contain a polynomial-time computable subset of arbitrarily large polynomial census.*

We ask whether or not Proposition 1 is optimal.

Definition 2 *A polynomial-time computable, length-increasing 1-1 function f is an* annihilating function *if every polynomial time decidable subset of the range of f is sparse.*

Polynomial-time computations can do little to invert an annihilating function. The definition, although originally intended as a tool to overthrow the Berman-Hartmanis isomorphism conjecture [BH77, KMR89], can be motivated on a purely cryptographic basis: To defeat a traffic analysis, two sites will send invalid messages to maintain a constant level of virtual traffic, irrespective of the actual traffic. If an eavesdropper could distinguish valid from invalid messages, this strategm would fail. The point behind the definition of an annihilating function is that a polynomial-time algorithm will not permit an eavesdropper to pick out enough valid messages upon which to base a traffic analysis.

We would like to know whether or not annihilating functions exist. It probably doesn't make sense to attack this question directly, as annihilating functions are 1-way functions in at least the Grollman-Selman sense, and so their existence would entail P $\neq$ UP and therefore P $\neq$ NP. As a surrogate, we obtain:

*The first author was supported in part by NSF Grant DCR-8602562

†The third author was supported in part by NSF Grant DCR-8602991

Theorem 3 *With probability 1 relative to a random oracle, annihilating functions exist.*

The instant reaction to Theorem 3 is to ask whether or not it gives us any meaningful insight into the unrelativized case. In general, we do not believe that it is reasonable to base one's intuitions about unrelativized computational world upon relativized worlds. After all, unrestricted relativizations can be used to produce conflicting "worlds."

Random relativizations, on the other hand, cannot conflict with one another. The "measure 1" relativized theory is consistent and well-defined. More importantly, the successful use of pseudo-random number generators in lieu of truly random numbers in probabilistic factoring algorithms makes it seem plausible that computational complexity theory relative to a random oracle is similar to unrelativized computation complexity theory. This intuition was formalized by Bennett and Gill [BG81] as the random oracle hypothesis. Although the formal hypothesis was refuted [Kur83], the informal hypothesis is still compelling, and remains a basis for assigning credibility to random relativizations.

This brings us to a crucial point: do we believe that annihilating functions exist? We are divided ourselves on this question, and await further evidence.

References

[Ber77] L. Berman. *Polynomial Reducibilities and Complete Sets.* PhD thesis, Cornell University, 1977.

[BG81] Charles H. Bennett and John Gill. Relative to a random oracle A, $\mathbf{P}^A \neq \mathbf{NP}^A \neq$ **co-NP**^{A} with probability 1. *SIAM Journal on Computing*, 10:96–113, February 1981.

[BH77] L. Berman and J. Hartmanis. On isomorphism and density of NP and other complete sets. *SIAM Journal on Computing*, 6:305–322, June 1977.

[CYa82] Andrew C. Yao. Theory and application of trapdoor functions. In *23rd Annual IEEE Symposium on the Foundations of Computer Science*, pages 80–91, 1982.

[KMR89] Stuart A. Kurtz, Stephen R. Mahaney, and James S. Royer. The isomorphism conjecture fails relative to a random oracle. In *Proceedings of the 21st Annual ACM Symposium on Theory of Computing*, pages 157–166, 1989.

[Kur83] Stuart A. Kurtz. On the random oracle hypothesis. *Information and Control*, 57:40–47, 1983.

"Practical IP" $\subseteq$ MA

Gilles BRASSARD †
Université de Montréal
Département IRO
C.P. 6128, Succ. "A"
Montréal (Québec)
CANADA H3C 3J7

Ivan Bjerre DAMGAARD ‡
Aarhus University, Mathematical Institute
Ny Munkegade
DK 8000 Aarhus C
DENMARK

Interactive protocols [GMR] and Arthur-Merlin games [B] have attracted considerable interest since their introduction a few years ago. These notions make it (probably) possible to extend the concept of what is "efficiently" provable to include, for instance, graph *non*-isomorphism [GMW]. In this short note, we assume that the reader is familiar with interactive protocols, Arthur-Merlin games, and the notion of zero-knowledge [GMR].

In the previous paragraph, we put quotes around "efficiently" because it is only the Verifier that is required to be efficient (i.e.: polynomial time). On the other hand, both interactive protocols and Arthur-Merlin games allow the Prover (or "Merlin") to be infinitely powerful. In fact, not only is the Prover *allowed* to be powerful but she is actually *required* to be so in many of the most interesting theorems concerning these notions [B, GS, F, etc.]. For instance, in the graph non-isomorphism protocol, the Prover must be capable of deciding graph isomorphism.

An important pair of results state that $MA \subseteq AM = IP[k]$ [B, GS], but again this requires the Prover to have considerable computing power *even if the original* MA *protocol is feasible*! From a practical point of view, this is silly in the sense that a polynomial-time Prover can run an MA protocol if only given a polynomial piece of advice, whereas it is not at all clear that she could run the corresponding AM protocol without additional power and/or information. (This is because the Prover must be able to satisfy exponentially many challenges in an AM setting.)

---

† Supported in part by Canada NSERC grant A4107
‡ Supported by the Danish Natural Science Research Council

For this reason, it is our opinion that **MA** is the natural extension of **NP** to randomness. This opinion is not new: it was already voiced in [BC]. However, here we claim that this is not merely an opinion but actually a theorem, *albeit* a rather trivial one. To achieve this goal, of course, we must be more precise on what we mean by "Practical **IP**": it is the class of languages that can be handled when both the Prover and the Verifier are restricted to being polynomial time.

This definition raises an important issue: if Prover and Verifier have similar computing abilities (and algorithmic knowledge), how did the Prover manage to obtain a hard enough proof to be of interest to the Verifier? (It is obviously *un*interesting if the Verifier can figure out the proof by himself.) One possible answer is that the Prover was lucky enough or that she worked hard enough to find it (this would presumably be the case for an eventual proof of FLT). A much more interesting answer, at least in cryptographic settings, is that the Prover obtained the statement of her claim *together with its proof*, as a result of running a probabilistic polynomial-time process (starting from some randomly chosen trap-door information). For instance, if the Prover wants a statement of the general form "the integer n is the product of exactly two distinct primes", she can simply choose the primes at random and multiply them. She then knows the factors of the result even though she is not better than the Verifier at factoring large integers. Read [AABFH] for a very nice theory on the efficient generation of solved hard instances of problems in **NP**.

Whatever is the origin of the information that allows the polynomial-time Prover to run her share of the interactive protocol, that information is necessarily polynomial in length. It is therefore reasonable to assert that "Practical **IP**" is *included* in "Polynomial-time **IP** with polynomial advice for the Prover" (**PIP/Poly**), where of course "polynomial-time" restricts both the Prover and the Verifier. (We are not willing to claim that "Practical **IP**" = **PIP/Poly** because in our view the really practical case for cryptography is when the advice comes from trap-door information rather than hard labour or luck.) Therefore, in order to prove the assertion given in the title of this paper, it suffices to prove that **PIP/Poly** $\subseteq$ **MA** (in fact, these classes are equal, but the reverse inclusion is irrelevant for our purpose).

Consider a language L in **PIP/Poly**, some x in L, and the polynomial-length advice a that the (polynomial-time) Prover could use through an **IP** to convince the Verifier that x is in L. The fact that L belongs to **MA** is obvious: given only x, an all-powerful Prover (Merlin) can figure out this advice a and simply give it to the Verifier (Arthur). The Verifier can then (in polynomial time) simulate the polynomial-time Prover and her interaction with him. This complete the proof that "Practical **IP**" $\subseteq$ **MA**. An open question is whether the inclusion is strict: in particular, is it possible in general to generate solved hard instances for every hard languages in **MA**? The reader is referred once more to [AABFH] for preliminary results concerning **NP**.

An interesting situation occurs if one is interested in zero-knowledge protocols [GMR]. It is shown in [BCC] (under cryptographic assumptions) that MA protocols can be carried out in zero-knowledge *by a polynomial-time Prover* provided she is given the corresponding piece of advice. This is in sharp contrast with the result of [GMW] in which an MA protocol must first be transformed into an AM protocol before it can be carried out in zero-knowledge, hence even a practical MA protocol requires a powerful Prover to be carried out in zero-knowledge if the technique of [GMW] is used. (This situation was already pointed out in [BCC].) In conclusion, [BCC] allows us to claim that

$$\text{"Practical IP"} = \text{"Practical zero-knowledge"},$$

which is the "practical" version of "everything provable is provable in zero-knowledge" [IY, BGGHKMR].

References

[AABFH]
Abadi, M., E. Allender, A. Broder, J. Feigenbaum and L. Hemachandra, "On generating hard instances of problems in NP", these *CRYPTO 88 Proceedings*.

[Ba] Babai, L., "Trading group theory for randomness", *Proceedings of the 17th Annual ACM Symposium on the Theory of Computing*, 1985, pp. 421-429.

[BGGHKMR]
Ben-Or, M., O. Goldreich, S. Goldwasser, J. Hastad, J. Kilian, S. Micali and P. Rogaway, "Everything provable is provable in zero-knowledge", these *CRYPTO 88 Proceedings*.

[BCC] Brassard, G., D. Chaum and C. Crépeau, "Minimum disclosure proofs of knowledge", *Journal of Computer and System Sciences*, to appear, 1988.

[BC] Brassard, G. and C. Crépeau, "Non-transitive transfer of confidence: a *perfect* zero-knowledge interactive protocol for SAT and beyond", *Proceedings of the 27th Annual IEEE Symposium on the Foundations of Computer Science*, October 1986, pp. 188-195.

[F] Fortnow, L., "The complexity of perfect zero-knowledge", *Proceedings of the 19th Annual ACM Symposium on the Theory of Computing*, May 1987, pp. 204-209.

[GMW] Goldreich, O., S. Micali and A. Wigderson, "Proofs that yield nothing but their validity and a methodology of cryptographic protocol design", *Proceedings of the 27th Annual IEEE Symposium on the Foundations of Computer Science*, October 1986, pp. 174-187.

[GMR] Goldwasser, S., S. Micali and C. Rackoff, "The knowledge complexity of interactive proof-systems", *Proceedings of the 17th Annual ACM Symposium on the Theory of Computing*, 1985, pp. 291-304.

[GS] Goldwasser, S. and M. Sipser, "Arthur-Merlin games versus interactive proof systems", *Proceedings of the 18th Annual ACM Symposium on the Theory of Computing*, May 1986, pp. 59-68.

[IY] Impagliazzo, R. and M. Yung, "Direct minimum-knowledge computations", *Proceedings of CRYPTO 87*, August 1987, pp. 40-51.

Zero-Knowledge Authentication Scheme with Secret Key Exchange (extended abstract)

Jørgen Brandt, Ivan Damgård, Peter Landrock, Torben Pedersen

University of Aarhus,
Department of Mathematics and Computer Science.

Abstract

In this note we first develop a new computationally zero-knowledge interactive proof system of knowledge, which then is modified into an authentication scheme with secret key exchange for subsequent conventional encryption. Implemented on a standard 32-bit chip or similar, the whole protocol, which involves mutual identification of two users, exchange of a random common secret key and verification of certificates for the public keys (RSA, 512 bits) takes less than 0.7 seconds.

1 Introduction

Recently, some very effective zero-knowledge identification-schemes have been constructed, such as Fiat-Shamir, [FS], Micali-Shamir, [MS], and Guillou-Quisquater, [GQ]. However, they only provide authentication, not confidentiality. Consequently, if one aims at implementing a hybrid system based on public keys to provide user authentication and secret key exchange, these fast schemes are of no help.

The protocol, we are going to describe, is designed for software implementation on a standard chip such as Intel 80386 or Motorola 68030 or perhaps on a DSP-chip with the extra requirement that the whole communication setup is based on 512 bits RSA-modulus and takes less than 2 seconds.

2 The Basic Authentication Scheme

We first describe an example of a more general construction of a computationally zero-knowledge interactive proof system of knowledge based on a public key system.

In the example we use RSA as the public key system, but the method would work in general under some reasonably weak assumptions.

For proof technical reasons we need two independent general public key pairs (one-way functions with trapdoor) $\tilde{n} = \tilde{p}\tilde{q}$, $\tilde{e}$ a public exponent prime to $\phi(\tilde{n})$, and $\bar{n} = \bar{p}\bar{q}$, $\bar{e}$ a public exponent prime to $\phi(\bar{n})$. $(\tilde{n}, \tilde{e})$ and $(\bar{n}, \bar{e})$ can be generated when the system is set up. Then $\tilde{p}, \tilde{q}, \bar{p}$ and $\bar{q}$ can be deleted and are therefore assumed unknown to all users. Thus the prover, P and the verifier, V, receive as auxiliary input $\tilde{n}$ and $\bar{n}$, which they must trust to be generated correctly. This is similar to the non-interactive zero-knowledge model where P and V share a bit-string which they must trust to be random [BFM].

In the final version of this paper, we remove the need for $\tilde{n}$. This requires a more complicated argument to prove the zero-knowledge property, but simplifies the protocol (see Section 4).

Moreover each prover computes his own public-key pair $n = pq$, e public exponent, d secret ($ed \equiv 1 \bmod \phi(n)$). The "knowledge" he is going to prove to possess is that he can provide arbitrary signatures $m^d \bmod n$. For convenience, we assume that $k = \log n = \log \tilde{n} = \log \bar{n}$.

Protocol:

1. V chooses arbitrary bit-strings Q, R of length $\frac{1}{2}\log n$, subject to $(Q \parallel R) < n$, where $(Q \parallel R)$ is the number represented by the concatenation, R in the least significant half.
   He also chooses S, S' of length $\frac{1}{2}\log \tilde{n}$, subject to $(S \parallel R), (S' \parallel Q) < \tilde{n}$. He computes $(Q \parallel R)^e \bmod n$, $(S \parallel R)^{\tilde{e}} \bmod \tilde{n}$ and $(S' \parallel Q)^{\tilde{e}} \bmod \tilde{n}$ and sends these numbers to P.

2. P recovers $(Q \parallel R)$, chooses T, T', arbitrary bit strings of length $\frac{1}{2}\log n$, subject to $(T \parallel R), (T' \parallel Q) < \bar{n}$, computes $(T \parallel R)^{\bar{e}} \bmod \bar{n}$ and $(T' \parallel Q)^{\bar{e}} \bmod \bar{n}$ and sends these numbers to V.

3. V sends S, S' to P.

4. Using what he received in step 2) and 3), P checks the values for $(S \parallel R)^{\tilde{e}} \bmod \tilde{n}$ and $(S' \parallel Q)^{\tilde{e}} \bmod \tilde{n}$ that he received in step 1).
   If OK, he sends T, T' to V. Otherwise, he halts.

5. V uses T, T' to check the value for $(T \parallel R)^{\bar{e}} \bmod \bar{n}$ and $(T' \parallel Q)^{\bar{e}} \bmod \bar{n}$ he received in step 2).
   If OK, he accepts. Otherwise he rejects.

We now informally sketch how to prove the correctness and security of this protocol. We will say that $f(x)$ is pseudorandom given $g(x)$ (where $x \in \{0,1\}^k$) if no polynomial time probabilistic algorithm can distinguish between pairs of the form $(g(x), f(x))$ and $(g(x), r)$, where r is chosen uniformly from $Im(f)$.

It has been shown earlier, [ACGS], that the least significant $\log k$ bits of x are pseudorandom given $x^e \bmod n$. Following Micali and Schnorr [MS] we stretch this a little to get

Assumption*:
The least significant $\frac{1}{2}k$ bits of x are pseudorandom given $x^e \bmod n$ if the factors of n are unknown.

When we speak of a proof system of knowledge, we mean that the following must hold for an arbitrary prover P^*: if P^* completes the protocol successfully with non-negligible probability, then there exists a probabilistic polynomial time algorithm, which with the help of P^* finds x from $x^e \bmod n$ with non-negligible probability.

This is a little twist of the notion defined by Fiat, Feige and Shamir in [FFS]: a prover does not prove his possession of a certain bit-string, but his ability to do something.

Theorem 1.
Under * the protocol is a zero-knowledge proof system of knowledge.

Proof:
Zero-knowledge:
Given an arbitrary verifier, V^*, we describe a simulator, M_{V^*}. We may assume that the algorithm used to generate $\tilde{n}$ is public. Therefore M_{V^*} can generate $\tilde{n}$ with known factorization and correct distribution and give this number to V^* as input. We may now proceed as follows:

1. Receive $(Q \parallel R)^e \bmod n$, $(S \parallel R)^{\tilde{e}} \bmod \tilde{n}$ and $(S' \parallel Q)^{\tilde{e}} \bmod \tilde{n}$ ¿from V^*.
2. Recover $(Q \parallel R)$, and check against $(Q \parallel R)^e \bmod n$. If OK, compute and send $(T \parallel R)^{\tilde{e}} \bmod \tilde{n}$ and $(T' \parallel Q)^{\tilde{e}} \bmod \tilde{n}$ as P would have done. Otherwise send $X^{\tilde{e}} \bmod \tilde{n}$, $Y^{\tilde{e}} \bmod \tilde{n}$ for random X, Y.

Now observe:

- If V^* sends correct messages in step 1), we can complete the protocol with exactly the right distribution of messages.
- If this is not the case, P would always stop in step 3), and the random $X^{\tilde{e}} \bmod \tilde{n}$, $Y^{\tilde{e}} \bmod \tilde{n}$ cannot be distinguished from the "real" $(T \parallel R)^{\tilde{e}} \bmod \tilde{n}$, $(T' \parallel Q)^{\tilde{e}} \bmod \tilde{n}$ - by *.

¿From this, it is clear that the simulation works.

Proof-system:
Completeness is obvious. As for soundness assume some P^* completes the protocol

with non-negligible probability. We now describe an algorithm M_{P^*} which will find x from $x^e \bmod n$ with non-negligible probability.
Choose $\bar{n}$ with known factorization and give this as input to P^*. Suppose we are given $x^e \bmod n$, x unknown. Then choose $y, z < \bar{n}$ and send $x^e \bmod n$, $y^{\bar{e}} \bmod \bar{n}$ and $z^{\bar{e}} \bmod \bar{n}$ to P^*.
Receive $(T \parallel R)^{\bar{e}} \bmod \bar{n}$ and $(T' \parallel Q)^{\bar{e}} \bmod \bar{n}$ ¿from P^*, recover $(Q \parallel R) = x$.
By *, P^* cannot distinguish the random $y^{\bar{e}} \bmod \bar{n}$, $z^{\bar{e}} \bmod \bar{n}$ ¿from the real $(S \parallel R)^{\bar{e}} \bmod \bar{n}$ and $(S' \parallel Q)^{\bar{e}} \bmod \bar{n}$, hence the recovered $(Q \parallel R)$ will be correct with essentially the same (i.e. non-negligible) probability as in an actual conversation with V. □

In a formal proof of this fact, what we get is a result saying that P^* finds x with the same probability when talking to V as when talking to M_{P^*} or the combined system (P^*, M_{P^*}) can invert $y^{\bar{e}} \bmod \tilde{n}$. But this is an empty statement if M_{P^*} already knows $(\tilde{p}, \tilde{q})$. This is why we cannot use $\bar{n} = \tilde{n}$.

3 Combining with Key-Exchange

For this, we can use exactly the same protocol as before, except that V will choose first $K < n$ and compute $(Q \parallel R) \equiv K^e \bmod n$. The protocol runs as before, but P can recover K as $K \equiv ((Q \parallel R)^e)^{d^2} \bmod n$. We can now use the least significant $\frac{1}{2}k$ bits of K as secret key.

Theorem 2.
The secret key exchanged as above is pseudorandom given the conversation between P and V, assuming *.

Proof:
By *, the key is pseudorandom given $(Q \parallel R)$, and given $(Q \parallel R)$, the rest of the conversation can be simulated exactly. □

4 Practical Implementation

As mentioned earlier, it is possible to redesign the protocol so that it does not use $\tilde{n}$. The resulting version is given below. By a somewhat complicated argument, given in the final version of this paper, also this version can be proved to be a zero-knowledge proof system. It is an open problem, whether the protocol remains correct if we use the tempting simplification of putting $\bar{n} = n$.

Our protocol is intended for use in a public network, where users wish to be convinced about each others identities before communicating secretly. Part of this goal is often achieved by letting some center, C, provide signatures (certificates) on public keys. It would be natural to let C provide $\bar{n}$ too. This does not constitute any serious problems: as mentioned, the factorization of $\bar{n}$ can simply be forgotten after setting up the system.

Thus the practical version of the protocol will be as follows:

1. V chooses $K < n$ randomly and computes Q and R as $(Q \parallel R) \equiv K^e \bmod n$. He then computes $(Q \parallel R)^e \bmod n$ and sends this number to P.

2. P recovers K, Q and R. As before P chooses T, T', computes $(T \parallel R)^{\bar{e}} \bmod \bar{n}$ and $(T' \parallel Q)^{\bar{e}} \bmod \bar{n}$ and sends these numbers to V.

3. V sends Q, R to P.

4. P checks the values for Q and R.
   If OK, he sends T, T' to V. Otherwise, he halts.

5. V checks the value for $(T \parallel R)^{\bar{e}} \bmod \bar{n}$ and $(T' \parallel Q)^{\bar{e}} \bmod \bar{n}$ he received in step 2).
   If OK, he accepts. Otherwise he rejects.

For efficiency, it is advisable to choose e and $\bar{e}$ small, i.e. ≈ 3. Then the only really time consuming part is finding $x^{d^2} \bmod n$ from x. Note that the version with key exchange does not require more time, if e is small. P can precompute $d^2 \bmod \phi(n)$, and the protocol will then only require 2 extra exponentiations to the e'th power, which is negligible. With a little care, it is possible to use $e = 2$, which will be even more efficient.

Finally note, that A can prove himself to B <u>while</u> B is proving himself to A. In particular we can ensure that "decryption" of $(Q \parallel R)^e$ takes place simultaneosly for A and B. Thus this will not take more time than the basic version of the protocol.

It will be natural to obtain the final key as the xor of the two produced keys, since this will ensure that the key is known to precisely A and B, and no one else.

The basic operations needed for this protocol have been implemented on a standard 16 MHz Intel 80386 processor. The results of this show that the whole protocol, including verification of public-key certificates can be completed in less than 0.7 seconds.

References

[ACGS] Alexi,W., Chor, B., Goldreich, O. and Schnorr, C.P.: "RSA and Rabin Functions: Certain Parts Are as Hard as the Whole". Proc. of the 25th FOCS, 1984, pp. 449-457.

[BFM] Blum, M., Feldman, P. and Micali, S.: "Proving Security Against Chosen Cyphertext Attack". These proceedings.

[FFS] Feige, U., Fiat, A. and Shamir, A.: "Zero Knowledge Proofs of Identity". Proc. of the 19th STOC, 1987, pp. 210-217.

[FS] Fiat, A. and Shamir, A.: "How to Prove Yourself: Practical Solution to Identification and Signature Problems". Advances in Cryptology - CRYPTO'86, Lecture Notes in Computer Science 263, 1987, pp. 186-199.

[GQ] Guillou, L. and Quisquater, J-J: "A "Paradoxical" Identity-Based Signature Scheme Resulting from Zero-Knowledge". These proceedings.

[MS] Micali, S. and Shamir, A.: "An Improvement of the Fiat-Shamir Identification and Signature Scheme". These proceedings.

Author Index

Lecture Notes in Computer Science

Vol. 296: R. Janßen (Ed.), Trends in Computer Algebra. Proceedings, 1987. V, 197 pages. 1988.

Vol. 297: E.N. Houstis, T.S. Papatheodorou, C.D. Polychronopoulos (Eds.), Supercomputing. Proceedings, 1987. X, 1093 pages. 1988.

Vol. 298: M. Main, A. Melton, M. Mislove, D. Schmidt (Eds.), Mathematical Foundations of Programming Language Semantics. Proceedings, 1987. VIII, 637 pages. 1988.

Vol. 299: M. Dauchet, M. Nivat (Eds.), CAAP '88. Proceedings, 1988. VI, 304 pages. 1988.

Vol. 300: H. Ganzinger (Ed.), ESOP '88. Proceedings, 1988. VI, 381 pages. 1988.

Vol. 301: J. Kittler (Ed.), Pattern Recognition. Proceedings, 1988. VII, 668 pages. 1988.

Vol. 302: D.M. Yellin, Attribute Grammar Inversion and Source-to-source Translation. VIII, 176 pages. 1988.

Vol. 303: J.W. Schmidt, S. Ceri, M. Missikoff (Eds.), Advances in Database Technology – EDBT '88. X, 620 pages. 1988.

Vol. 304: W.L. Price, D. Chaum (Eds.), Advances in Cryptology – EUROCRYPT '87. Proceedings, 1987. VII, 314 pages. 1988.

Vol. 305: J. Biskup, J. Demetrovics, J. Paredaens, B. Thalheim (Eds.), MFDBS 87. Proceedings, 1987. V, 247 pages. 1988.

Vol. 306: M. Boscarol, L. Carlucci Aiello, G. Levi (Eds.), Foundations of Logic and Functional Programming. Proceedings, 1986. V, 218 pages. 1988.

Vol. 307: Th. Beth, M. Clausen (Eds.), Applicable Algebra, Error-Correcting Codes, Combinatorics and Computer Algebra. Proceedings, 1986. VI, 215 pages. 1988.

Vol. 308: S. Kaplan, J.-P. Jouannaud (Eds.), Conditional Term Rewriting Systems. Proceedings, 1987. VI, 278 pages. 1988.

Vol. 309: J. Nehmer (Ed.), Experiences with Distributed Systems. Proceedings, 1987. VI, 292 pages. 1988.

Vol. 310: E. Lusk, R. Overbeek (Eds.), 9th International Conference on Automated Deduction. Proceedings, 1988. X, 775 pages. 1988.

Vol. 311: G. Cohen, P. Godlewski (Eds.), Coding Theory and Applications 1986. Proceedings, 1986. XIV, 196 pages. 1988.

Vol. 312: J. van Leeuwen (Ed.), Distributed Algorithms 1987. Proceedings, 1987. VII, 430 pages. 1988.

Vol. 313: B. Bouchon, L. Saitta, R.R. Yager (Eds.), Uncertainty and Intelligent Systems. IPMU '88. Proceedings, 1988. VIII, 408 pages. 1988.

Vol. 314: H. Göttler, H.J. Schneider (Eds.), Graph-Theoretic Concepts in Computer Science. Proceedings, 1987. VI, 254 pages. 1988.

Vol. 315: K. Furukawa, H. Tanaka, T. Fujisaki (Eds.), Logic Programming '87. Proceedings, 1987. VI, 327 pages. 1988.

Vol. 316: C. Choffrut (Ed.), Automata Networks. Proceedings, 1986. VII, 125 pages. 1988.

Vol. 317: T. Lepistö, A. Salomaa (Eds.), Automata, Languages and Programming. Proceedings, 1988. XI, 741 pages. 1988.

Vol. 318: R. Karlsson, A. Lingas (Eds.), SWAT 88. Proceedings, 1988. VI, 262 pages. 1988.

Vol. 319: J.H. Reif (Ed.), VLSI Algorithms and Architectures – AWOC 88. Proceedings, 1988. X, 476 pages. 1988.

Vol. 320: A. Blaser (Ed.), Natural Language at the Computer. Proceedings, 1988. III, 176 pages. 1988.

Vol. 321: J. Zwiers, Compositionality, Concurrency and Partial Correctness. VI, 272 pages. 1989.

Vol. 322: S. Gjessing, K. Nygaard (Eds.), ECOOP '88. European Conference on Object-Oriented Programming. Proceedings, 1988. VI, 410 pages. 1988.

Vol. 323: P. Deransart, M. Jourdan, B. Lorho, Attribute Grammars. IX, 232 pages. 1988.

Vol. 324: M.P. Chytil, L. Janiga, V. Koubek (Eds.), Mathematical Foundations of Computer Science 1988. Proceedings. IX, 562 pages. 1988.

Vol. 325: G. Brassard, Modern Cryptology. VI, 107 pages. 1988.

Vol. 326: M. Gyssens, J. Paredaens, D. Van Gucht (Eds.), ICDT '88. 2nd International Conference on Database Theory. Proceedings, 1988. VI, 409 pages. 1988.

Vol. 327: G.A. Ford (Ed.), Software Engineering Education. Proceedings, 1988. V, 207 pages. 1988.

Vol. 328: R. Bloomfield, L. Marshall, R. Jones (Eds.), VDM '88. VDM – The Way Ahead. Proceedings, 1988. IX, 499 pages. 1988.

Vol. 329: E. Börger, H. Kleine Büning, M.M. Richter (Eds.), CSL '87. 1st Workshop on Computer Science Logic. Proceedings, 1987. VI, 346 pages. 1988.

Vol. 330: C.G. Günther (Ed.), Advances in Cryptology – EUROCRYPT '88. Proceedings, 1988. XI, 473 pages. 1988.

Vol. 331: M. Joseph (Ed.), Formal Techniques in Real-Time and Fault-Tolerant Systems. Proceedings, 1988. VI, 229 pages. 1988.

Vol. 332: D. Sannella, A. Tarlecki (Eds.), Recent Trends in Data Type Specification. V, 259 pages. 1988.

Vol. 333: H. Noltemeier (Ed.), Computational Geometry and its Applications. Proceedings, 1988. VI, 252 pages. 1988.

Vol. 334: K.R. Dittrich (Ed.), Advances in Object-Oriented Database Systems. Proceedings, 1988. VII, 373 pages. 1988.

Vol. 335: F.A. Vogt (Ed.), CONCURRENCY 88. Proceedings, 1988. VI, 401 pages. 1988.

Vol. 336: B.R. Donald, Error Detection and Recovery in Robotics. XXIV, 314 pages. 1989.

Vol. 337: O. Günther, Efficient Structures for Geometric Data Management. XI, 135 pages. 1988.

Vol. 338: K.V. Nori, S. Kumar (Eds.), Foundations of Software Technology and Theoretical Computer Science. Proceedings, 1988. IX, 520 pages. 1988.

Vol. 339: M. Rafanelli, J.C. Klensin, P. Svensson (Eds.), Statistical and Scientific Database Management. Proceedings, 1988. IX, 454 pages. 1989.

Vol. 340: G. Rozenberg (Ed.), Advances in Petri Nets 1988. VI, 439 pages. 1988.

Vol. 341: S. Bittanti (Ed.), Software Reliability Modelling and Identification. VII, 209 pages. 1988.

Vol. 342: G. Wolf, T. Legendi, U. Schendel (Eds.), Parcella '88. Proceedings, 1988. 380 pages. 1989.

Vol. 343: J. Grabowski, P. Lescanne, W. Wechler (Eds.), Algebraic and Logic Programming. Proceedings, 1988. 278 pages. 1988.

Vol. 344: J. van Leeuwen, Graph-Theoretic Concepts in Computer Science. Proceedings, 1988. VII, 459 pages. 1989.

Vol. 345: R.T. Nossum (Ed.), Advanced Topics in Artificial Intelligence. VII, 233 pages. 1988 (Subseries LNAI).

Vol. 346: M. Reinfrank, J. de Kleer, M.L. Ginsberg, E. Sandewall (Eds.), Non-Monotonic Reasoning. Proceedings, 1988. XIV, 237 pages. 1989 (Subseries LNAI).

Vol. 347: K. Morik (Ed.), Knowledge Representation and Organization in Machine Learning. XV, 319 pages. 1989 (Subseries LNAI).

Vol. 348: P. Deransart, B. Lorho, J. Maluszyński (Eds.), Programming Languages Implementation and Logic Programming. Proceedings, 1988. VI, 299 pages. 1989.

Vol. 349: B. Monien, R. Cori (Eds.), STACS 89. Proceedings, 1989. VIII, 544 pages. 1989.

Vol. 350: A. Törn, A. Žilinskas, Global Optimization. X, 255 pages. 1989.

Vol. 351: J. Díaz, F. Orejas (Eds.), TAPSOFT '89. Volume 1. Proceedings, 1989. X, 383 pages. 1989.